W9-CRX-965

PLEASE STAMP DATE DUE, BOTH BELOW AND ON CARD

DATE DUE	DATE DUE	DATE DUE	DATE DUE
SEP 3 1990	DEC 1 8 1996		
12/15/90			
1/15/90			
9/30	MAR 2 2 2002		

C

MATRIX

COMPUTATIONS

JOHNS HOPKINS SERIES
IN THE MATHEMATICAL SCIENCES

MATRIX
COMPUTATIONS

GENE H. GOLUB

Department of Computer Science
Stanford University

CHARLES F. VAN LOAN

Department of Computer Science
Cornell University

THE JOHNS HOPKINS UNIVERSITY PRESS

BALTIMORE, MARYLAND

Originally published, 1983
Second printing, 1984

The Johns Hopkins University Press, Baltimore, Maryland 21218

The paper in this book is acid-free and meets the guidelines for permanence and durability of the Committee on Production Guidelines for Book Longevity of the Council on Library Resources.

LIBRARY OF CONGRESS CATALOGING IN PUBLICATION DATA

Golub, Gene H. (Gene Howard), 1932–
  Matrix computations.

  (Johns Hopkins series in the mathematical sciences; 3)
  Bibliography: p. 447.
  Includes index.
  1. Matrices—Data processing.   I. Van Loan, Charles F.
II. Title.   III. Series.
QA188.G65   1983   512.9′434   83–7897
ISBN 0-8018-3010-9
ISBN 0-8018-3011-7 Pbk

DEDICATED TO

ALSTON S. HOUSEHOLDER

AND

JAMES H. WILKINSON

Contents

11 Functions of Matrices 380

12 Special Topics 404

Preface

It can be argued that the "mission" of numerical analysis is to provide the scientific community with effective software tools. What makes this enterprise so interesting is that its participants require skills from both mathematics and computer science. Indeed, good software development demands a mathematical understanding of the problem to be solved, a flair for algorithmic expression, and an appreciation for finite precision arithmetic. The aim of this book is to provide the reader with these skills as they pertain to matrix computations.

Great progress has been made in this area since the mid-1950s. This is borne out by the existence of quality programs for many linear equations, least squares, and eigenvalue problems. Typical are the routines in EISPACK and LINPACK, whose widespread use has had the effect of elevating the level of algorithmic thought in various applied areas. By using the programs in these packages as building blocks, scientists and engineers can piece together more complicated software tools that are tailored specifically for their needs. This development encourages the writing of well-structured programs, a welcome trend since more and more scientific research is manifested in software.

The impact of numerical linear algebra is felt in other ways. The habits of the field—our reliance on orthogonal matrices, our appreciation of problem sensitivity, our careful consideration of roundoff—have spilled over into many areas of research. A prime example of this is the increased use of the singular value decomposition (SVD) as an *analytical* tool by many statisticians and control engineers. Workers in these areas are reformulating numerous theoretical concepts in the "language" of the SVD and as a result are finding it much easier to implement their ideas in the presence of roundoff error and inexact data.

Further evidence of the growing impact of numerical linear algebra is in the area of hardware design. Recent developments in floating point arithmetic and parallel processing have in no small measure been provoked by activity in the matrix computation field.

We have written this book in order to impart a sense of unity to this expanding and exciting field. Much has been accomplished since the publication in 1965 of Wilkinson's monumental treatise *The Algebraic Eigenvalue*

Problem. Many of these modern developments have been discussed in survey articles and in specialized volumes such as *Solving Least Squares Problems* by Lawson and Hanson and *The Symmetric Eigenvalue Problem* by Parlett. We feel that the time is appropriate for a synthesis of this material. In this regard we see *Matrix Computations* as a comprehensive, somewhat more advanced version of Stewart's *Introduction to Matrix Computations*, published nearly a decade ago.

We anticipate three categories of readers: graduate students in technical areas, computational scientists and engineers, and our colleagues in numerical analysis. We have included special features addressed to each group.

For students (and their instructors) we have included a large number of problems. Many of these are computational and can form the nucleus of a programming project. Our experience in teaching with the book is that EIS-PACK and LINPACK assignments greatly enliven its contents. C. B. Moler's MATLAB, an easy-to-use system for performing matrix calculations, has also been used successfully in conjunction with this text.

For practicing engineers and scientists who wish to use the volume as a reference book, we have tried to minimize the interconnections among chapters. We have also included an annotated bibliography at the end of almost every section in order to hasten the search for reference material on any given topic.

For our colleagues in numerical analysis we have sprinkled our algorithmic discussions with a generous amount of perturbation theory and error analysis. It is our intention to provide these readers with enough detail so that they can understand and solve the matrix problems that arise in their own work. Research in numerical linear algebra is frequently instigated by ongoing work in other areas of numerical analysis. For example, some of the best methods for solving sparse linear systems have been developed by researchers concerned with the numerical solution of partial differential equations. Similarly, it was the activity in the quasi-Newton area that prompted the development of techniques for updating various matrix factorizations.

This book took six years to write and has undergone several title changes: (1) *A Last Course in Matrix Computations*, (2) *Applied Matrix Computations*, and (3) *Advanced Matrix Computations*. The first title, aside from being "cutesy," is misleading. We do not pretend that our treatment of matrix computations is complete. In particular, many topics in the vibrant area of sparse matrix computation were excluded from the text simply because we did not wish to delve into graph theory and data structures. The second title was dismissed because we do not dwell at great length on applications. For the most part, the matrix problems we consider are treated as given—their origins are not chased down. We recognize this as a pedagogic shortcoming but one which can be offset by an experienced teacher. Moreover, we suspect that many of our readers will be experienced themselves, thereby obviating the need for excessive motivation. Finally, the last title was dispensed with because the book *does* contain introductory material, especially in the first four chapters. We chose to include elementary topics for the sake of completeness and because we think that our approach to the rudiments of the subject will interest teachers of undergraduate numerical analysis.

What, then, is the book about if it is incomplete, less than applied, and not entirely advanced? A brief synopsis of its contents should answer this question.

The first three chapters contain necessary background material. Matrix algebra is reviewed, and some key algorithms are established. The pace is rather brisk. Readers who struggle with the problems in these early chapters will no doubt struggle with the remainder of the book.

Chapter 4 presents and analyzes the algorithm of Gaussian elimination. Experts will probably find all portions of this chapter boring except perhaps the style of our error analysis and our discussion of the condition estimation problem.

Much current research in numerical linear algebra focusses on problems in which the matrices have special structure—e.g., are large and sparse. The art of exploiting structure is the central theme of Chapter 5, where various special-purpose linear equation solvers are described.

Chapter 6 picks up on another trend in the field—the increased reliance on orthogonal matrices. We discuss several orthogonalization methods and show how they can be applied to the least squares problem. Special attention is paid to the handling of rank deficiency.

The all-powerful QR algorithm for the unsymmetric eigenvalue problem is the centerpiece of Chapter 7. Our pedagogic derivation should help to demystify this important technique. We also comment on invariant subspace calculation and the generalized eigenvalue problem.

In Chapter 8 we continue our discussion of the eigenvalue problem by focussing on the important symmetric case. We first describe the symmetric QR algorithm and then proceed to show how symmetry permits several alternative computational procedures.

Up to this point in the book our treatment of sparsity is rather scattered. Banded linear system solvers are discussed in Chapter 5, simultaneous iteration is described in Chapter 7, Rayleigh quotient iteration in Chapter 8, and so on. Chapters 9 and 10, however, are entirely devoted to the solving of sparse matrix problems. The discussion revolves around the Lanczos method and its country cousin, the method of conjugate gradients. We show how various sparse eigenvalue, least squares, and linear equation problems can be solved using these important algorithms.

The purpose of the last two chapters in the book is to illustrate the wide applicability of the algorithms presented in Chapters 4–8. Chapter 11 deals with the problem of computing a function of a matrix, something that is frequently required in applications of control theory. Chapter 12 describes a selection of special matrix problems, several of which highlight the power of the singular value decomposition.

Indeed, perhaps the most recurring theme in the book is the practical and theoretical value of this matrix decomposition. Its algorithmic and mathematical properties have a key role to play in nearly every chapter. In many respects *Matrix Computations* is an embellishment of the manuscript "Everything You Wanted to Know about the Singular Value Decomposition (But Were Afraid to Ask)" authored by our colleague Alan Cline.

A word is in order about references to available software. We have concentrated on EISPACK and LINPACK, and just about every subroutine in these packages is alluded to in the text. In addition, numerous "tech report" references that we cite in the annotated bibliographies are in fact references to software. It should be stressed, however, that we have no direct experience with much of the referenced software aside from EISPACK and LINPACK. Caveat emptor.

Many people assisted in the production of the book. Richard Bartels helped to gather references and to revise the first draft of some early chapters. The writing of this book was prompted by his organization of the Workshop in Matrix Computations held at Johns Hopkins University in August 1977.

Bob Plemmons, John Dennis, Alan Laub, and Don Heller taught from various portions of the text and provided numerous constructive criticisms. George Cybenko generously helped us with the section on Toeplitz matrices, while Bo Kagstrom offered many intelligent comments on our treatment of invariant subspace computation. Per-Åke Wedin diligently read early versions of Chapters 2 and 6 and expertly guided our revisions. Uri Ascher and Roger Horn have our gratitude for independently suggesting the book's final title and so does an anonymous reviewer who made many valuable suggestions.

Last, but not least, we happily acknowledge the influence of our colleagues Cleve Moler and Pete Stewart. Their own work, which so perfectly captures the spirit of the field, has strongly affected the balance and style of our own presentation.

Using the Book

Abbreviations

The following references are frequently cited in the text:

SLE   G. E. Forsythe and C. B. Moler (1967). *Computer Solution of Linear Algebraic Systems*, Prentice-Hall, Englewood Cliffs.

SLS   C. L. Lawson and R. J. Hanson (1974). *Solving Least Squares Problems*, Prentice-Hall, Englewood Cliffs.

SEP   B. N. Parlett (1980). *The Symmetric Eigenvalue Problem*, Prentice-Hall, Englewood Cliffs.

IMC   G. W. Stewart (1973). *Introduction to Matrix Computations*, Academic Press, New York.

AEP   J. H. Wilkinson (1965). *The Algebraic Eigenvalue Problem*, Clarendon Press, Oxford.

Mnemonics are used to identify these "global" references—e.g., "Wilkinson (AEP, chap. 5)." All other bibliographic information associated with a given chapter or section appears locally. There is a master bibliography at the end of the book.

References to the software packages LINPACK and EISPACK are tacit references to the corresponding manuals:

EISPACK   B. T. Smith, J. M. Boyle, Y. Ikebe, V. C. Klema, and C. B. Moler (1970). *Matrix Eigensystem Routines: EISPACK Guide*, 2nd ed., Springer-Verlag, New York.

EISPACK 2 B. S. Garbow, J. M. Boyle, J. J. Dongarra, and C. B. Moler (1972). *Matrix Eigensystem Routines: EISPACK Guide Extension*, Springer-Verlag, New York.

LINPACK   J. Dongarra, J. R. Bunch, C. B. Moler, and G. W. Stewart (1978). *LINPACK Users Guide*, SIAM Publications, Philadelphia.

Thus, "subroutine *xyz* may be found in LINPACK (Chap. 2)" means that subroutine *xyz* is described in the second chapter of the LINPACK manual. ALGOL versions of many EISPACK and LINPACK subroutines are collected in

HACLA   J. H. Wilkinson and C. Reinsch, eds. (1971). *Handbook for Automatic Computation*, Vol. 2, *Linear Algebra*, Springer-Verlag, New York.

Section Interdependence

The reader is advised to check the appropriate "reading paths" before consulting an isolated section or chapter from the book. The reading paths appear at the beginning of Chapter 5-12, and they collectively indicate the logical dependence among all the sections in the volume. (Knowledge of Chapters 1-4 is assumed throughout.)

A Note to Instructors

The book can be used in several different types of courses. Here are some samples:

> Title: Introduction to Matrix Computations (1 semester)
> Syllabus: Chapters 2-4 and 5.1-5.3, 6.1-6.5, 7.1-7.6, 8.1-8.2

> Title: Matrix Computations (2 semesters)
> Syllabus: Chapters 1-12

> Title: Linear Equation and Least Square Problems (1 semester)
> Syllabus: Chapters 4-6, 9-10, and 12.1-12.4, 12.6

> Title: Eigenvalue Problems (1 semester)
> Syllabus: Chapters 7-9, 11, and 12.5

In each case we strongly recommend the inclusion of computing assignments that involve the use of state-of-the-art software packages like EISPACK and LINPACK. The subject can be further enlivened by using MATLAB, an easy-to-use-system for performing matrix computations. See

C. B. Moler (1980). "MATLAB User's Guide," Technical Report CS81-1, Department of Computer Science, University of New Mexico, Albuquerque, New Mexico, 87131.

MATRIX
COMPUTATIONS

Background Matrix Algebra

§1.1 Vectors and Matrices
§1.2 Independence, Orthogonality, Subspaces
§1.3 Special Matrices
§1.4 Block Matrices and Complex Matrices

This chapter reviews basic matrix algebra. Its primary purpose is to serve as a glossary of concepts and notation. Readers wishing a more leisurely introduction should consult the references that are given at the end of the chapter.

Sec. 1.1. Vectors and Matrices

$\mathbb{R}^{m \times n}$ denotes the vector space of all m-by-n real matrices:

$$A \in \mathbb{R}^{m \times n} \iff A = [a_{ij}] = \begin{bmatrix} a_{11} & \cdots & a_{1n} \\ \vdots & & \vdots \\ a_{m1} & \cdots & a_{mn} \end{bmatrix},$$

When a capital letter is used to denote a matrix (e.g., A, B, Δ), the corresponding lowercase letter with the subscript ij refers to the (i, j) component (e.g., a_{ij}, b_{ij}, δ_{ij}).

Some of the basic manipulations with matrices are addition ($\mathbb{R}^{m \times n} \times \mathbb{R}^{m \times n} \to \mathbb{R}^{m \times n}$),

$$C = A + B \qquad c_{ij} = a_{ij} + b_{ij},$$

multiplication by a scalar ($\mathbb{R} \times \mathbb{R}^{m \times n} \to \mathbb{R}^{m \times n}$),

$$C = \alpha A \qquad c_{ij} = \alpha a_{ij},$$

matrix-matrix multiplication ($\mathbb{R}^{m \times n} \times \mathbb{R}^{n \times p} \to \mathbb{R}^{m \times p}$),

$$C = AB \qquad c_{ij} = \sum_{k=1}^{n} a_{ik} b_{kj},$$

transposition ($\mathbb{R}^{m \times n} \to \mathbb{R}^{n \times m}$)

$$C = A^{\mathrm{T}} \qquad c_{ij} = a_{ji}$$

and differentiation ($\mathbb{R}^{m \times n} \rightarrow \mathbb{R}^{m \times n}$),

$$C = (c_{ii}(\alpha)) \qquad \dot{C} = \frac{d}{d\alpha} C = [\dot{c}_{ij}(\alpha)].$$

The *n*-by-*n* matrices are said to be square. The *n*-by-*n* *identity matrix* is denoted by I_n and its *k*-th column by $e_k^{(n)}$:

$$I_n = \begin{bmatrix} 1 & \cdots & 0 \\ \vdots & \ddots & \vdots \\ 0 & \cdots & 1 \end{bmatrix} \qquad e_k^{(n)} = (0, \ldots, 0, \underset{k}{1}, 0, \ldots, 0)^{\mathrm{T}}.$$

When the dimension is clear from context, we simply write I and e_k, respectively.

If A and B in $\mathbb{R}^{n \times n}$ satisfy $AB = I$, then B is the *inverse* of A and is denoted by A^{-1}. If A^{-1} exists, then A is said to be *nonsingular*; otherwise A is *singular*. $A^{-\mathrm{T}}$ denotes $(A^{-1})^{\mathrm{T}} = (A^{\mathrm{T}})^{-1}$.

If $A = (a) \in \mathbb{R}^{1 \times 1}$, then its *determinant* is given by $\det(A) = a$. For $A \in \mathbb{R}^{n \times n}$ we have

$$\det(A) = \sum_{j=1}^{n} (-1)^{j+1} a_{1j} \det(A_{1j})$$

where A_{1j} is an $(n-1)$-by-$(n-1)$ matrix obtained by deleting the first row and *j*-th column of A. Useful properties of the determinant include

(i) $\det(AB) = \det(A)\det(B)$ $A, B \in \mathbb{R}^{n \times n}$

(ii) $\det(A^{\mathrm{T}}) = \det(A)$ $A \in \mathbb{R}^{n \times n}$

(iii) $\det(cA) = c^n \det(A)$ $c \in \mathbb{R}, A \in \mathbb{R}^{n \times n}$

(iv) $\det(A) \neq 0 \iff A$ is nonsingular $A \in \mathbb{R}^{n \times n}$.

$\mathbb{R}^m$ denotes $\mathbb{R}^{m \times 1}$, and for these column vectors we customarily use lowercase letters and denote individual components with single subscripts. Thus, if $A \in \mathbb{R}^{m \times n}$, $x \in \mathbb{R}^n$, and $y = Ax$, then

$$y_i = \sum_{j=1}^{n} a_{ij} x_j. \qquad\qquad i = 1, \ldots, m$$

The *outer product* of $x \in \mathbb{R}^m$ and $y \in \mathbb{R}^n$ is given by

$$xy^{\mathrm{T}} = \begin{bmatrix} x_1 y_1 & \cdots & x_1 y_n \\ \vdots & & \vdots \\ x_m y_1 & \cdots & x_m y_n \end{bmatrix} \in \mathbb{R}^{m \times n}$$

and their *inner product* (if $m = n$) by

$$x^T y = \sum_{i=1}^{n} x_i y_i = y^T x.$$

Suppose $A \in \mathbb{R}^{n \times n}$ is nonsingular and that u and v are in $\mathbb{R}^n$. If $v^T A^{-1} u \neq -1$, then

(1.1-1) $$(A + uv^T)^{-1} = A^{-1} - \frac{A^{-1} u \, v^T A^{-1}}{1 + v^T A^{-1} u}.$$

This is known as the *Sherman-Morrison formula*.

A useful generalization of (1.1-1) is the *Sherman-Morrison-Woodbury formula*,

(1.1-2) $$(A + UV^T)^{-1} = A^{-1} - A^{-1} U(I + V^T A^{-1} U)^{-1} V^T A^{-1},$$

where $A \in \mathbb{R}^{n \times n}$, $U \in \mathbb{R}^{n \times k}$, $V \in \mathbb{R}^{n \times k}$, and both A and $(I + V^T A^{-1} U)$ are nonsingular.

If $A \in \mathbb{R}^{m \times n}$ and we write

$$A = [c_1, \ldots, c_n],$$

then $c_k \in \mathbb{R}^m$ is the k-th column of A. Likewise,

$$A = \begin{bmatrix} r_1^T \\ \vdots \\ r_m^T \end{bmatrix}$$

means that r_k^T is the k-th row of A. If

$$1 \leqslant i_1 < \cdots < i_r \leqslant m,$$

and

$$1 \leqslant j_1 < \ldots < j_s \leqslant n,$$

then the matrix $B = (b_{pq}) \in \mathbb{R}^{r \times s}$ defined by $b_{pq} = a_{i_p j_q}$ is a *submatrix* of A. If $r = s$ and $i_p = j_p$ for $p = 1, \ldots, r$, then B is a *principal submatrix*. If in addition, $i_p = j_p = p$ for $p = 1, \ldots, r$, then B is a *leading principal submatrix*.

Problems

P1.1-1. For $A \in \mathbb{R}^{m \times n}$ and $B \in \mathbb{R}^{n \times p}$, verify $(AB)^T = B^T A^T$.

P1.1-2. Show that the inverse of a nonsingular matrix is unique.

P1.1-3. Show that if $x, y \in \mathbb{R}^n$ then $(xy^T)^k = (x^T y)^{k-1} xy^T$.

P1.1-4. Show that if $A = [a_1, \ldots, a_n] \in \mathbb{R}^{m \times n}$ and $B = [b_1, \ldots, b_n] \in \mathbb{R}^{p \times n}$ then

$$AB^T = \sum_{i=1}^{n} a_i b_i^T.$$

P1.1-5. Suppose $A(\alpha) \in \mathbb{R}^{m \times n}$ and $B(\alpha) \in \mathbb{R}^{n \times p}$ are matrices whose entries are differentiable functions of the scalar α. Show

$$\frac{d}{d\alpha}[A(\alpha)B(\alpha)] = \left[\frac{d}{d\alpha}A(\alpha)\right]B(\alpha) + A(\alpha)\left[\frac{d}{d\alpha}B(\alpha)\right].$$

P1.1-6. Suppose $A(\alpha) \in \mathbb{R}^{n \times n}$ has entries which are differentiable functions of the scalar α. Assuming that $A(\alpha)$ is always nonsingular, show

$$\frac{d}{d\alpha}[A(\alpha)^{-1}] = -A(\alpha)^{-1}\left[\frac{d}{d\alpha}A(\alpha)\right]A(\alpha)^{-1}.$$

P1.1-7. Prove (1.1-2).

P1.1-8. Suppose $A \in \mathbb{R}^{n \times n}$, $b \in \mathbb{R}^n$, and that $\phi : \mathbb{R}^n \to \mathbb{R}$ is defined by $\phi(x) = \frac{1}{2}x^TAx - x^Tb$. Show that the gradient of ϕ is given by

$$\nabla\phi(x) = \frac{1}{2}(A^T + A)x - b.$$

Sec. 1.2. Independence, Orthogonality, Subspaces

A set of vectors $\{a_1, \ldots, a_n\}$ in $\mathbb{R}^m$ is *linearly independent* if

$$\sum_{j=1}^{n} \alpha_j a_j = 0 \Leftrightarrow \alpha_1 = \cdots = \alpha_n = 0.$$

Otherwise, a nontrivial combination of $a_1, \ldots, a_n$ is zero and $\{a_1, \ldots, a_n\}$ is said to be *linearly dependent*.

A *subspace* of $\mathbb{R}^m$ is a subset that is also a vector space. The set of all linear combinations of $a_1, \ldots, a_n \in \mathbb{R}^m$ is a subspace referred to as the *span* of $\{a_1, \ldots, a_n\}$:

$$\text{span}\{a_1, \ldots, a_n\} = \left\{\sum_j \beta_j a_j \mid \beta_1, \ldots, \beta_n \in \mathbb{R}\right\}$$

If $\{a_1, \ldots, a_n\}$ is independent and $b \in \text{span}\{a_1, \ldots, a_n\}$, then b is a unique linear combination of $a_1, \ldots, a_n$.

If $S_1, \ldots, S_k$ are subspaces of $\mathbb{R}^m$, then their sum S, defined by

$$S = \{a_1 + a_2 + \cdots + a_k \mid a_i \in S_i, i = 1, \ldots, k\}$$

is also a subspace. S is said to be a *direct sum* if each $v \in S$ has a *unique representation* $v = a_1 + \cdots + a_k, a_i \in S_i$. In this case we write $S = S_1 \oplus \cdots \oplus S_k$.

The intersection of a collection of subspaces is also a subspace, e.g., $S = S_1 \cap S_2 \cap \cdots \cap S_k$.

The subset $\{a_{i_1}, \ldots, a_{i_k}\}$ is a *maximal linearly independent subset* of $\{a_1, \ldots, a_n\}$ if $\{a_{i_1}, \ldots, a_{i_k}\}$ is linearly independent and is not properly contained in any linearly independent subset of $\{a_1, \ldots, a_n\}$. If $\{a_{i_1}, \ldots, a_{i_k}\}$ is maximal, then

$$\text{span}\{a_1, \ldots, a_n\} = \text{span}\{a_{i_l}, \ldots, a_{i_k}\}$$

and $\{a_{i_1}, \ldots, a_{i_k}\}$ is a *basis* for span $\{a_1, \ldots, a_n\}$. If $S \subset \mathbb{R}^m$ is a subspace, then there exist independent basic vectors $a_1, \ldots, a_k$ in S such that

$$S = \text{span}\{a_1, \ldots, a_k\}.$$

All bases for a subspace S have the same number of elements. This number is the *dimension* of S and is denoted by $\dim(S)$.

There are two important subspaces associated with a matrix A in $\mathbb{R}^{m \times n}$. The *range* of A is defined by

$$R(A) = \{y \in \mathbb{R}^m \mid y = Ax \text{ for some } x \in \mathbb{R}^n\}$$

and the *null space* of A by

$$N(A) = \{x \in \mathbb{R}^n \mid Ax = 0\}.$$

If $A = [a_1, \ldots, a_n]$ then

$$R(A) = \text{span}\{a_1, \ldots, a_n\}.$$

The *rank* of a matrix A is defined by

$$\text{rank}(A) = \dim[R(A)].$$

It can be shown that $\text{rank}(A) = \text{rank}(A^T)$, and thus, the rank of a matrix equals the maximal number of independent rows or columns.

For any $A \in \mathbb{R}^{m \times n}$, $\dim[N(A)] + \text{rank}(A) = n$. If $m = n$, then the following are equivalent:

(i) A is nonsingular

(ii) $N(A) = \{0\}$

(iii) $\text{rank}(A) = n$.

A set of vectors $\{x_1, \ldots, x_p\}$ in $\mathbb{R}^m$ is *orthogonal* if $x_i^T x_j = 0$ whenever $i \neq j$ and *orthonormal* if $x_i^T x_j = \delta_{ij}$. More generally, a collection of subspaces $S_1, \ldots, S_p$ of $\mathbb{R}^m$ is mutually orthogonal if $x^T y = 0$ whenever $x \in S_i$ and $y \in S_j$ for $i \neq j$.

The *orthogonal complement* of a subspace $S \subset \mathbb{R}^m$ is defined by

$$S^\perp = \{y \in \mathbb{R}^m \mid y^T x = 0 \text{ for all } x \in S\}.$$

It can be shown that $R(A)^\perp = N(A^T)$.

The vectors $v_1, \ldots, v_k$ form an *orthonormal basis* for a subspace $S \subset \mathbb{R}^m$ if they are orthonormal and span S. It is always possible to extend such a basis to a full orthonormal basis $\{v_1, \ldots, v_m\}$ for $\mathbb{R}^m$. Note that in this case,

$$S^\perp = \text{span}\{v_{k+1}, \ldots, v_m\}.$$

Problems

P1.2-1. Show that if $v_1, \ldots, v_k \in \mathbb{R}^m$ are orthonormal, then they are independent.

P1.2-2. Show that if $S \subset \mathbb{R}^m$ is a subspace, then $\dim(S) + \dim(S^\perp) = m$.

P1.2-3. Show that if $A \in \mathbb{R}^{m \times n}$, then $N(A^T) = R(A)^\perp$.

P1.2-4. Show that if $A \in \mathbb{R}^{m \times n}$ has rank p, then there exists an $X \in \mathbb{R}^{m \times p}$ and a $Y \in \mathbb{R}^{n \times p}$ such that $A = XY^T$, where $\text{rank}(X) = \text{rank}(Y) = p$.

Sec. 1.3. Special Matrices

Matrices with special patterns of zero entries, with special symmetries, and with other special properties are listed in this section for reference. Let us begin by classifying matrices according to their zero-nonzero structure. We say that $A \in \mathbb{R}^{m \times n}$ has *lower bandwidth* r and *upper bandwidth* s if $a_{ij} = 0$ whenever $i > j + r$ and $j > i + s$. If $r = s$, then A is simply said to have bandwidth r. Several classes of band matrices that frequently occur have special names. We say that A is

diagonal	if $a_{ij} = 0$ whenever $i \neq j$
tridiagonal	if $a_{ij} = 0$ whenever $|i - j| > 1$
upper bidiagonal	if $a_{ij} = 0$ whenever $i > j$ or $j > i + 1$
upper triangular	if $a_{ij} = 0$ whenever $i > j$
strictly upper triangular	if $a_{ij} = 0$ whenever $i \geq j$
upper Hessenberg	if $a_{ij} = 0$ whenever $i > j + 1$.

Analogous definitions hold for *lower bidiagonal, lower triangular, strictly lower triangular*, and *lower Hessenberg* matrices. (Just apply the above definitions to A^T.)

The adjectives "upper" and "lower" will usually be deleted wherever context permits. A triangular matrix is *unit triangular* if it has ones on its diagonal.

A special notation is convenient for diagonal matrices. If $A \in \mathbb{R}^{m \times n}$ and we write

$$A = \text{diag}(\alpha_1, \ldots, \alpha_k), \qquad\qquad k = \min\{m, n\}$$

then $A = (a_{ij})$ is diagonal and $a_{ii} = \alpha_i$ for $i = 1, \ldots, k$.

A matrix is *sparse* if it has relatively few nonzero entries. Banded matrices in $\mathbb{R}^{m \times n}$ whose bandwidths are much smaller than m and n are sparse.

The zero-nonzero structure of a matrix can be conveniently displayed by letting "x" denote an arbitrary nonzero scalar. Thus if $A \in \mathbb{R}^{5 \times 4}$ is upper bidiagonal, then

$$A = \begin{bmatrix} x & x & 0 & 0 \\ 0 & x & x & 0 \\ 0 & 0 & x & x \\ 0 & 0 & 0 & x \\ 0 & 0 & 0 & 0 \end{bmatrix}.$$

There are several important types of square matrices. We say that $A \in \mathbb{R}^{n \times n}$ is

symmetric	if $A^T = A$
skew-symmetric	if $A^T = -A$
positive definite	if $x^T A x > 0, 0 \neq x \in \mathbb{R}^n$
non-negative definite	if $x^T A x \geq 0, x \in \mathbb{R}^n$
indefinite	if $(x^T A x)(y^T A y) < 0$ for some $x, y \in \mathbb{R}^n$.
orthogonal	if $A^T A = I_n$
nilpotent	if $A^k = 0$ for some k.
idempotent	if $A^2 = A$.
positive	if $a_{ij} > 0$ for all i and j
non-negative	if $a_{ij} \geq 0$ for all i and j
diagonally dominant	if $|a_{ii}| > \Sigma_{j \neq i} |a_{ij}|$ for all i
permutation	if $A = [e_{s_1}, \ldots, e_{s_n}]$ where $(s_1, \ldots, s_n)$ is a permutation of $(1, 2, \ldots, n)$.

Let $A \in \mathbb{R}^{n \times n}$. If $X \in \mathbb{R}^{n \times n}$ is nonsingular, then we say that A and $X^T A X$ are *congruent*. We point out that the properties of symmetry, skew-symmetry, and definiteness are preserved under congruence transformations.

Problems

P1.3-1.   Suppose $A \in \mathbb{R}^{n \times n}$ has positive diagonal entires. Show that if both A and A^T are diagonally dominant then A is positive definite.

P1.3-2.   Show that $A = \begin{bmatrix} a & b \\ 0 & c \end{bmatrix}$ is positive definite if $a > 0, c > 0$, and $|b| < 2\sqrt{ac}$.

P1.3-3.   Show that the following subsets of $\mathbb{R}^{n \times n}$ are closed under matrix multiplication: (a) upper triangular matrices, (b) permutation matrices, and (c) orthogonal matrices.

P1.3-4.   Show that the following subsets of $\mathbb{R}^{n \times n}$ are closed under inversion: (a) positive definite matrices, (b) nonsingular symmetric matrices, (c) unit triangular matrices, and (d) nonsingular triangular matrices

P1.3-5. Show that if $A \in \mathbb{R}^{m \times n}$ is upper triangular and $B \in \mathbb{R}^{n \times m}$ is upper Hessenberg, then $C = AB$ is upper Hessenberg.

P1.3-6. Show that a strictly upper triangular square matrix is nilpotent.

P1.3-7. Suppose $A \in \mathbb{R}^{m \times m}$ and that $X \in \mathbb{R}^{m \times n}$ satisfies $\text{rank}(X) = n$. Show that the properties of symmetry, skew-symmetry, and positive definiteness are preserved by the transformation $X^T A X$.

P1.3-8. Show that $A \in \mathbb{R}^{n \times n}$ is positive definite if and only if its symmetric part, $(A + A^T)/2$, is positive definite.

P1.3-9. Show that a permutation matrix is orthogonal.

P1.3-10. Show that if S is skew-symmetric then $I - S$ is nonsingular and $(I - S)^{-1}(I + S)$ is orthogonal. (This is known as the *Cayley transform* of S.)

P1.3-11. Define $B = A^T A$ where $A \in \mathbb{R}^{m \times n}$. Show that B is non-negative definite. Show that B is positive definite if and only if $\text{rank}(A) = n$. Show that if A is upper triangular with bandwidth p, then B has bandwidth p.

P1.3-12. Show that a triangular orthogonal matrix is diagonal.

P1.3-13. Show that if the k-th diagonal entry of an upper triangular matrix is zero, then the first k columns of the matrix are dependent.

Sec. 1.4. Block Matrices and Complex Matrices

Let $\{m_1, \ldots, m_p\}$ and $\{n_1, \ldots, n_q\}$ be sets of positive integers. If

$$A_{ij} \in \mathbb{R}^{m_i \times n_j} \qquad\qquad i = 1, \ldots, p, \ j = 1, \ldots, q$$

then

$$A = [A_{ij}] = \begin{bmatrix} A_{11} & A_{12} & \cdots & A_{1q} \\ A_{21} & A_{22} & \cdots & A_{2q} \\ \vdots & \vdots & & \vdots \\ A_{p1} & A_{p2} & \cdots & A_{pq} \end{bmatrix}$$

is a p-by-q *block matrix* and A_{ij} is referred to as the (i, j) block.

Many of the properties described in the previous section extend to block matrices. For example,

$$A = \begin{bmatrix} A_{11} & A_{12} \\ 0 & A_{22} \end{bmatrix}$$

is a 2-by-2 block upper triangular matrix, while

$$\begin{bmatrix} A_{11} & A_{12} & 0 \\ A_{21} & A_{22} & A_{23} \\ 0 & A_{32} & A_{33} \end{bmatrix}$$

is a 3-by-3 block tridiagonal matrix. Other generalizations of the definitions in §1.3 are possible by simply substituting A_{ij} for a_{ij}.

A convenient way to indicate the block dimensions in a matrix is as follows:

$$A = \begin{bmatrix} A_{11} & A_{12} \\ A_{21} & A_{22} \end{bmatrix} \begin{matrix} r \\ s \end{matrix} \\ \quad\ \ u \quad\ v \quad .$$

With this notation, for example, A_{21} is an s-by-u block.

Block notation allows us to regard a given matrix in more than one way. For example, a $2k$-by-$2k$ five-diagonal matrix ($a_{ij} = 0$, $|i - j| > 2$) can be viewed as a k-by-k block tridiagonal matrix with 2-by-2 blocks. Such observations can sometimes suggest new algorithms.

Furthermore, it is sometimes useful to partition matrices and vectors conformably and observe how they interact at the block level. For example,

$$A = [B, \quad C, \quad D], w = \begin{bmatrix} x \\ y \\ z \end{bmatrix} \begin{matrix} p \\ q \\ r \end{matrix} \Rightarrow Aw = Bx + Cy + Dz.$$
$$\quad\ \ p \quad q \quad r$$

As with block matrices, the concepts in the previous sections also carry over to complex matrices. The set of m-by-n complex matrices will be denoted by $\mathbb{C}^{m \times n}$ and the set of complex n-vectors by $\mathbb{C}^n$. If $A \in \mathbb{C}^{m \times n}$, then its *conjugate transpose* A^H is defined by $A^H = (\overline{a_{ji}})$. The inner product of x and y in $\mathbb{C}^n$ thus has the form

$$x^H y = \sum_{i=1}^{n} \overline{x_i} y_i = \overline{y^H x}.$$

We say that $A \in \mathbb{C}^{n \times n}$ is *unitary* if $A^H A = I_n$, *Hermitian* if $A^H = A$, and *positive definite* if $x^H A x > 0$ for all nonzero $x \in \mathbb{C}^n$.

In general, the terminology for real matrices and vectors carries over to the complex case in obvious fashion.

Problems

P1.4-1. Under what conditions is a block triangular matrix triangular? Under what conditions is a block symmetric matrix symmetric?

P1.4-2. Show that if

$$A = \begin{bmatrix} A_{11} & A_{12} \\ A_{21} & A_{22} \end{bmatrix} \begin{matrix} p \\ q \end{matrix}$$
$$\quad\quad p \quad\ q$$

is positive definite, then A_{11} and A_{22} are also positive definite.

P1.4-3. Show that if $A \in \mathbb{C}^{n \times n}$ is Hermitian, then $A = P + iQ$, where $P^{\mathrm{T}} = P \in \mathbb{R}^{n \times n}$, $-Q^{\mathrm{T}} = Q \in \mathbb{R}^{n \times n}$, and $i^2 = -1$.

P1.4-4. Find a 2-by-2 real matrix A with the property that $x^{\mathrm{T}} A x > 0$ for all $0 \neq x \in \mathbb{R}^2$ but which is not positive definite when regarded as a member of $\mathbb{C}^{2 \times 2}$.

P1.4-5. Show that if $Q = Q_1 + iQ_2$ is unitary with $Q_1, Q_2 \in \mathbb{R}^{n \times n}$, then the $2n$-by-$2n$ real matrix

$$Z = \begin{bmatrix} Q_1 & -Q_2 \\ Q_2 & Q_1 \end{bmatrix}$$

is orthogonal.

Notes and References for Chap. 1

There are many introductory linear algebra texts but only a few that provide the necessary background for the beginning student of matrix computations. Among them, we have found the following very useful:

B. Noble and J. W. Daniel (1977). *Applied Linear Algebra*, Prentice-Hall, Englewood Cliffs.

P. R. Halmos (1958). *Finite Dimensional Vector Spaces*, 2nd ed., Van Nostrand-Reinhold, Princeton.

G. Strang (1976). *Linear Algebra and Its Applications*, Academic Press, New York.

S. J. Leon (1980). *Linear Algebra with Applications*. Macmillan, New York.

More encyclopedic treatments include

R. Bellman (1970). *Introduction to Matrix Analysis*, 2nd ed., McGraw-Hill, New York.

A. S. Householder (1964). *The Theory of Matrices in Numerical Analysis*, Ginn (Blaisdell), Boston.

F. R. Gantmacher (1959). *The Theory of Matrices, vols. 1 and 2*, Chelsea, New York.

Stewart (IMC, chap. 1) also provides an excellent review of matrix algebra for numerical analysts.

Measuring Vectors, Matrices, Subspaces, and Linear System Sensitivity

§2.1 Vector Norms

§2.2 Matrix Norms

§2.3 The Singular Value Decomposition

§2.4 Orthogonal Projections and the C-S Decomposition

§2.5 The Sensitivity of Square Linear Systems

The problem concerning us in subsequent chapters is how to find $x \in \mathbb{R}^n$ such that $Ax = b$ where $A \in \mathbb{R}^{m \times n}$ and $b \in \mathbb{R}^m$ are given. This problem has a solution if and only if $b \in R(A)$, and it is unique if and only if $N(A) = \{0\}$.

This complete characterization of the $Ax = b$ solution set is inadequate from the numerical point of view. Finite precision arithmetic and the reality of inexact data force us to ask the following questions:

(a) If A and b are perturbed by a "small" amount, how is x affected?

(b) What does it mean for A to be "nearly" rank deficient?

(c) If $b \notin R(A)$, then how can x be determined so that Ax is "close" to b?

To answer these quantitative questions we need a language for making precise such notions as "small perturbations," "near rank deficiency," and "distance" in a vector space. Norms provide this language.

In the first two sections we define vector and matrix norms and summarize some of their important properties. Using norms, we establish the very fundamental singular value decomposition in §2.3. A related decomposition called the C-S decomposition is then presented in §2.4. The C-S decomposition enables us to quantify the notion of distance between subspaces. We conclude in §2.5 with an analysis of the $Ax = b$ problem and introduce the important concept of condition.

We wish to stress the importance of having a facility with the material in this chapter. It will be used frequently throughout the book without explicit mention.

All references are given at the end of the chapter.

Sec. 2.1. Vector Norms

A *vector norm* on $\mathbb{R}^n$ is a function $f : \mathbb{R}^n \to \mathbb{R}$ with the following properties

(2.1-1) $f(x) \geqslant 0$ for all $x \in \mathbb{R}^n$ with equality if and only if $x = 0$.

(2.1-2) $f(x + y) \leqslant f(x) + f(y)$ for all $x, y \in \mathbb{R}^n$.

(2.1-3) $f(\alpha x) = |\alpha| \, f(x)$ for all $\alpha \in R, x \in \mathbb{R}^n$.

We shall denote such a function $f(x)$ by $\|x\|$ with subscripts on the double bar to distinguish between various norms.

A useful class of norms are the *Hölder* or *p-norms* defined by

(2.1-4) $$\|x\|_p = (|x_1|^p + \cdots + |x_n|^p)^{1/p} \qquad\qquad p \geqslant 1$$

of which

$$\|x\|_1 = |x_1| + \cdots + |x_n|$$
$$\|x\|_2 = (|x_1|^2 + \cdots + |x_n|^2)^{1/2} = (x^Tx)^{1/2}$$

and

$$\|x\|_\infty = \max_i |x_i|$$

are the most important. A classic result concerning *p-norms* is the *Hölder inequality*

(2.1-5) $$|x^Ty| \leqslant \|x\|_p \|y\|_q. \qquad\qquad \frac{1}{p} + \frac{1}{q} = 1$$

A very important special case of this is the *Cauchy-Schwartz inequality*:

(2.1-6) $$|x^Ty| \leqslant \|x\|_2 \|y\|_2.$$

Notice that the 2-norm is invariant under orthogonal transformation, for if $Q^TQ = I$, then

$$\|Qx\|_2^2 = x^TQ^TQx = x^Tx = \|x\|_2^2.$$

All norms on $\mathbb{R}^n$ are *equivalent*, i.e., if $\| \cdot \|_\alpha$ and $\| \cdot \|_\beta$ are norms on $\mathbb{R}^n$, then there exist positive constants c_1 and c_2 such that

(2.1-7) $$c_1\|x\|_\alpha \leqslant \|x\|_\beta \leqslant c_2\|x\|_\alpha$$

for all $x \in \mathbb{R}^n$. For example, for any $x \in \mathbb{R}^n$ we have

$$\|x\|_2 \leqslant \|x\|_1 \leqslant \sqrt{n}\,\|x\|_2$$

(2.1-8)
$$\|x\|_\infty \leqslant \|x\|_2 \leqslant \sqrt{n}\,\|x\|_\infty$$

$$\|x\|_\infty \leqslant \|x\|_1 \leqslant n\,\|x\|_\infty.$$

Norms serve the same purpose on vector spaces that absolute value does on the real line: they permit the concept of distance. More precisely, $\mathbb{R}^n$ together with a norm on $\mathbb{R}^n$ defines a metric space. Therefore, we have the familiar notions of neighborhood, open sets, convergence, and continuity when working with vectors and vector-valued functions.

Suppose $\hat{x} \in \mathbb{R}^n$ is an approximation to $x \in \mathbb{R}^n$. For a given vector norm $\|\cdot\|$ we say that

$$\epsilon_a = \|\hat{x} - x\|$$

is the *absolute error* and

$$\epsilon_r = \frac{\|\hat{x} - x\|}{\|x\|} \qquad\qquad x \neq 0$$

the *relative error*. Relative error in the ∞-norm can be translated into a statement about the number of correct significant digits in x. In particular, if

$$\frac{\|\hat{x} - x\|_\infty}{\|x\|_\infty} \cong 10^{-p}$$

then the largest component of $\hat{x}$ will have approximately p correct significant digits.

EXAMPLE 2.1-1. If $x = (1.234, .05674)^T$ and $\hat{x} = (1.235, .05128)^T$, then $\|\hat{x} - x\|_\infty / \|x\|_\infty \cong .0043 \cong 10^{-3}$. Note that $\hat{x}_1$ has about three significant digits that are correct while only one significant digit in $\hat{x}_2$ is right.

Problems

P2.1-1. Show that for $x \in \mathbb{R}^n$, $\lim_{p \to \infty} \|x\|_p = \max_{1 \leqslant i \leqslant n} |x_i|$.

P2.1-2. Prove the Cauchy-Schwartz inequality (2.1-6).

P2.1-3. Verify that $\|\cdot\|_1$, $\|\cdot\|_2$, and $\|\cdot\|_\infty$ are vector norms.

P2.1-4. Prove that each of the inequalities in (2.1-8) is correct. Give $n = 2$ examples illustrating how each upper bound can be attained.

P2.1-5. Show that in $\mathbb{R}^n$, $x^{(i)} \to x$ if and only if $x_k^{(i)} \to x_k$ for $k = 1, \ldots, n$.

P2.1-6. Show that any vector norm on $\mathbb{R}^n$ is uniformly continuous by verifying the inequality

$$|\,\|x\| - \|y\|\,| \leqslant \|x - y\|. \qquad\qquad x, y \in \mathbb{R}^n$$

P2.1-7. Show that the function $f(x) = (x^T A x)^{1/2}$ is a vector norm on $\mathbb{R}^n$ if and only if A is positive definite.

P2.1-8. Let $\| \cdot \|$ be a vector norm on $\mathbb{R}^m$ and assume $A \in \mathbb{R}^{m \times n}$. Show that if rank$(A) = n$ then $\|x\|_A = \|Ax\|$ is a vector norm on $\mathbb{R}^n$.

P2.1-9. Let x and y be in $\mathbb{R}^n$ and define $\phi: \mathbb{R} \to \mathbb{R}$ by $\phi(\alpha) = \|x - \alpha y\|_2$. Show that ϕ is minimized when $\alpha = x^T y / y^T y$.

Sec. 2.2. Matrix Norms

The analysis of matrix algorithms requires that we be able to assess the size of matrices. The notion of a matrix norm is useful for this purpose. Since $\mathbb{R}^{m \times n}$ is isomorphic to $\mathbb{R}^{mn}$, the definition of a matrix norm should be equivalent to the definition of a vector norm:

(2.2-1) $f(A) \geqslant 0$ for all $A \in \mathbb{R}^{m \times n}$ with equality if and only if $A = 0$.

(2.2-1) $f(A + B) \leqslant f(A) + f(B)$ for all $A, B \in \mathbb{R}^{m \times n}$.

(2.2-3) $f(\alpha A) = |\alpha| f(A)$ for all $\alpha \in \mathbb{R}, A \in \mathbb{R}^{m \times n}$.

Any function f satisfying these criteria is a matrix norm and will be denoted with the double bar notation.

The most frequently used matrix norms in numerical analysis are the F-norm (Frobenius norm),

$$(2.2\text{-}4) \qquad \|A\|_F = \left[\sum_{i=1}^{m} \sum_{j=1}^{n} |a_{ij}|^2 \right]^{1/2}$$

and the p-norms

$$(2.2\text{-}5) \qquad \|A\|_p = \sup_{x \neq 0} \frac{\|Ax\|_p}{\|x\|_p}.$$

The vector p-norms that are part of this definition are defined in (2.1-4). The verification that (2.2-4) and (2.2-5) are matrix norms is left as an exercise.

It is important to notice that (2.2-4) and (2.2-5) define families of norms—the 2-norm on $\mathbb{R}^{3 \times 2}$ is a different function from the 2-norm on $\mathbb{R}^{5 \times 6}$. Thus, the observation that

$$(2.2\text{-}6) \qquad \|AB\|_p \leqslant \|A\|_p \|B\|_p \qquad A \in \mathbb{R}^{m \times n}, B \in \mathbb{R}^{n \times q}$$

is really an observation about the relationship between three different norms. Formally, we say that norms f_1, f_2 and f_3 on $\mathbb{R}^{m \times q}$, $\mathbb{R}^{m \times n}$ and $\mathbb{R}^{n \times q}$, respectively, are *mutually consistent* if for all $A \in \mathbb{R}^{m \times n}$ and $B \in \mathbb{R}^{n \times q}$ we have $f_1(AB) \leqslant f_2(A) f_3(B)$.

Not all matrix norms satisfy (2.2-6). For example, if $\|A\|_\Delta = \max |a_{ij}|$ and

$$A = B = \begin{bmatrix} 1 & 1 \\ 1 & 1 \end{bmatrix},$$

then $\|AB\|_\Delta > \|A\|_\Delta \|B\|_\Delta$. Unless otherwise stated, we will always work with norms that satisfy $\|AB\| \leqslant \|A\| \|B\|$.

The p-norms have the important property that for every $A \in \mathbb{R}^{m \times n}$

$$\|Ax\|_p \leqslant \|A\|_p \|x\|_p. \qquad\qquad \forall x \in \mathbb{R}^n$$

More generally, for any vector norm $\|\cdot\|_\alpha$ on $\mathbb{R}^n$ and $\|\cdot\|_\beta$ on $\mathbb{R}^m$,

$$\|Ax\|_\beta \leqslant \|A\|_{\alpha,\beta} \|x\|_\alpha,$$

where $\|A\|_{\alpha,\beta}$ is a matrix norm defined by

(2.2-7) $$\|A\|_{\alpha,\beta} = \sup_{x \neq 0} \frac{\|Ax\|_\beta}{\|x\|_\alpha} = \sup_{\|x\|_\alpha = 1} \|Ax\|_\beta.$$

We say that $\|\cdot\|_{\alpha,\beta}$ is *subordinate* to the vector norms $\|\cdot\|_\alpha$ and $\|\cdot\|_\beta$. Since the set $\{x \in \mathbb{R}^n \mid \|x\|_\alpha = 1\}$ is compact and $\|\cdot\|_\beta$ is continuous, it follows that

(2.2-8) $$\|A\|_{\alpha,\beta} = \sup_{\|x\|_\alpha = 1} \|Ax\|_\beta = \|Ax^*\|_\beta$$

for some $x^* \in \mathbb{R}^n$ having unit α-norm.

The Frobenius and p-norms (especially $p = 1, 2, \infty$) satisfy certain inequalities that will be frequently used in the sequel without justification. Thus, it is useful to tabulate these relationships for future reference. For $A \in \mathbb{R}^{m \times n}$ we have

(2.2-9) $$\|A\|_2 \leqslant \|A\|_F \leqslant \sqrt{n} \|A\|_2$$

(2.2-10) $$\max |a_{ij}| \leqslant \|A\|_2 \leqslant \sqrt{mn} \max |a_{ij}|$$

(2.2-11) $$\|A\|_2 \leqslant \sqrt{\|A\|_1 \|A\|_\infty}$$

(2.2-12) $$\|A\|_1 = \max_j \sum_{i=1}^m |a_{ij}|$$

(2.2-13) $$\|A\|_\infty = \max_i \sum_{j=1}^n |a_{ij}|$$

(2.2-14) $$\frac{1}{\sqrt{n}} \|A\|_\infty \leqslant \|A\|_2 \leqslant \sqrt{m} \|A\|_\infty$$

(2.2-15) $$\frac{1}{\sqrt{m}} \|A\|_1 \leqslant \|A\|_2 \leqslant \sqrt{n} \|A\|_1.$$

An important property of the Frobenius norm and the 2-norm is that they are invariant with respect to orthogonal transformations; i.e., for all orthogonal Q and Z of appropriate dimensions we have

(2.2-16) $$\|QAZ\|_F = \|A\|_F$$

and

(2.2-17) $$\|QAZ\|_2 = \|A\|_2.$$

Problems

P2.2-1. Show $\|AB\|_p \leqslant \|A\|_p \|B\|_p$ $(1 \leqslant p \leqslant \infty)$.

P2.2-2. Let B be any submatrix of A. Show that $\|B\|_p \leqslant \|A\|_p$.

P2.2-3. Show that if $D = \text{diag}(\mu_1, \ldots, \mu_k) \in \mathbb{R}^{m \times n}$ with $k = \min\{m, n\}$, then $\|D\|_p = \max |\mu_i|$.

P2.2-4. Verify (2.2-9). (Hint: $\|a\|_2 \leqslant \|A\|_2$, where a is any column of A.)

P2.2-5. Verify (2.2-10).

P2.2-6. Verify (2.2-11).

P2.2-7. Verify (2.2-12) and (2.2-13).

P2.2-8. Verify (2.2-14) and (2.2-15).

P2.2-9. Verify (2.2-16) and (2.2-17).

P2.2-10. Show that if $\|\cdot\|_\alpha$ is a norm on $\mathbb{R}^{n \times n}$ satisfying

$$\|AB\|_\alpha \leqslant \|A\|_\alpha \|B\|_\alpha, \text{ then } \|A\|_\alpha < 1 \text{ implies } \lim_{k \to \infty} A^k = 0.$$

P2.2-11. Show that if $0 \neq s \in \mathbb{R}^n$ and $E \in \mathbb{R}^{n \times n}$, then

$$\left\| E\left(I - \frac{ss^{\mathrm{T}}}{s^{\mathrm{T}}s}\right) \right\|_F^2 = \|E\|_F^2 - \frac{\|Es\|_2^2}{s^{\mathrm{T}}s}.$$

P2.2-12. Suppose $u \in \mathbb{R}^m$ and $v \in \mathbb{R}^n$. Show that if $E = uv^{\mathrm{T}}$ then $\|E\|_F = \|E\|_2 = \|u\|_2 \|v\|_2$ and that $\|E\|_\infty = \|u\|_\infty \|v\|_1$.

P2.2-13. Suppose $A \in \mathbb{R}^{m \times n}$, $y \in \mathbb{R}^m$, and $0 \neq s \in \mathbb{R}^n$. Show $E = (y - As)s^{\mathrm{T}}/s^{\mathrm{T}}s$ has the smallest 2-norm among all $E \in \mathbb{R}^{m \times n}$ satisfying $(A + E)s = y$.

Sec. 2.3. The Singular Value Decomposition

The theory of norms developed in the previous two sections can be used to prove one of the most important decompositions in matrix computations—the singular value decomposition.

THEOREM 2.3-1: *Singular Value Decomposition (SVD).* If $A \in \mathbb{R}^{m \times n}$ then there exist orthogonal matrices

$$U = [u_1, \ldots, u_m] \in \mathbb{R}^{m \times m}$$

and

$$V = [v_1, \ldots, v_n] \in \mathbb{R}^{n \times n}$$

such that

$$U^{\mathrm{T}}AV = \text{diag}(\sigma_1, \ldots, \sigma_p) \qquad p = \min\{m, n\}$$

where

$$\sigma_1 \geqslant \sigma_2 \geqslant \cdots \geqslant \sigma_p \geqslant 0.$$

Proof. Let $x \in \mathbb{R}^n$ and $y \in \mathbb{R}^m$ be such that $\|x\|_2 = \|y\|_2 = 1$ and $Ax = \sigma y$ with $\sigma = \|A\|_2$. Let

$$V = [x, V_1] \in \mathbb{R}^{n \times n}$$

and

$$U = [y, U_1] \in \mathbb{R}^{m \times m}$$

be orthogonal. (Recall that it is always possible to extend an orthonormal set of vectors to an orthonormal basis for the whole space.) It follows that $U^T A V$ has the following structure:

$$A_1 \equiv U^T A V = \begin{bmatrix} \sigma & w^T \\ 0 & B \end{bmatrix} \begin{matrix} 1 \\ m\text{-}1 \end{matrix}$$
$$\quad\quad\quad\quad\quad\quad 1 \quad n\text{-}1$$

Since

$$\left\| A_1 \begin{pmatrix} \sigma \\ w \end{pmatrix} \right\|_2^2 \geqslant (\sigma^2 + w^T w)^2$$

it follows that $\|A_1\|_2^2 \geqslant \sigma^2 + w^T w$. But since $\sigma^2 = \|A\|_2^2 = \|A_1\|_2^2$, we must have $w = 0$. An obvious induction argument completes the proof. $\square$

The σ_i are the *singular values* of A and the vectors u_i and v_i are, respectively, the i-th *left singular vector* and the i-th *right singular vector*. It is easy to verify that

$$Av_i = \sigma_i u_i$$
$$\quad\quad\quad\quad\quad\quad\quad i = 1, \ldots, p$$
$$A^T u_i = \sigma_i v_i.$$

It will be convenient to have the following notation in conjunction with singular values:

$$\sigma_i(A) = \text{the } i\text{-th largest singular value of } A,$$
$$\sigma_{\max}(A) = \text{the largest singular value of } A,$$
$$\sigma_{\min}(A) = \text{the smallest singular value of } A.$$

The singular values of a matrix A are precisely the lengths of the semi-axes of the hyperellipsoid E defined by

$$E = \{y \mid y = Ax, \|x\|_2 = 1\}.$$

EXAMPLE 2.3-1.

$$\text{The matrix } A = \begin{bmatrix} .96 & 1.72 \\ 2.28 & .96 \end{bmatrix} \text{ has SVD}$$

$$A = U\Sigma V^{T} = \begin{bmatrix} .6 & -.8 \\ .8 & .6 \end{bmatrix} \begin{bmatrix} 3 & 0 \\ 0 & 1 \end{bmatrix} \begin{bmatrix} .8 & .6 \\ .6 & -.8 \end{bmatrix}^{T}$$

and the associated ellipsoid shown in Figure 2.3-1.

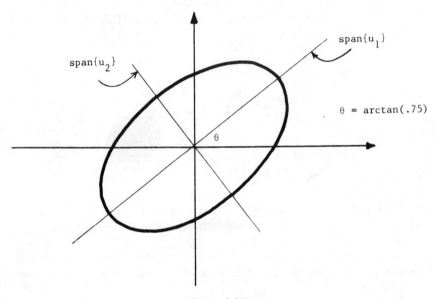

Figure 2.3-1.

The SVD reveals a great deal about the structure of a matrix as evidenced by the following corollary.

COROLLARY 2.3-2. If the SVD of A is given by Theorem 2.3-1 and

$$\sigma_1 \geqslant \cdots \geqslant \sigma_r > \sigma_{r+1} = \cdots = \sigma_p = 0,$$

then

(2.3-1) $\text{rank}(A) = r,$

(2.3-2) $N(A) = \text{span}\{v_{r+1}, \ldots, v_n\},$

(2.3-3) $R(A) = \text{span}\{u_1, \ldots, u_r\},$

$$(2.3\text{-}4) \qquad A = \sum_{i=1}^{r} \sigma_i u_i v_i^{\mathrm{T}} = U_r \Sigma_r V_r^{\mathrm{T}} \text{ where } U_r = [u_1, \ldots, u_r],$$

$$V_r = [v_1, \ldots, v_r], \qquad \text{and} \qquad \Sigma_r = \mathrm{diag}(\sigma_1, \ldots, \sigma_r),$$

$$(2.3\text{-}5) \qquad \|A\|_F^2 = \sigma_1^2 + \cdots + \sigma_p^2,$$

$$(2.3\text{-}6) \qquad \|A\|_2 = \sigma_1.$$

The proofs of these results are left as exercises.

One of the most valuable aspects of the SVD is that it enables us to deal sensibly with the concept of matrix rank. Numerous theorems in linear algebra have the form "if such-and-such a matrix has full rank, then such-and-such a property holds." While neat and aesthetic, results of this flavor fail to foreshadow the numerical difficulties frequently encountered in situations where near rank deficiency prevails. Rounding errors and fuzzy data make rank determination a nontrivial exercise. In this regard the SVD is particularly useful because it permits us to quantify the notion of near rank deficiency.

COROLLARY 2.3-3.   Let the SVD of $A \in \mathbb{R}^{m \times n}$ be given by Theorem 2.3-1. If $k < r = \mathrm{rank}(A)$ and

$$(2.3\text{-}7) \qquad A_k = \sum_{i=1}^{k} \sigma_i u_i v_i^{\mathrm{T}},$$

then

$$(2.3\text{-}8) \qquad \min_{\mathrm{rank}(B)=k} \|A - B\|_2 = \|A - A_k\|_2 = \sigma_{k+1}.$$

Proof.   Since $U^{\mathrm{T}} A_k V = \mathrm{diag}(\sigma_1, \ldots, \sigma_k, 0, \ldots, 0)$ it follows that $\mathrm{rank}(A_k) = k$ and that

$$\|A - A_k\|_2 = \|U^{\mathrm{T}}(A - A_k)V\|_2 = \|\mathrm{diag}(0, \ldots, 0, \sigma_{k+1}, \ldots, \sigma_p)\|_2 = \sigma_{k+1}.$$

Now suppose for some $B \in \mathbb{R}^{m \times n}$ we have $\mathrm{rank}(B) = k$. It follows that we can find orthonormal vectors $x_1, \ldots, x_{n-k}$ such that

$$N(B) = \mathrm{span}\{x_1, \ldots, x_{n-k}\}.$$

A dimension argument shows that

$$\mathrm{span}\{x_1, \ldots, x_{n-k}\} \cap \mathrm{span}\{v_1, \ldots, v_{k+1}\} \neq \{0\}.$$

Let z be a unit 2-norm vector in this intersection. Since $Bz = 0$ and

$$Az = \sum_{i=1}^{k+1} \sigma_i (v_i^{\mathrm{T}} z) u_i,$$

we have

$$\|A - B\|_2^2 \geqslant \|(A - B)z\|_2^2 = \|Az\|_2^2 = \sum_{i=1}^{k+1} \sigma_i^2 (v_i^T z)^2 \geqslant \sigma_{k+1}^2.$$

This completes the proof of the theorem. $\square$

Corollary 2.3-3 says that the smallest singular value of A is the 2-norm distance of A to the set of all rank-deficient matrices. It also follows from the corollary that the set of full rank matrices in $\mathbb{R}^{m \times n}$ is both open and dense.

Problems

P2.3-1. Prove Corollary 2.3-2.

P2.3-2. Prove that

$$\sigma_{max}(A) = \max_{\substack{x \in \mathbb{R}^n \\ y \in \mathbb{R}^m}} \frac{y^T A x}{\|x\|_2 \|y\|_2}.$$

P2.3-3. Use the SVD to show that if $A \in \mathbb{R}^{m \times n}$ then there exist $Q \in \mathbb{R}^{m \times n}$ and $P \in \mathbb{R}^{n \times n}$ such that $A = QP$, where $Q^T Q = I_n$ and P is symmetric and non-negative definite. (This decomposition is sometimes referred to as the *polar decomposition* because it is analogous to the complex number factorization $z = e^{i \arg(z)} |z|$.)

P2.3-4. Show that any matrix in $\mathbb{R}^{m \times n}$ is the limit of a sequence of full rank matrices.

P2.3-5. What can be said about the singular value decomposition of a symmetric positive definite matrix?

P2.3-6. Show that if $A \in \mathbb{R}^{m \times n}$ has rank n, then $\|A(A^T A)^{-1} A^T\|_2 = 1$.

P2.3-7. Use the SVD to establish the fundamental identity $\text{rank}(A) = \text{rank}(A^T)$.

P2.3-8. What is the nearest rank one matrix to $A = \begin{bmatrix} 1 & M \\ 0 & 1 \end{bmatrix}$ in the Frobenius norm?

Sec. 2.4. Orthogonal Projections and the C-S Decomposition

Let $S \subset \mathbb{R}^n$ be a subspace. $P \in \mathbb{R}^{n \times n}$ is the *orthogonal projection* onto S if

(2.4-1)
$$\begin{aligned} &\text{(a)} \quad R(P) = S \\ &\text{(b)} \quad P^2 = P \\ &\text{(c)} \quad P^T = P. \end{aligned}$$

From this definition it is easy to show that if $x \in \mathbb{R}^n$ then $Px \in S$ and $(I - P)x \in S^\perp$.

EXAMPLE 2.4-1. $P = vv^T / v^T v$ is the orthogonal projection onto $S = \text{span}\{v\}$, $v \in \mathbb{R}^n$. Pictorially we have Figure 2.4-1.

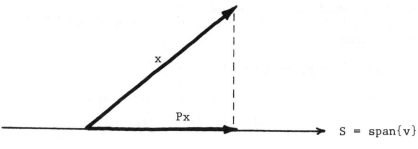

Figure 2.4-1.

If P_1 and P_2 are each orthogonal projections, then for any $z \in \mathbb{R}^n$ we have

$$(2.4\text{-}2) \qquad \| (P_1 - P_2)z \|_2^2 = (P_1 z)^{\mathrm{T}}(I - P_2)z + (P_2 z)^{\mathrm{T}}(I - P_1)z.$$

If in addition, $R(P_1) = R(P_2) = S$, then the right-hand side of this expression is zero. This shows that the orthogonal projection for a subspace is unique.

If the columns of $V = [v_1, \ldots, v_k]$ are an orthonormal basis for a subspace S, then it follows that

$$P = VV^{\mathrm{T}}$$

is the unique orthogonal projection onto S. It should be stressed that although P is unique, V is not.

There are several important orthogonal projections associated with the SVD

$$A = [U_r, \ \bar{U}_r] \, \Sigma \, [V_r, \ \bar{V}_r]^{\mathrm{T}}$$
$$\quad r \quad m\text{-}r \qquad r \quad n\text{-}r$$

where $r = \text{rank}(A)$. Namely,

$V_r V_r^{\mathrm{T}}$ is the orthogonal projection onto $N(A)^{\perp} = R(A^{\mathrm{T}})$,

$\bar{V}_r \bar{V}_r^{\mathrm{T}}$ is the orthogonal projection onto $N(A)$,

$U_r U_r^{\mathrm{T}}$ is the orthogonal projection onto $R(A)$,

$\bar{U}_r \bar{U}_r^{\mathrm{T}}$ is the orthogonal projection onto $R(A)^{\perp} = N(A^{\mathrm{T}})$.

The one-to-one correspondence between subspaces and orthogonal projections enables us to devise a notion of distance between subspaces. Suppose S_1 and S_2 are subspaces of $\mathbb{R}^n$ and that $\dim(S_1) = \dim(S_2)$. We define the *distance* between these two spaces by

$$(2.4\text{-}3) \qquad\qquad \text{dist}(S_1, S_2) = \| P_1 - P_2 \|_2,$$

where P_i is the orthogonal projection onto S_i.

By considering the case of one-dimensional subspaces in $\mathbb{R}^2$, we can obtain a geometrical interpretation of $\text{dist}(\cdot, \cdot)$. Suppose $S_1 = \text{span}\{x\}$ and $S_2 =$

span$\{y\}$, where x and y are unit 2-norm vectors in $\mathbb{R}^2$. Assume that $x^Ty = \cos(\theta)$ where $\theta \in [0, \frac{\pi}{2}]$, i.e., the diagram depicted in Figure 2.4-2. It follows that the difference between the projections onto these spaces satisfies

$$P_1 - P_2 = xx^T - yy^T = x[x - (x^Ty)y]^T - [y - (x^Ty)x]y^T.$$

If $\theta = 0$, then $\text{dist}(S_1, S_2) = \|P_1 - P_2\|_2 = \sin(\theta) = 0$. If $\theta \neq 0$ then

$$U_x = [u_1, u_2] = [x, -[y - (y^Tx)x]/\sin(\theta)]$$

and

$$V_x = [v_1, v_2] = [[x - (x^Ty)y]/\sin(\theta), y]$$

are defined and orthogonal. It follows that

$$P_1 - P_2 = U_x \, \text{diag}[\sin(\theta), \sin(\theta)] \, V_x^T$$

is the SVD of $P_1 - P_2$. Consequently, $\text{dist}(S_1, S_2) = \sin(\theta)$, the sine of the angle between the two subspaces.

This nice geometrical interpretation of the distance function can be extended to higher dimensions by invoking the following theorem:

THEOREM 2.4-1: *C-S Decomposition.* If

$$Q = \begin{bmatrix} Q_{11} & Q_{12} \\ Q_{21} & Q_{22} \end{bmatrix} \begin{matrix} k \\ j \end{matrix} \qquad k \geqslant j$$

$$\begin{matrix} k & j \end{matrix}$$

is orthogonal, then there exist orthogonal matrices U_1, $V_1 \in \mathbb{R}^{k \times k}$ and orthogonal matrices U_2, $V_2 \in \mathbb{R}^{j \times j}$ such that

$$(2.4\text{-}4) \quad \begin{bmatrix} U_1 & 0 \\ 0 & U_2 \end{bmatrix}^T \begin{bmatrix} Q_{11} & Q_{12} \\ Q_{21} & Q_{22} \end{bmatrix} \begin{bmatrix} V_1 & 0 \\ 0 & V_2 \end{bmatrix} = \begin{bmatrix} I & 0 & 0 \\ 0 & C & S \\ 0 & -S & C \end{bmatrix} \begin{matrix} k\text{-}j \\ j \\ j \end{matrix}$$

$$\begin{matrix} k\text{-}j & j & j \end{matrix}$$

where

$$C = \text{diag}(c_1, \ldots, c_j) \qquad c_i = \cos(\theta_i)$$

$$S = \text{diag}(s_1, \ldots, s_j) \qquad s_i = \sin(\theta_i)$$

and $0 \leqslant \theta_1 \leqslant \theta_2 \leqslant \cdots \leqslant \theta_j \leqslant \pi/2$.

Proof. See Davis and Kahan (1970) or Stewart (1977). The assumption $k \geqslant j$ is made for notational convenience only. □

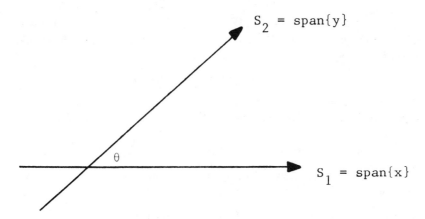

Figure 2.4-2.

Roughly speaking, the C-S decomposition amounts to a simultaneous diagonalization of the blocks of an orthogonal matrix.

EXAMPLE 2.4-2.   The matrices

$$Q = \begin{bmatrix} -.716 & .548 & .433 \\ -.698 & -.555 & -.451 \\ -.006 & -.626 & .780 \end{bmatrix}, \quad U = \begin{bmatrix} -.721 & -.692 & .000 \\ -.692 & .721 & .000 \\ .000 & .000 & 1 \end{bmatrix},$$

and

$$V = \begin{bmatrix} .999 & -.010 & .000 \\ -.010 & -.999 & .000 \\ .000 & .000 & 1 \end{bmatrix}$$

are orthogonal to three decimals and satisfy

$$U^T Q V = \begin{bmatrix} 1.000 & .000 & .000 \\ .000 & .780 & .625 \\ .000 & -.625 & .780 \end{bmatrix}.$$

Thus, in the language of Theorem 2.4-1, $k = 2, j = 1, c_1 = .780$ and $s_1 = .625$.

We now show how an appropriately chosen C-S decomposition reveals much about the "closeness" of a pair of equidimensional subspaces.

COROLLARY 2.4-2. Let $W = [W_1, W_2]$ and $Z = [Z_1, Z_2]$ be orthogonal matrices where $W_1, Z_1 \in \mathbb{R}^{n \times k}$ and $W_2, Z_2 \in \mathbb{R}^{n \times (n-k)}$. If $S_1 = R(W_1)$ and $S_2 = R(Z_1)$ then

$$\text{dist}(S_1, S_2) = \sqrt{1 - \sigma_{\min}^2(W_1^T Z_1)}.$$

Proof. Let $Q = W^T Z$ and assume that $k \geqslant j = n - k$. Let the C-S decomposition of Q be given by (2.4-4), $(Q_{ip} = W_i^T Z_p)$. It follows that

$$\| W_1^T Z_2 \|_2 = \| W_2^T Z_1 \|_2 = s_j = \sqrt{1 - c_j^2} = \sqrt{1 - \sigma_{\min}^2(W_1^T Z_1)}.$$

Since $W_1 W_1^T$ and $Z_1 Z_1^T$ are the orthogonal projections onto S_1 and S_2, respectively, we have

$$\text{dist}(S_1, S_2) = \| W_1 W_1^T - Z_1 Z_1^T \|_2 = \| W^T (W_1 W_1^T - Z_1 Z_1^T) Z \|_2$$

$$= \left\| \begin{pmatrix} 0 & W_1^T Z_2 \\ W_2^T Z_1 & 0 \end{pmatrix} \right\|_2 = s_j.$$

If $k < j$, then the above argument goes through merely by setting

$$Q = [W_2, W_1]^T [Z_2, Z_1]$$

and noting that $\sigma_{\min}(W_2^T Z_1) = \sigma_{\min}(W_1^T Z_2) = s_j$. $\square$

EXAMPLE 2.4-3. If $S_1 = \text{span}\{e_1, e_2\}$ and $S_2 = \text{span}\{e_2, e_3\}$, then the matrix Q in the corollary is given by

$$Q = [e_1, e_2, e_3, e_4, \ldots, e_n]^T [e_2, e_3, e_1, e_4, \ldots, e_n]$$

$$= [e_2, e_3, e_1, e_4, \ldots, e_n].$$

Thus, $\text{dist}(S_1, S_2) = 1$. This shows that subspace intersection is not equivalent to zero distance.

Problems

P2.4-1. Show that if P is an orthogonal projection, then $Q = I - 2P$ is orthogonal.

P2.4-2. What are the singular values of an orthogonal projection?

Sec. 2.5. The Sensitivity of Square Linear Systems

We now use the tools developed in the previous sections to analyze the linear system problem $Ax = b$ where $A \in \mathbb{R}^{n \times n}$ is nonsingular and $b \in \mathbb{R}^n$. Our aim is to examine how perturbations in A and b affect the solution x.

We begin by applying the singular value decomposition. If

$$A = \sum_{i=1}^{n} \sigma_i u_i v_i^T = U \Sigma V^T$$

is the SVD of A given by (2.3-4), then

$$(2.5\text{-}1) \qquad x = A^{-1}b = (U\Sigma V^{\mathrm{T}})^{-1}b = \sum_{i=1}^{n} \frac{u_i^{\mathrm{T}}b}{\sigma_i} v_i.$$

This expansion shows that small changes in A or b can induce relatively large changes in x if σ_n is small. Indeed, if $\cos(\theta) = |u_n^{\mathrm{T}}b| / \|b\|_2$ and

$$(A - \epsilon u_n v_n^{\mathrm{T}})y = b + \epsilon \operatorname{sign}(u_n^{\mathrm{T}}b)u_n, \qquad\qquad \sigma_n > \epsilon \geqslant 0$$

then it can be shown that

$$\|y - x\|_2 \geqslant \frac{\epsilon}{\sigma} \|x\|_2 \cos(\theta).$$

Thus, $O(\epsilon)$ perturbations can alter the solution by an amount ϵ/σ_n.

That the magnitude of σ_n should have a bearing on the sensitivity of the $Ax = b$ problem should come as no surprise when we recall from Corollary 2.3-3 that σ_n is the distance from A to the set of singular matrices. As we approach this set, it is intuitively clear that the solution x should be increasingly sensitive to perturbations.

A precise measure of linear system sensitivity can be obtained by considering the parameterized system

$$(A + \epsilon F)x(\epsilon) = b + \epsilon f, \qquad\qquad x(0) = x$$

where $F \in \mathbb{R}^{n \times n}$ and $f \in \mathbb{R}^n$. If A is nonsingular, then it is clear that $x(\epsilon)$ is differentiable in a neighborhood of zero and moreover,

$$\dot{x}(0) = A^{-1}(f - Fx).$$

Thus, the Taylor series expansion for $x(\epsilon)$ has the form

$$x(\epsilon) = x + \epsilon \dot{x}(0) + O(\epsilon^2).$$

Using any vector norm and consistent matrix norm we obtain

$$(2.5\text{-}2) \qquad \frac{\|x(\epsilon) - x\|}{\|x\|} \leqslant \epsilon \|A^{-1}\| \left\{ \frac{\|f\|}{\|x\|} + \|F\| \right\} + O(\epsilon^2).$$

Define the *condition number* $\kappa(A)$ by

$$(2.5\text{-}3) \qquad \kappa(A) = \|A\| \, \|A^{-1}\|.$$

Using the inequality $\|b\| \leqslant \|A\| \, \|x\|$ it follows from (2.5-2) that

$$(2.5\text{-}4) \qquad \frac{\|x(\epsilon) - x\|}{\|x\|} \leqslant \kappa(A)[\rho_A + \rho_b] + O(\epsilon^2),$$

where

$$\rho_A = \epsilon \frac{\|F\|}{\|A\|}$$

and

$$\rho_b = \epsilon \frac{\|f\|}{\|b\|}$$

represent the relative errors in A and b, respectively. Thus, the relative error in x can be $\kappa(A)$ times the relative error in A and b. In this sense, the condition number $\kappa(A)$ quantifies the sensitivity of the $Ax = b$ problem.

Note that $\kappa(\cdot)$ depends on the underlying norm. When this norm is to be stressed, we use subscripts, e.g.,

$$(2.5\text{-}5) \qquad \kappa_2(A) = \|A\|_2 \|A^{-1}\|_2 = \sigma_1(A)/\sigma_n(A).$$

Thus, the 2-norm condition of a matrix A measures the elongation of its associated hyperellipsoid (cf. Example 2.3-1).

We mention two other characterizations of the condition number. For p-norm condition numbers we have

$$(2.5\text{-}6) \qquad \kappa_p(A)^{-1} = \min_{A+E \text{ singular}} \frac{\|E\|_p}{\|A\|_p}.$$

This result may be found in Kahan (1966) and shows that $\kappa_p(A)$ measures the relative p-norm distance from A to the set of singular matrices.

For any norm we also have

$$(2.5\text{-}7) \qquad \kappa(A) = \lim_{\delta \to 0} \sup_{\|E\| \leqslant \delta \|A\|} \frac{\|(A + E)^{-1} - A^{-1}\|}{\delta} \frac{1}{\|A^{-1}\|}.$$

This imposing result merely says that the condition number is a normalized Frechet derivative of the map $A \to A^{-1}$. Further details may be found in Rice (1966). Recall that we were initially led to $\kappa(A)$ through differentiation.

If $\kappa(A)$ is large, then A is said to be *ill-conditioned*. Note this is a norm-dependent property. However, any two condition numbers $\kappa_\alpha(\cdot)$ and $\kappa_\beta(\cdot)$ on $\mathbb{R}^{n \times n}$ are equivalent in that constants c_1 and c_2 can be found for which

$$c_1 \kappa_\alpha(A) \leqslant \kappa_\beta(A) \leqslant c_2 \kappa_\alpha(A). \qquad\qquad A \in \mathbb{R}^{n \times n}$$

For example, on $\mathbb{R}^{n \times n}$ we have

$$\frac{1}{n} \kappa_2(A) \leqslant \kappa_1(A) \leqslant n \kappa_2(A)$$

$$(2.5\text{-}8) \qquad \frac{1}{n} \kappa_\infty(A) \leqslant \kappa_2(A) \leqslant n \kappa_\infty(A)$$

$$\frac{1}{n^2} \kappa_1(A) \leqslant \kappa_\infty(A) \leqslant n^2 \kappa_1(A).$$

Thus, if a matrix is ill-conditioned in the α-norm, it is ill-conditioned in the β-norm modulo the constants c_1 and c_2 above.

For any of the p-norms we have $\kappa_p(A) \geqslant 1$. Matrices with small condition numbers are said to be *well-conditioned*. In the 2-norm, orthogonal matrices are perfectly conditioned in that $\kappa_2(Q) = 1$ if Q is orthogonal.

It is natural to consider how well determinant size measures ill-conditioning. Unfortunately, there is little correlation between $\det(A)$ and the condition of $Ax = b$. For example, the matrices B_n defined by

$$(2.5\text{-}9) \qquad B_n = \begin{bmatrix} 1 & -1 & \cdots & -1 \\ & 1 & & -1 \\ & & \ddots & \vdots \\ \mathbf{0} & & & 1 \end{bmatrix} \in \mathbb{R}^{n \times n}$$

have determinant 1, but $\kappa_\infty(B_n) = n\, 2^{n-1}$. On the other hand, a very well-conditioned matrix can have a very small determinant. For example,

$$D_n = \text{diag}(10^{-1}, \ldots, 10^{-1}) \in \mathbb{R}^{n \times n}$$

satisfies $\kappa_p(D_n) = 1$ although $\det(D_n) = 10^{-n}$.

We conclude this section by presenting one final $Ax = b$ perturbation theorem. The earlier result (2.5-4) was useful to derive because it highlighted the connection between $\kappa(A)$ and the rate of change of $x(\epsilon)$ at $\epsilon = 0$. However, it is a little dissatisfying because it is contingent on ϵ being "small enough" and because it sheds no light on the size of the $O(\epsilon^2)$ term. In short, it is not a completely rigorous result. This is in contrast to the following:

THEOREM 2.5-1. Suppose $b \in \mathbb{R}^n$ and $A \in \mathbb{R}^{n \times n}$ is nonsingular. If $\Delta A \in \mathbb{R}^{n \times n}$ satisfies

$$\|\Delta A\|\, \|A^{-1}\| = r < 1,$$

then the perturbed system

$$(A + \Delta A)y = b + \Delta b \qquad\qquad \Delta b \in \mathbb{R}^n$$

is nonsingular. Moreover, if

$$\|\Delta A\| \leqslant \delta\|A\|$$

and

$$\|\Delta b\| \leqslant \delta\|b\|$$

for some $\delta \geqslant 0$, then

$$(2.5\text{-}10) \qquad \frac{\|x - y\|}{\|x\|} \leqslant 2\,\delta\,\kappa(A)/(1 - r). \qquad\qquad x \neq 0$$

Proof. See exercises P2.5-3, P2.5-4, and P2.5-5. $\square$

EXAMPLE 2.5-1. The $Ax = b$ problem

$$\begin{bmatrix} 1 & 0 \\ 0 & 10^{-6} \end{bmatrix} \begin{bmatrix} x_1 \\ x_2 \end{bmatrix} = \begin{bmatrix} 1 \\ 10^{-6} \end{bmatrix}$$

has solution $x = (1, 1)^T$ and condition $\kappa_\infty(A) = 10^6$. If $\Delta b = (10^{-6}, 0)^T$, $\Delta A = 0$, and $(A + \Delta A)y = b + \Delta b$, then the inequality (2.5-10) in the ∞-norm has the form

$$\frac{10^{-6}}{10^0} \leqslant 2 \times 10^{-6} \times 10^6.$$

Thus, the upper bound can be a gross overestimate of the error induced by the perturbation. On the other hand, if $\Delta b = (0, 10^{-6})^T$, $\Delta A = 0$, and $(A + \Delta A)y = b + \Delta b$, then the inequality says

$$\frac{10^0}{10^0} \leqslant 2 \times 10^{-6} \, 10^6.$$

Conclusion: there are perturbations for which the bound in (2.5-10) is essentially attained.

Problems

P2.5-1. Show that if $\|I\| \geqslant 1$ then $\kappa(A) \geqslant 1$.

P2.5-2. Show that for a given norm, $\kappa(AB) \leqslant \kappa(A)\,\kappa(B)$ and that $\kappa(\alpha A) = \kappa(A)$ for all nonzero α.

P2.5-3. (*Banach lemma*) Show that if both A and $A + E$ are nonsingular, then

$$\|(A + E)^{-1} - A^{-1}\| \leqslant \|E\| \, \|A^{-1}\| \, \|(A + E)^{-1}\|.$$

P2.5-4. Show that if $\|F\| < 1$ in any matrix norm, then $(I - F)^{-1} = \sum_{k=0}^{\infty} F^k$ and

$$\|(I - F)^{-1}\| \leqslant 1/(1 - \|F\|).$$

P2.5-5. Prove Theorem 2.5-1.

Notes and References for Chap. 2

Stewart (IMC, pp. 160–84) has a complete review of norms. See also

J. M. Ortega (1972). *Numerical Analysis: A Second Course*, Academic Press, New York.

A. S. Householder (1974). *The Theory of Matrices in Numerical Analysis*, Dover Publications, New York.

Forsythe and Moler (SLE) offer a good account of the SVD's role in the analysis of the $Ax = b$ problem. Their proof of the decomposition is more traditional than ours in that it makes use of the eigenvalue theory for symmetric matrices. Some early SVD references include

E. Beltrami (1873). "Sulle Funzioni Bilineari," *Giornale di Mathematiche 11*, 98–106.
C. Eckart and G. Young (1939). "A Principal Axis Transformation for Non-Hermitian Matrices," *Bull. Amer. Math. Soc. 45*, 118–21.

For generalizations of the SVD to infinite dimensional Hilbert space, see

I. C. Gohberg and M. G. Krein (1969). *Introduction to the Theory of Linear Non Self-Adjoint Operators*, Amer. Math. Soc., Providence, R.I.
F. Smithies (1970). *Integral Equations*, Cambridge University Press, Cambridge.

The concept condition is thoroughly investigated in

J. Rice (1966). "A Theory of Condition," *SIAM J. Num. Anal. 3*, 287–310.
W. Kahan (1966). "Numerical Linear Algebra," *Canadian Math. Bull. 9*, 757–801.

In some applications, it is necessary to examine the sensitivity of the solution to $Ax = b$ subject to structured perturbations. For example, we may wish to know how x varies as each element of b varies over some known interval. A paper concerned with this type of more refined sensitivity analysis is

J. E. Cope and B. W. Rust (1979). "Bounds on solutions of systems with inaccurate data," *SIAM J. Num. Anal. 16*, 950–63.

A proof of the C-S decomposition (Theorem 2.4-1) appears in

G. W. Stewart (1977). "On the Perturbation of Pseudo-Inverses, Projections and Linear Least Squares Problems," *SIAM Review 19*, 634–62.

The following papers discuss other aspects of this important decomposition:

C. Davis and W. Kahan (1970). "The Rotation of Eigenvectors by a Perturbation III," *SIAM J. Num. Anal. 7*, 1–46.
C. C. Paige and M. Saunders (1981). "Toward a Generalized Singular Value Decomposition," *SIAM J. Num. Anal. 18*, 398–405.

Numerical Matrix Algebra

§3.1 **Matrix Algorithms**
§3.2 **Rounding Errors**
§3.3 **Householder Transformations**
§3.4 **Givens Transformations**
§3.5 **Gauss Transformations**

In this chapter we develop more tools for constructing, describing, and analyzing the matrix algorithms that appear subsequently in the book. In §3.1 we familiarize the reader with our somewhat informal style of presenting algorithms. Next, we consider the effects of finite precision arithmetic and develop techniques that can be used to simplify the quantification of roundoff error. In the last three sections we describe the basic tools for zeroing components of vectors and matrices: Householder, Givens, and Gauss transformations. Most algorithms in the book are based on these important zeroing transformations.

All references are given at the end of the chapter.

Sec. 3.1 Matrix Algorithms

Consider the problem of computing $C = AB$ where both A and B are n-by-n upper triangular matrices. The entries of C are defined as follows:

(3.1-1)
$$c_{ij} = \begin{cases} \sum\limits_{k=i}^{j} a_{ik} b_{kj} & 1 \leqslant i \leqslant j \leqslant n \\ 0 & 1 \leqslant j < i \leqslant n \end{cases}.$$

Although complete mathematically, (3.1-1) is inadequate as an algorithm. For example, in what order are the c_{ij} computed? The question is critical if we want to overwrite A with the product AB.

To resolve the ambiguity inherent in statements like (3.1-1) we follow Stewart (IMC, pp. 83–93) and express our algorithms in a programming-like language that is precise enough to convey the important algorithmic concepts, but informal enough to permit the suppression of cumbersome details.

By way of illustration, here is a wholly adequate way of describing the algorithm for triangular matrix multiplication with overwriting:

ALGORITHM 3.1-1.   Given n-by-n upper triangular matrices A and B, the following algorithm computes the product $C = AB$ and stores the result in A.

> For $i = n, n - 1, \ldots, 1$
> > For $j = n, n - 1, \ldots, i$
> >
> > $$a_{ij} := \sum_{k=i}^{j} a_{ik} b_{kj}$$

The symbol ":=" denotes arithmetic assignment. A statement of the form "$a := b$" should be interpreted to mean "a becomes b."

There are, of course, remaining ambiguities in Algorithm 3.1-1. For example, how is the summation evaluated? This is an important question but somewhat distracting if, for instance, our aim is to illustrate how A can be overwritten by AB. Consequently, the summation expression is preferable to the following more detailed description of the inner product calculation:

> $s := 0$
> For $k = i, \ldots, j$
> > $s := s + a_{ik} b_{kj}$
> $a_{ij} := s$

Let us investigate the amount of work involved in Algorithm 3.1-1. In matrix computations it is customary to count multiplicative operations ($\times, \div$) since the number of additive operations ($+, -$) is approximately the same. Note that c_{ij} ($i \leqslant j$) requires ($j - i + 1$) multiplications. Using the identities

$$\sum_{p=1}^{q} p = q(q + 1)/2$$

and

$$\sum_{p=1}^{q} p^2 = \tfrac{1}{3}q^3 + \tfrac{1}{2}q^2 + \tfrac{1}{6}q$$

and ignoring low-order terms since we are interested only in order of magnitude, we find that Algorithm 3.1-1 requires $n^3/6$ multiplicative operations:

$$\sum_{i=n}^{1} \sum_{j=n}^{i} (j - i + 1) \cong \sum_{i=n}^{1} \sum_{j=1}^{n-i} j \cong \sum_{i=n}^{1} \frac{(n - i)^2}{2} \cong \tfrac{1}{2} \sum_{i=1}^{n} i^2 \cong n^3/6.$$

This means of quantifying work is necessarily crude when applied to computer programs, for it ignores subscripting, paging, and the countless other operations that go on during execution. As a way of acknowledging these "background" operations we shall use C. B. Moler's concept of a *flop*. A flop is more or less the amount of work associated with the statement

$$s := s + a_{ik}b_{kj},$$

i.e., in FORTRAN:

$$S = S + A(I, K) * B(K, J).$$

That is, a flop roughly constitutes the effort of doing a floating point add, a floating point multiply, and a little subscripting. Thus, Algorithm 3.1-1 requires $n^3/6$ flops. This terminology serves to remind us of the numerous overheads associated with floating point arithmetic.

In assessing the amount of storage that an algorithm needs, we are likewise concerned with order of magnitude. As written, Algorithm 3.1-1 requires $2n^2$ storage. However, an n-by-n upper triangular matrix T can be stored in an $n(n + 1)/2$ linear array t as follows:

(3.1-2) $$t = (t_{11}, t_{12}, t_{22}, t_{13}, \ldots, t_{1n}, \ldots, t_{nn}).$$

Storing T in this fashion halves memory requirements at the expense of more cumbersome subscripting; t_{ij} is in $t_{j+(i-1)(n-i/2)}$. With this technique, Algorithm 3.1-1 can be rewritten in such a way that only $n^2/2$ storage is required for each triangular matrix.

Problems

P3.1-1. Rewrite Algorithm 3.1-1 so that the product AB overwrites B as it is formed.

P3.1-2. Rewrite Algorithm 3.1-1 for the case when A and B are stored in linear $n(n + 1)/2$ arrays as suggested by (3.1-2).

P3.1-3. Suppose $X \in \mathbb{R}^{n \times p}$ and $A \in \mathbb{R}^{n \times n}$, with A symmetric. Specify an algorithm for computing $Y = X^{T}AX$ which requires $pn(n + p/2)$ flops. Do not overwrite A with Y.

Sec. 3.2. Rounding Errors

When calculations are performed on a computer, each arithmetic operation is generally affected by *roundoff error*. This error arises because the machine hardware can only represent a subset of the real numbers. We denote this subset by F and refer to its elements as *floating point numbers*. Following the conventions set forth in Forsythe, Malcolm, and Moler (1977, pp. 10–29), the floating point number system on a particular computer is characterized by four intergers: the *machine base* β, the *precision* t, the *underflow limit* L, and the *overflow limit* U. In particular, F consists of all numbers f of the form

$$f = \pm .d_1d_2 \cdots d_t \times \beta^e \qquad 0 \leq d_i < \beta, d_1 \neq 0, L \leq e \leq U$$

together with zero. Notice that if $0 \neq f \in F$, then $m \leqslant |f| \leqslant M$, where

$$m = \beta^{L-1}$$

and

$$M = \beta^{U}(1 - \beta^{-t}).$$

As an example, if $\beta = 2, t = 3, L = -1$, and $U = 2$, then the non-negative elements of F are represented by hash marks on the axis of Figure 3.2-1. Notice that the floating point numbers are not equally spaced.

Some typical values of (β, t, L, U) are $(16, 6, -64, 63)$ for IBM 370 short precision, $(16, 14, -64, 63)$ for IBM 370 long precision, $(2, 48, -976, 1070)$ for the CDC 6600, and $(2, 48, -16384, 8191)$ for the Cray-1.

To make general pronouncements about the effect of rounding errors on a given algorithm, it is necessary to have a model of computer arithmetic on F. To this end define the set G by

$$G = \{x \in \mathbb{R} \mid m \leqslant |x| \leqslant M\} \cup \{0\}$$

and the operator $fl : G \to F$ by

$$fl(x) = \begin{cases} \text{nearest } c \in F \text{ to } x \text{ if } rounded\ arithmetic \text{ is used. (If tie,} \\ \text{then round away from zero.)} \\ \text{nearest } c \in F \text{ to } x \text{ satisfying } |c| \leqslant |x| \text{ if } chopped \\ arithmetic \text{ is used.} \end{cases}$$

Let a and b be any two floating point numbers and let $\square$ denote any of the four arithmetic operations $+, -, *, /$. If $|a \square b| \notin G$, then an *arithmetic fault* occurs implying either *overflow* ($|a \square b| > M$) or *underflow* ($0 < |a \square b| < m$). If $|a \square b| \in G$, then in our model of floating point arithmetic we assume that *the computed version of* a $\square$ b *is given by* fl(a $\square$ b).

An arithmetic fault usually implies program termination, although with some machines and compilers, underflows are set to zero. If an arithmetic fault does not occur, then an important question to be answered concerns the accuracy of $fl(\text{a} \square \text{b})$.

The fl operator can be shown to satisfy

$$(3.2\text{-}1) \qquad\qquad fl(x) = x(1 + \epsilon), \qquad\qquad |\epsilon| \leqslant \mathbf{u}$$

where $\mathbf{u}$ is the *unit roundoff* defined by

$$(3.2\text{-}2) \qquad\qquad \mathbf{u} = \begin{cases} \frac{1}{2}\beta^{1-t} \text{ if rounded arithmetic is used.} \\ \beta^{1-t} \text{ if chopped arithmetic is used.} \end{cases}$$

It follows that $fl(a \square b) = (a \square b)(1 + \epsilon)$, $|\epsilon| < \mathbf{u}$. Thus,

$$(3.2\text{-}3) \qquad\qquad \frac{|fl(a \square b) - (a \square b)|}{|a \square b|} \leqslant \mathbf{u}, \qquad\qquad a \square b \neq 0$$

Figure 3.2-1.

showing that there is small relative error associated with individual arithmetic operations. It is important to realize, however, that this is not necessarily the case when a sequence of operations is involved.

EXAMPLE 3.2-1. If $\beta = 10$, $t = 3$ chopped arithmetic is used, then

$$fl[fl(10^{-3} + 1) - 1] = 0$$

implying a relative error of 1. On the other hand

$$fl[10^{-3} + fl(1 - 1)] = 10^{-3},$$

the exact answer. The example shows that floating point arithmetic is not always associative.

Another important aspect of finite precision arithmetic is the phenomenon of *catastrophic cancellation*. Roughly speaking, this term refers to the extreme loss of correct significant digits when small numbers are additively computed from large numbers. A well-known example is the computation of $e^{-a}(a > 0)$ via Taylor series. The roundoff error associated with this method is approximately **u** times the largest partial sum. For large a, this error can actually be larger than the correct exponential, and there will be no correct digits in the answer no matter how many terms in the series are summed. On the other hand, if enough terms in the Taylor series for e^a are added and the result reciprocated, then an estimate of e^{-a} to full precision is attained. See Forsythe, Malcolm and Moler (1977, pp. 14–16).

Accounting for the rounding errors in a matrix computation can be a combinatoric headache. To illustrate this consider the rounding errors that result in the following algorithm.

ALGORITHM 3.2-1. Given x and y in $\mathbb{R}^n$, the following algorithm computes $x^T y$ and stores the result in s.

$$s := 0$$
$$\text{For } k = 1, \ldots, n$$
$$s := s + x_k y_k$$

This algorithm requires n flops.

In trying to express the rounding errors of this algorithm, we are immediately confronted with a notational problem: the distinction between computed and exact quantities. When the underlying computations are clear, we shall use the $fl(\cdot)$ operator to signify computed quantities. Thus, $fl(x^Ty)$ denotes the computed output of Algorithm 3.2-1. Let us bound $|fl(x^Ty) - x^Ty|$.

If

$$s_p = fl\left[\sum_{k=1}^{p} x_k y_k\right]$$

then

$$s_1 = x_1 y_1(1 + \delta_1), \qquad\qquad |\delta_1| \leqslant \mathbf{u}$$

and for $p = 2, \ldots, n$

$$s_p = fl[s_{p-1} + fl(x_p y_p)]$$
$$= [s_{p-1} + x_p y_p(1 + \delta_p)](1 + \epsilon_p). \qquad |\delta_p|, |\epsilon_p| \leqslant \mathbf{u}$$

A little algebra shows that

$$fl(x^Ty) = s_n = \sum_{k=1}^{n} x_k y_k(1 + \gamma_k),$$

where

$$(1 + \gamma_k) = (1 + \delta_k) \prod_{j=k}^{n} (1 + \epsilon_j), \qquad\qquad \epsilon_1 \equiv 0$$

and thus

(3.2-4) $$|fl(x^Ty) - x^Ty| \leqslant \sum_{k=1}^{n} |x_k y_k| \, |\gamma_k|.$$

To proceed further, we must be able to bound the quantities $|\gamma_k|$ in terms of $\mathbf{u}$. The following result is useful for this purpose.

LEMMA 3.2-1.   If $(1 + \alpha) = \Pi_{k=1}^{n}(1 + \alpha_k)$ where $|\alpha_k| \leqslant \mathbf{u}$ and $n\mathbf{u} \leqslant .01$, then $|\alpha| \leqslant 1.01 \, n\mathbf{u}$.

Proof.   See Forsythe and Moler (SLE, p. 92). □

Applying this result to (3.2-4) under the "reasonable" assumption $n\mathbf{u} \leqslant .01$ gives

(3.2-5) $$|fl(x^Ty) - x^Ty| \leqslant 1.01 \, n\mathbf{u} |x|^T |y|,$$

where the absolute value notation is defined by $|z| = (|z_1|, \ldots, |z_n|)^T$.

Notice that if $|x^Ty| \ll |x|^T|y|$ then the relative error in $fl(x^Ty)$ may not be small.

An easier but less rigorous way of bounding α in Lemma 3.2-1 is to say

$$|\alpha| \leqslant n\mathbf{u} + O(\mathbf{u}^2).$$

With this convention we have

(3.2-6) $$|fl(x^Ty) - x^Ty| \leqslant n\mathbf{u}|x|^T|y| + O(\mathbf{u}^2).$$

Other ways of expressing the same result are

(3.2-7) $$|fl(x^Ty) - x^Ty| \leqslant \phi(n)\mathbf{u}|x|^T|y|$$

and

(3.2-8) $$|fl(x^Ty) - x^Ty| \leqslant cn\mathbf{u}|x|^T|y|,$$

where in (3.2-7) $\phi(n)$ is a "modest" function of n and in (3.2-8) c is a constant of order unity.

We shall not express a preference for any of the error bounding styles shown in (3.2-5) through (3.2-8). This spares us the necessity of translating the roundoff results that appear in the literature into a fixed format. Moreover, paying overly close attention to the details of an error bound is inconsistent with the "philosophy" of roundoff analysis. As Wilkinson (1971, p. 567) says,

> There is still a tendency to attach too much importance to the precise error bounds obtained by an a priori error analysis. In my opinion, the bound itself is usually the least important part of it. The main object of such an analysis is to expose the potential instabilities, if any, of an algorithm so that hopefully from the insight thus obtained one might be led to improved algorithms. Usually the bound itself is weaker than it might have been because of the necessity of restricting the mass of detail to a reasonable level and because of the limitations imposed by expressing the errors in terms of matrix norms. A priori bounds are not, in general, quantities that should be used in practice. Practical error bounds should usually be determined by some form of a posteriori error analysis, since this takes full advantage of the statistical distribution of rounding errors and of any special features, such as sparseness, in the matrix.

It is for these reasons that roundoff error analysis has a different flavor than "pure math" error estimation.

We conclude this section by reviewing the rounding errors associated with matrix manipulation. These results will be useful in subsequent discussion and require use of the absolute value notation,

$$E \in \mathbb{R}^{m \times n} \Rightarrow |E| = (|e_{ij}|)$$

and the convention that

$$|E| \leqslant |F| \iff |e_{ij}| \leqslant |f_{ij}| \text{ for all } i \text{ and } j.$$

It is easy to show that if A and B are floating point matrices and $\alpha \in F$, then

(3.2-9) $$fl(\alpha A) = \alpha A + E \qquad\qquad |E| \leqslant \mathbf{u}\,|\alpha A|$$

and

(3.2-10) $$fl(A + B) = (A + B) + \mathbf{E} \qquad\qquad |E| \leqslant \mathbf{u}\,|A + B|.$$

Moreover, if $A \in \mathbb{R}^{m \times n}$ and $B \in \mathbb{R}^{n \times p}$, then from the inner product result (3.2-6) we have

(3.2-11) $$fl(AB) = AB + E \qquad\qquad |E| \leqslant n\,\mathbf{u}\,|A|\,|B| + O(\mathbf{u}^2)$$

Notice that matrix multiplication does not necessarily give small relative error, since $|AB|$ may be much smaller than $|A|\,|B|$. For this reason, the notion of *inner product accumulation* is important. Roughly speaking, this means computing inner products in *double precision* arithmetic ($\mathbf{u} = \beta^{-2t}$) and truncating the result to t digits. Since the multiplication of two t-digit floating point numbers can be stored exactly in a double precision variable, the roundoff associated with accumulated inner product is very small. Whenever the hardware permits, it is always recommended that inner product accumulation be used for matrix-matrix multiplication.

The error bound in (3.2-11) shows how wrong the computed matrix product can be. This is an example of *forward error analysis*. An alternative style of characterizing the roundoff errors in an algorithm is accomplished through a technique known as *inverse error analysis*. Here, the rounding errors are related to the data of the problem rather than to its solution. By way of illustration, consider the $n = 2$ version of Algorithm 3.1-1, triangular matrix multiplication. It can be shown that

$$fl(AB) = \begin{bmatrix} a_{11}b_{11}(1 + \epsilon_1) & [a_{11}b_{12}(1 + \epsilon_2) + a_{12}b_{22}(1 + \epsilon_3)](1 + \epsilon_4) \\ 0 & a_{22}b_{22}(1 + \epsilon_5) \end{bmatrix},$$

where $|\epsilon_i| \leqslant \mathbf{u}$, $i = 1, \ldots, 5$. If

$$\hat{A} = \begin{bmatrix} a_{11} & a_{12}(1 + \epsilon_3)(1 + \epsilon_4) \\ 0 & a_{22}(1 + \epsilon_5) \end{bmatrix}$$

and

$$\hat{B} = \begin{bmatrix} b_{11}(1 + \epsilon_1) & b_{12}(1 + \epsilon_2)(1 + \epsilon_4) \\ 0 & b_{22} \end{bmatrix}$$

then it can easily be verified that

$$fl(AB) = \hat{A}\hat{B}$$

where

$$\hat{A} = A + E \qquad\qquad |E| \leqslant 2\mathbf{u}|A| + O(\mathbf{u}^2)$$
$$\hat{B} = B + F. \qquad\qquad |F| \leqslant 2\mathbf{u}|B| + O(\mathbf{u}^2)$$

In other words, the computed product is the exact product of slightly perturbed A and B.

This technique of "throwing" an algorithm's errors back to the data it acts upon is very useful and is frequently used in the sequel.

Problems

P3.2-1. Show that if Algorithm 3.2-1 is applied with $y = x$, then $fl(x^Tx) = x^Tx(1 + \alpha)$, where $|\alpha| \leqslant n\mathbf{u} + O(\mathbf{u}^2)$.

P3.2-2. Prove (3.2-11).

P3.2-3. Show that if $E \in \mathbb{R}^{m \times n}$ ($m \geqslant n$), then

$$\| \, |E| \, \|_2 \leqslant \sqrt{n} \, \|E\|_2.$$

P3.2-4. Assume the existence of a square root function satisfying $fl(\sqrt{x}) = \sqrt{x}(1 + \epsilon)$, $|\epsilon| \leqslant \mathbf{u}$. Give an algorithm for computing $\|x\|_2$ and bound the rounding errors.

P3.2-5. Show through a 2-by-2 example that matrix multiplication in finite precision arithmetic need not have small relative error.

P3.2-6. Using (3.2-11), show that if B is nonsingular then

$$\frac{\| fl(AB) - AB \|_F}{\|AB\|_F} \leqslant n\mathbf{u}\kappa_F(B) + O(\mathbf{u}^2).$$

P3.2-7. Show that $fl(1 - m) = 1 - \beta^{-t}$ when chopped arithmetic is used. (This example illustrates a shortcoming of our chopped arithmetic model; the result of subtracting m from 1 is 1 on all chopped arithmetic computers that we know. Although it is possible to remove this gap between our model and reality, the added complexity that would be imposed on our discussion of floating point computation would cloud the more fundamental issues.)

Sec. 3.3. Householder Transformations

Let $v \in \mathbb{R}^n$ be nonzero. An n-by-n matrix P of the form

$$(3.3-1) \qquad\qquad P = I - 2vv^T/v^Tv$$

is known as a *Householder transformation*. (Synonyms: Householder matrix, Householder reflection.) When a vector x is multiplied by P, it is reflected in the hyperplane span$\{v\}^\perp$. (See Fig. 3.3-1.) Householder matrices are symmetric and orthogonal. They are important because of their ability to zero specified entries in a matrix or vector.

In particular, given any nonzero $x \in \mathbb{R}^n$ it is easy to construct v in (3.3-1) such that Px is a multiple of e_1, the first column of I. Noting that

$$Px = \left[I - \frac{2vv^T}{v^Tv} \right] x = x - \frac{2v^Tx}{v^Tv} v,$$

we see that the requirement $Px \in \text{span}\{e_1\}$ implies $v \in \text{span}\{x, e_1\}$. Setting $v = x + \alpha e_1$ gives

$$v^Tx = x^Tx + \alpha x_1$$

and

$$v^Tv = x^Tx + 2\alpha x_1 + \alpha^2,$$

and therefore

$$Px = \left[1 - 2\frac{x^Tx + \alpha x_1}{x^Tx + 2\alpha x_1 + \alpha^2} \right] x - 2\alpha \frac{v^Tx}{v^Tv} e_1.$$

In order for the coefficient of x to be zero, we need only set $\alpha = \pm \|x\|_2$. Thus, if $v = x \pm \|x\|_2 e_1$ in (3.3-1), then $Px = \pm \|x\|_2 e_1$. It is this simple determination of v that makes the Householder matrix such a powerful tool.

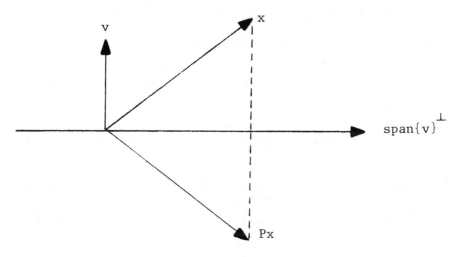

Figure 3.3-1.

EXAMPLE 3.1-1.   If $x = (3, 1, 5, 1)^T$ and $v = (9, 1, 5, 1)^T$, then

$$P = I - 2vv^T/v^Tv = \tfrac{1}{54} \begin{bmatrix} -27 & -9 & -45 & -9 \\ -9 & 53 & -5 & -1 \\ -45 & -5 & 29 & -5 \\ -9 & -1 & -5 & 53 \end{bmatrix}$$

has the property that $Px = (-6, 0, 0, 0)^T$.

There are a number of important practical details associated with the determination of a Householder matrix. One concern is the actual choice of α's sign. If x is close to a multiple of e_1, then $v = x - \text{sign}(x_1)\|x\|_2 e_1$ has small norm. Consequently, large relative error can be expected in the factor $\beta = 2/v^T v$. This difficulty can be avoided merely by choosing the sign of α to be the same as the sign of x's first component:

$$v = x + \text{sign}(x_1)\|x\|_2 e_1.$$

This ensures that $\|v\|_2 \geq \|x\|_2$ and guarantees near-perfect orthogonality in the computed P (see below).

An easy way to avoid overflow and underflow in the calculation of $\|x\|_2$ is to determine v from the vector $x/\|x\|_\infty$. The resulting Householder matrix is identical, although v will generally have a different norm.

Householder matrices can also be used to zero *any* contiguous block of vector components. For example, if $1 \leq k \leq j \leq n$ and

$$v^T = [0, \ldots, 0, x_k + \text{sign}(x_k)\alpha, x_{k+1}, \ldots, x_j, 0, \ldots, 0],$$

where $\alpha^2 = x_k^2 + \cdots + x_j^2$, then the Householder matrix $P = I - 2vv^T/v^T v$ has the property that

$$Px = [x_1, \ldots, x_{k-1}, -\text{sign}(x_k)\alpha, 0, \ldots, 0, x_{j+1}, \ldots, x_n]^T.$$

Note that P has the form $P = \text{diag}(I_{k-1}, \bar{P}, I_{n-j})$ where $\bar{P}$ is a Householder matrix defined by v's nontrivial portion.

Taking into account all of the above observations we obtain the following basic algorithm.

ALGORITHM 3.3-1. Given $x \in \mathbb{R}^n$ and indices k and j satisfying $1 \leq k \leq j \leq n$, this algorithm computes $v^T = (0, \ldots, v_k, \ldots, v_j, 0, \ldots, 0)$ and $\beta = 2/v^T v$ such that components $k + 1$ through j of $(I - \beta vv^T)x$ are zero. It is assumed that $(x_k, \ldots, x_j) \neq 0$.

$m := \max\{|x_k|, \ldots, |x_j|\}$

$\alpha := 0$

For $i = k$ to j

$\quad v_i := x_i/m$

$\quad \alpha := \alpha + v_i^2$

$\alpha := \sqrt{\alpha}$

$\beta := 1/(\alpha(\alpha + |v_k|))$

$v_k := v_k + \text{sign}(v_k)\alpha$

This algorithm requires about $2(j - k)$ flops.

It is critical to exploit structure when performing calculations with Householder matrices. Indeed, a careful look at the update

$$PA = (I - \beta vv^T)A = A - \beta v(A^Tv)^T \qquad\qquad A \in \mathbb{R}^{n \times n}$$

reveals that A is altered by a rank-one matrix when premultiplied by a Householder matrix. Moreover, P need not ever be formed explicitly provided β and v are available. This point is obvious in the following algorithm.

ALGORITHM 3.3-2. Given $A \in \mathbb{R}^{n \times q}$, $v^T = (0, \ldots, v_k, \ldots, v_j, 0, \ldots, 0)$, and $\beta = 2/v^Tv$, this algorithm overwrites A with $(I - \beta vv^T)A$.

> For $p = 1, \ldots, q$
> $\quad s := v_k a_{kp} + \cdots + v_j a_{jp}$
> $\quad s := \beta s$
> $\quad$ For $i = k, \ldots, j$
> $\qquad a_{ip} := a_{ip} - sv_i$

This algorithm requires $2q(j - k + 1)$ flops. An analogous algorithm exists for performing updates of the form $A := A(I - \beta vv^T)$.

The roundoff properties of Algorithms 3.3-1 and 3.3-2 are very favorable. Suppose $\hat{v}$ and $\hat{\beta}$ are the computed versions of v and β in Algorithm 3.3-1 and define

$$\hat{P} = I - \hat{\beta}\hat{v}\hat{v}^T.$$

Wilkinson (AEP, pp. 152–62) shows that if $P = I - \beta vv^T$, the exact P, then

$$\|P - \hat{P}\|_2 \leqslant 10\mathbf{u}.$$

Moreover, if $\hat{\beta}$ and $\hat{v}$ are used in Algorithm 3.3-2, then the resulting update $fl(\hat{P}A)$ satisfies

$$fl(\hat{P}A) = P(A + E),$$

where

$$\|E\|_2 \leqslant c(j - k + 1)^2\mathbf{u}\|A\|_2$$

and c is a constant of order unity. Thus, a computed Householder update of a matrix A is an exact Householder update of some matrix close to A.

We conclude the section with a discussion about keeping Householder matrices in factored form and about modified Householder matrices.

Frequently, an algorithm will determine an orthogonal matrix as a product of Householder matrices, e.g.,

$$(3.3\text{-}2) \qquad\qquad Q = P_1P_2 \cdots P_r \qquad\qquad P_i = I - \beta_i v^{(i)}[v^{(i)}]^T.$$

Instead of explicitly forming Q, it is often more economical to leave it in factored form and access it through the β_i and $v^{(i)}$. For example, to overwrite $A \in \mathbb{R}^{n \times q}$ with QA we execute the following loop:

(3.3-4)
$$\begin{aligned} &\text{For } i = r, \ldots, 1 \\ &\quad A := P_i A \quad \text{(via Algorithm 3.3-2)} \end{aligned}$$

This requires $2nrq$ flops. On the other hand, if we compute QA as follows,

(3.3-5)
$$\begin{aligned} &Q := I \\ &\text{For } i = 1, \ldots, r \\ &\quad Q := Q P_i \\ &A := QA \end{aligned}$$

then $2n^2r + n^2q$ flops are needed.

Of course, in some applications it is necessary to explicitly form Q (or parts of it). In these situations it is typically the case that the Householder vectors $v^{(i)}$ in (3.3-2) have the property that their first $i - 1$ components are zero. Because of this, it is more economical to accumulate the P_i in reverse order:

(3.3-6)
$$\begin{aligned} &Q := I \\ &\text{For } i = r, \ldots, 1 \\ &\quad Q := P_i Q \end{aligned}$$

This enables us to exploit the fact that

$$P_{i+1} \cdots P_r = \operatorname{diag}(I_i, \hat{Q})$$

where $\hat{Q} \in \mathbb{R}^{(n-i) \times (n-i)}$. Note that if we accumulate via (3.3-5) then Q is full after the first pass through the loop.

Another computational shortcut involving Householder matrices arises whenever the Householder vector v in $P = I - 2vv^T/v^Tv$ has only two or three nonzero elements. Suppose Algorithm 3.3-1 produces

$$v^T = (0, \ldots, 0, v_k, \ldots, v_j, 0, \ldots 0)$$

and $\beta = 2/v^Tv$. Since $|v_k| = \|v\|_\infty$, the vectors

$$\begin{aligned} y^T &= (0, \ldots, 0, y_k, \ldots, y_j, 0, \ldots, 0) & y_i &= v_i/v_k \\ w^T &= (0, \ldots, 0, w_k, \ldots, w_j, 0, \ldots, 0) & w_i &= \beta v_k v_i \end{aligned}$$

are defined, can be stably computed, and satisfy $P = I - wy^T$. Since $y_k = 1$, the update

$$PA = (I - wy^T)A = A - w(A^Ty)^T \qquad A \in \mathbb{R}^{n \times q}$$

can be executed in $q[2(j - k) + 1]$ flops, q flops fewer than Algorithm 3.3-2. This is a noteworthy reduction only if $j = k + 1$ or $k + 2$, for then the work counts become $3q$ and $5q$, respectively, instead of $4q$ and $6q$.

Householder matrices P represented in the form $P = I - wy^T$, where the first nonzero component in y is unity, are referred to as *modified* Householder matrices. We will tacitly assume their use whenever appropriate.

EXAMPLE 3.3-2.   If $v = (3, 4)^T$ then

$$P = I - 2vv^T/v^Tv = \tfrac{1}{25}\begin{bmatrix} 7 & -24 \\ -24 & -7 \end{bmatrix} = I - \begin{pmatrix} 18/25 \\ 24/25 \end{pmatrix}(1, 4/3).$$

Problems

P3.3-1. Execute Algorithm 3.3-1 with $x = (1, 7, 2, 3, -1)^T$.

P3.3-2. Adapt (3.3-6) so that it computes the first p ($p < n$) columns of Q. Assume that the first $i - 1$ components of $v^{(i)}$ are zero.

P3.3-3. Let x and y be nonzero vectors in $\mathbb{R}^n$. Give an algorithm for determining a Householder matrix P such that Px is a multiple of y.

P3.3-4. Give an algorithm for the update $A := AP$, where $A \in \mathbb{R}^{n \times q}$, $P \in \mathbb{R}^{q \times q}$, and P is a Householder matrix.

P3.3-5. Suppose $x \in \mathbb{C}^n$ and that $x_1 = |x_1| e^{i\theta}$ with $\theta \in \mathbb{R}$. Assume $x \neq 0$ and define $u = x + e^{i\theta} \|x\|_2 e_1$. Show that

$$P = I - \frac{2}{u^H u} uu^H$$

is unitary and that $Px = -e^{i\theta} \|x\|_2 e_1$.

P3.3-6. Show that if x and y are n-vectors, then $\det(I_n + xy^T) = 1 + x^T y$.

Sec. 3.4. Givens Transformations

Householder reflections are exceedingly useful for introducing zeros on a grand scale, e.g., the annihilation of all but the first component of a vector. However, in many computations it is necessary to zero elements more selectively. *Givens rotations* are the tool for doing this. These are rank-two corrections to the identity of the form

$$J(i, k, \theta) = \begin{bmatrix} 1 & \vdots & & \vdots & \\ \cdots & c & \cdots & s & \cdots \\ & \vdots & & \vdots & \\ \cdots & -s & \cdots & c & \cdots \\ & \vdots & & \vdots & 1 \end{bmatrix} \begin{matrix} i \\ \\ k \end{matrix} ,$$
$$\qquad\qquad\qquad i \qquad k$$

where $c = \cos(\theta)$ and $s = \sin(\theta)$ for some θ. Givens rotations are clearly orthogonal.

It is useful to contrast Householder reflections and Givens rotations in the $n = 2$ case. Suppose $Q = I - 2vv^T/v^Tv$ is a 2-by-2 Householder matrix. Without loss of generality we may assume that $v_1 = \cos(\theta)$ and $v_2 = \sin(\theta)$ for some θ. It follows that

$$Q = \begin{bmatrix} 1 - 2v_1^2 & -2v_1v_2 \\ -2v_1v_2 & 1 - 2v_2^2 \end{bmatrix} = \begin{bmatrix} \cos(2\theta) & -\sin(2\theta) \\ -\sin(2\theta) & -\cos(2\theta) \end{bmatrix}.$$

Thus, if $Q \in \mathbb{R}^{2 \times 2}$ is orthogonal, then either $Q = \begin{bmatrix} c & s \\ s & -c \end{bmatrix}$ is Householder or $Q = \begin{bmatrix} c & s \\ -s & c \end{bmatrix}$ is Givens.

More generally, if $Q \in \mathbb{R}^{n \times n}$ is an orthogonal matrix that agrees with the identity everywhere except possibly in some 2-by-2 principal submatrix, then either Q is a Givens rotation or else Q has the form

$$H(i, k, \theta) = \begin{bmatrix} 1 & \vdots & & \vdots & \\ \cdots c \cdots & & s \cdots \\ & \vdots & & \vdots & \\ \cdots s \cdots & & -c \cdots \\ & \vdots & & \vdots & 1 \end{bmatrix} \begin{matrix} i \\ k \end{matrix},$$

$$i \quad k$$

where $c = \cos(\theta)$ and $s = \sin(\theta)$. Sometimes the matrices $H(i, k, \theta)$ are referred to as Givens reflections. We shall not be concerned with them further, since we can get along nicely with just the Givens rotations.

Premultiplication by $J(i, k, \theta)$ amounts to a rotation of θ degrees in the (i, k) coordinate plane. Indeed, if $x \in \mathbb{R}^n$ and $y = J(i, k, \theta)x$, then

$$y_i = cx_i + sx_k$$
$$y_k = -sx_i + cx_k$$
$$y_j = x_j. \qquad\qquad j \neq i \text{ or } k$$

From these formulae it is clear that we can force y_k to be zero by setting

$$(3.4\text{-}1) \qquad c = x_i/(x_i^2 + x_k^2)^{1/2} \qquad s = x_k/(x_i^2 + x_k^2)^{1/2}.$$

Thus, it is a simple matter to zero a specified entry in a vector by using a Givens rotation.

EXAMPLE 3.4-1. If $x = (1, 2, 3, 4)^T$, $\cos(\theta) = 2/\sqrt{5}$, and $\sin(\theta) = 1/\sqrt{5}$, then $J(2, 4, \theta)x = (1, \sqrt{20}, 3, 0)^T$.

In determining c and s, it is preferable to use the following algorithm rather than the formulae (3.4-1).

ALGORITHM 3.4-1.   Given $x \in \mathbb{R}^n$ and indices i and k that satisfy $1 \leqslant i < k \leqslant n$, the following algorithm computes $c = \cos(\theta)$ and $s = \sin(\theta)$ such that the k-th component of $J(i, k, \theta)x$ is zero.

> If $x_k = 0$
>> then
>>> $c := 1$ and $s := 0$
>> else
>>> if $|x_k| \geqslant |x_i|$
>>>> then
>>>>> $t := x_i/x_k, s := 1/(1 + t^2)^{1/2}, c := st$
>>>> else
>>>>> $t := x_k/x_i, c := 1/(1 + t^2)^{1/2}, s := ct$

This algorithm requires four flops and a single square root. Note that it does *not* compute θ.

It is critical that the simple structure of $J(i, k, \theta)$ be exploited when computing products of the form $J(i, k, \theta)A$. Note that only rows i and k are affected by the premultiplication. Moreover, the elements in these rows are combined in an elemental way, as the following algorithm shows.

ALGORITHM 3.4-2.   Given $A \in \mathbb{R}^{n \times q}$, $c = \cos(\theta)$, $s = \sin(\theta)$, and indices i and k that satisfy $1 \leqslant i < k \leqslant n$, the following algorithm overwrites A with $J(i, k, \theta)A$.

> For $j = 1, \ldots, q$
>> $v := a_{ij}$
>> $w := a_{kj}$
>> $a_{ij} := cv + sw$
>> $a_{kj} := -sv + cw$

This algorithm requires $4q$ flops. An analogous algorithm exists for computing products of the form $A\,J(i, k, \theta)$.

The numerical properties of Givens rotations are as favorable as those for Householder reflections. In particular, it can be shown that the computed c and s in Algorithm 3.4-1 satisfy

$$\hat{c} = c(1 + \epsilon_c) \qquad\qquad \epsilon_c = O(\mathbf{u})$$
$$\hat{s} = s(1 + \epsilon_s) \qquad\qquad \epsilon_s = O(\mathbf{u}).$$

If $\hat{c}$ and $\hat{s}$ are subsequently used in Algorithm 3.4-2 and $fl[\hat{J}(i,\ k,\ \theta)A]$ denotes the computed update of A, then it is possible to show that

$$fl[\hat{J}(i,\ k,\ \theta)A] = J(i,\ k,\ \theta)A + E,$$

where

$$\frac{\|E\|_2}{\|A\|_2} = O(\mathbf{u}).$$

A detailed error analysis of Givens rotations may be found in Wilkinson (AEP, pp. 131–39).

Suppose $Q = J_1 \cdots J_N$ where each J_i is an n-by-n Givens rotation. In some applications it is more economical to keep the orthogonal matrix Q in factored form than to compute explicitly the product of the rotations. Using a technique demonstrated by Stewart (1976), it is possible to do this in a very compact way. The idea is to associate a single floating point number with each rotation. Specifically, if

$$Z = \begin{bmatrix} c & s \\ -s & c \end{bmatrix} \qquad\qquad c^2 + s^2 = 1$$

then we define the scalar ρ by

(3.4-2)
$$\rho = \begin{cases} 1 & \text{if } c = 0 \\ \tfrac{1}{2}\,\text{sign}(c)s & \text{if } |s| < |c|. \\ 2\,\text{sign}(s)/c & \text{if } |c| \leqslant |s| \end{cases}$$

The matrix $\pm Z$ can then be retrieved as follows:

if $\rho = 1$ *then* $c := 0$ and $s := 1$

(3.4-3)
if $|\rho| < 1$
  then $s := 2\rho$ and $c := \sqrt{(1 - s^2)}$
  else $c := 2/\rho$ and $s := \sqrt{(1 - c^2)}$

That $-Z$ may be generated is usually of no consequence (cf. §6.3). The reason for essentially storing the smaller of c and s is that the formula $\sqrt{(1 - x^2)}$ renders poor results if x is near unity. More details may be found in Stewart (1976).

We conclude with some remarks about the propagation of roundoff error in algorithms that involve sequences of Householder/Givens updates of a given matrix. To be precise, suppose $A = A_0 \in \mathbb{R}^{m \times n}$ is given and that matrices $A_1, \ldots, A_p = B$ are generated via the recursion

$$A_k = fl(\hat{Q}_k A_{k-1} \hat{Z}_k) \qquad\qquad k = 1, \ldots, p$$

Assume that the $\hat{Q}_k$ and $\hat{Z}_k$ are computed via Algorithms 3.3-1 and 3.4-1 and that the updates are performed via Algorithms 3.3-2 and 3.4-2. Let Q_k and Z_k be the orthogonal matrices that would be produced in the absence of round-off. It can be shown that

$$B = (Q_p \cdots Q_1)(A + E)(Z_1 \cdots Z_p),$$

where $\|E\|_2 \leqslant c\mathbf{u}\|A\|_2$ and c is a constant that depends mildly on n, m, and p. In plain English, B is an *exact* orthogonal update of a matrix near to A.

Problems

P3.4-1. Suppose $x^H = (\bar{x}_1, \bar{x}_2)$ is complex. Give an algorithm for determining a unitary matrix of the form

$$Q = \begin{bmatrix} c & \bar{s} \\ -s & c \end{bmatrix} \qquad c \in \mathbb{R}, c^2 + |s|^2 = 1$$

such that the second component of Qx is zero.

P3.4-2. Suppose x and y are unit vectors in $\mathbb{R}^n$. Give an algorithm using Givens transformations which computes an orthogonal Q such that $Qx = y$.

P3.4-3. Derive analogs of Algorithms 3.4-1 and 3.4-2 that are based on the matrices $H(i, k, \theta)$.

P3.4-4. Determine $c = \cos(\theta)$ and $s = \sin(\theta)$ such that $\begin{bmatrix} c & s \\ -s & c \end{bmatrix}\begin{bmatrix} 5 \\ 12 \end{bmatrix}$ is a multiple of $\begin{bmatrix} 1 \\ 1 \end{bmatrix}$.

P3.4-5. Why is the reciprocal of the cosine used in (3.4-2)?

Sec. 3.5. Gauss Transformations

Orthogonal transformations are not the only means for zeroing entries in a vector. For example, if $x_1 \neq 0$ and $\alpha = -x_2/x_1$, then

$$\begin{bmatrix} 1 & 0 \\ \alpha & 1 \end{bmatrix}\begin{bmatrix} x_1 \\ x_2 \end{bmatrix} = \begin{bmatrix} x_1 \\ 0 \end{bmatrix}.$$

More generally, suppose $x \in \mathbb{R}^n$ with $x_k \neq 0$. If

$$\alpha_i = x_i/x_k \qquad\qquad i = k + 1, \ldots, n$$

and

$$M_k = I - \alpha e_k^T, \qquad \alpha^T = (\underbrace{0, \ldots, 0}_{k \text{ times}}, \alpha_{k+1}, \ldots, \alpha_n)$$

then

$$M_k \begin{bmatrix} x_1 \\ \vdots \\ x_k \\ x_{k+1} \\ \vdots \\ x_n \end{bmatrix} = \begin{bmatrix} x_1 \\ \vdots \\ x_k \\ 0 \\ \vdots \\ 0 \end{bmatrix}.$$

The matrix M_k is said to be a *Gauss transformation*. (Synonym: elementary transformation.) The vector α which defines the "spike" in the Gauss transformation is referred to as the *Gauss vector*. The components of α are referred to as *multipliers*.

ALGORITHM 3.5-1. Given $x \in \mathbb{R}^n$ with $x_k \neq 0$, the following algorithm computes $\alpha = (0, \ldots, 0, \alpha_{k+1}, \ldots, \alpha_n)^T$ such that components $k + 1, \ldots, n$ of $(I - \alpha e_k^T)x$ are zero.

For $i = k + 1, \ldots, n$

   $\alpha_i := x_i/x_k$

This algorithm requires $n - k$ flops.

Multiplication by a Gauss transformation is particularly simple. If $x \in \mathbb{R}^n$ and $M = I - \alpha e_k^T$ is a Gauss transformation, then $Mx = x - x_k\alpha$. The update of a matrix is equally trivial:

ALGORITHM 3.5-2. Given $A \in \mathbb{R}^{n \times q}$ and $\alpha = (0, \ldots, 0, \alpha_{k+1}, \ldots, \alpha_n)^T$, this algorithm overwrites A with $(I - \alpha e_k^T)A$.

For $j = 1, \ldots, q$
   For $i = k + 1, \ldots, n$
      $a_{ij} := a_{ij} - a_{kj}\alpha_i$

This algorithm requires $(n - k)q$ flops.

EXAMPLE 3.5-1. If $x = (1, 2, 3, 4)^T$ and $\alpha = (0, 0, \frac{3}{2}, 2)^T$, then $(I - \alpha e_2^T)x = (1, 2, 0, 0)^T$.

Note that Algorithm 3.5-2 involves half the work of the comparable algorithm for Householder updating (Algorithm 3.3-2). The price paid for this efficiency is loss of stability unless the Gauss vector is suitably bounded. Specifically, if $\hat{\alpha}$ denotes the computed version of α in Algorithm 3.5-2, then it is easy to verify that

$$\hat{\alpha} = \alpha + e. \qquad\qquad |e| \leqslant \mathbf{u}|\alpha|$$

If α is used in Algorithm 3.5-2 and $fl(I - \hat{\alpha}e_k^T)A)$ denotes the resulting update, then

$$fl[(I - \hat{\alpha}e_k^T)A] = (I - \alpha e_k^T)A + E,$$

where

$$|E| \leqslant 3\mathbf{u}[|A| + |\alpha| \ |a_k^T|] + O(\mathbf{u}^2)$$

and $a_k^T = (a_{kl}, \ldots, a_{kp})$. Clearly, if α is big then the errors in the update may be large in comparison to $|A|$. For this reason care must be exercised when Gauss transformations are employed (cf. §4.4).

We close with some useful remarks about products of Gauss transformations. Suppose

$$M = M_k \cdots M_1,$$

where

$$M_i = I_n - \alpha^{(i)}e_i^T$$

with

$$\alpha^{(i)} = (0, \ldots, 0, -l_{i+1,i}, \ldots, -l_{ni})^T.$$

To begin with, M is an n-by-n unit lower triangular matrix. Rather than form this matrix explicitly, it is often more convenient to leave it in factored form since this requires less storage for $k < n$, e.g.,

$$[-\alpha^{(1)}, -\alpha^{(2)}] = \begin{bmatrix} 1 & 0 \\ l_{21} & 1 \\ l_{31} & l_{32} \\ \vdots & \vdots \\ l_{nl} & l_{n2} \end{bmatrix}. \qquad k = 2$$

An interesting aspect of representing M in this fashion is that it is tantamount to storing the first k columns of $M^{-1} = M_1^{-1} \cdots M_k^{-1}$. This follows because $M_i^{-1} = I + \alpha^{(i)}e_i^T$ and $e_j^T\alpha^{(i)} = 0$ for $j < i$. Thus,

$$M^{-1} = (I + \alpha^{(1)}e_1^T) \cdots (I + \alpha^{(k)}e_k^T) = I + \sum_{i=1}^{k} \alpha^{(i)}e_i^T.$$

Most algorithms that use Gauss transformations exploit this property.

Problems

P3.5-1. Show that if $M_k = I - \alpha e_k^T$ is a Gauss transformation, then $M_k^{-1} = I + \alpha e_k^T$.

P3.5-2. Suppose $\alpha^{(k)} = (0, \ldots, 0, \alpha_{k+1,k}, \ldots, \alpha_{nk})^{\mathrm{T}} \in \mathbb{R}^n$ for $k = 1, \ldots,$ $n - 1$. Show that if $M_k = I - \alpha^{(k)} e_k^{\mathrm{T}}$ and $L = (M_{n-1} \cdots M_1)^{-1}$, then $l_{ij} = \alpha_{ij}$ for all $i > j$.

P3.5-3. Determine a 3-by-3 Gauss transformation M such that

$$M \begin{bmatrix} 2 \\ 3 \\ 4 \end{bmatrix} = \begin{bmatrix} 2 \\ 7 \\ 8 \end{bmatrix}.$$

Notes and References for Chap. 3

The most complete treatment of roundoff error analysis is Wilkinson (AEP, chap. 3). The treatments in Forsythe and Moler (SLE, pp. 87–97) and Stewart (IMC, pp. 69–82) are also excellent. For a general introduction to the effects of roundoff error we highly recommend chapter 2 of

G. E. Forsythe, M. A. Malcom, and C. B. Moler (1977). *Computer Methods for Mathematical Computations*, Prentice-Hall, Englewood Cliffs,

as well as the classical reference

J. H. Wilkinson (1963). *Rounding Errors in Algebraic Processes*, Prentice-Hall, Englewood Cliffs.

A philosophical and historical overview of roundoff analysis was given in the 1970 John von Neumann Lecture delivered by J. H. Wilkinson and appears in

J. H. Wilkinson (1971). "Modern Error Analysis," *SIAM Review 13*, 548–68.

This paper gives a critique of the early papers on error analysis authored by Von Neumann and Goldstine, Turing, and Givens.

More recent developments in error analysis involve interval analysis, the building of statistical models of roundoff error, and the automating of the analysis itself. See

T. E. Hull and J. R. Swenson (1966). "Tests of Probabilistic Models for Propagation of Roundoff Errors," *Comm. Assoc. Comp. Mach. 9*, 108–13.

J. Larson and A. Sameh (1978). "Efficient Calculation of the Effects of Roundoff Errors," *ACM Trans. Math. Soft. 4*, 228–36.

J. M. Yohe (1979). "Software for Interval Arithmetic: A Reasonable Portable Package," *ACM Trans. Math. Soft. 5*, 50–63.

W. Miller and D. Spooner (1978). "Software for Roundoff Analysis, II," *ACM Trans. Math. Soft. 4*, 369–90.

Anyone engaged in serious software development needs a thorough understanding of floating point arithmetic. A good way to begin acquiring knowledge in this direction is to read

D. Stevenson (1981). "A Proposed Standard for Binary Floating Point Arithmetic," *Computer 14 (March)*, 51–62.

The subleties associated with the development of high-quality software, even for "simple" problems, are immense. A good example is the design of a subroutine to compute 2-norms:

J. M. Blue (1978). "A Portable FORTRAN Program to Find the Euclidean Norm of a Vector," *ACM Trans. Math. Soft. 4*, 15-23.

Householder matrices are named after A. S. Householder, who popularized their use in numerical analysis. However, others have been interested in their properties for quite some time. See

H. W. Turnbull and A. C. Aitken (1961). *An Introduction to the Theory of Canonical Matrices*, Dover Publications, New York, pp. 102-5.

Other references concerned with Householder transformations include

A. R. Gourlay (1970). "Generalization of Elementary Hermitian Matrices," *Comp. J. 13*, 411-12.
N. K. Tsao (1975). "A Note on Implementing the Householder Transformation," *SIAM J. Num. Anal. 12*, 53-58.
B. N. Parlett (1971). "Analysis of Algorithms for Reflections in Bisectors," *SIAM Review 13*, 197-208.

A detailed error analysis of Householder transformations is given in Lawson and Hanson (SLS, pp. 83-89).

The transformations that we have discussed in this chapter are extensively used as building blocks in more complicated algorithms. Thus, there is a premium on doing the computations right at this elementary level. See

C. L. Lawson, R. J. Hanson, F. T. Krogh, and D. R. Kincaid (1979). "Basic Linear Algebra Subprograms for FORTRAN Usage," *ACM Trans. Math. Soft. 5*, 308-23.

The basic linear algebra subroutines, or BLAS, are described in detail in LINPACK.

Givens rotations, named after W. Givens, are also referred to as Jacobi rotations. Jacobi devised a symmetric eigenvalue algorithm based on these transformations in 1846. See §8.4.

The storage scheme discussed in the text is detailed in

G. W. Stewart (1976). "The Economical Storage of Plane Rotations," *Numer. Math. 25*, 137-38.

Gaussian Elimination

§4.1 **Triangular Systems**

§4.2 **Computing the L-U Decomposition**

§4.3 **Roundoff Error Analysis of Gaussian Elimination**

§4.4 **Pivoting**

§4.5 **Improving and Estimating Accuracy**

The problem of solving a system of linear equations $Ax = b$ is central to the field of matrix computations. In this chapter we focus on the method of Gaussian elimination, the algorithm of choice when A is square, dense, and unstructured. When A does not fall into this category, then the algorithms of Chapters 5, 6, and 10 become of interest.

We motivate Gaussian elimination in §4.1 by discussing the ease with which triangular systems can be solved. The conversion of a general system to triangular form via Gauss transformations is then presented in §4.2. Unfortunately, the derived method behaves very poorly on a nontrivial class of problems. Our error analysis in §4.3 pinpoints the difficulty and motivates §4.4, where the concept of pivoting is introduced. Partial pivoting and complete pivoting are each described. In the final section we offer some remarks on the important practical issues associated with scaling, iterative improvement, and condition estimation.

Sec. 4.1. Triangular Systems

Consider the following 2-by-2 lower triangular system:

$$\begin{bmatrix} l_{11} & 0 \\ l_{21} & l_{22} \end{bmatrix} \begin{bmatrix} y_1 \\ y_2 \end{bmatrix} = \begin{bmatrix} b_1 \\ b_2 \end{bmatrix}.$$

If $l_{11}l_{22} \neq 0$, then the unknowns can be determined sequentially:

$$y_1 = b_1/l_{11}$$

$$y_2 = (b_2 - l_{21}y_1)/l_{22}.$$

This is the 2-by-2 version of an algorithm known as *forward elimination*:

ALGORITHM 4.1-1: *Forward Elimination.* Given an n-by-n nonsingular lower triangular matrix L and $b \in \mathbb{R}^n$, this algorithm finds $y \in \mathbb{R}^n$ such that $Ly = b$

> For $i = 1, \ldots, n$
>> $y_i := b_i$
>> For $j = 1, \ldots, i - 1$
>>> $y_i := y_i - l_{ij}y_j$
>> $y_i := y_i/l_{ii}$

This algorithm requires $n^2/2$ flops. The computed solution $\hat{y}$ satisfies

$$(4.1\text{-}1) \qquad (L + F)\hat{y} = b. \qquad\qquad |F| \leqslant n\mathbf{u}\,|L| + O(\mathbf{u}^2)$$

This result is established in Forsythe and Moler (SLE, pp. 104–5). It says that the computed solution is the exact solution of some slightly perturbed system. Moreover, each entry in F is small relative to the corresponding element of L.

The analogous algorithm for upper triangular systems is called *back-substitution*:

ALGORITHM 4.1-2: *Back-Substitution.* Given an n-by-n nonsingular upper triangular matrix U and $y \in \mathbb{R}^n$, this algorithm finds $x \in \mathbb{R}^n$ such that $Ux = y$.

> For $i = n, \ldots, 1$
>> $x_i := y_i$
>> For $j = i + 1, \ldots, n$
>>> $x_i := x_i - u_{ij}x_j$
>> $x_i := x_i/u_{ii}$

This algorithm requires $n^2/2$ flops. The computed solution $\hat{x}$ obtained by the algorithm can be shown to satisfy

$$(4.1\text{-}2) \qquad (U + F)\hat{x} = y. \qquad\qquad |F| \leqslant n\mathbf{u}\,|U| + O(\mathbf{u}^2)$$

The problem of solving nonsquare triangular systems deserves some mention. We consider two cases:

$$(4.1\text{-}3) \qquad \begin{bmatrix} L_{11} \\ L_{21} \end{bmatrix} x = \begin{bmatrix} b_1 \\ b_2 \end{bmatrix} \qquad \begin{matrix} L_{11} \in \mathbb{R}^{n \times n}, b_1 \in \mathbb{R}^n \\ L_{21} \in \mathbb{R}^{(m-n) \times n}, b_2 \in \mathbb{R}^{m-n} \end{matrix}$$

and

$$(4.1\text{-}4) \qquad [U_{11}, U_{12}] \begin{bmatrix} y_1 \\ y_2 \end{bmatrix} = b. \qquad \begin{matrix} U_{11} \in \mathbb{R}^{n \times n}, U_{12} \in \mathbb{R}^{n \times (n-m)} \\ y_1 \in \mathbb{R}^n, y_2 \in \mathbb{R}^{n-m} \end{matrix}$$

Assume that L_{11} is lower triangular, U_{11} is upper triangular, and that both are nonsingular. The system (4.1-3) can be solved as follows

(i) Apply forward elimination to $L_{11}x = b_1$.

(ii) If $L_{21}(L_{11}^{-1}b_1) \neq b_2$ then (4.1-3) has no solution.

For the upper triangular system (4.1-4) we have

(i) Apply back substitution to $U_{11}y_1 = b$.

(ii) Set $y_2 = 0$.

Further details about solving nonsquare systems may be found in §6.7. However, readers interested in this area are advised to be familiar with §§4.2–4.4, where we develop Gaussian elimination for rectangular matrices.

Problems

P4.1-1. Change Algorithms 4.1-1 and 4.1-2 so that the right-hand side is overwritten with the solution.

P4.1-2. Interchange the order of the loops in Algorithms 4.1-1 and 4.1-2.

P4.1-3. Give an algorithm for computing a nonzero $z \in \mathbb{R}^n$ such that $Uz = 0$ where $U \in \mathbb{R}^{n \times n}$ is upper triangular with the property that $u_{nn} = 0$ and $u_{11} \cdots u_{n-1,n-1} \neq 0$.

P4.1-4. Discuss how the determinant of a square triangular matrix could be computed with minimum risk of overflow and underflow.

P4.1-5. Rewrite Algorithm 4.1-2 given that U is stored in an array w as follows:

$$w = (u_{11}, u_{12}, u_{13}, u_{22}, u_{23}, u_{33}). \qquad n = 3$$

Notes and References for Sec. 4.1

FORTRAN codes for solving triangular linear systems may be found in LINPACK (chap. 6).

Sec. 4.2. Computing the L-U Decomposition

Assume that $A \in \mathbb{R}^{n \times n}$ is nonsingular and that we wish to solve the linear system $Ax = b$. In this section we show how Gauss transformations $M_1, \ldots, M_{n-1}$ can almost always be found such that

$$M_{n-1} \cdots M_2 M_1 A = U$$

is upper triangular. The original $Ax = b$ problem is then equivalent to the upper triangular system

$$Ux = (M_{n-1} \cdots M_2 M_1)b$$

which can be solved via back-substitution (Algorithm 4.1-2).

EXAMPLE 4.2-1.   If

$$A = \begin{bmatrix} 1 & 4 & 7 \\ 2 & 5 & 8 \\ 3 & 6 & 11 \end{bmatrix}, \qquad b = \begin{bmatrix} 1 \\ 1 \\ 1 \end{bmatrix}$$

and

$$M_1 = \begin{bmatrix} 1 & 0 & 0 \\ -2 & 1 & 0 \\ -3 & 0 & 1 \end{bmatrix}, \qquad M_2 = \begin{bmatrix} 1 & 0 & 0 \\ 0 & 1 & 0 \\ 0 & -2 & 1 \end{bmatrix},$$

then $(M_2 M_1 A)x = (M_2 M_1 b)$ has the form

$$\begin{bmatrix} 1 & 4 & 7 \\ 0 & -3 & -6 \\ 0 & 0 & 2 \end{bmatrix} \begin{bmatrix} x_1 \\ x_2 \\ x_3 \end{bmatrix} = \begin{bmatrix} 1 \\ -1 \\ 0 \end{bmatrix}.$$

In deriving the algorithm for computing the M_i, we shall assume that A is rectangular. The added generality is painless and enriches the discussion of underdetermined systems in §6.7.

Suppose, then, that $A \in \mathbb{R}^{m \times n}$ and that for some $k < \min\{m, n\}$ we have determined Gauss transformations $M_1, \ldots, M_{k-1} \in \mathbb{R}^{m \times m}$ such that

$$A^{(k-1)} \equiv M_{k-1} \cdots M_1 A = \begin{bmatrix} A_{11}^{(k-1)} & A_{12}^{(k-1)} \\ 0 & A_{22}^{(k-1)} \end{bmatrix} \begin{matrix} k-1 \\ m-k+1 \end{matrix}$$
$$\qquad\qquad\qquad\qquad\quad k-1 \quad\ n-k+1$$

where $A_{11}^{(k-1)}$ is upper triangular. If

$$A_{22}^{(k-1)} = \begin{bmatrix} a_{kk}^{(k-1)} & \cdots & a_{kn}^{(k-1)} \\ \vdots & & \vdots \\ a_{mk}^{(k-1)} & \cdots & a_{mn}^{(k-1)} \end{bmatrix}$$

and $a_{kk}^{(k-1)}$ is nonzero, then the *multipliers*

$$l_{ik} = a_{ik}^{(k-1)}/a_{kk}^{(k-1)} \qquad\qquad i = k+1, \ldots, m$$

are defined. It follows that if

$$M_k = I - \alpha^{(k)} e_k^T$$

where

$$\alpha^{(k)} = (0, \ldots, 0, l_{k+1,k}, \ldots, l_{mk})^T,$$

then

$$A^{(k)} \equiv M_k A^{(k-1)} = \begin{bmatrix} A_{11}^{(k)} & A_{12}^{(k)} \\ 0 & A_{22}^{(k)} \end{bmatrix} \begin{matrix} k \\ m-k \end{matrix}$$
$$\begin{matrix} k & n-k \end{matrix}$$

with $A_{11}^{(k)}$ upper triangular. This illustrates the k-th step of *Gaussian elimination*. Recall from §3.5 that

$$(M_k \cdots M_1)^{-1} = M_1^{-1} \cdots M_k^{-1} = \prod_{i=1}^{k} (I_m + \alpha^{(i)} e_i^T) = I_m + \sum_{i=1}^{k} \alpha^{(i)} e_i^T,$$

and so we have

(4.2-1) $A = \begin{bmatrix} L_{11}^{(k)} & 0 \\ L_{21}^{(k)} & I_{m-k} \end{bmatrix} \begin{bmatrix} I_k & 0 \\ 0 & A_{22}^{(k)} \end{bmatrix} \begin{bmatrix} A_{11}^{(k)} & A_{12}^{(k)} \\ 0 & I_{n-k} \end{bmatrix},$

where

$$(M_k \cdots M_1)^{-1} \equiv \begin{bmatrix} L_{11}^{(k)} & 0 \\ L_{21}^{(k)} & I_{m-k} \end{bmatrix} = I_m + [\alpha^{(1)}, \ldots, \alpha^{(k)}, 0, \ldots, 0]$$

is unit lower triangular. Using this result we can establish the following important theorem.

THEOREM 4.2-1: *L-U Decomposition.*   Let A_k denote the leading k-by-k principal submatrix of $A \in \mathbb{R}^{m \times n}$. If A_k is nonsingular for $k = 1, \ldots, s = \min\{m-1, n\}$, then there exists a unit lower triangular matrix $L \in \mathbb{R}^{m \times m}$ and an upper triangular matrix $U \in \mathbb{R}^{m \times n}$ such that $A = LU$. Moreover, $\det(A_k) = u_{11} \cdots u_{kk}$ for $k = 1, \ldots, \min\{m, n\}$.

Proof.   Comparing $(1,1)$ blocks in (4.2-1) we conclude that $A_k = L_{11}^{(k)} A_{11}^{(k)}$. Since

$$\det[A_{11}^{(k)}] = a_{kk}^{(k-1)} \det[A_{11}^{(k-1)}],$$

it follows that

$$\det(A_k) = a_{11}^{(0)} \cdots a_{kk}^{(k-1)}. \qquad\qquad a_{11}^{(0)} \equiv a_{11}$$

The assumption that all the leading principal submatrices are nonsingular implies that the elimination process can continue until $k = s \equiv \min\{m-1, n\}$. At that stage, $U = A^{(s)} \in \mathbb{R}^{m \times n}$ is upper triangular, $L = M_1^{-1} \cdots M_s^{-1} \in \mathbb{R}^{m \times m}$ is unit lower triangular, and $A = LU$. □

The quantities $a_{kk}^{(k-1)}$ which have such a prominent role to play in Gaussian elimination are called *pivots*. Thus, if the conditions of Theorem 4.2-1 are satisfied, no zero pivots are encountered during the elimination.

The L-U decomposition is a "high-level" description of the quantities produced by Gaussian elimination. Expressing the outcome of a matrix algorithm in the "language" of matrix decompositions is very worthwhile. It facilitates generalization and highlights connections between algorithms that may appear very different at the scalar level.

Once the L-U decomposition of a square nonsingular matrix A has been computed, we are poised to solve $Ax = b$ using the algorithms of §4.1. In particular, if we use forward elimination to solve $Ly = b$ and apply back substitution to $Ux = y$, we find that $Ax = (LU)x = L(Ux) = Ly = b$.

EXAMPLE 4.2-2. If

$$
A = \begin{bmatrix} 1 & 4 & 7 \\ 2 & 5 & 8 \\ 3 & 6 & 11 \end{bmatrix} \quad \text{and} \quad b = \begin{bmatrix} 1 \\ 1 \\ 1 \end{bmatrix}
$$

then

$$
A = LU = \begin{bmatrix} 1 & 0 & 0 \\ 2 & 1 & 0 \\ 3 & 2 & 1 \end{bmatrix} \begin{bmatrix} 1 & 4 & 7 \\ 0 & -3 & -6 \\ 0 & 0 & 2 \end{bmatrix},
$$

$y = (1, -1, 0)^{\mathrm{T}}$ solves $Ly = b$, and $x = (-\frac{1}{3}, \frac{1}{3}, 0)^{\mathrm{T}}$ solves $Ux = y$.

In a computer program, the entries in A can be overwritten with the corresponding entries of L and U as they are produced. For example, after two steps in the reduction of a 5-by-4 matrix, the array A would look like this:

$$
\begin{bmatrix}
u_{11} & u_{12} & u_{13} & u_{14} \\
l_{21} & u_{22} & u_{23} & u_{24} \\
l_{31} & l_{32} & a_{33}^{(2)} & a_{34}^{(2)} \\
l_{41} & l_{42} & a_{43}^{(2)} & a_{44}^{(2)} \\
l_{51} & l_{52} & a_{53}^{(2)} & a_{54}^{(2)}
\end{bmatrix}.
$$

With this storage scheme we obtain

ALGORITHM 4.2-1: *Gaussian Elimination.* Given $A \in \mathbb{R}^{m \times n}$, the following algorithm computes the factorization $A = LU$ described in Theorem 4.2-1. The element a_{ij} is overwritten by l_{ij} if $i > j$ and $j \leq s = \min\{n, m-1\}$, and by u_{ij} otherwise. If A has a singular leading principal submatrix, then the algorithm may terminate prematurely.

For $k = 1, \ldots, s = \min\{m - 1, n\}$
  If $a_{kk} = 0$
    then
      quit
    else
      $w_j := a_{kj} \quad (j = k + 1, \ldots, n)$
      For $i = k + 1, \ldots, m$
        $\eta := a_{ik}/a_{kk}$
        $a_{ik} := \eta$
        For $j = k + 1, \ldots, n$
          $a_{ij} := a_{ij} - \eta w_j$

This algorithm requires $mns - (m + n)s^2/2 + s^3/3$ flops.

Unfortunately, Gaussian elimination can fail on some very simple matrices, e.g.,

$$A = \begin{bmatrix} 0 & 1 \\ 1 & 0 \end{bmatrix}.$$

While A has perfect 2-norm condition, it fails to have an L-U decomposition because it has a singular leading principal submatrix.

Clearly, modifications of Algorithm 4.2-1 are necessary if it is to be the cornerstone of an effective linear equation solver. The error analysis in the following section will suggest precisely what the needed modifications must achieve in order to ensure success.

Problems

P4.2-1. Show that if $A \in \mathbb{R}^{n \times n}$ has an L-U decomposition and is nonsingular, then L and U are unique.

P4.2-2. Suppose the entries of $A(\epsilon) \in \mathbb{R}^{n \times n}$ are continuously differentiable functions of the scalar ϵ. Assume that $A \equiv A(0)$ and all its principal submatrices are nonsingular. Show that for sufficiently small ϵ, the matrix $A(\epsilon)$ has an L-U decomposition $A(\epsilon) = L(\epsilon)U(\epsilon)$ and that $L(\epsilon)$ and $U(\epsilon)$ are both continuously differentiable.

P4.2-3. Suppose

$$A = \begin{bmatrix} A_{11} & A_{12} \\ A_{21} & A_{22} \end{bmatrix} \begin{matrix} k \\ m-k \end{matrix}$$
$$\,k \quad\;\; n-k$$

and that A_{11} is nonsingular. The matrix

$$S = A_{22} - A_{21}A_{11}^{-1}A_{12}$$

is called the *Schur complement* of A_{11} in A. Show that if A_{11} has an

L-U decomposition, then after k steps of Gaussian elimination S equals the matrix $A_{22}^{(k)}$ in (4.2-1).

P4.2-4. Suppose $A \in \mathbb{R}^{n \times n}$ has an L-U decomposition. Show how $Ax = b$ can be solved without accumulating L by applying Gaussian elimination to the n-by-$(n + 1)$ matrix $[A, b]$. How many flops are required?

P4.2-5. Obtain a "column oriented" version of Gaussian elimination by interchanging the order of the two innermost loops in Algorithm 4.2-1.

P4.2-6. Suppose

$$
B_k = \begin{matrix} \begin{bmatrix} U_{11}^{(k)} & U_{12}^{(k)} \\ v^{\mathrm{T}} & w^{\mathrm{T}} \end{bmatrix} & \begin{matrix} k \\ 1 \end{matrix} \\ \begin{matrix} k \quad\ \ n-k \end{matrix} \end{matrix}
$$

and that $U_{11}^{(k)}$ is nonsingular and upper triangular. Show how $y \in \mathbb{R}^k$ can be determined so that

$$
\begin{bmatrix} I_k & 0 \\ y^{\mathrm{T}} & 1 \end{bmatrix} \begin{bmatrix} U_{11}^{(k)} & U_{12}^{(k)} \\ v^{\mathrm{T}} & w^{\mathrm{T}} \end{bmatrix}
$$

is upper triangular. Show how this idea can be used to solve $Ax = b$ ($A \in \mathbb{R}^{n \times n}$) with only $n^2/2$ storage. (Hint: Read one row of $[A, b]$ at a time and update the triangular form.)

P4.2-7. Describe a variant of Gaussian elimination which introduces zeros into the columns of A in the order $n, n - 1, \ldots, 2$ and which produces the decomposition

$$
A = UL,
$$

where U is unit upper triangular and L is lower triangular.

P4.2-8. Matrices in $\mathbb{R}^{n \times n}$ of the form $N(y, k) = I + ye_k^{\mathrm{T}}$ where $y \in \mathbb{R}^n$ are said to be *Gauss-Jordan transformations*. (a) Give a formula for $N(y, k)^{-1}$ assuming it exists. (b) Given $x \in \mathbb{R}^n$, under what conditions can y be found so $N(y, k)x = e_k$? (c) Give an algorithm using Gauss-Jordan transformations that overwrites A with A^{-1}. What conditions on A ensure the success of your algorithm?

Notes and References for Sec. 4.2

Schur complements (P4.2-3) arise in many applications. For a survey of both practical and theoretical interest, see

R. W. Cottle (1974). "Manifestations of the Schur Complement." *Lin. Alg. & Its Applic.* 8, 189–211.

Schur complements are known as "Gauss transforms" in some application areas. The use of Gauss-Jordan transformations (P4.2-8) is detailed in

L. Fox (1964). *An Introduction to Numerical Linear Algebra*, Oxford University Press, Oxford.

Sec. 4.3. Roundoff Error Analysis of Gaussian Elimination

We now assess the effect of rounding errors when Algorithms 4.2-1, 4.1-1, and 4.1-2 are used to solve the linear system $Ax = b$. Before we proceed with the analysis, it is useful to consider the nearly "ideal" situation in which *no* roundoff occurs during the entire solution process except when A and b are stored. If $fl(b) = b + e$ and $fl(A) = A + E$ is nonsingular, then these assumptions imply that the computed solution $\hat{x}$ satisfies

(4.3-1) $(A + E)\hat{x} = (b + e)$. $\|E\|_\infty \leqslant \mathbf{u}\|A\|_\infty$, $\|e\|_\infty \leqslant \mathbf{u}\|b\|_\infty$

That is, $\hat{x}$ solves a "nearby" system exactly. Moreover, if $\mathbf{u}\kappa_\infty(A) \leqslant \frac{1}{2}$ (say), then by using Theorem 2.5-1 it can be shown that

(4.3-2) $$\frac{\|x - \hat{x}\|_\infty}{\|x\|_\infty} \leqslant 4\mathbf{u}\kappa_\infty(A).$$

The bounds (4.3-1) and (4.3-2) are "best possible." No general ∞-norm error analysis of a linear equation solver that requires the storage of A and b can render sharper bounds. As a consequence, we cannot justifiably criticize an algorithm for returning an inaccurate $\hat{x}$ if A is ill-conditioned relative to the machine precision, e.g., $\mathbf{u}\kappa_\infty(A) \cong 1$.

Let us see how the error bounds for Gaussian elimination compare with the ideal bounds above. We work with the infinity norm for convenience. Our first challenge is to quantify the roundoff errors associated with the computed triangular factors.

LEMMA 4.3-1.   Assume $A \in \mathbb{R}^{m \times n}$ is a matrix of floating point numbers. If no zero pivots are encountered during the execution of Gaussian elimination, then the computed triangular matrices $\hat{L}$ and $\hat{U}$ satisfy

$$\hat{L}\hat{U} = A + H$$

(4.3-3)

$$|H| \leqslant 3s\mathbf{u}\{|A| + |\hat{L}| \, |\hat{U}|\} + O(\mathbf{u}^2)$$

where $s = \min\{m - 1, n\}$.

Proof.   The proof is by induction on s. The lemma obviously holds for $s = 1$. Assume it holds for all $(m - 1)$-by-$(n - 1)$ floating point matrices. If

$$A = \begin{bmatrix} \alpha & w^T \\ v & B \end{bmatrix} \quad \begin{matrix} 1 \\ m-1 \end{matrix}$$
$$\begin{matrix} 1 & n-1 \end{matrix}$$

then $\hat{z} = fl(v/\alpha)$ and $\hat{A}_1 = fl(B - \hat{z}w^T)$ are computed in the first step of Gaussian elimination. We therefore have

(4.3-4) $$\hat{z} = \frac{1}{\alpha} v + f \qquad\qquad |f| \leqslant \frac{\mathbf{u}}{\alpha} |v|$$

and

(4.3-5) $\hat{A}_1 = B - \hat{z}w^T + F.$ $\quad |F| \leqslant 2\mathbf{u}[|B| + |\hat{z}| |w|^T] + O(\mathbf{u}^2)$

The algorithm now proceeds to calculate the L-U decomposition of $\hat{A}_1$. By induction the resulting triangular matrices $\hat{L}_1$ and $\hat{U}_I$ satisfy

(4.3-6)
$$\hat{L}_1\hat{U}_1 = \hat{A}_1 + H_1$$
$$|H_1| \leqslant 3(s - 1)\mathbf{u}[|\hat{A}_1| + |\hat{L}_1| |\hat{U}_1|] + O(\mathbf{u}^2).$$

Thus,

$$\hat{L}\hat{U} \equiv \begin{bmatrix} 1 & 0 \\ \hat{z} & \hat{L}_1 \end{bmatrix} \begin{bmatrix} \alpha & w^T \\ 0 & \hat{U}_1 \end{bmatrix} = A + \begin{bmatrix} 0 & 0 \\ \alpha f & H_1 + F \end{bmatrix} \equiv A + H.$$

From (4.3-5) it follows that

$$|A_1| \leqslant (1 + 2\mathbf{u})[|B| + |\hat{z}| |w|^T] + O(\mathbf{u}^2),$$

and therefore by using (4.3-6) we have

$$|H_1 + F| \leqslant 3s\mathbf{u}[|B| + |\hat{z}| |w|^T + |\hat{L}_1| |\hat{U}_1|] + O(\mathbf{u}^2).$$

Since $|\alpha f| \leqslant \mathbf{u}|v|$ it is easy to verify that

$$|H| \leqslant 3s\mathbf{u}\left\{ \begin{bmatrix} |\alpha| & |w|^T \\ |v| & |B| \end{bmatrix} + \begin{bmatrix} 1 & 0 \\ |\hat{z}| & |\hat{L}_1| \end{bmatrix} \begin{bmatrix} |\alpha| & |w|^T \\ 0 & |\hat{U}_1| \end{bmatrix} \right\}$$
$$+ O(\mathbf{u}^2),$$

thereby proving the lemma. $\square$

At this point we assume $m = n$ and proceed to examine the effect of round-off error when $\hat{L}$ and $\hat{U}$ are used in Algorithms 4.1-1 and 4.1-2 to solve $Ax = b$.

THEOREM 4.3-2. Let $\hat{L}\hat{U}$ be the computed L-U decomposition of the floating point matrix $A \in \mathbb{R}^{n \times n}$ obtained by Gaussian Elimination (Algorithm 4.2-1). If $\hat{y}$ is the computed solution to $\hat{L}y = b$ found via Algorithm 4.1-1 and $\hat{x}$ is the computed solution to $\hat{U}x = \hat{y}$ obtained from Algorithm 4.1-2, then $\hat{x}$ satisfies

(4.3-7) $\quad (A + E)\hat{x} = b.$ $\quad |E| \leqslant n\mathbf{u}[3|A| + 5|\hat{L}| |\hat{U}|] + O(\mathbf{u}^2)$

Proof. From (4.1-1) and (4.1-2) we have

$$(\hat{L} + F)\hat{y} = b \qquad\qquad |F| \leqslant n\mathbf{u}|\hat{L}| + O(\mathbf{u}^2)$$
$$(\hat{U} + G)\hat{x} = \hat{y} \qquad\qquad |G| \leqslant n\mathbf{u}|\hat{U}| + O(\mathbf{u}^2)$$

and thus

$$(\hat{L}\hat{U} + F\hat{U} + \hat{L}G + FG)\hat{x} = b.$$

Now

$$\hat{L}\hat{U} = A + H,$$

with $|H| \leqslant 3n\mathbf{u}[\,|A| + |\hat{L}|\,|\hat{U}|\,] + O(\mathbf{u}^2)$, and so by defining

$$E = H + F\hat{U} + \hat{L}G + FG$$

we find $(A + E)\hat{x} = b$ and

$$|E| \leqslant 3n\mathbf{u}[\,|A| + |\hat{L}|\,|\hat{U}|\,] + 2n\mathbf{u}[\,|\hat{L}|\,|\hat{U}|\,] + O(\mathbf{u}^2). \;\square$$

Were it not for the possibility of a large $|\hat{L}|\,|\hat{U}|$ term, (4.3-7) would compare favorably with the ideal bound in (4.3-1). (The factor n is of no consequence; cf. the Wilkinson quotation in §3.2.) Such a possibility exists, for there is nothing in Gaussian elimination to rule out the appearance of small pivots. If a small pivot is encountered, then we can expect large numbers to be present in $\hat{L}$ and $\hat{U}$.

EXAMPLE 4.3-1.   Suppose $\beta = 10$, $t = 3$, chopped arithmetic is used to solve

$$\begin{bmatrix} .001 & 1.00 \\ 1.00 & 2.00 \end{bmatrix} \begin{bmatrix} x_1 \\ x_2 \end{bmatrix} = \begin{bmatrix} 1.00 \\ 3.00 \end{bmatrix}.$$

Applying Gaussian elimination we get

$$\hat{L} = \begin{bmatrix} 1 & 0 \\ 1000 & 1 \end{bmatrix} \qquad \hat{U} = \begin{bmatrix} .001 & 1 \\ 0 & -1000 \end{bmatrix}$$

and a calculation shows

$$\hat{L}\hat{U} = \begin{bmatrix} .001 & 1 \\ 1 & 2 \end{bmatrix} + \begin{bmatrix} 0 & 0 \\ 0 & -2 \end{bmatrix} \equiv A + H.$$

Moreover, $6\begin{bmatrix} 10^{-6} & 10^{-3} \\ 10^{-3} & 1.001 \end{bmatrix}$ is the bounding matrix in (4.3-3), not a severe overestimate of $|H|$. If we go on to solve the problem via Algorithms 4.1-1 and 4.1-2 using the same precision arithmetic, then we obtain a computed solution $\hat{x} = (0, 1)^{\mathrm{T}}$. This is in contrast to the exact solution $x = (1.002\ldots, .998\ldots)^{\mathrm{T}}$.

We stress that there is no correlation between small (or vanishing) pivots and ill-conditioning. The example $A = \begin{bmatrix} 0 & 1 \\ 1 & 0 \end{bmatrix}$ makes this obvious. Thus, Gaussian elimination can give arbitrarily poor results, even for well-conditioned problems, i.e., the method is unstable.

In order to repair this shortcoming of the algorithm, it is necessary to introduce row and/or column interchanges during the elimination process with the intention of keeping the numbers that arise during the calculation suitably bounded. This idea is pursued in the next section.

Problems

P4.3-1. Establish (4.3-2).

P4.3-2. Show that if we drop the assumption that A is floating point in Lemma 4.3-1, then (4.3-3) holds with the coefficient "3" replaced by "4."

P4.3-3. Assume A is m-by-n. How many flops are required to compute $|\hat{L}| \, |\hat{U}|$? Show $fl(|\hat{L}| \, |\hat{U}|) \leqslant (1 + 2n\mathbf{u})|\hat{L}| \, |\hat{U}| + O(\mathbf{u}^2)$.

P4.3-4. Suppose $x = A^{-1}b$. If $e = x - \hat{x}$ (the error) and $r = b - A\hat{x}$ (the residual) then

$$\frac{\|r\|}{\|A\|} \leqslant \|e\| \leqslant \|A^{-1}\| \, \|r\|.$$

Assume consistency between the matrix and vector norm.

P4.3-5. Using $\beta = 10, t = 2$, chopped arithmetic, compute the L-U factorization of

$$A = \begin{bmatrix} 7 & 6 \\ 9 & 8 \end{bmatrix}.$$

For this example, what is the matrix H in (4.3-3)?

Notes and References for Sec. 4.3

Our style of inverse error analysis is patterned after

C. deBoor and A. Pinkus (1977). "A Backward Error Analysis for Totally Positive Linear Systems," *Numer. Math. 27*, 485–90.

The original roundoff analysis of Gaussian elimination appears in

J. H. Wilkinson (1961). "Error Analysis of Direct Methods of Matrix Inversion," *J. Assoc. Comp. Mach. 8*, 281–330.

Various improvements in the bounds and simplifications in the analysis have occurred over the years. See

J. K. Reid (1971). "A Note on the Stability of Gaussian Elimination," *J. Inst. Math. Applic. 8*, 374–75.

C. C. Paige (1973). "An Error Analysis of a Method for Solving Matrix Equations," *Math. Comp. 27*, 355–59.

H. H. Robertson (1977). "The Accuracy of Error Estimates for Systems of Linear Algebraic Equations," *J. Inst. Math. Applic. 20*, 409–14.

Sec. 4.4. Pivoting

Gaussian elimination is unstable because of the possibility of arbitrarily small pivots. This problem can be alleviated, however, by interchanging rows and columns during the elimination. To be precise, given $A \in \mathbb{R}^{m \times n}$ suppose that Gauss transformations $M_1, \ldots, M_{k-1} \in \mathbb{R}^{m \times m}$, permutation matrices $P_1, \ldots, P_{k-1} \in \mathbb{R}^{m \times m}$, and permutation matrices $\Pi_1, \ldots, \Pi_{k-1} \in \mathbb{R}^{n \times n}$ have been determined such that

$$(4.4\text{-}1) \qquad A^{(k-1)} = M_{k-1} P_{k-1} \cdots M_1 P_1 A \Pi_1 \cdots \Pi_{k-1}$$

$$= \begin{bmatrix} A_{11}^{(k-1)} & A_{12}^{(k-1)} \\ 0 & A_{22}^{(k-1)} \end{bmatrix} \begin{matrix} k-1 \\ m-k+1 \end{matrix}$$
$$ \begin{matrix} k-1 & n-k+1 \end{matrix}$$

where $A_{11}^{(k-1)}$ is upper triangular and

$$A_{22}^{(k-1)} = \begin{bmatrix} a_{kk}^{(k-1)} & \cdots & a_{kn}^{(k-1)} \\ \vdots & & \vdots \\ a_{mk}^{(k-1)} & \cdots & a_{mn}^{(k-1)} \end{bmatrix}.$$

If we compute M_k to zero entries $(k+1, k), \ldots, (m, k)$ of $A_{22}^{(k-1)}$ as in Gaussian elimination, then we risk generation of large multipliers whenever $a_{kk}^{(k-1)}$ is small.

Instead, we determine permutation matrices $\tilde{P}_k$ and $\tilde{\Pi}_k$ such that the leading entry in the matrix

$$(4.4\text{-}2) \qquad \tilde{P}_k A_{22}^{(k-1)} \tilde{\Pi}_k = \begin{bmatrix} \tilde{a}_{kk}^{(k-1)} & \cdots & \tilde{a}_{kn}^{(k-1)} \\ \vdots & & \vdots \\ \tilde{a}_{mk}^{(k-1)} & \cdots & \tilde{a}_{mn}^{(k-1)} \end{bmatrix}$$

is "acceptably large." (Details will follow.) Having done this we proceed as in ordinary Gaussian elimination and compute a Gauss transformation M_k such that if $P_k = \text{diag}(I_{k-1}, \tilde{P}_k)$ and $\Pi_k = \text{diag}(I_{k-1}, \tilde{\Pi}_k)$, then

$$A^{(k)} = M_k(P_k A^{(k-1)} \Pi_k) = \begin{bmatrix} A_{11}^{(k)} & A_{12}^{(k)} \\ 0 & A_{22}^{(k)} \end{bmatrix} \begin{matrix} k \\ m-k \end{matrix},$$
$$ \begin{matrix} k & n-k \end{matrix}$$

with $A_{11}^{(k)}$ is upper triangular. Note that if

$$M_k = I_m - \alpha^{(k)} e_k^{\mathrm{T}}$$

with

(4.4-3) $$\alpha^{(k)} = (0, \ldots, 0, \tilde{l}_{k+1,k}, \ldots, \tilde{l}_{mk})^{\mathrm{T}},$$

then

(4.4-4) $$\tilde{l}_{ik} = \tilde{a}_{ik}^{(k-1)}/\tilde{a}_{kk}^{(k-1)}. \qquad\qquad i = k + 1, \ldots, m$$

Thus, if we are to guard against large numbers arising during the elimination, we must choose the permutations $\tilde{P}_k$ and $\tilde{\Pi}_k$ in equation (4.4-2) so that the multipliers $\tilde{l}_{ik}$ are suitably bounded.

One strategy for determining P_k and Π_k is called *complete pivoting* and involves choosing these permutations such that

(4.4-5) $$|\tilde{a}_{kk}^{(k-1)}| = \max_{\substack{k \leqslant i \leqslant m \\ k \leqslant j \leqslant n}} |\tilde{a}_{ij}^{(k-1)}|.$$

This requires searching the submatrix $A_{22}^{(k-1)}$ for its largest entry. Note that if this number is zero then we may conclude from (4.4-1) that $\text{rank}(A) = k - 1$. It also follows from (4.4-5) that the multipliers $\tilde{l}_{ij}$ in (4.4-4) satisfy

(4.4-6) $$|\tilde{l}_{ij}| \leqslant 1.$$

In summary, Gaussian elimination with complete pivoting computes the decomposition

(4.4-7) $$M_s P_s \cdots M_1 P_1 A \Pi_1 \cdots \Pi_s = U = \begin{bmatrix} U_{11} & U_{12} \\ 0 & 0 \end{bmatrix} \begin{matrix} r \\ m-r \end{matrix}$$
$$\begin{matrix} r & n-r \end{matrix}$$

where $r = \text{rank}(A)$, $s = \min\{m - 1, r\}$, U_{11} is upper triangular and nonsingular, and the multipliers in the Gauss transformations $M_1, \ldots, M_s$ are each bounded by unity in modulus.

When we incorporate complete pivoting in Algorithm 4.2-1 we obtain

ALGORITHM 4.4-1: *Gaussian Elimination with Complete Pivoting.* Given $A \in \mathbb{R}^{m \times n}$, the following algorithm computes the decomposition (4.4-7) using the strategy of complete pivoting. The element a_{ij} is overwritten by u_{ij} whenever $i \leqslant j$ and by the multiplier $\tilde{l}_{ij}$ for $i > j$. The permutations $P_1, \ldots, P_s$ and $\Pi_1, \ldots, \Pi_s$ are represented by a pair of integer vectors $(r_1, \ldots, r_s)$ and $(c_1, \ldots, c_s)$ where $s = \min\{m - 1, \text{rank}(A)\}$. P_k is obtained by interchanging row k and row r_k of I_m. Π_k is obtained by interchanging column k and column c_k of I_n.

For $k = 1, \ldots, s = \min\{m - 1, \text{rank}(A)\}$
  Determine indices $p \in \{k, \ldots, m\}$ and $q \in \{k, \ldots, n\}$ such that

$$|a_{pq}| = \max_{\substack{k \leqslant i \leqslant m \\ k \leqslant j \leqslant n}} |a_{ij}|$$

  $r_k := p$
  $c_k := q$
  Swap a_{kj} and a_{pj} $(j = k, \ldots, n)$.
  Swap a_{ik} and a_{iq} $(i = 1, \ldots, m)$.
  $w_j := a_{kj}$ $(j = k + 1, \ldots, n)$
  For $i = k + 1, \ldots, m$
    $\eta := a_{ik}/a_{kk}$
    $a_{ik} := \eta$
    For $j = k + 1, \ldots, n$
      $a_{ij} := a_{ij} - \eta w_j$

This algorithm requires $mns - (m + n)s^2/2 + s^3/3$ flops and an equal number of comparisons. For the case $m = n = \text{rank}(A)$, this amounts to $n^3/3$ flops and comparisons.

  In principle. Gaussian elimination with complete pivoting can be used to determine the rank of a matrix. Yet roundoff errors make the probability of encountering an exactly zero pivot remote. In practice one would have to "declare" A to have rank k if the pivot element in step $k + 1$ was sufficiently small. The advisability of determining matrix rank in this fashion is discussed in §6.4.
  As in the case of (ordinary) Gaussian elimination, it is possible to couch the outcome of Algorithm 4.4-1 in the language of matrix decompositions. Define the permutation matrices P and Π by

$$P = P_s \cdots P_1 \qquad\qquad s = \min\{m - 1, \text{rank}(A)\}$$

$$\Pi = \Pi_1 \cdots \Pi_s$$

and set

$$L = P(M_s P_s \cdots M_1 P_1)^{-1}.$$

By carefully considering the structure of the M_i and the P_i it is possible to show that L is unit lower triangular. The subdiagonal entries in its first s columns are "made up" of the multipliers $\tilde{l}_{ij}$. It follows from (4.4-7) that Gaussian elimination with complete pivoting computes the L-U decomposition of a permuted version of A, namely,

(4.4-8) $P A \Pi = LU.$

We now examine the effect of roundoff error when this decomposition is used to solve the square nonsingular system $Ax = b$:

$$Ly = Pb$$

(4.4-9)
$$Uz = y$$

$$x = \Pi z.$$

The key thing to bear in mind is that there are no rounding errors associated with permutation. Consequently, it is not hard to show, using Theorem 4.3-2, that the computed solution $\hat{x}$ satisfies

(4.4-10) $(A + E)\hat{x} = b, \ |E| \leqslant n\mathbf{u}\{3|A| + 5\hat{P}^T|\hat{L}| \ |\hat{U}|\hat{\Pi}^T\} + O(\mathbf{u}^2)$

where $\hat{P}$, $\hat{\Pi}$, $\hat{L}$, and $\hat{U}$ are the computed analogs of P, Π, L, and U. Pivoting implies that the elements of $\hat{L}$ are bounded by one. Thus, $\|\hat{L}\|_\infty \leqslant n$ and we obtain the bound

(4.4-11) $$\|E\|_\infty \leqslant n\mathbf{u}\{3\|A\|_\infty + 5n\|\hat{U}\|_\infty\} + O(\mathbf{u}^2).$$

The problem now is to bound $\|\hat{U}\|_\infty$. Define the *growth factor* ρ by

(4.4-12) $$\rho = \max_{i,j,k} \frac{|\hat{a}_{ij}^{(k)}|}{\|A\|_\infty}$$

where $\hat{A}^{(k)}$ is the computed version of the matrix $A^{(k)}$ defined by equation (4.4-1). It follows that

(4.4-13) $$\|E\|_\infty \leqslant 8n^3\rho\|A\|_\infty\mathbf{u} + O(\mathbf{u}^2).$$

Whether or not this compares favorably with the "ideal" bound (4.3-1) hinges on the size of the growth factor ρ. (The factor n^3 is very pessimistic and may be ignored.) The growth factor measures how large the numbers become during the process of elimination. Wilkinson (1961) has shown that in exact arithmetic the elements of the matrix $A^{(k)}$ defined by (4.4-1) satisfy

$$|a_{ij}^{(k)}| \leqslant k^{1/2}(2\cdot3^{1/2} \cdots k^{1/k-1})^{1/2}\max|a_{ij}|.$$

The upper bound is a rather slow-growing function of k. This fact coupled with vast empirical evidence suggesting that ρ is always modestly sized (e.g. $\rho = 10$) permit us to conclude that *Gaussian elimination with complete pivoting is stable*. Inequality (4.4-13) is essentially no different from (4.3-1). The method solves a "nearby" linear system $(A + E)x = b$ exactly.

EXAMPLE 4.4-1. If Gaussian elimination with complete pivoting is applied to the problem

$$\begin{bmatrix} .001 & 1.00 \\ 1.00 & 2.00 \end{bmatrix} \begin{bmatrix} x_1 \\ x_2 \end{bmatrix} = \begin{bmatrix} 1.00 \\ 3.00 \end{bmatrix}$$

with $\beta = 10$, $t = 3$, chopped arithmetic, then

$$P = \begin{bmatrix} 0 & 1 \\ 1 & 0 \end{bmatrix}, \quad \Pi = \begin{bmatrix} 0 & 1 \\ 1 & 0 \end{bmatrix}, \quad \hat{L} = \begin{bmatrix} 1.00 & 0.00 \\ .500 & 1.00 \end{bmatrix}, \quad \hat{U} = \begin{bmatrix} 2.00 & 1.00 \\ 0.00 & .499 \end{bmatrix}$$

and $\hat{x} = (1.00, 1.00)^T$. This is correct to two significant digits in each component. Compare with Example 4.3-1.

The numerical stability obtained by introducing complete pivoting is not without cost. Roughly speaking, there are as many arithmetic comparisons performed as flops. This represents a nontrivial overhead and explains why a much cheaper pivoting strategy called *partial pivoting* is preferred. In partial pivoting only row interchanges are performed, i.e., all the permutations Π_k in Algorithm 4.4-1 are set to the identity. This means that we search only the first column of the matrix $A_{22}^{(k-1)}$ in (4.4-1) for its largest entry. Notice that if this number is zero then we may conclude that the first k columns of the original matrix are dependent.

Since only one-dimensional searches are involved when partial pivoting is used, the number of arithmetic comparisons is reduced by an order of magnitude.

EXAMPLE 4.4-2. If Gaussian elimination with partial pivoting is applied to the problem

$$\begin{bmatrix} .001 & 1.00 \\ 1.00 & 2.00 \end{bmatrix} \begin{bmatrix} x_1 \\ x_2 \end{bmatrix} = \begin{bmatrix} 1.00 \\ 3.00 \end{bmatrix}$$

with $\beta = 10$, $t = 3$, chopped arithmetic, then

$$P = \begin{bmatrix} 0 & 1 \\ 1 & 0 \end{bmatrix}, \quad \hat{L} = \begin{bmatrix} 1.00 & 0 \\ .001 & 1.00 \end{bmatrix}, \quad \hat{U} = \begin{bmatrix} 1.00 & 2.00 \\ 0.00 & 1.00 \end{bmatrix},$$

and $\hat{x} = (1.00, .996)^T$, which is correct to three digits.

The complete method for square nonsingular matrices strives to compute permutations $P_1, \ldots, P_{n-1}$ and Gauss transformations $M_1, \ldots, M_{n-1}$ such that

$$(4.4\text{-}14) \qquad M_{n-1}P_{n-1} \cdots M_1 P_1 A = U$$

is upper triangular. The detailed algorithm is as follows:

ALGORITHM 4.4-2: *Gaussian Elimination with Partial Pivoting.* Given $A \in \mathbb{R}^{n \times n}$ (nonsingular), the following algorithm computes the decomposition (4.4-14) using the strategy of partial pivoting. The element a_{ij} is overwritten by u_{ij} whenever $i \leqslant j$ and by the multiplier $\tilde{l}_{ij}$ otherwise ($\tilde{l}_{ij}$ is the (i, j) entry of

$-M_j$). The permutations $P_1, \ldots, P_{n-1}$ are represented by an integer vector $(r_1, \ldots, r_{n-1})$. P_k is obtained by interchanging row k and row r_k of I_n.

> For $k = 1, \ldots, n - 1$
> > Determine $p \in \{k, k + 1, \ldots, n\}$ so $|a_{pk}| = \max_{k \leqslant i \leqslant n} |a_{ik}|$.
> > $r_k := p$
> > Swap a_{kj} and a_{pj} $(j = k, \ldots, n)$
> > $w_j := a_{kj}$ $(j = k + 1, \ldots, n)$
> > For $i = k + 1, \ldots, n$
> > > $\eta := a_{ik}/a_{kk}$
> > > $a_{ik} := \eta$
> > > For $j = k + 1, \ldots, n$
> > > > $a_{ij} := a_{ij} - \eta w_j$

This algorithm requires $n^3/3$ flops and $O(n^2)$ comparisons.

If $P = P_{n-1} \cdots P_1$ and $L = P(M_{n-1}P_{n-1} \cdots M_1 P_1)^{-1}$, then it follows from (4.4-14) that

$$PA = LU.$$

As in the case of Gaussian elimination with complete pivoting, it is possible to show that L is unit lower triangular, its lower triangular entries being comprised of the multipliers $\tilde{l}_{ij}$. Moreover, when we solve $Ax = b$ via $Ly = Pb$, $Ux = y$, we obtain a computed solution $\hat{x}$ that solves a nearby problem $(A + E)\hat{x} = b$ in the sense that E satisfies (4.4-13). The growth factor associated with partial pivoting usually has the same order of magnitude as the growth factor for complete pivoting, although it is possible to contrive examples where the former has size 2^n. Most numerical analysts, however, regard serious element growth in Gaussian elimination with partial pivoting as highly unlikely to occur in practical problems. The method can be used with confidence.

Since Gaussian elimination with partial pivoting is the standard method for solving general square linear systems, it is convenient to assume hereafter that "Gaussian elimination" means "Gaussian elimination with partial pivoting."

Problems

P4.4-1. Let $A = LU$ be the L-U decomposition of $A \in \mathbb{R}^{n \times n}$ with $|l_{ij}| \leqslant 1$. Let a_i^T and u_i^T denote the i-th rows of A and U, respectively. Verify the equation

$$u_i^T = a_i^T - \sum_{j=1}^{i-1} l_{ij} u_j^T$$

and use it to show that $\|U\|_\infty \leqslant 2^{n-1} \|A\|_\infty$. (Hint: Take norms and use induction.)

P4.4-2. Let $A \in \mathbb{R}^{n \times n}$ be defined by

$$a_{ij} = \begin{cases} 1 & \text{if } i = j \text{ or } j = n \\ -1 & \text{if } i > j \\ 0 & \text{otherwise} \end{cases}$$

Show that A has an L-U decomposition with $|l_{ij}| \leqslant 1$ and $u_{nn} = 2^{n-1}$.

P4.4-3. Show that if $PA\Pi = LU$ is obtained via Gaussian elimination with complete pivoting, then $|u_{ii}| \geqslant |u_{ij}|$ $(i = 1, \ldots, n; j = i, \ldots, n)$.

P4.4-4. Show that if $A^T \in \mathbb{R}^{n \times n}$ is diagonally dominant, then A has a decomposition $A = LU$ with $|l_{ij}| \leqslant 1$.

P4.4-5. Assume that $A \in \mathbb{R}^{n \times n}$ is nonsingular. (a) Show how Gaussian elimination with partial pivoting can be used to solve the p linear systems $Ax = b^{(1)}, \ldots, Ax = b^{(p)}$ in $n^3/3 + pn^2$ flops. (b) Give an algorithm for computing A^{-1}. (c) Is your method for part (a) cheaper than forming A^{-1} explicitly and then computing $A^{-1}b^{(i)}$ for $i = 1, \ldots, p$?

P4.4-6. Suppose $A \in \mathbb{R}^{n \times n}$ has an L-U decomposition and that L and U are known. Give an algorithm which can compute the (i, j) entry of A^{-1} in approximately $\frac{1}{2}(n - j)^2 + \frac{1}{2}(n - i)^2$ flops.

P4.4-7. Show that if $\hat{X}$ is the computed inverse in P4.4-5(b), then $A\hat{X} = I + F$, where

$$\|F\|_1 \leqslant n\mathbf{u}[3 + 5n^2\rho]\|A\|_\infty\|\hat{X}\|_1.$$

P4.4-8. Suppose Algorithm 4.4-1 is applied to a nonsingular matrix $A \in \mathbb{R}^{n \times n}$. Detail how the output can be used to solve the system $Ax = b$.

P4.4-9. Consider the matrices

$$A = \begin{bmatrix} 4 & 1 & 1 & 1 \\ 1 & 3 & 0 & 0 \\ 1 & 0 & 2 & 0 \\ 1 & 0 & 0 & 1 \end{bmatrix} \quad \text{and} \quad \tilde{A} = \begin{bmatrix} 1 & 0 & 0 & 1 \\ 0 & 2 & 0 & 1 \\ 0 & 0 & 3 & 1 \\ 1 & 1 & 1 & 4 \end{bmatrix}.$$

If Gaussian elimination is carried out in $\beta = 10, t = 3$, arithmetic, verify that accurate factors L and U are produced from each matrix. (Note that all of the zero elements of $\tilde{A}$ are preserved in the resulting factors L and U. This suggests that for sparse matrices it may be useful at times to pick a pivot which appears to be numerically inferior in order to preserve sparsity.)

P4.4-10. What happens in Algorithm 4.4-2 if A is singular?

Notes and References for Sec. 4.4

FORTRAN codes for solving general linear systems may be found in LINPACK (Chapter 1) and ALGOL versions in

H. J. Bowdler, R. S. Martin, G. Peters, and J. H. Wilkinson (1966). "Solution of Real and Complex Systems of Linear Equations," *Numer. Math. 8*, 217–34. See also HACLA, pp. 93–110.

The conjecture that $|a_{ij}^{(k)}| \leqslant n \max |a_{ij}|$ when complete pivoting is used has been proven in the real $n = 4$ case in

C. W. Cryer (1968). "Pivot Size in Gaussian Elimination," *Numer. Math. 12*, 335–45.

Other papers concerned with element growth and pivoting are

P. A. Businger (1971). "Monitoring the Numerical Stability of Gaussian Elimination," *Numer. Math. 16*, 360–61.

A. M. Cohen (1974). "A Note on Pivot Size in Gaussian Elimination," *Lin. Alg. & Its Applic. 8*, 361–68.

A. M. Erisman and J. K. Reid (1974). "Monitoring the Stability of the Triangular Factorization of a Sparse Matrix," *Numer. Math. 22*, 183–86.

J. K. Reid (1971). "A Note on the Stability of Gaussian Elimination," *J. Inst. Math. Applics. 8*, 374–75.

J. H. Wilkinson (1961). "Error Analysis of Direct Methods of Matrix Inversion," *J. Assoc. Comp. Mach. 8*, 281–330.

The designers of sparse Gaussian elimination codes are interested in the topic of element growth because multipliers greater than unity are sometimes tolerated for the sake of minimizing fill-in.

Sec. 4.5. Improving and Estimating Accuracy

Suppose Gaussian elimination (with partial or complete pivoting) is used to solve the n-by-n system $Ax = b$. The results of the previous section when coupled with Theorem 2.5-1 permit us to conclude that if t-digit, base β floating point arithmetic is used, then the computed solution $\hat{x}$ satisfies

$$(4.5\text{-}1) \qquad (A + E)\hat{x} = b \qquad \qquad \frac{\|E\|_\infty}{\|A\|_\infty} \cong \beta^{-t}$$

and

$$(4.5\text{-}2) \qquad \frac{\|\hat{x} - x\|_\infty}{\|x\|_\infty} \cong \beta^{-t}\kappa_\infty(A).$$

(In this section we rely exclusively on the infinity norm, since it is useful in practice and very handy when analyzing roundoff error.) Suppose $\kappa_\infty(A) \cong \beta^q$. Since (4.5-1) and (4.5-2) imply

$$\|b - A\hat{x}\|_\infty = \|E\hat{x}\|_\infty \cong \beta^{-t}\|A\|_\infty\|\hat{x}\|_\infty$$

and

$$\frac{\|\hat{x} - x\|_\infty}{\|x\|_\infty} \cong \beta^{q-t},$$

respectively, we have

Heuristic I: Gaussian elimination produces a solution $\hat{x}$ that renders a relatively small residual.

Heuristic II: Gaussian elimination produces a solution $\hat{x}$ that has about $t \log_{10}\beta - \log_{10}[\kappa_\infty(A)]$ correct decimal digits.

As an illustration of these heuristics, consider the system

$$\begin{bmatrix} .986 & .579 \\ .409 & .237 \end{bmatrix} \begin{bmatrix} x_1 \\ x_2 \end{bmatrix} = \begin{bmatrix} .235 \\ .107 \end{bmatrix}$$

in which $\kappa_\infty(A) \cong 700$ and $x = (2, -3)^T$. Table 4.5-1 summarizes the effect of roundoff error when Gaussian elimination with partial pivoting is used for various machine precisions. It confirms the heuristics.

Whether or not one is content with the computed solution $\hat{x}$ depends on the requirements of the underlying source problem. In many applications accuracy is not important but small residuals are. In such a situation, $\hat{x}$ is probably adequate. On the other hand, if the correct number of digits in $\hat{x}$ is an issue, then the following questions take on particular importance:

—Can we "preprocess" A so that less roundoff is generated during the elimination?

—Can we improve the accuracy of $\hat{x}$ by iterative means?

—Can we efficiently estimate $\kappa_\infty(A)$ and thereby obtain a computable approximation to the relative error in $\hat{x}$?

As will be subsequently shown in the remainder of this section, the answer to each of these questions is a qualified "yes."

Scaling

Let β be the machine base and define the diagonal matrices D_1 and D_2 by

$$D_1 = \text{diag}(\beta^{r_1}, \ldots, \beta^{r_n})$$

and

$$D_2 = \text{diag}(\beta^{c_1}, \ldots, \beta^{c_n}).$$

The solution to the n-by-n linear system $Ax = b$ can be found by solving the scaled system $(D_1^{-1}AD_2)y = D_1^{-1}b$:

$$P(D_1^{-1}AD_2) = LU \qquad \text{(Gaussian elimination)}$$

$$Lw = P(D_1^{-1}b) \qquad \text{(forward elimination)}$$

$$Uy = w \qquad \text{(back substitution)}$$

$$x = D_2 y.$$

The scalings of A, b, and y require only $O(n^2)$ flops and may be accomplished without roundoff. Note that D_1 scales equations and D_2 scales unknowns.

It follows from Heuristic II that if $\hat{x}$ and $\hat{y}$ are the computed versions of x and y, then

$$(4.5\text{-}3) \qquad \frac{\|D_2^{-1}(\hat{x} - x)\|_\infty}{\|D_2^{-1}x\|_\infty} = \frac{\|\hat{y} - y\|_\infty}{\|y\|_\infty} \cong \beta^{-t}\kappa_\infty(D_1^{-1}AD_2).$$

Thus, if $\kappa_\infty(D_1^{-1}AD_2)$ can be made considerably smaller than $\kappa_\infty(A)$, then we might expect a correspondingly more accurate $\hat{x}$, provided errors are measured in the "D_2 norm":

$$\|e\|_{D_2} = \|D_2^{-1}e\|_\infty.$$

This is the objective of scaling. Note that it encompasses two issues: the condition of the scaled problem and the appropriateness of appraising error in the D_2-norm.

An interesting but very difficult mathematical problem concerns the exact minimization of $\kappa_p(D_1^{-1}AD_2)$ for general diagonal D_i and various p. What results there are in this direction are not very practical. This is hardly discouraging, however, when we recall that (4.5-3) is heuristic and it makes little sense to minimize exactly a heuristic bound. What we seek is a fast, approximate method for improving the quality of the computed solution $\hat{x}$.

One technique of this variety is *simple row scaling*. In this scheme D_2 is the identity and D_1 is chosen so that each row in $D_1^{-1}A$ has approximately the same ∞-norm. Row scaling minimizes the likelihood of adding a very small number to a very large number during elimination—an event that can greatly diminish accuracy.

Table 4.5-1

β	t	$\hat{x}_1$	$\hat{x}_2$	$\dfrac{\|\hat{x} - x\|_\infty}{\|x\|_\infty}$	$\dfrac{\|b - A\hat{x}\|}{\|A\|_\infty\|\hat{x}\|_\infty}$
10	3	2.11	−3.17	5×10^{-2}	2.0×10^{-3}
10	4	1.986	−2.975	8×10^{-3}	1.5×10^{-4}
10	5	2.0019	−3.0032	1×10^{-3}	2.1×10^{-6}
10	6	2.00025	−3.00094	3×10^{-4}	4.2×10^{-7}

EXAMPLE 4.5-1 (Forsythe and Moler [SLE, pp. 34, 40]). If

$$\begin{bmatrix} 10. & 100{,}000 \\ 1. & 1. \end{bmatrix} \begin{bmatrix} x_1 \\ x_2 \end{bmatrix} = \begin{bmatrix} 100{,}000 \\ 2. \end{bmatrix}$$

and the equivalent row-scaled problem

$$\begin{bmatrix} .0001 & 1. \\ 1. & 1. \end{bmatrix} \begin{bmatrix} x_1 \\ x_2 \end{bmatrix} = \begin{bmatrix} 1. \\ 2. \end{bmatrix}$$

are each solved using $\beta = 10$, $t = 3$, rounded arithmetic, then approximate solutions $\hat{x} = (0.00, 1.00)^T$ and $\hat{x} = (1.00, 1.00)^T$ are respectively computed. The exact solution is $x = (1.0001\ldots, .9999\ldots)^T$.

Slightly more complicated than simple row scaling is *row-column equili-bration*. Here, the object is to choose D_1 and D_2 so that the ∞-norm of each row and column of $D_1^{-1}AD_2$ belongs to the interval $[\frac{1}{\beta}, 1]$. An efficient proce-dure for doing this is given in McKeeman (1962).

It cannot be stressed too much that simple row scaling and row-column equilibration do not "solve" the scaling problem. Indeed, either technique can render a worse $\hat{x}$ than if no scaling whatever is used. The ramifications of this point are thoroughly discussed in Forsythe and Moler (SLE, chap. 11). The basic recommendation is that the scaling of equations and unknowns must proceed on a problem-by-problem basis. General scaling strategies are unreliable. It is best to scale (if at all) on the basis of what the source problem proclaims about the significance of each a_{ij}. Measurement units and data error may have to be considered.

Iterative Improvement

Suppose $Ax = b$ has been solved via $PA = LU$ and that we wish to improve the accuracy of the computed solution $\hat{x}$. If we calculate

$$r = b - A\hat{x}$$

$$Ly = Pr$$

(4.5-4)

$$Uz = y$$

$$x_{new} = \hat{x} + z$$

then in exact arithmetic

$$Ax_{new} = A(\hat{x} + z) = (b - r) + Az = b.$$

Unfortunately, the naive floating point execution of (4.5-4) renders an x_{new} that is no more accurate than $\hat{x}$. This is to be expected since

$$\hat{r} = \text{fl}(b - A\hat{x})$$

has few, if any, correct significant digits. (Recall Heuristic I.) Consequently, $\hat{z} = \text{fl}(A^{-1}r) \cong A^{-1} \cdot \text{noise} \cong \text{noise}$ is a very poor correction.

For (4.5-4) to produce an effective correction it is necessary to compute the residual $b - Ax$ with extended precision floating point arithmetic. Typically this means that if t-digit arithmetic is used to compute $PA = LU, x, y$, and z, then $2t$-digit arithmetic is used to form $b - A\hat{x}$. The process can be iterated giving

$$x := 0 \qquad (t\text{-digit})$$

$$PA = LU \qquad (t\text{-digit})$$

Repeat:

(4.5-5)
$$r := b - Ax \qquad (2t\text{-digit})$$

$$\text{Solve } Ly = Pr \text{ for } y. \qquad (t\text{-digit})$$

$$\text{Solve } Uz = y \text{ for } z. \qquad (t\text{-digit})$$

$$x := x + z \qquad (t\text{-digit})$$

This is referred to as *iterative improvement*. We stress that the original A must be used in the computation of r.

The basic result concerning the performance of (4.5-5) is summarized in the following heuristic:

Heuristic III: If a base β machine is used in (4.5-5) and $\kappa_\infty(A) \cong \beta^q$, then after k passes through the loop, x will have approximately

$$\min\{t, k(t - q)\}$$

correct significant base β digits.

Roughly speaking, if $\mathbf{u}\kappa_\infty(A) \leqslant 1$, then iterative improvement can ultimately produce a solution that is correct to full precision. Note that the process is relatively cheap. Each improvement costs $O(n^2)$, to be compared with the original $O(n^3)$ investment in the decomposition $PA = LU$. Of course, no improvement may result if A is badly enough conditioned with respect to the machine precision.

EXAMPLE 4.5-2. If (4.5-5) is applied to the system

$$\begin{bmatrix} .986 & .579 \\ .409 & .237 \end{bmatrix} \begin{bmatrix} x_1 \\ x_2 \end{bmatrix} = \begin{bmatrix} .235 \\ .107 \end{bmatrix}$$

with $\beta = 10$ and $t = 3$, then

$$\hat{x} = \begin{bmatrix} 2.11 \\ -3.17 \end{bmatrix}, \begin{bmatrix} 1.99 \\ -2.99 \end{bmatrix}, \begin{bmatrix} 2.00 \\ -3.00 \end{bmatrix}, \ldots$$

The exact solution is $x = (2, -3)^T$.

The primary drawback of iterative improvement is that its implementation is somewhat machine-dependent. This discourages its use in software that is intended for wide distribution. The need for retaining an original copy of A is another aggravation associated with the method.

On the other hand, iterative improvement is usually very easy to implement on a given machine that has provision for the accumulation of inner products, i.e., provision for the $2t$-digit calculation of inner products between the rows of A and x. In a "small t" computing environment the presence of an iterative improvement routine can significantly widen the class of solvable $Ax = b$ problems.

Condition Estimation

Suppose that we have solved $Ax = b$ via $PA = LU$ and that we now wish to ascertain the number of correct digits in the computed solution $\hat{x}$. It follows from Heuristic II that in order to do this we will need an estimate of the condition $\kappa_\infty(A) = \|A\|_\infty \|A^{-1}\|_\infty$. Computing $\|A\|_\infty$ poses no problem; we merely invoke the formula

$$\|A\|_\infty = \max_i \sum_j |a_{ij}|.$$

The challenge is with respect to the factor $\|A^{-1}\|_\infty$. Conceivably we could estimate this quantity by $\|\hat{X}\|_\infty$, where $\hat{X} = [\hat{x}_1, \ldots, \hat{x}_n]$ and $\hat{x}_i$ is the computed solution to $Ax_i = e_i$. The trouble with this approach is its expense; $\hat{\kappa}_\infty = \|A\|_\infty \|\hat{X}\|_\infty$ costs three times as much as $\hat{x}$.

The central problem of *condition estimation* is how to estimate the condition number in $O(n^2)$ flops assuming the availability of $PA = LU$ (or some other apt decomposition). An approach described in Forsythe and Moler (SLE, p. 51) is based on iterative improvement and the heuristic

$$\beta^{-t}\kappa_\infty(A) \cong \|z\|_\infty / \|\hat{x}\|_\infty$$

where z is the correction to $\hat{x}$ in (4.5-4). While the resulting condition estimator is $O(n^2)$, it suffers from the shortcomings of iterative improvement, namely, machine dependency.

Cline, Moler, Stewart, and Wilkinson (1979) have proposed a condition estimator without this flaw. It is based on exploitation of the implication

$$Ay = d \Rightarrow \|A^{-1}\|_\infty \geq \|y\|_\infty / \|d\|_\infty.$$

The idea behind their estimator is to choose d so that the solution y is large in norm and then set

$$\hat{\kappa}_\infty = \|A\|_\infty \|y\|_\infty / \|d\|_\infty.$$

The success of this method hinges on how close the ratio $\|y\|_\infty / \|d\|_\infty$ is to its maximum value $\|A^{-1}\|_\infty$.

Consider the case when $A = T$ is upper triangular. The relation between d and y is completely specified by the following "column version" of back substitution:

$$p_i := 0 \qquad (i = 1, \ldots, n)$$

(4.5-6)

For $k = n, \ldots, 1$

$$y_k := (d_k - p_k)/t_{kk}$$

$$p_i := p_i + t_{ik} y_k \qquad (i = 1, \ldots, k - 1)$$

Normally we use this algorithm to solve a *given* triangular system $Ty = d$. Now, however, we are free to pick the right-hand side d subject to the "constraint" that y is large relative to d.

One way to encourage growth in y is to choose d_k from the set $\{-1, +1\}$ according to whether $(1 - p_k)/t_{kk}$ or $(-1 - p_k)/t_{kk}$ is larger. In other words, (4.5-6) is invoked with y_k being assigned as follows:

$$y_k := [-\text{sign}(p_k) - p_k]/t_{kk}.$$

Since d is then a vector of the form $d = (\pm 1, \ldots, \pm 1)^T$, we obtain the estimator $\hat{\kappa}_\infty = \|T\|_\infty \|y\|_\infty$.

A more complicated estimator results if $d_k \in \{-1, +1\}$ is chosen so as to encourage growth both in y_k *and* the updated running sums $p_1, \ldots, p_{k-1}$:

ALGORITHM 4.5-1.   Let $T \in \mathbb{R}^{n \times n}$ be nonsingular and upper triangular and let $w_1, \ldots, w_n$ be a given set of weights. The following algorithm computes $y \in \mathbb{R}^n$ such that $\|y\|_\infty$ approximates $\|T^{-1}\|_\infty$.

$$p_i := 0 \qquad (i = 1, \ldots, n)$$

For $k = n, \ldots, 1$

$$y_k^+ := (1 - p_k)/t_{kk}$$

$$y_k^- := (-1 - p_k)/t_{kk}$$

$$s_+ := |y_k^+| + \sum_{i=1}^{k-1} w_i |p_i + t_{ik} y_k^+|$$

$$s^- := |y_k^-| + \sum_{i=1}^{k-1} w_i |p_i + t_{ik} y_k^-|$$

If $s^+ \geqslant s^-$

then

$$y_k := y_k^+$$

else

$$y_k := y_k^-$$

$$p_i := p_i + t_{ik} y_k \qquad (i = 1, \ldots, k - 1)$$

This algorithm requires $5n^2/2$ flops. A reasonable way to choose the weights is to set $w_i = 1/|t_{ii}|$ for $i = 1, \ldots, n$. An obvious lower triangular version of the procedure can be formulated.

EXAMPLE 4.5-3. If Algorithm 4.5-1 with $w_i \equiv 1$ is applied to B_n given in (2.5-9) then $y = (2^{n-1}, 2^{n-2}, \ldots, 2, 1)^T$ and $\|y\|_\infty = \|B_n^{-1}\|_\infty$.

We are now in a position to describe a procedure for estimating the condition of a square nonsingular matrix A whose $PA = LU$ decomposition we know:

(i) Apply the lower triangular version of Algorithm 4.5-1 to U^T to obtain a large norm solution to $U^T y = d$.

(ii) Solve the triangular systems

$$L^T r = y$$
$$Lw = Pr$$
$$Uz = w.$$

(iii) $\hat{\kappa}_\infty = \|A\|_\infty \|z\|_\infty / \|r\|_\infty$.

Note that $\|z\|_\infty \leq \|A^{-1}\|_\infty \|r\|_\infty$. The method is based on several heuristics. First, if A is ill-conditioned and $PA = LU$, then it is usually the case that U is correspondingly ill-conditioned. The lower triangle L tends to be fairly well-conditioned. Thus, it is more profitable to apply Algorithm 4.5-1 to U than to L. The vector r, because it solves $A^T P^T r = d$, tends to be rich in the direction of the left singular vector associated with $\sigma_{min}(A)$. Righthand sides with this property render large solutions to the problem $Az = r$.

In practice it is found that (i)–(iii) above produces good order-of-magnitude estimates of the actual condition number. See LINPACK.

Problems

P4.5-1 Show by example that there may be more than one way to equilibrate a matrix.

P4.5-2. Using $\beta = 10$, $t = 2$, chopped arithmetic, solve

$$\begin{bmatrix} 11 & 15 \\ 5 & 7 \end{bmatrix} \begin{bmatrix} x_1 \\ x_2 \end{bmatrix} = \begin{bmatrix} 7 \\ 3 \end{bmatrix}$$

using Gaussian elimination with partial pivoting. Do one step of iterative improvement using $t = 4$ arithmetic to compute the residual. (Don't forget to truncate the computed residual to two digits.)

P4.5-3. Suppose $P(A + E) = \hat{L}\hat{U}$, where P is a permutation, $\hat{L}$ is lower triangular with $|\hat{l}_{ij}| \leq 1$, and U is upper triangular. Show that

$$\kappa_\infty(A) \geq \frac{\|A\|_\infty}{\|E\|_\infty + \min_i |\hat{u}_{ii}|}.$$

Conclude that if a small pivot is encountered when Gaussian elimination with pivoting is applied to A, then A is ill-conditioned. The converse is not true. (Let $A = B_n$.)

P4.5-4. (Kahan 1966) The system $Ax = b$ where

$$A = \begin{bmatrix} 2 & -1 & 1 \\ -1 & 10^{-10} & 10^{-10} \\ 1 & 10^{-10} & 10^{-10} \end{bmatrix} \quad b = \begin{bmatrix} 2(1 + 10^{-10}) \\ -10^{-10} \\ 10^{-10} \end{bmatrix}$$

has solution $x = (10^{-10}, -1, 1)^{\mathrm{T}}$. (a) Show that if $(A + E)y = b$ and $|E| \leqslant 10^{-8}|A|$, then $|x - y| \leqslant 10^{-7}|x|$. That is, small relative changes in A's entries do not induce large changes in x even though $\kappa_\infty(A) = 10^{10}$. (b) Define $D = \mathrm{diag}(10^{-5}, 10^5, 10^5)$. Show $\kappa_\infty(DAD) \leqslant 5$.

P4.5-5. Apply Algorithm 4.5-1 with weights $w_1 = \cdots = w_n = 0$ to the matrix

$$T = \begin{bmatrix} 1 & 0 & M & -M \\ 0 & 1 & -M & M \\ 0 & 0 & 1 & 0 \\ 0 & 0 & 0 & 1 \end{bmatrix} \qquad M \in \mathbb{R}$$

Repeat with weights $w_1 = \cdots = w_n = 1$. What is $\kappa_\infty(T)$?

Notes and References for Sec. 4.5

The following papers are concerned with the scaling of $Ax = b$ problems:

F. L. Bauer (1963). "Optimally Scaled Matrices," *Numer. Math. 5*, 73–87.

P. A. Businger (1968). "Matrices Which Can be Optimally Scaled," *Numer. Math. 12*, 346–48.

T. Fenner and G. Loizou (1974). "Some New Bounds on the Condition Numbers of Optimally Scaled Matrices," *J. Assoc. Comp. Mach. 1*, 514–24.

G. H. Golub and J. M. Varah (1974). "On a Characterization of the Best 1_2-scaling of a Matrix," *SIAM J. Num. Anal. 11*, 472–79.

C. McCarthy and G. Strang (1973). "Optimal Conditioning of Matrices," *SIAM J. Num. Anal. 10*, 370–88.

R. Skeel (1981). "Effect of Equilibration on Residual Size for Partial Pivoting," *SIAM J. Num. Anal. 18*, 449–55.

A. van der Sluis (1969). "Condition Numbers and Equilibration Matrices," *Numer. Math. 14*, 14–23.

A. van der Sluis (1970). "Condition, Equilibration, and Pivoting in Linear Algebraic Systems," *Numer. Math. 15*, 74–86.

Part of the difficulty in scaling concerns the selection of a norm in which to measure errors. An interesting discussion of this frequently overlooked point appear in

W. Kahan (1966). "Numerical Linear Algebra," *Canadian Math. Bull. 9*, 757–801.

For a rigorous analysis of iterative improvement see

C. B. Moler (1967). "Iterative Refinement in Floating Point," *J. Assoc. Comp. Mach. 14*, 316-71.

The condition estimator that we described is given in

A. K. Cline, C. B. Moler, G. W. Stewart, and J. H. Wilkinson (1979). "An Estimate for the Condition Number of a Matrix," *SIAM J. Num. Anal. 16*, 368-75.

and is incorporated in LINPACK (pp. 1.10-1.13). Other references concerned with the condition estimation problem include

C. G. Broyden (1973). "Some Condition Number Bounds for the Gaussian Elimination Process," *J. Inst. Math. Applic. 12*, 273-86.

F. Lemeire (1973). "Bounds for Condition Numbers of Triangular and Trapezoid Matrices," *BIT 15*, 58-64.

J. M. Varah (1975). "A Lower Bound for the Smallest Singular Value of a Matrix," *Lin. Alg. & Its Applic. 11*, 1-2.

R. S. Varga (1976). "On Diagonal Dominance Arguments for Bounding $\|A^{-1}\|$," *Lin. Alg. & Its Applic. 14*, 211-17.

G. W. Stewart (1980). "The Efficient Generation of Random Orthogonal Matrices with an Application to Condition Estimators," *SIAM J. Num. Anal. 17*, 403-9.

A. K. Cline, A. R. Conn, and C. Van Loan (1982). "Generalizing the LINPACK Condition Estimator," in *Numerical Analysis*, ed. J. P. Hennart, Lecture Notes in Mathematics no. 909, Springer-Verlag, New York.

D. P. O'Leary (1980). "Estimating Matrix Condition Numbers," *SIAM J. Sci. & Stat. Comp. 1*, 205-9.

R. G. Grimes and J. G. Lewis (1981). "Condition Number Estimation for Sparse Matrices," *SIAM J. Sci. & Stat. Comp. 2*, 384-88.

Special Linear Systems

It is a basic tenet of numerical analysis that structure should be exploited whenever solving a problem. In numerical algebra this translates into an expectation that algorithms for general matrix problems can be streamlined in the presence of such properties as symmetry, definiteness, and sparsity. This is the central theme of the current chapter, where our principal aim is to devise special algorithms for computing special variants of the L-U decomposition.

We begin by pointing out the connection between the triangular factors L and U when A is symmetric. This is achieved by examining the L-D-M^T decomposition in §5.1. We then turn our attention to the important case when A is both symmetric and positive definite, deriving the stable Cholesky decomposition in §5.2. Unsymmetric positive definite systems are also investigated in this section. In §5.3 banded versions of Gaussian elimination and other factorization methods are discussed. We then examine the interesting situation when A is symmetric but indefinite. Our treatment of this problem in §5.4 highlights the numerical analyst's ambivalence towards pivoting. We love pivoting for the stability it induces but despise it for the structure that it can destroy. Fortunately, there is a happy resolution to this conflict in the symmetric indefinite problem.

Any block banded matrix is also banded and so the methods of §5.3 are applicable. Yet there are occasions when it pays not to adopt this point of view. To illustrate this we consider the important case of block tridiagonal systems in §5.5.

In the final two sections we examine some very interesting $O(n^2)$ algorithms that can be used to solve Vandermonde and Toeplitz systems.

81

Reading Path

$$§5.6 \qquad\qquad\qquad §5.4$$

$$\text{Chapters 1-4} \xrightarrow{\ \nearrow\ } §5.1 \rightarrow §5.2 \rightarrow §5.3 \xrightarrow[\searrow]{\nearrow} §5.5$$

$$§5.7$$

Sec. 5.1. The L-D-M^T and L-D-L^T Decompositions

In order to specialize Gaussian elimination to the case when A is symmetric, it is convenient to establish a slight variation of the L-U decomposition.

THEOREM 5.1-1: *L-D-M^T Decomposition.* If all the leading principal sub-matrices of $A \in \mathbb{R}^{n \times n}$ are nonsingular, then there exist unit lower triangular matrices L and M and a diagonal matrix $D = \text{diag}(d_1, \ldots, d_n)$ such that $A = LDM^T$.

Proof. By Theorem 4.2-1 we have $A = LU$, where L is unit lower triangular and U is upper triangular. Define $D = \text{diag}(d_1, \ldots, d_n)$, where $d_i = u_{ii}$ for $i = 1, \ldots, n$. Notice that D is nonsingular and that $M^T = D^{-1}U$ is unit upper triangular. Since $A = LU = LD(D^{-1}U) = LDM^T$, the proof is complete. $\square$

Once the L-D-M^T decomposition of A is obtained, the solution to $Ax = b$ may be found by solving the following three systems:

$$(5.1\text{-}1) \qquad \begin{array}{ll} Ly = b & n^2/2 \text{ flops} \\ Dz = y & n \text{ flops} \\ M^Tx = z & n^2/2 \text{ flops.} \end{array}$$

The L-D-M^T decomposition can be found by computing $A = LU$ via Gaussian elimination and then determining D and M from $U = DM^T$. However, an interesting alternative algorithm can be derived by comparing entries in the matrix equation $A = LDM^T$. In particular, by comparing the (k,k) entries we find

$$a_{kk} = \sum_{p=1}^{k-1} l_{kp}d_p m_{kp} + d_k,$$

while comparison of the (i,k) and (k,i) entries $(i > k)$ gives

$$a_{ik} = \sum_{p=1}^{k-1} l_{ip}d_p m_{kp} + l_{ik}d_k$$

$$a_{ki} = \sum_{p=1}^{k-1} l_{kp}d_p m_{ip} + d_k m_{ik}.$$

If we solve these three equations for d_k, l_{ik}, and m_{ik}, respectively, and then properly sequence the resulting expressions, we obtain the following method for the determination of L, D, and M:

For $k = 1, \ldots, n$

$$d_k = a_{kk} - \sum_{p=1}^{k-1} l_{kp} d_p m_{kp}$$

(5.1-2)   For $i = k + 1, \ldots, n$

$$l_{ik} = \left(a_{ik} - \sum_{p=1}^{k-1} l_{ip}(d_p m_{kp}) \right) \Big/ d_k$$

$$m_{ik} = \left(a_{ki} - \sum_{p=1}^{k-1} (l_{kp} d_p) m_{ip} \right) \Big/ d_k$$

Example 5.1-1.

$$A = \begin{bmatrix} 10 & 10 & 20 \\ 20 & 25 & 40 \\ 30 & 50 & 61 \end{bmatrix} = \begin{bmatrix} 1 & 0 & 0 \\ 2 & 1 & 0 \\ 3 & 4 & 1 \end{bmatrix} \begin{bmatrix} 10 & 0 & 0 \\ 0 & 5 & 0 \\ 0 & 0 & 1 \end{bmatrix} \begin{bmatrix} 1 & 1 & 2 \\ 0 & 1 & 0 \\ 0 & 0 & 1 \end{bmatrix}$$

An operation count reveals that $2n^3/3$ flops are required in (5.1-2)—twice the work in Gaussian elimination. However, it should be observed that the quantities

$$r_{kp} = d_p m_{kp} \qquad\qquad k = 1, \ldots, n - 1$$

and

$$w_{kp} = l_{kp} d_p \qquad\qquad k = 1, \ldots, n - 1$$

are independent of the inner loop index i. By computing these quantities outside the inner loop, an algorithm requiring $n^3/3$ flops results.

ALGORITHM 5.1-1. Given $A \in \mathbb{R}^{n \times n}$, this algorithm computes unit lower triangular matrices L and M and a diagonal matrix $D = \text{diag}(d_1, \ldots, d_n)$ such that $A = LDM^T$. The entry a_{ij} is overwritten with l_{ij} if $i > j$ and with m_{ji} if $i < j$. The algorithm terminates prematurely if no such decomposition exists.

For $k = 1, \ldots, n$
    For $p = 1, 2, \ldots, k - 1$
        $r_p := d_p a_{pk}$
        $w_p := a_{kp} d_p$
    $d_k := a_{kk} - \sum_{p=1}^{k-1} a_{kp} r_p$

If $d_k = 0$
  then
      quit
  else
      For $i = k + 1, \ldots, n$

$$a_{ik} := \left(a_{ik} - \sum_{p=1}^{k-1} a_{ip} r_p \right) \bigg/ d_k$$

$$a_{ki} := \left(a_{ki} - \sum_{p=1}^{k-1} w_p a_{pi} \right) \bigg/ d_k$$

This algorithm requires $n^3/3$ flops.

The computed solution $\hat{x}$ to $Ax = b$ obtained via this algorithm and (5.1-1) can be shown to satisfy

$$(A + E)\hat{x} = b,$$

where

(5.1-3) $|E| \leqslant n\mathbf{u}\,[3\,|A| + 5\,|\hat{L}|\,|\hat{D}|\,|\hat{M}^{\mathrm{T}}|] + O(\mathbf{u}^2)$

and $\hat{L}$, $\hat{D}$, and $\hat{M}$ are the computed versions of L, D, and M, respectively.

As in the case of the L-U decomposition considered in §4.2, the upper bound in (5.1-3) is without limit unless some form of pivoting is done. Hence, for Algorithm 5.1-1 to be a practical procedure, it must be modified so as to compute a decomposition of the form $PA = LDM^{\mathrm{T}}$, where P is a permutation matrix chosen so that the entries in L satisfy $|l_{ij}| \leqslant 1$. The details of this will not be pursued, since they are straightforward and since our main object for introducing the L-D-M^{T} factorization is to motivate special methods for symmetric systems.

THEOREM 5.1-2: *L-D-L^T Decomposition*.   If $A = LDM^{\mathrm{T}}$ is the L-D-M^{T} decomposition of a nonsingular symmetric matrix A, then $L = M$.

Proof.   The matrix $M^{-1}AM^{-\mathrm{T}} = M^{-1}LD$ is both symmetric and lower triangular and therefore diagonal. Since D is nonsingular, this implies that $M^{-1}L$ is also diagonal. But $M^{-1}L$ is unit lower triangular and so $M^{-1}L = I$. $\square$

In view of this result, it is possible to halve the work in Algorithm 5.1-2 when it is applied to a symmetric matrix.

ALGORITHM 5.1-2.   Given a symmetric $A \in \mathbb{R}^{n \times n}$, this algorithm computes a unit lower triangular matrix L and a diagonal matrix $D = \text{diag}(d_1, \ldots, d_n)$ such that $A = LDL^{\mathrm{T}}$. The entry a_{ij} is overwritten with l_{ij} if $i > j$. The algorithm may terminate prematurely if the decomposition fails to exist.

For $k = 1, \ldots, n$
    For $p = 1, \ldots, k - 1$
        $r_p := d_p a_{kp}$
    $d_k := a_{kk} - \sum\limits_{p=1}^{k-1} a_{kp} r_p$
    If $d_k = 0$
        then
            quit
        else
            For $i = k + 1, \ldots, n$
$$a_{ik} := \left(a_{ik} - \sum_{p=1}^{k-1} a_{ip} r_p \right) \Big/ d_k$$

This algorithm requires $n^3/6$ flops.

EXAMPLE 5.1-2.

$$A = \begin{bmatrix} 10 & 20 & 30 \\ 20 & 45 & 80 \\ 30 & 80 & 171 \end{bmatrix} = \begin{bmatrix} 1 & 0 & 0 \\ 2 & 1 & 0 \\ 3 & 4 & 1 \end{bmatrix} \begin{bmatrix} 10 & 0 & 0 \\ 0 & 5 & 0 \\ 0 & 0 & 1 \end{bmatrix} \begin{bmatrix} 1 & 2 & 3 \\ 0 & 1 & 4 \\ 0 & 0 & 1 \end{bmatrix}$$

In the next section we show that if A is both symmetric and positive definite, then Algorithm 5.1-2 not only runs to completion, but is extremely stable.

Problems

P5.1-1. Show that the L-D-M^T decomposition of a nonsingular A is unique if it exists.

P5.1-2. At the k-th stage in Algorithm 5.1-1, d_k and the k-th columns of L and M are determined. Derive an equivalent algorithm in which d_k and the k-th *rows* of L and M are found at the k-th stage.

P5.1-3. (*Doolittle reduction*) Let $A = LU$ be the L-U decomposition $A \in \mathbb{R}^{n \times n}$. Derive the algorithm

    For $k = 1, 2, \ldots, n$
        For $i = k, \ldots, n$
$$u_{ki} = a_{ki} - \sum_{p=1}^{k-1} l_{kp} u_{pi}$$
        For $i = k + 1, \ldots, n$
$$l_{ik} = \left(a_{ik} - \sum_{p=1}^{k-1} l_{ip} u_{pk} \right) \Big/ u_{kk}$$

P5.1-4. (Symmetric storage) Suppose the n-by-n symmetric matrix $A = (a_{ij})$ is stored in an $n(n + 1)/2$ vector c as follows:

$$c = (a_{11}, a_{21}, a_{22}, a_{31}, \ldots, a_{n1}, a_{n2}, \ldots, a_{nn}).$$

Rewrite Algorithm 5.1-2 with A stored in this fashion. (Notice that a_{ij} $(i \geqslant j)$ is stored in component $j + i(i - 1)/2$ of c. Do as much indexing outside the inner loops as possible.)

P5.1-5. Change Algorithm 5.1-1 so that it computes a factorization of the form $PA = LDM^T$, where L and M are both unit lower triangular and P is a permutation that is chosen so $|l_{ij}| \leqslant 1$.

Notes and References for Sec. 5.1

Algorithm 5.1-1 and the Doolittle reduction (P5.1-3) are examples of "compact" elimination schemes. Another such scheme is the Crout elimination method, whereby a lower triangular L and a unit upper triangular U are determined such that $A = LU$. The Crout and Doolittle techniques were popular during the days of desk calculators because there are far fewer intermediate results than in Gaussian elimination. These methods still have attraction because they can be implemented with accumulated inner products. For remarks along these lines see chapter 4 of

L. Fox (1964). *An Introduction to Numerical Linear Algebra*, Oxford University Press, Oxford.

as well as Stewart (IMC, pp. 131–39). An ALGOL procedure may be found in

H. J. Bowdler, R. S. Martin, G. Peters, and J. H. Wilkinson (1966). "Solution of Real and Complex Systems of Linear Equations," *Numer. Math 8*, 217–34. See also HACLA, pp. 93–110.

See also

G. E. Forsythe (1960). "Crout with Pivoting," *Comm. Assoc. Comp. Mach. 3*, 507–8.

W. M. McKeeman (1962). "Crout with Equilibration and Iteration," *Comm. Assoc. Comp. Mach. 5*, 553–55.

Sec. 5.2. Positive Definite Systems

A very important class of "special" $Ax = b$ problems are those in which $A \in \mathbb{R}^{n \times n}$ is positive definite, i.e. $x^T A x > 0$ for all nonzero $x \in \mathbb{R}^n$. Our primary goal in this section is to develop special algorithms for solving symmetric positive definite systems. However, for purposes of motivation and because they constitute an important class in their own right, we first consider unsymmetric positive definite systems. The starting point in the analysis is the following existence theorem.

THEOREM 5.2-1. If $A \in \mathbb{R}^{n \times n}$ is positive definite, then A has an L-D-M^T decomposition and the diagonal entries of D are positive.

Proof. All principle submatrices of a positive definite matrix are positive definite and therefore nonsingular. Hence, by Theorem 5.1-1 there exist unit lower triangular matrices L and M and $D = \text{diag}(d_1, \ldots, d_n)$ such that $A = LDM^T$. Since $S = DML^{-T} = L^{-1}AL^{-T}$ is (a) positive definite and (b) upper triangular with $s_{ii} = d_i$, it follows that the d_i are positive. $\square$

Now the mere existence of an L-D-M^T decomposition does not mean that its computation is advisable, because the resulting factors may have unacceptably large elements. For example, the matrix

$$A = \begin{bmatrix} \epsilon & m \\ -m & \epsilon \end{bmatrix} \qquad m \gg \epsilon > 0$$

is positive definite and has an L-D-M^T decomposition given by

$$A = \begin{bmatrix} 1 & 0 \\ -m/\epsilon & 1 \end{bmatrix} \begin{bmatrix} \epsilon & 0 \\ 0 & \epsilon+m^2/\epsilon \end{bmatrix} \begin{bmatrix} 1 & m/\epsilon \\ 0 & 1 \end{bmatrix}.$$

Because $m/\epsilon \gg 1$, pivoting is recommended for this example.

The following result suggests when to expect element growth when computing the L-D-M^T decomposition of a positive definite matrix.

THEOREM 5.2-2. Let $A \in \mathbb{R}^{n \times n}$ be positive definite, and define the matrices T and S by

$$T = (A + A^T)/2 \qquad S = (A - A^T)/2.$$

If $A = LDM^T$, then

(5.2-1) $\| \, |L| \; |D| \; |M^T| \, \|_F \leqslant n[\|T\|_2 + \|ST^{-1}S\|_2]$

Proof. See Golub and Van Loan (1979). $\square$

The theorem can be used to suggest when it is safe not to pivot when solving positive definite systems. Assume that the computed factors $\hat{L}$, $\hat{D}$, and $\hat{M}$ satisfy

(5.2-2) $\| \, |\hat{L}| \; |\hat{D}| \; |\hat{M}^T| \, \|_F \leqslant c \| \, |L| \; |D| \; |M^T| \, \|_F,$

where c is some constant, usually of modest size. It follows from (5.2-1) and the analysis in §4.3 that if these factors are used to compute a solution to $Ax = b$, then the computed solution satisfies $(A + E)\hat{x} = b$ with

(5.2-3) $\|E\|_F \leqslant \mathbf{u}\{3\,n\|A\|_F + 5\,cn^2\,(\|T\|_2 + \|ST^{-1}S\|_2)\} + 0(\mathbf{u}^2).$

Since $\|T\|_2 \leqslant \|A\|_2$, it follows that it is safe not to pivot provided that

(5.2-4) $$\frac{\|ST^{-1}S\|_2}{\|A\|_2} = \Omega$$

is not too large. In other words, the norm of the skew part has to be modest relative to the condition of the symmetric part. Fortunately, this condition is always satisfied whenever A is symmetric and positive definite. Symmetric positive definite systems are the most frequently occurring class of structured linear systems. They can be solved by computing the following variant of the L-D-L^T decomposition:

THEOREM 5.2-3: *Cholesky Decomposition.*   If $A \in \mathbb{R}^{n \times n}$ is symmetric positive definite, then there exists a lower triangular $G \in \mathbb{R}^{n \times n}$ with positive diagonal entries such that $A = GG^T$.

Proof.   From Theorem 5.1-2, there exists a unit lower triangular L and a diagonal $D = \text{diag}(d_1, \ldots, d_n)$ such that $A = LDL^T$. Since the d_k are positive, the matrix $G = L \, \text{diag}(d_1^{1/2}, \ldots, d_n^{1/2})$ is real, lower triangular, has positive diagonal entries, and satisfies $A = GG^T$. $\square$

The decomposition $A = GG^T$ is known as the *Cholesky decomposition* and G is referred to as the *Cholesky triangle*.

EXAMPLE 5.2-1.   The matrix

$$A = \begin{bmatrix} 2 & -2 \\ -2 & 5 \end{bmatrix} = \begin{bmatrix} 1 & 0 \\ -1 & 1 \end{bmatrix} \begin{bmatrix} 2 & 0 \\ 0 & 3 \end{bmatrix} \begin{bmatrix} 1 & -1 \\ 0 & 1 \end{bmatrix}$$

is positive definite and has a Cholesky decomposition $A = GG^T$ where

$$G = \begin{bmatrix} \sqrt{2} & 0 \\ -\sqrt{2} & \sqrt{3} \end{bmatrix}.$$

The proof of the Cholesky decomposition is constructive. However, a more effective method for computing the Cholesky triangle can be derived by comparing entries in the equation $A = GG^T$. For $i \geq k$ we have

$$a_{ik} = \sum_{p=1}^{k} g_{ip} g_{kp}.$$

Rearranging this equation we obtain

$$g_{ik} = \left(a_{ik} - \sum_{p=1}^{k-1} g_{ip} g_{kp} \right) \Big/ g_{kk} \qquad\qquad i > k$$

(5.2-5)

$$g_{kk} = \left(a_{kk} - \sum_{p=1}^{k-1} g_{kp}^2 \right)^{1/2}$$

Properly sequenced, these equations can be used to compute G a column at a time or a row at a time. In the former case we have

ALGORITHM 5.2-1: *Cholesky Decomposition (Column Version).* Given a symmetric positive definite $A \in \mathbb{R}^{n \times n}$, the following algorithm computes a lower triangular $G \in \mathbb{R}^{n \times n}$ such that $A = GG^T$. The entry a_{ij} is overwritten by $g_{ij} \ (i \geqslant j)$.

For $k = 1, 2, \ldots, n$

$$a_{kk} := \left(a_{kk} - \sum_{p=1}^{k-1} a_{kp}^2 \right)^{1/2}$$

For $i = k + 1, \ldots, n$

$$a_{ik} := \left(a_{ik} - \sum_{p=1}^{k-1} a_{ip}a_{kp} \right) \bigg/ a_{kk}$$

This algorithm requires $n^3/6$ flops.

The numerical stability of this algorithm is easily deduced from the inequality

$$g_{ij}^2 \leqslant \sum_{p=1}^{i} g_{ip}^2 = a_{ii}$$

This result shows that the Cholesky triangle is nicely bounded. The same conclusion can be reached from the equality $\| G \|_2^2 = \| A \|_2$.

The roundoff errors associated with Algorithm 5.2-1 have been extensively studied in a classical paper by Wilkinson (1968). Using the results in this paper it can be shown that if $\hat{x}$ is the computed solution to $Ax = b$, obtained by computing $A = GG^T$ and then solving $Gy = b$ and $G^Tx = y$, then

$$(A + E)\hat{x} = b,$$

where

$$\| E \|_2 \leqslant c_n \mathbf{u} \| A \|_2$$

and c_n is a small constant depending upon n. Moreover, Wilkinson shows that if

$$q_n \mathbf{u} \| A \|_2 \| A^{-1} \|_2 \leqslant 1,$$

where q_n is another small constant, then the Cholesky process runs to completion, i.e., no square roots of negative numbers arise.

EXAMPLE 5.2-2.   If Algorithm 5.2-1 is applied to the positive definite matrix

$$A = \begin{bmatrix} 100 & 15 & .01 \\ 15 & 2.3 & .01 \\ .01 & .01 & 1.00 \end{bmatrix}$$

and $\beta = 10$, $t = 2$, rounded arithmetic used, then $\hat{g}_{11} = 10$, $\hat{g}_{21} = 1.5$, $\hat{g}_{31} = .001$ and $\hat{g}_{22} = 0.0$. The algorithm then breaks down trying to compute g_{32}.

Problems

P5.2-1. Show that if

$$A = \begin{bmatrix} \alpha & w^T \\ v & B \end{bmatrix} \begin{matrix} 1 \\ n-1 \end{matrix}$$
$$\quad\quad 1 \quad n-1$$

is positive definite, then so is $B - vw^T/\alpha$.

P5.2-2. Show that if $A \in \mathbb{R}^{n \times n}$ is symmetric and positive definite, then

$$|a_{ij}| \leq (a_{ii} + a_{jj})/2 \qquad\qquad i \neq j$$

and

$$|a_{ij}| \leq \sqrt{a_{ii} a_{jj}}.$$

P5.2-3. Suppose

$$A = \begin{bmatrix} \alpha & v^T \\ v & B \end{bmatrix} \begin{matrix} 1 \\ n-1 \end{matrix}$$
$$\quad\quad 1 \quad n-1$$

is symmetric and positive definite. Note that

$$A = \begin{bmatrix} 1 & 0 \\ v/\alpha & I_{n-1} \end{bmatrix} \begin{bmatrix} \alpha & 0 \\ 0 & B-vv^T/\alpha \end{bmatrix} \begin{bmatrix} 1 & v^T/\alpha \\ 0 & I_{n-1} \end{bmatrix}.$$

(a) Observing that $B - vv^T/\alpha$ is positive definite and symmetric, derive an $n^3/6$ algorithm for computing the LDL^T decomposition of A which involves $n - 1$ repetitions of the above decomposition. (b) Incorporate complete pivoting in your algorithm of part (a). (Note from P5.2-2 that the largest element in a positive definite matrix occurs on the diagonal.) (c) Suppose A is nonnegative definite. Show that after $r = rank(A)$ steps, your algorithm in part (b) has computed the decomposition

$$P^T A P = L D L^T,$$

where P is a permutation, $D = \text{diag}(d_1, \ldots, d_r) \in \mathbb{R}^{r \times r}$, and $L \in \mathbb{R}^{n \times r}$ is unit lower triangular.

P5.2-4. Suppose that $A + iB$ is Hermitian and positive definite with A, $B \in \mathbb{R}^{n \times n}$. (a) Show that

$$C = \begin{bmatrix} A & -B \\ B & A \end{bmatrix}$$

is symmetric and positive definite. (b) Formulate a $4n^3/3$ algorithm for solving $(A + iB)(x + iy) = (b + ic)$, where b, c, x, and y are in $\mathbb{R}^n$. How much storage is required?

P5.2-5. Show that the Cholesky decomposition is unique for positive definite symmetric matrices.

P5.2-6. Suppose $A \in \mathbb{R}^{n \times n}$ is symmetric and positive definite. Give an algorithm for computing an upper triangular matrix $R \in \mathbb{R}^{n \times n}$ such that $A = RR^T$.

P5.2-7. Let $A \in \mathbb{R}^{n \times n}$ be positive definite and set $T = (A + A^T)/2$ and $S = (A - A^T)/2$.
(a) Show $\|A^{-1}\|_2 \leqslant \|T^{-1}\|_2$.
(b) Show $x^T A^{-1} x \leqslant x^T T^{-1} x$, where $x \in \mathbb{R}^n$.
(c) Show that if $A = LDM^T$, then

$$d_k \geqslant 1/\|T^{-1}\|_2. \qquad\qquad k = 1, \ldots, n$$

P5.2-8. Formulate a row version of Algorithm 5.2-1.

P5.2-9. Show that the following matrix is not positive definite:

$$A = \begin{bmatrix} 36 & 30 & 18 \\ 30 & 41 & 23 \\ 18 & 23 & 12 \end{bmatrix}.$$

Notes and References for Sec. 5.2

The definiteness of the quadratic form $x^T A x$ can frequently be established by considering the mathematics of the underlying problem. For example, the discretization of certain partial differential operators gives rise to "provably" positive definite matrices. The question of whether pivoting is necessary when solving positive definite systems is considered in

G. H. Golub and C. Van Loan (1979). "Unsymmetric Positive Definite Linear Systems," *Lin. Alg. & Its Applic. 28*, 85–98.

The results in this paper were motivated by the earlier work on linear systems of the form $(I + S)x = b$, where S is skew symmetric:

A. Buckley (1974). "A Note on Matrices $A = 1 + H$, H Skew-Symmetric," *Z. Angew. Math. Mech 54*, 125–26.
A. Buckley (1977). "On the Solution of Certain Skew-Symmetric Linear Systems," *SIAM J. Num. Anal. 14*, 566–70.

Symmetric positive definite systems constitute the most important class of special $Ax = b$ problems. ALGOL programs for both the G-G^T and L-D-L^T factorizations are given in

R. S. Martin, G. Peters, and J. H. Wilkinson (1965). "Symmetric Decomposition of a Positive Definite Matrix," *Numer. Math.* 7, 362–83. See also HACLA, pp. 9–30.

The technique of iterative improvement which we discussed in §4.5 in connection with the L-U decomposition, can also be implemented with the decompositions G-G^T and L-D-L^T. See

R. S. Martin, G. Peters, and J. H. Wilkinson (1966). "Iterative Refinement of the Solution of a Positive Definite System of Equations," *Numer. Math 8*, 203–16. See also HACLA, pp. 31–44.

The roundoff errors associated with the Cholesky decomposition are discussed in

J. H. Wilkinson (1968). "A Priori Error Analysis of Algebraic Processes," *Proc. International Congress Math.* (Moscow: Izdat. Mir, 1968), pp. 629–39.

The question of how the Cholesky triangle G changes when $A = GG^T$ is perturbed is analyzed in

G. W. Stewart (1977b). "Perturbation Bounds for the QR Factorization of a Matrix," *SIAM J. Num. Anal. 14*, 509–18.

An ALGOL procedure for inverting a symmetric positive definite matrix without any additional storage is given in

F. L. Bauer and C. Reinsch (1970). "Inversion of Positive Definite Matrices by the Gauss-Jordan Method," in HACLA, pp. 45–49.

FORTRAN programs for solving symmetric positive definite systems are in LINPACK, chaps. 3 and 8.

Sec. 5.3. Banded Systems

Frequently the matrix of coefficients in a linear system is banded. Recall that $A = (a_{ij})$ has upper bandwidth q if $a_{ij} = 0$ whenever $j > i + q$ and lower bandwidth p if $a_{ij} = 0$ whenever $i > j + p$. This is the case whenever the equations can be ordered so that each unknown x_i appears in only a few equations in a "neighborhood" of the i-th equation. Great economies can be realized when solving banded systems because the triangular factors in LU, GG^T, LDM^T, etc., are also banded:

THEOREM 5.3-1. Suppose $A \in \mathbb{R}^{n \times n}$ has an L-U decomposition $A = LU$. If A has upper bandwidth q and lower bandwidth p, then U has upper bandwidth q and L has lower bandwidth p.

Proof.   The proof is by induction on n. Writing

$$
A = \begin{bmatrix} \alpha & w^T \\ v & B \end{bmatrix} = \begin{bmatrix} 1 & 0 \\ v/\alpha & I_{n-1} \end{bmatrix} \begin{bmatrix} 1 & 0 \\ 0 & B - vw^T/\alpha \end{bmatrix} \begin{bmatrix} \alpha & w^T \\ 0 & I_{n-1} \end{bmatrix}
$$

it is clear that $B - vw^T/\alpha$ has upper bandwidth q and lower bandwidth p. This is because only the first q components of w and the first p components of v are nonzero. Let $L_1 U_1$ be the L-U decomposition of this matrix. Using the induction hypothesis and the sparsity of w and v, it follows that

$$
L = \begin{bmatrix} 1 & 0 \\ v/\alpha & L_1 \end{bmatrix} \quad \text{and } U = \begin{bmatrix} \alpha & w^T \\ 0 & U_1 \end{bmatrix}
$$

have the desired bandwidth properties and satisfy $A = LU$. $\square$

The specialization of Gaussian elimination to banded matrices having an L-U decomposition is straightforward.

ALGORITHM 5.3-1.   Given $A \in \mathbb{R}^{n \times n}$ with upper bandwidth q and lower bandwidth p, the following algorithm computes the decomposition $A = LU$, assuming it exists. The entry a_{ij} is overwritten by l_{ij} if $i > j$ and by u_{ij} otherwise.

> For $k = 1, \ldots, n - 1$
> > For $i = k + 1, \ldots, \min\{k + p, n\}$
> > > $a_{ik} := a_{ik}/a_{kk}$
> > For $i = k + 1, \ldots, \min\{k + p, n\}$
> > > For $j = k + 1, \ldots, \min\{k + q, n\}$
> > > > $a_{ij} := a_{ij} - a_{ik} a_{kj}$

This algorithm requires $w(p, q)$ flops, where

$$
w(p, q) = \begin{cases} npq - \frac{1}{2} pq^2 - \frac{1}{6} p^3 + pn & p \leqslant q \\ npq - \frac{1}{2} qp^2 - \frac{1}{6} q^3 + qn & p > q \end{cases}
$$

Similar results hold for the L-D-M^T decomposition. Note that $w(p, q) \ll n^3/3$ whenever p or q is much smaller than n.

Analogous savings can also be made when solving banded triangular systems:

ALGORITHM 5.3-2.   Let $L \in \mathbb{R}^{n \times n}$ be a nonsingular unit lower triangular matrix having lower bandwidth p. Given $b \in \mathbb{R}^n$, the following algorithm overwrites b with the solution to $Ly = b$.

For $i = 1, 2, \ldots, n$

$$b_i := b_i - \sum_{j=max\{1,i-p\}}^{i-1} l_{ij} b_j$$

This algorithm requires $np - p^2/2$ flops

ALGORITHM 5.3-3.   Let $U \in \mathbb{R}^{n \times n}$ be a nonsingular upper triangular matrix having upper bandwidth q. Given $b \in \mathbb{R}^n$, the following algorithm overwrites b with the solution to $Ux = b$.

For $i = n, n - 1, \ldots, 2, 1$

$$b_i := \left(b_i - \sum_{j=i+1}^{min\{i+q,n\}} u_{ij} b_j \right) \Big/ u_{ii}$$

This algorithm requires $n(q+1) - q^2/2$ flops.

Gaussian elimination with partial pivoting can also be specialized to exploit band structure in A. If, however, $PA = LU$, then the band properties of L and U are not quite so simple. For example, if A is tridiagonal and the first two rows are interchanged at the very first step of the algorithm, then u_{13} is non-zero. Consequently, row interchanges expand bandwidth. Precisely how the band enlarges is the subject of the following theorem.

THEOREM 5.3-4.   Suppose $A \in \mathbb{R}^{n \times n}$ is nonsingular and has upper and lower bandwidths q and p, respectively. If Gaussian elimination with partial pivoting is used to compute Gauss transformations

$$M_j = I - \alpha^{(j)} e_j^T \qquad\qquad j = 1, \ldots, n - 1$$

and permutations $P_1, \ldots, P_{n-1}$ such that

$$M_{n-1} P_{n-1} \ldots M_1 P_1 A = U$$

is upper triangular, then U has upper bandwidth $p + q$ and $\alpha_i^{(j)} = 0$ whenever $i \leqslant j$ or $i > j + p$.

Proof.   Let $PA = LU$ be the decomposition computed by Gaussian elimination with partial pivoting and recall that $P = P_{n-1} \ldots P_1$. Write $P^T = [e_{s_1}, \ldots, e_{s_n}]$, where $\{s_1, \ldots, s_n\}$ is a permutation of $\{1, 2, \ldots, n\}$. If $s_i > i + p$ then it follows that the leading i-by-i principal submatrix of PA is singular, since

$$(PA)_{ij} = a_{s_i,j} = 0 \qquad j = 1, \ldots, s_i - p - 1$$

and $s_i - p - 1 \geqslant i$. This implies that U and A are singular, a contradiction. Thus, $s_i \leqslant i + p$ for $i = 1, \ldots, n$ and therefore, PA has upper bandwidth $p + q$. It follows from Theorem 5.3-1 that U has upper bandwidth $p + q$.

The assertion about the $\alpha^{(j)}$ can be verified by observing that only elements $(j + 1, j), \ldots, (j + p, j)$ of the partially reduced matrix

$$M_{j-1} P_{j-1} \cdots M_1 P_1 A$$

need to be zeroed by M_j. $\square$

Thus, pivoting destroys band structure in the sense that U becomes "fatter" than A's upper triangle, while nothing at all can be said about L's bandwidth. However, since the j-th column of L is a permutation of the Gauss vector $\alpha^{(j)}$, it follows that L has at most $p + 1$ nonzeros per column.

As an example of an unsymmetric band matrix computation, we illustrate how Gaussian elimination with partial pivoting can be applied to factor an upper Hessenberg matrix $H(h_{ij} = 0, i > j + 1)$. After $k - 1$ steps of Algorithm 4.4-2 we are left with an upper Hessenberg matrix of the form

$$M_{k-1} P_{k-1} \ldots P_2 M_1 P_1 A = \begin{bmatrix} x & x & x & x & x & x \\ 0 & x & x & x & x & x \\ 0 & 0 & x & x & x & x \\ 0 & 0 & x & x & x & x \\ 0 & 0 & 0 & x & x & x \\ 0 & 0 & 0 & 0 & x & x \end{bmatrix} \qquad k = 3, n = 6.$$

By virtue of the special structure of this matrix, we see that P_k is either the identity or the identity with rows k and $k + 1$ interchanged, and that M_k is a Gauss transformation with a single nonzero multiplier in the $(k + 1, k)$ position. This illustrates the k-th step of the following algorithm.

ALGORITHM 5.3-4. Given an upper Hessenberg matrix $H \in \mathbb{R}^{n \times n}$, the following algorithm computes the factorization $M_{n-1} P_{n-1} \ldots M_1 P_1 A = U$ where each P_i is a permutation, each M_i is a Gauss transformation whose entries are bounded by unity, and U is upper triangular. The entry h_{ij} is overwritten with u_{ij} if $i \leqslant j$, and by the sole multiplier of M_j if $i = j + 1$. If $P_k = I$, then $p_k = 0$. If P_k interchanges rows k and $k + 1$, then $p_k = 1$.

For $k = 1, \ldots, n - 1$
    If $|h_{kk}| < |h_{k+1,k}|$
        then
            Interchange h_{kj} and $h_{k+1,j}, j = k, \ldots, n$.
            $p_k := 1$
        else

$$p_k := 0$$
$$\text{if } h_{kk} \neq 0$$
$$\text{then}$$
$$t := h_{k+1,k}/h_{kk}$$
$$\text{For } j = k + 1, \ldots, n$$
$$h_{k+1,j} := h_{k+1,j} - t h_{kj}$$
$$h_{k+1,k} := t$$

This algorithm requires $n^2/2$ flops.

The rest of this section is devoted to banded $Ax = b$ problems where the matrix A is also symmetric positive definite. The fact that pivoting is unnecessary for such matrices leads to some very compact, elegant algorithms. In particular, it follows from Theorem 5.3-1 that if $A = GG^T$ is the Cholesky decomposition of A, then G has the same bandwidth as A. This leads to the following banded version of the Cholesky decomposition, Algorithm 5.2-1.

ALGORITHM 5.3-5. Given a symmetric positive definite $A \in \mathbb{R}^{n \times n}$ with bandwidth p, the following algorithm computes a lower triangular matrix G with lower bandwidth p such that $A = GG^T$. The entry a_{ij} is overwritten by g_{ij} ($i \geq j$).

$$\text{For } i = 1, \ldots, n$$
$$\text{For } j = \max\{1, i-p\}, \ldots, i - 1$$
$$a_{ij} := \left(a_{ij} - \sum_{k=\max\{1,i-p\}}^{j-1} a_{ik} a_{jk} \right) \Big/ a_{jj}$$

$$a_{ii} := \left(a_{ii} - \sum_{k=\max\{1,i-p\}}^{i-1} a_{ik}^2 \right)^{1/2}$$

This algorithm requires $(np^2/2) - (p^3/3) + (3/2)(np - p^2)$ flops and n square roots.

Thus for $p \ll n$, the Cholesky decomposition requires $n(p^2 + 3p)/2$ flops and n square roots. Moreover, only an $n \times (p + 1)$ array is required if A is stored in the following fashion:

$$\begin{bmatrix} 0 & 0 & a_{11} \\ 0 & a_{21} & a_{22} \\ a_{31} & a_{32} & a_{33} \\ a_{42} & a_{43} & a_{44} \\ \vdots & \vdots & \vdots \\ a_{n,n-2} & a_{n,n-1} & a_{nn} \end{bmatrix} \qquad p = 2$$

Storing A (and G) in this way means that the (i,j) entry of the matrix is found in the $(i,j-i+p+1)$ entry of the storage array. When this data structure is used, the subscript computations must be made as efficient as possible.

If we add up the work involved in solving a symmetric positive definite $Ax = b$ problem with bandwidth p ($p \ll n$), we find that a total of $n[p^2/2 + 7p/2 + 2]$ flops and n square roots are required. (The triangular systems $Gy = b$ and $G^Tx = y$ involve $n(p + 1)$ flops each, while $A = GG^T$ requires $np^2/2 + 3np/2$ flops and n square roots.)

Now if the work involved in computing a square root is equivalent to three or four flops, then for small p, the n square roots are a significant portion of the computation. For this reason, a banded version of the L-D-L^T decomposition is preferred for narrow bandwidth problems. Indeed, a careful flop count of the steps $A = LDL^T$, $Ly = b$, $Dz = y$, and $L^Tx = z$ reveals that $n[p^2/2 + 7p/2 + 1]$ flops and *no* square roots are needed.

As an illustration, we present a complete symmetric positive definite tridiagonal system solver. Setting

$$
L = \begin{bmatrix}
1 & & & & \mathbf{0} \\
e_1 & 1 & & & \\
 & e_2 & 1 & & \\
 & & & \ddots & \\
\mathbf{0} & & & e_{n-1} & 1
\end{bmatrix}
$$

and $D = \mathrm{diag}(d_1, \ldots, d_n)$, we deduce from the equation $A = LDL^T$ that

$$a_{11} = d_1$$

$$a_{k,k-1} = e_{k-1}d_{k-1} \qquad\qquad k = 2, 3, \ldots, n$$

$$a_{kk} = d_k + e_{k-1}^2 d_{k-1} = d_k + e_{k-1}a_{k,k-1} \qquad k = 2, \ldots, n$$

Thus,

$$d_1 = a_{11}$$
For $k = 2, \ldots, n$
$$e_{k-1} = a_{k,k-1}/d_{k-1}$$
$$d_k = a_{kk} - e_{k-1}a_{k,k-1}$$

The systems $Ly = b$, $Dz = y$, and $L^Tx = z$ can each solved in n flops. The complete algorithm is as follows:

ALGORITHM 5.3-6. Given an n-by-n symmetric, tridiagonal, positive definite matrix A and $b \in \mathbb{R}^n$, the following algorithm overwrites b with the solution to $Ax = b$. It is assumed that the diagonal of A is stored in $(d_1, \ldots, d_n)$ and the superdiagonal in $(e_1, \ldots, e_{n-1})$.

For $k = 2, \ldots, n$
$\quad t := e_{k-1}$
$\quad e_{k-1} := t/d_{k-1}$
$\quad d_k := d_k - te_{k-1}$
For $k = 2, \ldots, n$
$\quad b_k := b_k - e_{k-1}b_{k-1}$
For $k = 1, \ldots, n$
$\quad b_k := b_k/d_k$
For $k = n - 1, \ldots, 1$
$\quad b_k := b_k - e_k b_{k+1}$

This algorithm requires $5n$ flops.

Problems

P5.3-1. Derive a banded version of Algorithm 5.1-1 (L-D-M^T) similar to Algorithm 5.3-1.

P5.3-2. Show how the output of Algorithm 5.3-3 can be used to solve the upper Hessenberg system $Hx = b$.

P5.3-3. Give an algorithm for solving unsymmetric tridiagonal systems $Ax = b$ that uses Gaussian elimination with partial pivoting and which requires only 4 n-vectors of floating point storage.

P5.3-4. For $C \in \mathbb{R}^{n \times n}$ define the *profile indices* $m(c, i) = \min\{j \,|\, c_{ij} \neq 0\}$, where $i = 1, 2, \ldots, n$. Show that if $A = GG^T$ is the Cholesky factorization of A, then $m(A, i) = m(G, i)$ for $i = 1, 2, \ldots, n$. (We say that G has the same "profile" as A.)

P5.3-5. Suppose $A \in \mathbb{R}^{n \times n}$ is symmetric positive definite with profile indices $m_i = m(A, i)$ where $i = 1, \ldots, n$. Assume that A is stored in a one-dimensional array v as follows:

$$v = (a_{11}, a_{2,m_2}, \ldots, a_{22}, a_{3,m_3}, \ldots, a_{33}, \ldots, a_{n,m_n}, \ldots, a_{nn}).$$

Write an algorithm which overwrites v with the corresponding entries of the Cholesky factor G and then uses this factorization to solve $Ax = b$. How many flops are required?

P5.3-6. For $C \in \mathbb{R}^{n \times n}$ define $p(C, i) = \max\{j \,|\, c_{ij} \neq 0\}$. Suppose that $A \in \mathbb{R}^{n \times n}$ has an L-U factorization $A = LU$ and that

$$m(A, 1) \leqslant m(A, 2) \leqslant \ldots \leqslant m(A, n)$$
$$p(A, 1) \leqslant p(A, 2) \leqslant \ldots \leqslant p(A, n).$$

Show that $m(A, i) = m(L, i)$ and $p(A, i) = p(U, i)$ for $i = 1, \ldots, n$. Recall the definition of $m(A, i)$ from P5.3-4.

Notes and References for Sec. 5.3

We wish to stress again that our flop counts are meant only to guide our appraisals of work. The reader should not assign too much meaning to their precise value, especially in the band matrix area where so much depends on the cleverness of the implementation.

FORTRAN codes for banded linear systems may be found in LINPACK, chaps. 2, 4, and 7. The literature concerned with banded systems is immense. Some representative papers include

E. L. Allgower (1973). "Exact Inverses of Certain Band Matrices," *Numer. Math. 21*, 279-84.

Z. Bohte (1975). "Bounds for Rounding Errors in the Gaussian Elimination for Band Systems," *J. Inst. Math. Applic. 16*, 133-42.

R. S. Martin and J. H. Wilkinson (1965). "Symmetric Decomposition of Positive Definite Band Matrices," *Numer. Math. 7*, 355-61. See also HACLA, pp. 50-56.

R. S. Martin and J. H. Wilkinson (1967). "Solution of Symmetric and Unsymmetric Band Equations and the Calculation of Eigenvalues of Band Matrices," *Numer. Math. 9*, 279-301. See also HACLA, pp. 70-92.

I. S. Duff (1977). "A Survey of Sparse Matrix Research," *Proc. IEEE 65*, 500-535.

A topic of considerable interest in the area of banded matrices deals with methods for reducing the width of the band. See

E. Cuthill (1972). "Several Strategies for Reducing the Bandwidth of Matrices," in *Sparse Matrices and Their Applications*, ed. D. J. Rose and R. A. Willoughby, Plenum Press, New York.

N. E. Gibbs, W. G. Poole, Jr., and P. K. Stockmeyer (1976). "An Algorithm for Reducing the Bandwidth and Profile of a Sparse Matrix," *SIAM J. Num. Anal. 13*, 236-50.

N. E. Gibbs, W. G. Poole, and P. K. Stockmeyer (1976). "A Comparison of Several Bandwidth and Profile Reduction Algorithms," *ACM Trans. Math. Soft 2*, 322-30.

As we mentioned, tridiagonal systems arise with particular frequency. Thus, it is not surprising that a great deal of attention has been focussed on special methods for this class of banded problems:

C. Fischer and R. A. Usmani (1969). "Properties of Some Tridiagonal Matrices and Their Application to Boundary Value Problems," *SIAM J. Num. Anal. 6*, 127-42.

J. Lambiotte and R. G. Voigt (1975). "The Solution of Tridiagonal Linear Systems of the CDC-STAR 100 Computer," *ACM Trans. Math. Soft. 1*, 308-29.

M. A. Malcolm and J. Palmer (1974). "A Fast Method for Solving a Class of Tridiagonal Systems of Linear Equations," *Comm. Assoc. Comp. Mach. 17*, 14-17.

D. J. Rose (1969). "An Algorithm for Solving a Special Class of Tridiagonal Systems of Linear Equations," *Comm. Assoc. Comp. Mach. 12*, 234-36.

H. S. Stone (1973). "An Efficient Parallel Algorithm for the Solution of a Tridiagonal Linear System of Equations," *J. Assoc. Comp. Mach. 20*, 27-38.

H. S. Stone (1975). "Parallel Tridiagonal Equation Solvers," *ACM Trans. Math. Soft. 1*, 289-307.

Chapter 4 of

J. A. George and J. W. Liu (1981). *Computer Solution of Large Sparse Positive Definite Systems*, Prentice-Hall, Englewood Cliffs,

contains a nice survey of band methods for positive definite systems.

Sec. 5.4. Symmetric Indefinite Systems

Suppose we wish to solve $Ax = b$, where $A \in \mathbb{R}^{n \times n}$ is symmetric but indefinite. Although A may have an L-D-L^T factorization, the entries in the factors can have arbitrary magnitude.

EXAMPLE 5.4-1.

$$\begin{bmatrix} \epsilon & 1 \\ 1 & \epsilon \end{bmatrix} = \begin{bmatrix} 1 & 0 \\ 1/\epsilon & 1 \end{bmatrix} \begin{bmatrix} \epsilon & 0 \\ 0 & \epsilon - 1/\epsilon \end{bmatrix} \begin{bmatrix} 1 & 0 \\ 1/\epsilon & 1 \end{bmatrix}^T \qquad 1 \gg \epsilon > 0$$

Hence, the unmodified use of this factorization cannot be recommended for symmetric indefinite systems. Some form of pivoting must be introduced. The aim of this section is to show how this can be accomplished without essentially altering the $n^3/6$ flop count for L-D-L^T.

In order to obtain an algorithm this fast it is necessary to preserve symmetry. This rules out partial pivoting, since row interchanges alone destroy symmetry. This prompts us to consider how a permutation P might be determined so that

$$PAP^T = LDL^T$$

might be safely computed. Example 5.4-1 shows that this may not be possible. This is because

$$P \begin{bmatrix} \epsilon & 1 \\ 1 & \epsilon \end{bmatrix} P^T = \begin{bmatrix} \epsilon & 1 \\ 1 & \epsilon \end{bmatrix}$$

for any 2-by-2 permutation P. The challenge, it seems, is how to involve the off-diagonal entries in the pivoting process.

We will present an $n^3/6$ method due to Aasen (1971) that does this by computing the factorization

(5.4-1) $$PAP^T = LTL^T,$$

where $L = (l_{ij})$ is unit lower triangular, P is a permutation chosen such that $|l_{ij}| \leqslant 1$, and T is tridiagonal. Note that if we solve the systems $Lw = Pb$, $Tz = w$, $L^Tv = z$, and $Px = v$, then $Ax = b$. Since the computation of (5.4-1) dominates the overall computational effort, it does not matter if we ignore symmetry and use Gaussian elimination with partial pivoting to solve the symmetric indefinite tridiagonal system $Tz = w$. (The property of being definite or indefinite is preserved under congruence.)

Our discussion of Aasen's method can be considerably simplified if we first outline a somewhat less efficient method for calculating (5.4-1) that was demonstrated by Parlett and Reid (1970). Their algorithm is sufficiently illustrated by displaying the $k = 2$ step for the case $n = 5$. At the beginning of this step the matrix A has been transformed to

$$A^{(1)} = M_1 P_1 A P_1^T M_1^T = \begin{bmatrix} \alpha_1 & \beta_2 & 0 & 0 & 0 \\ \beta_2 & \alpha_2 & v_3 & v_4 & v_5 \\ 0 & v_3 & x & x & x \\ 0 & v_4 & x & x & x \\ 0 & v_5 & x & x & x \end{bmatrix},$$

where P_1 is a permutation that has been chosen so that the entries in the Gauss transformation M_1 are bounded by unity in modulus. Scanning the vector $(v_3, v_4, v_5)^T$ for its largest entry, we now determine a 3-by-3 permutation $\tilde{P}_2$ such that

$$\tilde{P}_2 \begin{bmatrix} v_3 \\ v_4 \\ v_5 \end{bmatrix} = \begin{bmatrix} \tilde{v}_3 \\ \tilde{v}_4 \\ \tilde{v}_5 \end{bmatrix}, \quad |\tilde{v}_3| = \max \{ |v_3|, |v_4|, |v_5| \}.$$

If this maximal element is zero, we set $M_2 = P_2 = I$ and proceed to the next step. Otherwise, we set $P_2 = \text{diag}(I_2, \tilde{P}_2)$ and

$$M_2 = I_5 - \frac{1}{\tilde{v}_3} (0, 0, 0, \tilde{v}_4, \tilde{v}_5)^T e_3^T$$

and observe that

$$A^{(2)} = M_2 P_2 A^{(1)} P_2^T M_2^T = \begin{bmatrix} \alpha_1 & \beta_2 & 0 & 0 & 0 \\ \beta_2 & \alpha_2 & \tilde{v}_3 & 0 & 0 \\ 0 & \tilde{v}_3 & x & x & x \\ 0 & 0 & x & x & x \\ 0 & 0 & x & x & x \end{bmatrix}.$$

In general, the process continues for $n - 2$ steps, leaving us with a tridiagonal matrix

(5.4-2) $T = A^{(n-2)} = (M_{n-2} P_{n-2} \cdots M_1 P_1) A (M_{n-2} P_{n-2} \cdots M_1 P_1)^T.$

It can be shown that (5.4-1) holds with

$$P = P_{n-2} \cdots P_1$$

and

$$L = (M_{n-2}P_{n-2} \cdots M_1P_1P^T)^{-1}.$$

Analysis of L reveals that its first column is e_1 and that its subdiagonal entries in column k $(k > 1)$ are "made up" of the multipliers in M_{k-1}.

EXAMPLE 5.4-2. If the Parlett-Reid algorithm is applied to

$$A = \begin{bmatrix} 0 & 1 & 2 & 3 \\ 1 & 2 & 2 & 2 \\ 2 & 2 & 3 & 3 \\ 3 & 2 & 3 & 4 \end{bmatrix},$$

then

$$P_1 = [e_1, e_4, e_3, e_2]$$
$$M_1 = I_4 - (0, 0, \tfrac{2}{3}, \tfrac{1}{3})^T e_2^T$$
$$P_2 = [e_1, e_2, e_4, e_3]$$
$$M_2 = I_4 - (0, 0, 0, \tfrac{1}{2})^T e_3^T$$

and $PAP^T = LTL^T$, where

$$P = \begin{bmatrix} 1 & 0 & 0 & 0 \\ 0 & 0 & 0 & 1 \\ 0 & 1 & 0 & 0 \\ 0 & 0 & 1 & 0 \end{bmatrix}, \quad L = \begin{bmatrix} 1 & 0 & 0 & 0 \\ 0 & 1 & 0 & 0 \\ 0 & \tfrac{1}{3} & 1 & 0 \\ 0 & \tfrac{2}{3} & \tfrac{1}{2} & 1 \end{bmatrix}, \quad T = \begin{bmatrix} 0 & 3 & 0 & 0 \\ 3 & 4 & \tfrac{2}{3} & 0 \\ 0 & \tfrac{2}{3} & \tfrac{10}{9} & 0 \\ 0 & 0 & 0 & \tfrac{1}{2} \end{bmatrix}.$$

The efficient implementation of the Parlett-Reid method requires care when computing the update

(5.4-3) $$A^{(k)} = M_k(P_k A^{(k-1)} P_k^T) M_k^T.$$

To see what is involved with a minimum of notation, suppose $B = B^T$ has order $n - k$ and that we wish to form

$$B_+ = (I - we_1^T) B (I - we_1^T)^T,$$

where w and e_1 are in R^{n-k}. Such a calculation is at the heart of (5.4-3). If we set

$$u = Be_1 - \frac{b_{11}}{2} w,$$

then the symmetric matrix

$$B_+ = B - wu^T - uw^T$$

can be formed in $(n - k)^2$ flops. Summing this quantity from $k = 1$ to $k = n - 2$ indicates that the Parlett-Reid procedure requires $n^3/3$ flops—twice what we would like.

The $n^3/6$ Aasen method can be derived by reconsidering the calculation of the Gauss transformations $M_1, \ldots, M_{n-2}$. For clarity, we temporarily ignore pivoting.

Suppose that as above we have determined Gauss transformations $M_1, \ldots, M_{k-1}$ such that

$$(M_{k-1} \ldots M_1)A(M_{k-1} \ldots M_1)^T = \begin{bmatrix} T_{11} & \begin{array}{c} 0 \\ \hline v^T \end{array} \\ \hline 0 \quad v & T_{22} \end{bmatrix} \begin{array}{c} k-1 \\ 1 \\ n-k \end{array}$$
$$\quad\quad\quad\quad\quad\quad\quad\quad\quad k-1 \quad 1 \quad n-k$$

where

$$T_{11} = \begin{bmatrix} \alpha_1 & \beta_2 & & \mathbf{0} \\ \beta_2 & \alpha_2 & \ddots & \\ & \ddots & \ddots & \beta_k \\ \mathbf{0} & & \beta_k & \alpha_k \end{bmatrix}.$$

Assume that we know all of T_{11} except α_k. In Aasen's method, the goal of the k-th step is to determine M_k, α_k, and β_{k+1}.

Note that

$$M_1^{-1} \ldots M_{k-1}^{-1} = (l_{ij}) = \begin{bmatrix} L_{11} & 0 \\ L_{21} & I \end{bmatrix} \begin{array}{c} k \\ n-k \end{array}$$
$$\quad\quad\quad\quad\quad\quad\quad\quad k \quad n-k$$

is unit lower triangular with first column e_1. This follows because each M_i has the form

$$M_i = I - \underbrace{(0, \ldots, 0, x, \ldots, x)}_{i+1}^T e_{i+1}^T.$$

(See §3.5.) Since

(5.4-4)
$$A = \begin{bmatrix} L_{11} & 0 \\ L_{21} & I \end{bmatrix} \begin{bmatrix} H_{11} & H_{12} \\ \hline 0 & v & H_{22} \end{bmatrix},$$

where

(5.4-5)
$$\begin{bmatrix} T_{11} & 0 \\ \hline & v^T \\ \hline 0 & v & T_{22} \end{bmatrix} \begin{bmatrix} L_{11}^T & L_{12}^T \\ 0 & I \end{bmatrix} = \begin{bmatrix} H_{11} & H_{12} \\ \hline 0 & v & H_{22} \end{bmatrix} \quad \begin{matrix} k \\ \\ n-k \end{matrix}$$
$$\qquad\qquad\qquad\qquad\qquad\qquad\qquad\qquad k-1 \quad 1 \quad n-k$$

it follows that

(5.4-6)
$$v = \begin{bmatrix} v_{k+1} \\ \vdots \\ v_n \end{bmatrix} = \begin{bmatrix} a_{k+1,k} \\ \vdots \\ a_{nk} \end{bmatrix} - L_{21}H_{11}e_k.$$

Thus, once we know

$$H_{11}e_k \equiv \begin{bmatrix} h_1 \\ \vdots \\ h_k \end{bmatrix}$$

we can compute

(5.4-7)
$$v_i = a_{ik} - \sum_{j=1}^{k} l_{ij}h_j \qquad\qquad i = k+1, \ldots, n$$

and thereby obtain

$$M_k = I - \frac{1}{v_{k+1}}(0, \ldots, 0, v_{k+2}, \ldots, v_n)^T \, e_{k+1}^T.$$

The key to Aasen's method is to recognize from (5.4-5) that $H_{11} = T_{11} L_{11}^T$ and so

(5.4-8) $h_1 = \beta_2 l_{k2}$

(5.4-9) $h_i = \beta_i l_{k,i-1} + \alpha_i l_{ki} + \beta_{i+1} l_{k,i+1} \qquad\qquad i = 2, \ldots, k-1$

(5.4-10) $h_k = \beta_k l_{k,k-1} + \alpha_k.$

These are not quite "working formulae," since α_k is unknown. To circumvent this problem we use the alternative equation

(5.4-10')
$$h_k = a_{kk} - \sum_{i=2}^{k-1} l_{ki}h_i,$$

which can be derived by comparing (k,k) entries in (5.4-4). Thus, the multipliers in M_k can be determined by invoking (5.4-8), (5.4-9), and (5.4-10') followed by (5.4-7). The k-th step is completed by setting

$$(5.4\text{-}11) \qquad \beta_{k+1} = v_{k+1} \quad \text{and} \quad \alpha_k = \begin{cases} a_{11} & k = 1 \\ h_k - \beta_k l_{k,k-1} & k > 1 \end{cases}$$

The incorporation of pivots in Aasen's method poses no serious difficulty. The vector v is calculated as above. Its largest entry is then moved into the pivot position by a permutation $\tilde{P}_k$ of order $n - k$. Of course, this permutation must be suitably applied to the unreduced portion of A; i.e., $T_{22} := \tilde{P}_k T_{22} \tilde{P}_k^T$. The previously computed Gauss vectors, which may be stored in the "sub-tridiagonal" portion of A, must be premultiplied by $\tilde{P}_k$ in order to emerge with the proper L matrix. The overall process is as follows.

ALGORITHM 5.4-1: *Aasen's Method.* Given a symmetric matrix $A \in \mathbb{R}^{n \times n}$, the following algorithm computes permutations $P_1, \ldots, P_{n-2}$, a unit lower triangular matrix $L = (l_{ij})$ having first column e_1, and a tridiagonal matrix $T = (t_{ij})$ such that if $P = P_{n-2} \ldots P_1$ then $PAP^T = LTL^T$. The permutations are chosen such that $|l_{ij}| \leq 1$. The entry l_{ij} $(j = 2, \ldots, n - 2, i = j + 1, \ldots, n)$ overwrites $a_{i,j-1}$. The diagonal of T is stored in $\alpha_1, \ldots, \alpha_n$ and its subdiagonal in $\beta_2, \ldots, \beta_n$. For $k = 1, \ldots, n - 2$, the integer p_k indicates that the permutation P_k is the identity with rows $k + 1$ and p_k interchanged.

For $k = 1, \ldots, n$
$\quad l_1 := 0$
$\quad l_i := a_{k,i-1} \qquad (i = 2, \ldots, k - 1)$
$\quad l_k := 1$
$\quad h_i := \beta_i l_{i-1} + \alpha_i l_i + \beta_{i+1} l_{i+1} \qquad (i = 2, \ldots, k - 1)$
$\quad h_k := a_{kk} - \sum_{i=2}^{k-1} l_i h_i$
$\quad$ if $k = 1$
$\qquad$ *then* $\alpha_1 := a_{11}$
$\qquad$ *else* $\alpha_k := h_k - \beta_k l_{k-1}$
$\quad$ if $k < n$
$\qquad$ *then*
$\qquad\qquad$ For $i = k + 1, \ldots, n$
$$v_i := a_{ik} - \sum_{j=2}^{k} a_{i,j-1} h_j$$
$\qquad\qquad$ Find q $(k+1 \leq q \leq n)$ so
$$|v_q| = \max \{ |v_{k+1}|, \ldots, |v_n| \}$$
$\qquad\qquad p_k := q$

> Interchange v_{k+1} and v_q, $a_{k+1,j}$ and a_{qj} ($j = 1, \ldots, n$),
> and $a_{j,k+1}$ and a_{jq} ($j = k + 1, \ldots, n$).
> $\beta_{k+1} := v_{k+1}$
> If $v_q \neq 0$ and $k \leqslant n - 2$
>   then
> $$a_{ik} := v_i / v_{k+1} \qquad (i = k + 2, \ldots, n)$$

This algorithm requires $n^3 / 6$ flops.

Aasen's method is stable in the same sense that Gaussian elimination with partial pivoting is stable. That is, the exact factorization of a matrix near A is obtained provided $\| \hat{T} \| / \| A \| \approx 1$, where $\hat{T}$ is the computed version of the tridiagonal matrix T. In general, this is almost always the case.

Rather than compute (5.4-1), Bunch and others propose the calculation of

$$(5.4\text{-}12) \qquad PAP^T = LDL^T,$$

where D is block diagonal with 1-by-1 and 2-by-2 blocks, $L = (l_{ij})$ is unit lower triangular, and P is a permutation matrix that is determined so that $|l_{ij}| \leqslant 1$. Methods for computing a factorization of this type are called *diagonal pivoting methods*. They differ in how P is generated and in how the dimensions of D's blocks are determined.

We first describe the strategy of Bunch and Parlett (1971). Suppose

$$\Pi A \Pi^T = \begin{bmatrix} E & C^T \\ C & B \end{bmatrix} \begin{matrix} s \\ n-s \end{matrix}$$
$$\phantom{\Pi A \Pi^T = \begin{bmatrix} E & C^T \end{bmatrix}} \begin{matrix} s & n-s \end{matrix}$$

where Π is a permutation matrix and $s = 1$ or 2. If A is nonzero, then it is always possible to choose these quantities so that E is nonsingular, thereby enabling us to write

$$(5.4\text{-}13) \quad \Pi A \Pi^T = \begin{bmatrix} I_s & 0 \\ CE^{-1} & I_{n-s} \end{bmatrix} \begin{bmatrix} E & 0 \\ 0 & B - CE^{-1}C^T \end{bmatrix} \begin{bmatrix} I_s & E^{-1}C^T \\ 0 & I_{n-s} \end{bmatrix}.$$

For the sake of stability, the s-by-s "pivot" E should be chosen so that the entries in

$$(5.4\text{-}14) \qquad \tilde{A} = (\tilde{a}_{ij}) = B - CE^{-1}C^T$$

are suitably bounded. To this end, let $\alpha \in (0, 1)$ be given and define

$$\mu_0 = \max_{i,j} |a_{ij}| \qquad \mu_1 = \max_i |a_{ii}|.$$

The Bunch-Parlett pivot strategy is as follows:

If $\mu_1 \geqslant \alpha\mu_0$, then $s = 1$ and choose Π so $|e_{11}| = \mu_1$.

If $\mu_1 < \alpha\mu_0$, then $s = 2$ and choose Π so $|e_{21}| = \mu_0$.

It is easy to verify from (5.4-14) that if $s = 1$ then

(5.4-15) $$|\tilde{a}_{ij}| \leqslant (1 + \alpha^{-1})\mu_0,$$

while $s = 2$ implies

(5.4-16) $$|\tilde{a}_{ij}| \leqslant \frac{3 - \alpha}{1 - \alpha}\mu_0.$$

By equating $(1 + \alpha^{-1})^2$, the growth factor associated with two $s = 1$ steps, and $(3 - \alpha)/(1 - \alpha)$, the corresponding $s = 2$ factor, Bunch and Parlett conclude that $\alpha = (1 + \sqrt{17})/8$ is optimum from the standpoint of minimizing the bound on element growth. The process is then repeated on the $(n - s)$ order symmetric matrix $\tilde{A}$. A simple induction argument establishes that the factorization (5.4-12) exists and that $n^3/6$ flops are required if the work associated with pivot determination is ignored.

The resulting algorithm is shown by Bunch (1971a) to be as stable as Gaussian elimination with complete pivoting. Unfortunately, the overall process requires between $n^3/12$ and $n^3/6$ comparisons, since a two-dimensional search (for μ_0) must be performed at each stage of the reduction. The actual number of comparisons depends on the total number of 2-by-2 pivots. As a consequence, the Bunch-Parlett method for computing (5.4-12) is considerably slower than the technique of Aasen. See Barwell and George [1976].

This is not the case with the diagonal pivoting method of Bunch and Kaufman (1977). In their scheme it is only necessary to scan two columns at each stage of the reduction:

Let r $(2 \leqslant r \leqslant n)$ be such that $|a_{r1}| = \max_{2 \leqslant i \leqslant n} |a_{i1}|$.

Let p $(2 \leqslant p \leqslant n)$ be such that $p \neq r$ and $|a_{pr}| = \max_{\substack{2 \leqslant i \leqslant n \\ i \neq r}} |a_{ir}|$.

If $|a_{11}| \geqslant \alpha\,|a_{r1}|$ then $\Pi = I$ and $s = 1$.

Otherwise, if $|a_{11}|\,|a_{pr}| \geqslant \alpha\,|a_{r1}|^2$ then $\Pi = I$ and $s = 1$.

Otherwise, if $|a_{rr}| \geqslant \alpha\,|a_{pr}|$ then $s = 1$ and choose Π so that $e_{11} = a_{rr}$.

Otherwise, $s = 2$ and choose Π such that $E = \begin{bmatrix} a_{11} & a_{pr} \\ a_{rp} & a_{rr} \end{bmatrix}$.

Again, in order to minimize the chance of element growth, $\alpha = (1 + \sqrt{17})/8$. Overall, the Bunch-Kaufman algorithm requires $n^3/6$ flops, $O(n^2)$ comparisons, and, like all the methods of this section, $n^2/2$ storage. Its speed and roundoff properties are similar to those of Aasen's method, although in the latter algorithm it is possible to accumulate inner products.

EXAMPLE 5.4-3. If the Bunch-Kaufman algorithm is applied to

$$A = \begin{bmatrix} 0 & 1 & 2 & 3 \\ 1 & 2 & 2 & 2 \\ 2 & 2 & 3 & 3 \\ 3 & 2 & 3 & 4 \end{bmatrix}$$

then in one step we obtain $\Pi = [e_1, e_4, e_3, e_2]$,

$$D = \begin{bmatrix} 0 & 3 & 0 & 0 \\ 3 & 4 & 0 & 0 \\ 0 & 0 & \frac{7}{9} & \frac{5}{9} \\ 0 & 0 & \frac{5}{9} & \frac{10}{9} \end{bmatrix},$$

and

$$L = \begin{bmatrix} 1 & 0 & 0 & 0 \\ 0 & 1 & 0 & 0 \\ \frac{1}{9} & \frac{2}{3} & 1 & 0 \\ \frac{2}{9} & \frac{1}{3} & 0 & 1 \end{bmatrix}$$

and $\Pi A \Pi^T = LDL^T$.

Problems

P5.4-1. Show that if all the 2-by-2 principal submatrices of an n-by-n symmetric matrix A are singular, then A is singular.

P5.4-2. Give an algorithm for solving $Ax = b$ assuming that Algorithm 5.4-1 has been executed.

P5.4-3. Establish (5.4-11).

P5.4-4. Apply Aasen's method to

$$A = \begin{bmatrix} 0 & 4 & 1 & 2 \\ 4 & 2 & 5 & 3 \\ 1 & 5 & 0 & 1 \\ 2 & 3 & 1 & 2 \end{bmatrix}.$$

P5.4-5. Apply the Bunch-Kaufman diagonal pivoting method to the example in P5.4-4.

P5.4-6. Show that no 2-by-2 pivots can arise in the Bunch-Kaufman algorithm if A is positive definite.

P5.4-7. Verify (5.4-15) and (5.4-16).

Notes and References for Sec. 5.4

The basic references for computing (5.4-1) are

B. N. Parlett and J. K. Reid (1970). "On the Solution of a System of Linear Equations Whose Matrix is Symmetric but not Definite," *BIT 10*, 386–97.

J. O. Aasen (1971). "On the Reduction of a Symmetric Matrix to Tridiagonal Form," *BIT 11*, 233–42.

The diagonal pivoting literature includes

J. R. Bunch and B. N. Parlett (1971). "Direct Methods for Solving Symmetric Indefinite Systems of Linear Equations," *SIAM J. Num. Anal. 8*, 639–55.

J. R. Bunch (1971a). "Analysis of the Diagonal Pivoting Method," *SIAM J. Num. Anal. 8*, 656–80.

J. R. Bunch (1974). "Partial Pivoting Strategies for Symmetric Matrices," *SIAM J. Num. Anal. 11*, 521–28.

J. R. Bunch and L. Kaufman (1977). "Some Stable Methods for Calculating Inertia and Solving Symmetric Linear Systems," *Math. Comp. 31*, 162–79.

J. R. Bunch, L. Kaufman, and B. N. Parlett (1976). "Decomposition of a Symmetric Matrix," *Numer. Math. 27*, 95–109.

The last reference contains an ALGOL version of the diagonal pivoting method contained in LINPACK, chap. 5.

The question of whether Aasen's method is to be preferred to the Bunch-Kaufman algorithm is studied in

V. Barwell and J. A. George (1976). "A Comparison of Algorithms for Solving Symmetric Indefinite Systems of Linear Equations," *ACM Trans. Math. Soft. 2*, 242–51.

They suggest that the two algorithms behave similarly with perhaps a slight edge to Aasen when n is larger than 200. The performance data in this paper is the basis of an interesting statistical analysis in

D. Hoaglin (1977). "Mathematical Software and Exploratory Data Analysis," in *Mathematical Software III*, ed. John Rice, Academic Press, New York, pp. 139–59.

The small advantage of Aasen's method is perhaps due to its simpler pivot strategy.

Another idea for a cheap pivoting strategy utilizes error bounds based on more liberal interchange criteria, an idea borrowed from some work done in the area of sparse elimination methods. See

R. Fletcher (1976). "Factorizing Symmetric Indefinite Matrices," *Lin. Alg. & Its. Applic. 14*, 257-72.

We also mention the paper

A. Dax and S. Kaniel (1977). "Pivoting Techniques for Symmetric Gaussian Elimination," *Numer. Math. 28*, 221-42,

in which diagonal pivot entries are "built up" if necessary by using *upper* triangular multiplier matrices. Unfortunately, the technique appears to require $O(n^3)$ comparisons.

Before using any symmetric $Ax = b$ solver, it may be advisable to equilibrate A. An $O(n^2)$ algorithm for accomplishing this task is given in

J. R. Bunch (1971*b*). "Equilibration of Symmetric Matrices in the Max-Norm," *J. Assoc. Comp. Mach. 18*, 566-72.

Sec. 5.5. Block Tridiagonal Systems

The numerical solution of many mathematical problems involves matrices that have exploitable block structure. For example, in constrained optimization linear systems of the form

$$(5.5\text{-}1) \qquad \begin{bmatrix} A & B \\ B^\mathrm{T} & 0 \end{bmatrix} \begin{bmatrix} y \\ z \end{bmatrix} = \begin{bmatrix} c \\ d \end{bmatrix}$$

must frequently be solved where A is symmetric positive definite and B has full column rank. In this situation, it pays to exploit this structure rather than to treat (5.5-1) as "just another" symmetric indefinite system. See Heath (1978) for a detailed discussion.

The art of how to take advantage of block structure is perhaps most greatly advanced by researchers who are concerned with the numerical solution of partial differential equations. Discretization in this area frequently leads to block tridiagonal systems such as

$$(5.5\text{-}2) \qquad Ax \equiv \begin{bmatrix} D_1 & F_1 & & & \\ E_2 & D_2 & F_2 & & \\ & \ddots & \ddots & \ddots & \\ & & \ddots & \ddots & F_{n-1} \\ & & & E_n & D_n \end{bmatrix} \begin{bmatrix} x_1 \\ x_2 \\ \vdots \\ \\ x_n \end{bmatrix} = \begin{bmatrix} b_1 \\ b_2 \\ \vdots \\ \\ b_n \end{bmatrix}.$$

$$D_i, E_i, F_i \in \mathbb{R}^{q \times q}$$
$$x_i, b_i \in \mathbb{R}^q$$

By discussing several methods for solving this specific problem a general impression about the nature and value of block matrix calculations is achieved.

We begin by considering a block L-U factorization for the matrix in (5.5-2). Comparing blocks in

(5.5-3)

$$
\begin{bmatrix}
D_1 & F_1 & & & \mathbf{0} \\
E_2 & D_2 & F_2 & & \\
& \ddots & \ddots & \ddots & \\
& & & \ddots & F_{n-1} \\
\mathbf{0} & & & E_n & D_n
\end{bmatrix}
=
$$

$$
\begin{bmatrix}
I & & & \mathbf{0} \\
L_2 & I & & \\
& \ddots & \ddots & \\
\mathbf{0} & & L_n & I
\end{bmatrix}
\begin{bmatrix}
U_1 & F_1 & & \mathbf{0} \\
& U_2 & F_2 & \\
& & \ddots & F_{n-1} \\
\mathbf{0} & & & U_n
\end{bmatrix}
$$

we formally obtain the following algorithm for the L_i and U_i:

(5.5-4)

$$
\begin{aligned}
& U_1 = D_1 \\
& \text{For } i = 2, \ldots, n \\
& \qquad \text{Solve } L_i U_{i-1} = E_i \text{ for } L_i. \\
& \qquad U_i = D_i - L_i F_{i-1}
\end{aligned}
$$

The procedure is defined so long as $U_1, \ldots, U_{n-1}$ are nonsingular. This can be assured, for example, if the matrices

(5.5-5)

$$
\begin{bmatrix}
D_1 & F_1 & & & \mathbf{0} \\
E_2 & D_2 & F_2 & & \\
& \ddots & \ddots & F_{k-1} \\
\mathbf{0} & & E_k & D_k
\end{bmatrix}
\qquad k = 1, \ldots, n-1
$$

are nonsingular.

Having computed the above factorization, the vector x in (5.5-2) can be obtained via block forward elimination and back substitution:

(5.5-6)

$$
\begin{aligned}
& \text{For } i = 1, \ldots, n \\
& \qquad y_i = b_i - L_i y_{i-1} \qquad (L_1 y_0 \equiv 0) \\
& \text{For } i = n, \ldots, 1 \\
& \qquad U_i x_i = y_i - F_i x_{i+1} \qquad (F_n x_{n+1} \equiv 0)
\end{aligned}
$$

To carry out both (5.5-4) and (5.5-6), each U_i must be factored, since linear systems involving these submatrices must be solved. This could be done using Gaussian elimination with pivoting. However, this does not guarantee the stability of the overall process. (Just consider the case when the block size q is unity.) In order to obtain satisfactory bounds on the L_i and U_i it is necessary to make additional assumptions about the underlying block matrix. A sample result along these lines is the following:

THEOREM 5.5-1.   If the blocks in (5.5-2) satisfy

$$\|D_i^{-1}\|_1(\|F_{i-1}\|_1 + \|E_{i+1}\|_1) < 1 \quad i = 1, \ldots, n, \quad E_{n+1} \equiv F_0 \equiv 0$$

then the factorization (5.5-3) exists and the blocks in the factors satisfy

$$\|L_i\|_1 \leqslant 1 \qquad\qquad i = 2, \ldots, n$$

and

$$\|U_i\|_1 \leqslant \|A\|_1 \qquad\qquad i = 1, \ldots, n - 1$$

Proof.   The proof is by induction on n, the order of the block matrix A. It is clear that the $n = 1$ holds. To establish the inductive step, we factor A as follows:

$$(5.5\text{-}7) \quad A = \begin{bmatrix} I & \vdots & 0 & & 0 \\ \hline L_2 & \vdots & I & & \\ 0 & \vdots & & \ddots & \\ 0 & \vdots & & & I \end{bmatrix} \begin{bmatrix} I & \vdots & 0 & \cdots & 0 \\ \hline 0 & \vdots & & & \\ \vdots & \vdots & & C & \\ 0 & \vdots & & & \end{bmatrix} \begin{bmatrix} U_1 & \vdots & F_1 & \cdots & 0 \\ \hline 0 & \vdots & I & & \\ \vdots & \vdots & & \ddots & \\ 0 & \vdots & & & I \end{bmatrix},$$

where $U_1 = D_1$, $L_2 = E_2 D_1^{-1}$, and

$$(5.5\text{-}8) \quad C = \begin{bmatrix} \bar{D}_2 & F_2 & & & \text{\Large 0} \\ E_3 & D_3 & F_3 & & \\ & \ddots & \ddots & \ddots & \\ & & & & F_{n-1} \\ \text{\Large 0} & & & E_n & D_n \end{bmatrix}, \qquad \bar{D}_2 = D_2 - E_2 D_1^{-1} F_1.$$

From the hypothesis we have

$$\|D_2^{-1}E_2 D_1^{-1}F_1\| \leqslant (\|D_1^{-1}\|\,\|E_2\|)(\|D_2^{-1}\|\,\|F_1\|) < 1 - \|D_2^{-1}\|\,\|E_3\|.$$

Since $\bar{D}_2 = D_2(I - D_2^{-1}E_2 D_1^{-1}F_1)$, it follows by taking norms that $\|\bar{D}_2^{-1}\|\,\|E_3\| < 1$. (All norms in the proof are 1-norms.) By induction we then know that C can be factored as

$$(5.5\text{-}9) \quad C = \begin{bmatrix} I & & & \text{\Large 0} \\ L_3 & I & & \\ & \ddots & \ddots & \\ \text{\Large 0} & & L_n & I \end{bmatrix} \begin{bmatrix} U_2 & F_2 & & \text{\Large 0} \\ & \ddots & \ddots & \\ & & \ddots & F_{n-1} \\ \text{\Large 0} & & & U_n \end{bmatrix}$$

where $\|L_i\| \leqslant 1$ for $i = 3, \ldots, n$ and $\|U_i\| \leqslant \|C\|$ for $i = 2, \ldots, n - 1$. We leave as an exercise the inequality

(5.5-10)
$$\|C\| \leqslant \|A\|.$$

The theorem now follows by substituting (5.5-9) into (5.5-7) and observing that $\|L_2\| \leqslant 1$ and $\|U_1\| \leqslant \|A\|$. $\square$

The theorem requires that A^T be diagonally dominant in the block sense. Formally, a k-by-k block matrix $C = (C_{ij})$ with square invertible diagonal blocks is *block diagonally dominant* if

$$\|C_{ii}^{-1}\| \sum_{\substack{j=1 \\ j \neq i}}^{k} \|C_{ij}\| < 1 \qquad\qquad i = 1, \ldots, k$$

for some norm $\|\cdot\|$. Using the theorem we can assert that if A^T has this property and if the linear equations in both (5.5-4) and (5.5-6) are solved stably, then the entire method for computing x is stable.

At this point it is reasonable to ask why we do not simply regard the matrix A in (5.5-2) as a qn-by-qn matrix having scalar entries and bandwidth $2q - 1$. Band Gaussian elimination as described in §5.3 could be applied. The effectiveness of this course of action depends on such things as the dimensions of the blocks and the sparsity patterns within each block.

For example, suppose we wish to solve

(5.5-10)
$$\begin{bmatrix} D_1 & F_1 \\ E_2 & D_2 \end{bmatrix} \begin{bmatrix} x_1 \\ x_2 \end{bmatrix} = \begin{bmatrix} b_1 \\ b_2 \end{bmatrix},$$

where D_1 and D_2 are diagonal and F_1 and E_2 are tridiagonal. Assume that each of these blocks is n-by-n and that it is "safe" to solve (5.5-10) via (5.5-4) and (5.5-6). Note that

$$U_1 = D_1 \qquad\qquad \text{(diagonal)}$$
$$L_2 = E_2 U_1^{-1} \qquad\qquad \text{(tridiagonal)}$$
$$U_2 = D_2 - E_2 D_1^{-1} F_1 \qquad \text{(pentadiagonal)}$$

and

$$y_1 = b_1$$
$$y_2 = b_2 - E_2(D_1^{-1} y_1)$$
$$(D_2 - E_2 D_1^{-1} F_1)x_2 = y_2$$
$$D_1 x_1 = y_1 - F_1 x_2.$$

Consequently, some very simple n-by-n calculations with the original banded blocks renders the solution.

On the other hand, the naive application of band Gaussian elimination to the system (5.5-10) would entail a great deal of unnecessary work and storage. (The system has bandwidth $n + 1$). However, by permuting the rows and columns of the system via the permutation $P = [e_1, e_{n+1}, e_2, \ldots, e_n, e_{2n}]$ we find

$$
PAP^T =
\begin{bmatrix}
x & x & 0 & x & 0 & 0 & 0 & 0 & 0 & 0 \\
x & x & x & 0 & 0 & 0 & 0 & 0 & 0 & 0 \\
0 & x & x & x & 0 & x & 0 & 0 & 0 & 0 \\
x & 0 & x & x & x & 0 & 0 & 0 & 0 & 0 \\
0 & 0 & 0 & x & x & x & 0 & x & 0 & 0 \\
0 & 0 & x & 0 & x & x & x & 0 & 0 & 0 \\
0 & 0 & 0 & 0 & 0 & x & x & x & 0 & x \\
0 & 0 & 0 & 0 & x & 0 & x & x & x & 0 \\
0 & 0 & 0 & 0 & 0 & 0 & 0 & x & x & x \\
0 & 0 & 0 & 0 & 0 & 0 & x & 0 & x & x
\end{bmatrix}
\qquad n = 5
$$

Applying band Gaussian elimination to this matrix is not nearly so disastrous. (As a scalar matrix, it has bandwidth 3.)

We refer the reader to Varah (1972) and George (1974) for further details concerning the solving of such block systems.

We next describe a method for solving the block tridiagonal system (5.5-2) that is called *block cyclic reduction*. For simplicity, we assume that A has the form

$$
(5.5\text{-}11) \qquad A =
\begin{bmatrix}
D & F & & & \mathbf{0} \\
F & D & F & & \\
 & \ddots & \ddots & \ddots & \\
 & & \ddots & \ddots & F \\
\mathbf{0} & & & F & D
\end{bmatrix}
\in \mathbb{R}^{qn \times qn},
$$

where F and D are q-by-q matrices that satisfy $DF = FD$ and where $n = 2^k - 1$. These conditions frequently hold in practice. For example, when Poisson's equation is discretized on a rectangle,

$$D = \begin{bmatrix} 4 & -1 & & & 0 \\ -1 & 4 & -1 & & \\ & -1 & 4 & \ddots & \\ & & \ddots & \ddots & -1 \\ 0 & & & -1 & 4 \end{bmatrix}$$

and $F = -I_q$. The integer n is determined by the size of the mesh and can often be chosen to be of the form $n = 2^k - 1$. (Sweet [1977] shows how to proceed when the dimension is not of this form.)

The basic idea behind cyclic reduction is to halve the dimension of the problem on hand repeatedly until we are left with a single q-by-q system for the unknown subvector $x_{2^{k-1}}$. This system is then solved by standard means. The previously eliminated x_i are found by a back-substitution process.

The general procedure is adequately motivated by considering the case $n = 7$:

(5.5-12)
$$\begin{aligned} Dx_1 + Fx_2 & & & = b_1 \\ Fx_1 + Dx_2 + Fx_3 & & & = b_2 \\ Fx_2 + Dx_3 + Fx_4 & & & = b_3 \\ Fx_3 + Dx_4 + Fx_5 & & & = b_4 \\ Fx_4 + Dx_5 + Fx_6 & & & = b_5 \\ Fx_5 + Dx_6 + Fx_7 & = b_6 \\ Fx_6 + Dx_7 & = b_7. \end{aligned}$$

For $i = 2, 4, 6$ we multiply equations $i - 1, i$, and $i + 1$ by $F, -D$, and F, respectively, and add the resulting equations to obtain

$$\begin{aligned} (2F^2 - D^2)x_2 + F^2x_4 & & & = F(b_1 + b_3) - Db_2 \\ F^2x_2 + (2F^2 - D^2)x_4 + F^2x_6 & & & = F(b_3 + b_5) - Db_4 \\ F^2x_4 + (2F^2 - D^2)x_6 & = F(b_5 + b_7) - Db_6. \end{aligned}$$

Thus, with this tactic we have removed the odd-indexed x_i and are left with a reduced block tridiagonal system of the form

$$\begin{aligned} D^{(1)}x_2 + F^{(1)}x_4 & & & = b_2^{(1)} \\ F^{(1)}x_2 + D^{(1)}x_4 + F^{(1)}x_6 & = b_4^{(1)} \\ F^{(1)}x_4 + D^{(1)}x_6 & = b_6^{(1)}, \end{aligned}$$

where $D^{(1)} = 2F^2 - D^2$ and $F^{(1)} = F^2$ commute. Applying the same elimination strategy as above, we multiply these three equations respectively by $F^{(1)}$,

$-D^{(1)}$, and $F^{(1)}$. When these transformed equations are added together, we obtain the single equation

$$(2[F^{(1)}]^2 - [D^{(1)}]^2)x_4 = F^{(1)}(b_2^{(1)} + b_6^{(1)}) - D^{(1)}b_4^{(1)},$$

which we write as

$$D^{(2)}x_4 = b_4^{(2)}.$$

This completes the cyclic reduction. We now solve this (small) q-by-q system for x_4. The vectors x_2 and x_4 are then found by solving the systems

$$D^{(1)}x_2 = b_2^{(1)} - F^{(1)}x_4$$

and

$$D^{(1)}x_6 = b_6^{(1)} - F^{(1)}x_4.$$

Finally, we use the first, third, fifth, and seventh equations in (5.5-12) to compute x_1, x_3, x_5, and x_7, respectively.

For general $n = 2^k - 1$ we set $D^{(0)} = D$, $F^{(0)} = F$, and $b_i^{(0)} = b_i$ for $i = 1, \ldots, n$ and compute:

(5.5-13)
$$
\begin{aligned}
&\text{For } p = 1, \ldots, k - 1 \\
&\quad D^{(p)} = 2[F^{(p-1)}]^2 - [D^{(p-1)}]^2 \\
&\quad F^{(p)} = [F^{(p-1)}]^2 \\
&\quad r = 2^p \\
&\quad \text{For } j = 1, \ldots, 2^{k-p} - 1 \\
&\qquad b_{jr}^{(p)} = F^{(p-1)}(b_{jr-1}^{(p-1)} + b_{jr+1}^{(p-1)}) - D^{(p-1)}b_{jr}^{(p-1)}
\end{aligned}
$$

The x_i are then computed as follows:

(5.5-14)
$$
\begin{aligned}
&\text{Solve } D^{(k-1)}x_{2^{k-1}} = b_1^{(k-1)} \qquad \text{for } x_{2^{k-1}}. \\
&\text{For } p = k - 2, \ldots, 0 \\
&\quad r = 2^p \\
&\quad \text{For } j = 1, \ldots, 2^{k-p-1} \\
&\qquad \text{Solve} \\
&\qquad D^{(p)}x_{(2j-1)r} =
\begin{cases}
b_{(2j-1)r}^{(p)} - F^{(p)}x_{2jr} & (j = 1) \\
b_{(2j-1)r}^{(p)} - F^{(p)}x_{(2j-2)r} & (j = 2^{k-p-1}) \\
b_{(2j-1)r}^{(p)} - F^{(p)}x_{2jr} + x_{(2j-2)r} & (\text{otherwise})
\end{cases} \\
&\qquad \text{for } x_{(2j-1)r}.
\end{aligned}
$$

The amount of work required to perform these recursions depends greatly upon the sparsity of the $D^{(p)}$ and $F^{(p)}$. In the worst case when these matrices are full, the overall flop count has order $\log_2(n)q^3$. Care must be exercised in order to ensure stability during the reduction. See Buneman (1969).

Example 5.5-1.   Suppose $q = 1$, $D = (4)$, and $F = (-1)$ in (5.5-11) and that we wish to solve

$$\begin{bmatrix} 4 & -1 & 0 & 0 & 0 & 0 & 0 \\ -1 & 4 & -1 & 0 & 0 & 0 & 0 \\ 0 & -1 & 4 & -1 & 0 & 0 & 0 \\ 0 & 0 & -1 & 4 & -1 & 0 & 0 \\ 0 & 0 & 0 & -1 & 4 & -1 & 0 \\ 0 & 0 & 0 & 0 & -1 & 4 & -1 \\ 0 & 0 & 0 & 0 & 0 & -1 & 4 \end{bmatrix} \begin{bmatrix} x_1 \\ x_2 \\ x_3 \\ x_4 \\ x_5 \\ x_6 \\ x_7 \end{bmatrix} = \begin{bmatrix} 2 \\ 4 \\ 6 \\ 8 \\ 10 \\ 12 \\ 22 \end{bmatrix}. \quad n = 2^3 - 1$$

By executing (5.5-13) we obtain the reduced systems

$$\begin{bmatrix} -14 & 1 & 0 \\ 1 & -14 & 1 \\ 0 & 1 & -14 \end{bmatrix} \begin{bmatrix} x_2 \\ x_4 \\ x_6 \end{bmatrix} = \begin{bmatrix} -24 \\ -48 \\ -80 \end{bmatrix} \qquad p = 1$$

and

$$[-194][x_4] = [-776]. \qquad p = 2$$

The x_i are then determined via (5.5-14):

$$x_4 = 4$$
$$x_2 = 2, x_6 = 6 \qquad\qquad p = 1$$
$$x_1 = 1, x_3 = 3, x_5 = 5, x_7 = 7. \qquad p = 0$$

Problems

P5.5-1. Show that if the matrices in (5.5-5) are nonsingular, then (5.5-4) is defined.

P5.5-2. Prove inequality (5.5-10).

P5.5-3. Show that a block diagonally dominant matrix is nonsingular.

Notes and References for Sec. 5.5

A discussion of the symmetric indefinite system mentioned at the beginning of the section may be found in

M. T. Heath (1978). "Numerical Algorithms for Nonlinearly Constrained Optimization." Report STAN-CS 78-656, Department of Computer Science, Stanford University (Ph.D. thesis).

The following papers provide insight into the various nuances of block matrix computations:

J. A. George (1974). "On Block Elimination for Sparse Linear Systems," *SIAM J. Num. Anal. 11*, 585–603.

J. M. Varah (1972). "On the Solution of Block-Tridiagonal Systems Arising from Certain Finite-Difference Equations," *Math. Comp. 26*, 859–68.

The property of block diagonal dominance and its various implications is the central theme in

D. G. Feinold and R. S. Varga (1962). "Block Diagonally Dominant Matrices and Generalizations of the Gershgorin Circle Theorem," *Pacific J. Math. 12*, 1241–50.

Early methods that involve the idea of cyclic reduction are described in

B. L. Buzbee, G. H. Golub, and C. W. Nielson (1970). "On Direct Methods for Solving Poisson's Equations," *SIAM J. Num. Anal. 7*, 627–56.

R. W. Hockney (1965). "A Fast Direct Solution of Poisson's Equation Using Fourier Analysis," *J. Assoc. Comp. Mach. 12*, 95–113.

The accumulation of the right-hand sides must be done with great care, for otherwise there would be a significant loss of accuracy. A stable way of doing this is discribed in

O. Buneman (1969). "A Compact Non-Iterative Poisson Solver," Report 294, Stanford University Institute for Plasma Research, Stanford, Calif.

More recent literature concerned with cyclic reduction includes

B. L. Buzbee, F. W. Dorr, J. A. George, and G. H. Golub (1971). "The Direct Solution of the Discrete Poisson Equation on Irregular Regions," *SIAM J. Num. Anal. 8*, 722–36.

B. L. Buzbee and F. W. Dorr (1974). "The Direct Solution of the Biharmonic Equation on Rectangular Regions and the Poisson Equation on Irregular Regions," *SIAM J. Num. Anal. 11*, 753–63.

P. Concus and G. H. Golub (1973). "Use of Fast Direct Methods for the Efficient Numerical Solution of Nonseparable Elliptic Equations," *SIAM J. Num. Anal. 10*, 1103–20.

F. W. Dorr (1970). "The Direct Solution of the Discrete Poisson Equation on a Rectangle," *SIAM Review 12*, 248–63.

F. W. Dorr (1973). "The Direct Solution of the Discrete Poisson Equation in $O(n^2)$ Operations," *SIAM Review 15*, 412–415.

Various generalizations and extensions to cyclic reduction have been proposed to handle the problem of irregular boundaries:

M. A. Diamond and D. L. V. Ferreira (1976). "On a Cyclic Reduction Method for the Solution of Poisson's Equation," *SIAM J. Num. Anal. 13*, 54–70;

the problem of arbitrary dimension:

R. A. Sweet (1974). "A Generalized Cyclic Reduction Algorithm," *SIAM J. Num. Anal. 11*, 506–20;

R. A. Sweet (1977). "A Cyclic Reduction Algorithm for Solving Block Tridiagonal Systems of Arbitrary Dimension," *SIAM J. Num. Anal. 14*, 706-20;

and the problem of periodic end conditions:

P. N. Swarztrauber and R. A. Sweet (1973). "The Direct Solution of the Discrete Poisson Equation on a Disk," *SIAM J. Num. Anal. 10*, 900-907.

A good overview of cyclic reduction is

D. Heller (1976). "Some Aspects of the Cyclic Reduction Algorithm for Block Tridiagonal Linear Systems," *SIAM J. Num. Anal. 13*, 484-96.

For certain matrices that arise in conjunction with elliptic partial differential equations, block elimination corresponds to rather natural operations on the underlying mesh. A classical example of this is the method of nested dissection described in

A. George (1973). "Nested Dissection of a Regular Finite Element Mesh," *SIAM J. Num. Anal. 10*, 345-63.

Finally, we mention the general survey in

J. R. Bunch (1976). "Block Methods for Solving Sparse Linear Systems," in *Sparse Matrix Computations*, ed. J. R. Bunch and D. J. Rose, Academic Press, New York.

Sec. 5.6. Vandermonde Systems

Suppose $x_0, \ldots, x_n \in \mathbb{R}$. A matrix $V \in R^{(n+1) \times (n+1)}$ of the form

$$V = V(x_0, \ldots, x_n) = \begin{bmatrix} 1 & 1 & \cdots & 1 \\ x_0 & x_1 & \cdots & x_n \\ \vdots & \vdots & & \vdots \\ x_0^n & x_1^n & \cdots & x_n^n \end{bmatrix}$$

is said to be a *Vandermonde matrix*. In this section we show how the systems

(5.6-1) $$V^T a = f = (f_0, \ldots, f_n)^T$$

and

(5.6-2) $$Vz = b = (b_0, \ldots, b_n)^T$$

can be solved in $O(n^2)$ flops.

Vandermonde systems arise in many approximation and interpolation problems. Indeed, the key to obtaining a fast Vandermonde solver is to recognize that solving (5.6-1) is equivalent to polynomial interpolation. This follows because if $V^T(a_0, \ldots, a_n)^T = f$ and

$$(5.6\text{-}3) \qquad\qquad p(x) = \sum_{j=0}^{n} a_j x^j,$$

then $p(x_i) = f_i$ for $i = 0, \ldots, n$.

Recall that if the x_i are distinct then there is a unique polynomial of degree n that interpolates $(x_0, f_0), \ldots, (x_n, f_n)$. Consequently, $V(x_0, \ldots, x_n)$ is nonsingular so long as the x_i are distinct. We will assume this throughout the section.

The first step in computing the a_j of (5.6-3) is to calculate the Newton representation

$$(5.6\text{-}4) \qquad\qquad p(x) = \sum_{k=0}^{n} c_k \prod_{i=0}^{k-1} (x - x_i).$$

The constants c_k are divided differences and may be determined as follows:

$$(5.6\text{-}5) \qquad
\begin{aligned}
& c_i := f_i \qquad (i = 0, \ldots, n) \\
& \text{For } k = 0, \ldots, n - 1 \\
& \qquad \text{For } i = n, \ldots, k + 1 \\
& \qquad\qquad c_i := (c_i - c_{i-1})/(x_i - x_{i-k-1})
\end{aligned}$$

See Conte and de Boor (1980, chap. 2).

The next task is to generate the a_i from the c_i. Define the polynomials $p_n(x), \ldots, p_0(x)$ by the iteration

$$
\begin{aligned}
& p_n(x) = c_n \\
& \text{For } k = n - 1, n - 2, \ldots, 1, 0 \\
& \qquad p_k(x) = c_k + (x - x_k)p_{k+1}(x)
\end{aligned}$$

and observe that $p_0(x) = p(x)$, the desired interpolant of (5.5-3). Writing

$$p_k(x) = a_k^{(k)} + a_{k+1}^{(k)}x + \cdots + a_n^{(k)}x^{n-k}$$

and equating like powers of x in the equation $p_k = c_k + (x - x_k)p_{k+1}$ gives the following recursion for the coefficients $a_i^{(k)}$:

$$
\begin{aligned}
& a_n^{(n)} = c_n \\
& \text{For } k = n - 1, \ldots, 0 \\
& \qquad a_k^{(k)} = c_k - x_k a_{k+1}^{(k+1)} \\
& \qquad \text{For } i = k + 1, \ldots, n - 1 \\
& \qquad\qquad a_i^{(k)} = a_i^{(k+1)} - x_k a_{i+1}^{(k+1)} \\
& \qquad a_n^{(k)} = a_n^{(k+1)}
\end{aligned}$$

Consequently, the coefficients $a_i = a_i^{(0)}$ can be calculated as follows:

$$(5.6\text{-}6) \qquad
\begin{aligned}
& a_i := c_i \ (i = 0, \ldots, n) \\
& \text{For } k = n - 1, \ldots, 0 \\
& \qquad \text{For } i = k, \ldots, n - 1 \\
& \qquad\qquad a_i := a_i - x_k a_{i+1}
\end{aligned}$$

Combining this iteration with (5.6-5) renders the following algorithm.

ALGORITHM 5.6-1.   Given distinct real scalars $x_0, \ldots, x_n$ and $f = (f_0, \ldots, f_n)^{\mathrm{T}}$ $\in \mathbb{R}^{n+1}$, the following algorithm overwrites f with the solution $(a_0, \ldots, a_n)^{\mathrm{T}}$ to the Vandermonde system $V(x_0, \ldots, x_n)^{\mathrm{T}} a = f$.

> For $k = 0, \ldots, n - 1$
> > For $i = n, \ldots, k + 1$
> > > $f_i := (f_i - f_{i-1})/(x_i - x_{i-k-1})$
> For $k = n - 1, \ldots, 0$
> > For $i = k, \ldots, n - 1$
> > > $f_i := f_i - f_{i+1} x_k$

This algorithm requires n^2 flops.

EXAMPLE 5.6-1.   Suppose Algorithm 5.6-1 is used to solve

$$
\begin{bmatrix} 1 & 1 & 1 & 1 \\ 1 & 2 & 4 & 8 \\ 1 & 3 & 9 & 27 \\ 1 & 4 & 16 & 64 \end{bmatrix}
\begin{bmatrix} a_0 \\ a_1 \\ a_2 \\ a_3 \end{bmatrix}
=
\begin{bmatrix} 10 \\ 26 \\ 58 \\ 112 \end{bmatrix}.
$$

The first "k-loop" computes the Newton representation of $p(x)$:

$$ p(x) = 10 + 16(x - 1) + 8(x - 1)(x - 2) + (x - 1)(x - 2)(x - 3). $$

The second k-loop computes $a = (4, 3, 2, 1)^{\mathrm{T}}$ from $(10, 16, 8, 1)^{\mathrm{T}}$.

Now consider the system $Vz = b$. To derive an efficient algorithm for this problem we express the above calculations in matrix-vector language. Define the lower bidiagonal matrix $L_k(\alpha) \in \mathbb{R}^{(n+1) \times (n+1)}$ by

$$
L_k(\alpha) =
\left[
\begin{array}{c|cccc}
I_k & & & 0 & \\
\hline
& 1 & & & 0 \\
0 & -\alpha & 1 & & \\
& & \ddots & \ddots & \\
& 0 & & -\alpha & 1
\end{array}
\right]
$$

and the diagonal matrix D_k by

$$D_k = \text{diag}(\underbrace{1, \ldots, 1}_{k+1}, x_{k+1} - x_0, \ldots, x_n - x_{n-k-1}).$$

With these definitions it is easy to verify from (5.6-5) that if $f = (f_0, \ldots, f_n)^{\mathrm{T}}$ and $c = (c_0, \ldots, c_n)^{\mathrm{T}}$ then

$$c = U^{\mathrm{T}}f,$$

where U is the upper triangular matrix defined by

$$U^{\mathrm{T}} = D_{n-1}^{-1}L_{n-1}(1) \cdots D_0^{-1}L_0(1).$$

Similarly, from (5.6-6) we have

$$a = L^{\mathrm{T}}c,$$

where L is the unit lower triangular matrix defined by

$$L^{\mathrm{T}} = L_0(x_0)^{\mathrm{T}} \cdots L_{n-1}(x_{n-1})^{\mathrm{T}}.$$

Thus, $a = L^{\mathrm{T}}U^{\mathrm{T}}f$ where $V^{-\mathrm{T}} = L^{\mathrm{T}}U^{\mathrm{T}}$. In other words, Algorithm 5.6-1 solves $V^{\mathrm{T}}a = f$ by tacitly computing the "U-L" decomposition of V^{-1}.

Consequently, the solution to the system $Vz = b$ is given by

$$z = V^{-1}b = U(Lb)$$
$$= [L_0(1)^{\mathrm{T}}D_0^{-1} \cdots L_{n-1}(1)^{\mathrm{T}}D_{n-1}^{-1}][L_{n-1}(x_{n-1}) \cdots L_0(x_0)b]$$

This observation gives rise to the following algorithm.

ALGORITHM 5.6-2.   Given distinct real scalars $x_0, \ldots, x_n$ and $b = (b_0, \ldots, b_n)^{\mathrm{T}}$ $\in \mathbb{R}^{n+1}$, the following algorithm overwrites b with the solution $z = (z_0, \ldots, z_n)^{\mathrm{T}}$ to the Vandermonde system $Vz = b$.

> For $k = 0, \ldots, n - 1$
> > For $i = n, \ldots, k + 1$
> > > $b_i := b_i - x_k b_{i-1}$
> > For $k = n - 1, \ldots, 0$
> > > For $i = k + 1, \ldots, n$
> > > > $b_i := b_i / (x_i - x_{i-k-1})$
> > > For $i = k, \ldots, n - 1$
> > > > $b_i := b_i - b_{i+1}$

This algorithm requires n^2 flops.

EXAMPLE 5.6-2.   Suppose Algorithm 5.6-2 is used to solve

$$
\begin{bmatrix} 1 & 1 & 1 & 1 \\ 1 & 2 & 3 & 4 \\ 1 & 4 & 9 & 16 \\ 1 & 8 & 27 & 64 \end{bmatrix}
\begin{bmatrix} z_0 \\ z_1 \\ z_2 \\ z_3 \end{bmatrix}
=
\begin{bmatrix} 0 \\ -1 \\ 3 \\ 35 \end{bmatrix} .
$$

The first k-loop computes the vector

$$
\begin{bmatrix} 0 \\ -1 \\ 6 \\ 6 \end{bmatrix}
= L_3(3)\, L_2(2)\, L_1(1)
\begin{bmatrix} 0 \\ -1 \\ 3 \\ 35 \end{bmatrix}
$$

The second k-loop then calculates

$$
\begin{bmatrix} 3 \\ -4 \\ 0 \\ 1 \end{bmatrix}
= L_0(1)^{\mathrm{T}} D_0^{-1} L_1(1)^{\mathrm{T}} D_1^{-1} L_2(1)^{\mathrm{T}} D_2^{-1}
\begin{bmatrix} 0 \\ -1 \\ 3 \\ 35 \end{bmatrix} .
$$

Algorithms 5.6-1 and 5.6-2 are discussed and analyzed in Bjorck and Pereyra (1970). Their experience is that these algorithms frequently produce surprisingly accurate solutions, even when V is ill-conditioned. They also show how to update the solution when a new coordinate pair (x_{n+1}, f_{n+1}) is added to the set of points to be interpolated, and how to solve *confluent Vandermonde systems*, i.e., systems involving matrices like

$$
V = V(x_0, x_1, x_1, x_3) =
\begin{bmatrix}
1 & 1 & 0 & 1 \\
x_0 & x_1 & 1 & x_3 \\
x_0^2 & x_1^2 & 2x_1^2 & x_3^2 \\
x_0^3 & x_1^3 & 3x_1^2 & x_3^3
\end{bmatrix} .
$$

Problems

P5.6-1. Show that if $V = V(x_0, \ldots, x_n)$ then

$$
\det(V) = \prod_{n \geqslant i > j \geqslant 0} (x_i - x_j).
$$

P5.6-2. (Gautschi 1975a) Verify the following inequality for the $n = 1$ case:

$$\| V(x_0, \ldots, x_n)^{-1} \|_\infty \leqslant \max_{0 \leqslant k \leqslant n} \prod_{\substack{i=0 \\ i \neq k}}^{n} \frac{1 + |x_i|}{|x_k - x_i|}.$$

(Equality results if the x_i are all on the same ray in the complex plane.)

P5.6-3. Use Algorithm 5.6-2 to solve the Vandermonde system

$$V(1, 2, 3, 4)z = (1, 2, 3, 4)^T.$$

Notes and References for Sec. 5.6

Our discussion of Vandermonde linear systems is drawn from the papers

A. Bjorck and V. Pereyra (1970). "Solution of Vandermonde Systems of Equations," *Math. Comp. 24*, 893–903.

A. Bjorck and T. Elfving (1973). "Algorithms for Confluent Vandermonde Systems," *Numer. Math. 21*, 130–37.

The latter reference includes an Algol procedure. See also

G. Galimberti and V. Pereyra (1970). "Numerical Differentiation and the Solution of Multidimensional Vandermonde Systems," *Math. Comp. 24*, 357–64.

G. Galimberti and V. Pereyra (1971). "Solving Confluent Vandermonde Systems of Hermite Type," *Numer. Math. 18*, 44–60.

H. Van de Vel (1977). "Numerical Treatment of a Generalized Vandermonde System of Equations," *Lin. Alg. & Its Applic. 17*, 149–74.

Interesting theoretical results concerning the condition of Vandermonde systems may be found in

W. Gautschi (1975a). "Norm Estimates for Inverses of Vandermonde Matrices," *Numer. Math. 23*, 337–47.

W. Gautschi (1975b). "Optimally Conditioned Vandermonde Matrices," *Numer. Math. 24*, 1–12.

In

G. H. Golub and W. P. Tang (1981). "The Block Decomposition of a Vandermonde Matrix and Its Applications," *BIT 21*, 505–17.

a block Vandermonde algorithm is given that enables one to circumvent complex arithmetic in certain interpolation problems.

The divided difference computations we discussed are detailed in chapter 2 of

S. D. Conte and C. de Boor (1980). *Elementary Numerical Analysis: An Algorithmic Approach*, 3rd ed., McGraw-Hill, New York.

Sec. 5.7. Toeplitz Systems

Matrices whose entries are constant along each diagonal arise in many applications and are called *Toeplitz matrices*. Formally, $T \in \mathbb{R}^{n \times n}$ is Toeplitz if there exist scalars $r_{-n+1}, \ldots, r_0, \ldots, r_{n-1}$ such that $a_{ij} = r_{j-i}$ for all i and j. Thus,

$$
T = \begin{bmatrix}
r_0 & r_1 & r_2 & r_3 \\
r_{-1} & r_0 & r_1 & r_2 \\
r_{-2} & r_{-1} & r_0 & r_1 \\
r_{-3} & r_{-2} & r_{-1} & r_0
\end{bmatrix} \qquad n = 4
$$

is Toeplitz.

Toeplitz matrices belong to the larger class of *persymmetric matrices*. We say that $B \in \mathbb{R}^{n \times n}$ is persymmetric if it is symmetric about its northeast-southwest diagonal, i.e., if $b_{ij} = b_{n-j+1,n-i+1}$ for all i and j. This is equivalent to requiring

$$
B = EB^T E,
$$

where

$$
E = [e_n, \ldots, e_1]
$$

is the *n-by-n exchange matrix*. Note that (a) Toeplitz matrices are persymmetric and (b) the inverse of a nonsingular Toeplitz matrix is persymmetric. In this section we show how the careful exploitation of (b) can enable one to solve Toeplitz systems in $O(n^2)$ time. The discussion will be restricted to the important case when T is also symmetric and positive definite.

Assume, then, that we have scalars $r_1, \ldots, r_{n-1}$ such that the matrices

$$
T_k = \begin{bmatrix}
1 & r_1 & r_2 & \cdots & r_{k-1} \\
r_1 & 1 & r_1 & & \\
\vdots & & \ddots & \ddots & \vdots \\
& & & \ddots & \\
r_{k-1} & & \cdots & & 1
\end{bmatrix} \qquad k = 1, \ldots, n
$$

are all positive definite. (There is no loss of generality in normalizing the diagonal.) Three algorithms will be described:

   (i) Durbin's algorithm for the *Yule-Walker problem* $T_n y = -(r_1, \ldots, r_n)^T$.

   (ii) Levinson's algorithm for the general r.h.s. $T_n x = b$ problem.

   (iii) Trench's algorithm for computing $B = T_n^{-1}$.

In deriving these methods, we will denote the k-by-k exchange matrix by E_k, i.e., $E_k = [e_k^{(k)}, \ldots, e_1^{(k)}]$.

We begin by presenting Durbin's algorithm for the Yule-Walker equations which arise in conjunction with certain linear prediction problems. Suppose for some k that satisfies $1 \leqslant k \leqslant n - 1$ we have solved the k-th order Yule-Walker system

$$T_k y = -(r_1, \ldots, r_k)^T = -r^T.$$

We now show how the $(k + 1)$-st order system

$$\begin{bmatrix} T_k & E_k r \\ r^T E_k & 1 \end{bmatrix} \begin{bmatrix} z \\ \alpha \end{bmatrix} = \begin{bmatrix} -r \\ -r_{k+1} \end{bmatrix}$$

can be solved in $O(k)$ flops. First observe that

$$z = T_k^{-1}(-r - \alpha E_k r) = y - \alpha T_k^{-1} E_k r$$

and

$$\alpha = -r_{k+1} - r^T E_k z.$$

Since T_k^{-1} is persymmetric, $T_k^{-1} E_k = E_k T_k^{-1}$ and thus,

$$z = y + \alpha E_k y.$$

By substituting this into the above expression for α we find

$$\alpha = -r_{k+1} - r^T E_k (y + \alpha E_k y) = -(r_{k+1} + r^T E_k y)/(1 + r^T y).$$

The denominator is positive because T_{k+1} is positive definite and because

$$\begin{bmatrix} I & E_k y \\ 0 & 1 \end{bmatrix}^T \begin{bmatrix} T_k & E_k r \\ r^T E_k & 1 \end{bmatrix} \begin{bmatrix} I & E_k y \\ 0 & 1 \end{bmatrix} = \begin{bmatrix} T_k & 0 \\ 0 & 1 + r^T y \end{bmatrix}.$$

We have illustrated the k-th step of an algorithm proposed by Durbin (1960). It proceeds by generating solutions to the Yule-Walker systems

$$T_k y^{(k)} = -r^{(k)} = -(r_1, \ldots, r_k)^T$$

for $k = 1, \ldots, n$:

(5.7-1)

$$y^{(1)} = -r_1$$
$$\text{For } k = 1, \ldots, n - 1$$
$$\beta_k = 1 + r^{(k)T} y^{(k)}$$
$$\alpha_k = -(r_{k+1} + r^{(k)T} E_k y^{(k)})/\beta_k$$
$$z^{(k)} = y^{(k)} + \alpha_k E_k y^{(k)}$$
$$y^{(k+1)} = \begin{bmatrix} z^{(k)} \\ \alpha_k \end{bmatrix}$$

As it stands, this algorithm would require $\frac{3}{2} n^2$ flops to generate $y = y^{(n)}$. It is possible, however, to reduce the amount of work even further by noting that

$$\beta_k = 1 + r^{(k)\mathrm{T}} y^{(k)}$$

$$= 1 + [r^{(k-1)\mathrm{T}}, r_k] \begin{bmatrix} y^{(k-1)} + \alpha_{k-1} E_{k-1} y^{(k-1)} \\ \\ \alpha_{k-1} \end{bmatrix}$$

$$= 1 + r^{(k-1)\mathrm{T}} y^{(k-1)} + \alpha_{k-1} (r^{(k-1)\mathrm{T}} E_{k-1} y^{(k-1)} + r_k)$$

$$= \beta_{k-1} + \alpha_{k-1} (-\beta_{k-1}\alpha_{k-1}) = (1 - \alpha_{k-1}^2)\beta_{k-1}.$$

Using this recursion we obtain the following algorithm.

ALGORITHM 5.7-1 (*Durbin 1960*).   Given real numbers $1 = r_0, r_1, \ldots, r_n$ and that $T = (r_{|i-j|}) \in \mathbb{R}^{n \times n}$ is positive definite, the following algorithm computes $y \in \mathbb{R}^n$ such that $Ty = -(r_1, \ldots, r_n)^\mathrm{T}$.

$$y_1 := -r_1$$
$$\beta := 1$$
$$\alpha := -r_1$$
For $k = 1, \ldots, n - 1$
$$\qquad \beta := (1 - \alpha^2)\beta$$
$$\qquad \alpha := -\left(r_{k+1} + \sum_{i=1}^{k} r_{k+1-i} y_i\right)\bigg/ \beta$$
$$\qquad \text{For } i = 1, \ldots, k$$
$$\qquad\qquad z_i := y_i + \alpha y_{k+1-i}$$
$$\qquad y_i := z_i \qquad (i = 1, \ldots, k)$$
$$\qquad y_{k+1} := \alpha$$

This algorithm requires n^2 flops. We have included an auxilliary vector z for clarity, but it can be dispensed with in practice.

EXAMPLE 5.7-1.   Suppose we wish to solve the Yule-Walker system

$$\begin{bmatrix} 1 & .5 & .2 \\ .5 & 1 & .5 \\ .2 & .5 & 1 \end{bmatrix} \begin{bmatrix} y_1 \\ y_2 \\ y_3 \end{bmatrix} = - \begin{bmatrix} .5 \\ .2 \\ .1 \end{bmatrix}$$

using the Durbin algorithm. After one pass through the loop we obtain

$$\alpha = \tfrac{1}{15}, \qquad \beta = \tfrac{3}{4}, \quad \text{and} \quad y = \begin{bmatrix} -\tfrac{8}{15} \\ \tfrac{1}{15} \end{bmatrix}.$$

We then compute

$$\beta := (1 - \alpha^2)\beta = \tfrac{56}{75}$$

$$\alpha := -(r_3 + r_2 y_1 + r_1 y_2)/\beta = -\tfrac{1}{28}$$

$$z_1 := y_1 + \alpha y_2 = -\tfrac{225}{420}$$

$$z_2 := y_2 + \alpha y_1 = -\tfrac{36}{420},$$

giving the final solution $y = \tfrac{1}{140}(-75, 12, -5)^{\mathrm{T}}$.

With a marginal amount of extra work, it is possible to solve a symmetric positive definite Toeplitz system having an arbitrary right-hand side. Suppose for some k satisfying $1 \leqslant k < n$ that we have solved the system

$$(5.7\text{-}2) \qquad T_k x = b = (b_1, \ldots, b_k)^{\mathrm{T}}$$

and that we now wish to solve

$$(5.7\text{-}3) \qquad \begin{bmatrix} T_k & E_k r \\ r^{\mathrm{T}} E_k & 1 \end{bmatrix} \begin{bmatrix} v \\ \mu \end{bmatrix} = \begin{bmatrix} b \\ b_{k+1} \end{bmatrix}.$$

Here, $r = (r_1, \ldots, r_k)^{\mathrm{T}}$ as above. Assume that we also have at our disposal the solution to the k-th order Yule-Walker system $T_k y = -r$. Since

$$v = T_k^{-1}(b - \mu E_k r) = x + \mu E_k y,$$

it follows that

$$\mu = b_{k+1} - r^{\mathrm{T}} E_k v = b_{k+1} - r^{\mathrm{T}} E_k x - \mu r^{\mathrm{T}} y = (b_{k+1} - r^{\mathrm{T}} E_k x)/(1 + r^{\mathrm{T}} y).$$

Consequently, we can effect the transition from (5.7-2) to (5.7-3) in $O(k)$ flops.

This suggests that we can efficiently solve the system $T_n x = b$ by solving the systems

$$T_k x^{(k)} = b^{(k)} = (b_1, \ldots, b_k)^{\mathrm{T}}$$

and

$$T_k y^{(k)} = -r^{(k)} = -(r_1, \ldots, r_k)^{\mathrm{T}}$$

"in parallel" for $k = 1, \ldots, n$. This is the gist of the following algorithm.

ALGORITHM 5.7-2 (*Levinson 1947*).   Given $b \in \mathbb{R}^n$, scalars $1 = r_0, r_1, \ldots, r_{n-1}$, and that $T = (r_{|i-j|}) \in \mathbb{R}^{n \times n}$ is positive definite, the following algorithm computes $x \in \mathbb{R}^n$ such that $Tx = b$.

$$y_1 := -r_1$$
$$x_1 := b_1$$
$$\beta := 1$$
$$\alpha := -r_1$$

For $k = 1, \ldots, n - 1$

$\quad \beta := (1 - \alpha^2)\beta$

$$\mu := \left(b_{k+1} - \sum_{i=1}^{k} r_i x_{k+1-i}\right) \Big/ \beta$$

$\quad v_i := x_i + \mu y_{k+1-i} \qquad (i = 1, \ldots, k)$

$\quad x_i := v_i \qquad (i = 1, \ldots, k)$

$\quad x_{k+1} := \mu$

$\quad$ If $k < n - 1$

$\qquad$ *then*

$$\alpha := -\left(r_{k+1} + \sum_{i=1}^{k} r_i y_{k+1-i}\right)\Big/\beta$$

$$z_i := y_i + \alpha y_{k+1-i} \qquad (i = 1, \ldots, k)$$

$$y_i := z_i \qquad (i = 1, \ldots, k)$$

$$y_{k+1} := \alpha$$

This algorithm requires $2n^2$ flops. The vectors z and v are for clarity and may be dispensed with.

EXAMPLE 5.7-2.   Suppose we wish to solve the symmetric positive definite Toeplitz system

$$\begin{bmatrix} 1 & .5 & .2 \\ .5 & 1 & .5 \\ .2 & .5 & 1 \end{bmatrix} \begin{bmatrix} x_1 \\ x_2 \\ x_3 \end{bmatrix} = \begin{bmatrix} 4 \\ -1 \\ 3 \end{bmatrix}$$

using the above algorithm. After one pass through the loop we obtain

$$\alpha = \tfrac{1}{15}, \qquad \beta = \tfrac{3}{4}, \qquad y = \begin{bmatrix} -\frac{8}{15} \\ \frac{1}{15} \end{bmatrix}, \quad \text{and} \quad x = \begin{bmatrix} 6 \\ -4 \end{bmatrix}.$$

We then compute

$$\beta := (1 - \alpha^2)\beta = \tfrac{56}{75}$$

$$\mu := (b_3 - r_1 x_2 - r_2 x_1)/\beta = \tfrac{285}{56}$$

$$v_1 := x_1 + \mu y_2 = \tfrac{355}{56}$$

$$v_2 := x_2 + \mu y_1 = -\tfrac{376}{56}$$

giving the final solution $x = \tfrac{1}{56}(355, -376, 285)^{\mathrm{T}}$.

One of the most surprising properties of a symmetric positive definite Toeplitz matrix T_n is that its inverse can be calculated in $O(n^2)$ flops. To derive the algorithm for doing this, partition T_n^{-1} as follows

$$
(5.7-4) \qquad T_n^{-1} = \begin{bmatrix} A & Er \\ r^T E & 1 \end{bmatrix}^{-1} = \begin{bmatrix} B & v \\ v^T & \gamma \end{bmatrix} ,
$$

where $A = T_{n-1}$, $E = E_{n-1}$, and $r = (r_1, \ldots, r_{n-1})^T$. From the equation

$$
\begin{bmatrix} A & Er \\ r^T E & 1 \end{bmatrix} \begin{bmatrix} v \\ \gamma \end{bmatrix} = \begin{bmatrix} 0 \\ 1 \end{bmatrix}
$$

it follows that

$$
Av = -\gamma Er = -\gamma E(r_1, \ldots, r_{n-1})^T
$$

and

$$
\gamma = 1 - r^T E v.
$$

If y solves the $(n - 1)$-st Yule-Walker system $Ay = -r$, then these expressions imply that

$$
\gamma = 1/(1 + r^T y)
$$

$$
v = \gamma E y.
$$

Thus, the last row and column of T_n^{-1} are readily obtained.

It remains for us to develop working formulae for the entries of the submatrix B in (5.7-4). Since

$$
AB + Erv^T = I_{n-1},
$$

it follows that

$$
B = A^{-1} - (A^{-1}Er)v^T = A^{-1} + \frac{vv^T}{\gamma} .
$$

Now since $A = T_{n-1}$ is nonsingular and Toeplitz, its inverse is persymmetric. Thus,

$$
b_{ij} = (A^{-1})_{ij} + \frac{v_i v_j}{\gamma} = (A^{-1})_{n-j,n-i} + \frac{v_i v_j}{\gamma}
$$

$$
(5.7-5) \qquad = b_{n-j,n-i} - \frac{v_{n-j} v_{n-i}}{\gamma} + \frac{v_i v_j}{\gamma}
$$

$$
= b_{n-j,n-i} + \frac{1}{\gamma} (v_i v_j - v_{n-j} v_{n-i}).
$$

This indicates that although B is not persymmetric, we can readily compute an element b_{ij} from its reflection across B's northeast-southwest axis. Coupling this with the fact that A^{-1} is persymmetric enables us to determine B from its "edges" to its "interior."

Because the order of operations is rather cumbersome to describe, we preview the formal specification of the algorithm pictorially. To this end, assume that we know the last column and row of A^{-1}:

$$
A^{-1} = \left[\begin{array}{ccccc|c}
u & u & u & u & u & k \\
u & u & u & u & u & k \\
u & u & u & u & u & k \\
u & u & u & u & u & k \\
u & u & u & u & u & k \\
\hline
k & k & k & k & k & k
\end{array}\right],
$$

where $u \equiv$ unknown, $k \equiv$ known, and $n = 6$. Alternately exploiting the persymmetry of A^{-1} and the recursion (5.7-5), we can compute B as follows:

$$
\left[\begin{array}{ccccc|c}
k & k & k & k & k & k \\
k & u & u & u & u & k \\
k & u & u & u & u & k \\
k & u & u & u & u & k \\
k & u & u & u & u & k \\
\hline
k & k & k & k & k & k
\end{array}\right]
\rightarrow
\left[\begin{array}{ccccc|c}
k & k & k & k & k & k \\
k & u & u & u & k & k \\
k & u & u & u & k & k \\
k & u & u & u & k & k \\
k & k & k & k & k & k \\
\hline
k & k & k & k & k & k
\end{array}\right]
\rightarrow
\left[\begin{array}{ccccc|c}
k & k & k & k & k & k \\
k & k & k & k & k & k \\
k & k & u & u & k & k \\
k & k & u & u & k & k \\
k & k & k & k & k & k \\
\hline
k & k & k & k & k & k
\end{array}\right]
$$

$$
\rightarrow
\left[\begin{array}{ccccc|c}
k & k & k & k & k & k \\
k & k & k & k & k & k \\
k & k & u & k & k & k \\
k & k & k & k & k & k \\
k & k & k & k & k & k \\
\hline
k & k & k & k & k & k
\end{array}\right]
\rightarrow
\left[\begin{array}{ccccc|c}
k & k & k & k & k & k \\
k & k & k & k & k & k \\
k & k & k & k & k & k \\
k & k & k & k & k & k \\
k & k & k & k & k & k \\
\hline
k & k & k & k & k & k
\end{array}\right].
$$

Of course, when one is computing a matrix that is both symmetric and persymmetric, such as A^{-1}, it is only necessary to compute the "upper wedge" of the matrix—e.g.,

$$\begin{matrix} \text{x x x x x x} \\ \text{x x x x} \\ \text{x x} \end{matrix} \qquad n = 6$$

With this last observation, we are ready to present the overall algorithm.

ALGORITHM 5.7-3 (*Trench 1964*).   Given scalars $1 = r_0, r_1, \ldots, r_{n-1}$ with the property that $T_n = (r_{|i-j|})$ is positive definite, the following algorithm computes $B = T_n^{-1}$. Only those b_{ij} for which $i \leq j$ and $i + j \leq n + 1$ are computed.

Use Algorithm 5.7-1 to solve $T_{n-1} y = -(r_1, \ldots, r_{n-1})^T$.

$$\gamma := 1 \Big/ \left(1 + \sum_{k=1}^{n-1} r_i y_i \right)$$
$v_i := \gamma y_{n-i} \qquad (i = 1, \ldots, n-1)$
$b_{11} := \gamma$
$b_{1j} := v_{n+1-j} \qquad (j = 2, \ldots, n)$
For $i = 2, \ldots,$ floor$[(n-1)/2] + 1$
  For $j = i, \ldots, n - i + 1$
    $b_{ij} := b_{i-1,j-1} + (v_{n+1-j} v_{n+1-i} - v_{i-1} v_{j-1})/\gamma$

This algorithm requires $\frac{7}{4} n^2$ flops.

EXAMPLE 5.7-3.   If the above algorithm is applied to compute the inverse B of the positive definite Toeplitz matrix

$$T_3 = \begin{bmatrix} 1 & .5 & .2 \\ .5 & 1 & .5 \\ .2 & .5 & 1 \end{bmatrix},$$

we obtain $\gamma := \frac{75}{56}$, $b_{11} := \frac{75}{56}$, $b_{12} := -\frac{5}{7}$, $b_{13} := \frac{5}{56}$, and $b_{22} := \frac{12}{7}$.

Error analyses for the above algorithms have been performed by Cybenko (1978), and we briefly report on some of his findings.

The key quantities turn out to be the α_k in (5.7-1). In exact arithmetic these scalars satisfy

(5.7-6) $|\alpha_k| < 1$

and can be used to bound $\| T_n^{-1} \|_1$:

$$(5.7\text{-}7) \quad \max \left\{ \frac{1}{\prod_{1}^{n-1} (1 - \alpha_j^2)}, \frac{1}{\prod_{1}^{n-1} (1 - \alpha_j)} \right\} \leq \| T_n^{-1} \|_1 \leq \prod_{1}^{n-1} \frac{1 + |\alpha_j|}{1 - |\alpha_j|}$$

Moveover, the solution to the Yule-Walker equations $T_n y = -(r_1, \ldots, r_n)$ satisfies

(5.7-8)
$$\| y \|_1 = \prod_{k=1}^{n-1} (1 + \alpha_k) - 1,$$

provided all the α_k are non-negative.

Now if $\hat{x}$ is the computed Durbin solution to the Yule-Walker equations then $r_D = T_n x + (r_1, \ldots, r_n)^T$ can be bounded as follows:

$$\| r_D \| \cong \mathbf{u} \prod_{k=1}^{n} (1 + |\hat{\alpha}_k|),$$

where $\hat{\alpha}_k$ is the computer version of α_k. By way of comparison, since each $|r_i|$ is bounded by unity, it follows that

$$\| r_C \| \cong \mathbf{u} \| y \|_1,$$

where r_C is the residual associated with the computed solution obtained via Cholesky. Note that the two residuals are of comparable magnitude provided (5.7-8) holds. Experimental evidence suggests that this is the case even if some of the α_k are negative. Similar comments apply to the numerical behavior of the Levinson algorithm.

For the Trench method, the computed inverse $\hat{B}$ of T_n^{-1} can be shown to satisfy

$$\frac{\| T_n^{-1} - \hat{B} \|_1}{\| T_n^{-1} \|} \cong \mathbf{u} \prod_{1}^{n-1} \frac{1 + |\hat{\alpha}_k|}{1 - |\hat{\alpha}_k|}.$$

In light of (5.7-7) we see that the right-hand side is an approximate upper bound for $\mathbf{u} \| T_n^{-1} \|_1$. This, in turn, is approximately the size of the relative error when T_n^{-1} is calculated using the Cholesky decomposition.

Problems

P5.6-1. For any $v \in \mathbb{R}^n$, define the vectors v_+ and v_- by
$$v_+ = \tfrac{1}{2}(v + E_n v)$$
$$v_- = \tfrac{1}{2}(v - E_n v).$$

Suppose $A \in \mathbb{R}^{n \times n}$ is symmetric and persymmetric. Show that if $Ax = b$ then $Ax_+ = b_+$ and $Ax_- = b_-$.

P5.7-2. Under what conditions is a Householder matrix persymmetric?

P5.7-3. Suppose $z \in \mathbb{R}^n$ and that $S = [e_2, \ldots, e_n, e_1] \in \mathbb{R}^{n \times n}$. Show that if $X = [z, Sz, \ldots, S^{n-1}z]$ then $X^T X$ is Toeplitz.

P5.7-4. Suppose the matrix A in Algorithm 5.3-6 is Toeplitz. Show that $\lim_{n \to \infty} d_n$ and $\lim_{n \to \infty} e_n$ exist.

P5.7-5. Show that the product of two lower triangular Toeplitz matrices is Toeplitz.

P5.7-6. Give an algorithm for determining $\mu \in \mathbb{R}$ such that $T_n + \mu(e_n e_1^T + e_1 e_n^T)$ is singular. Assume $T_n = (r_{|i-j|})$ is positive definite, where $r_0 = 1$.

P5.7-7. Rewrite Algorithm 5.7-2 so that it does not require the vectors z and v.

P5.7-8. Give an algorithm for computing $\kappa_\infty(T_k)$, where $k = 1, \ldots, n$.

P5.7-9. Let $U \in \mathbb{R}^{n \times n}$ be a unit upper triangular matrix with the property that

$$
U e_k = \begin{bmatrix} E_{k-1} y^{(k-1)} \\ 0 \end{bmatrix}, \qquad k = 2, \ldots, n
$$

where $y^{(k)}$ is defined by (5.7-1). Show $U^T T_n U = \text{diag}(1, \alpha_1, \ldots, \alpha_{n-1})$.

Notes and References for Sec. 5.7

The original references for the three algorithms described in this section are

J. Durbin (1960). "The Fitting of Time Series Models," *Rev. Inst. Int. Stat. 28*, 233-43.

N. Levinson (1947). "The Weiner RMS Error Criterion in Filter Design and Prediction," *J. Math. Phys. 25*, 261-78.

W. F. Trench (1964). "An Algorithm for the Inversion of Finite Toeplitz Matrices," *J. SIAM 12*, 515-22.

A more detailed description of the nonsymmetric Trench algorithm is given in

S. Zohar (1969). "Toeplitz Matrix Inversion: The Algorithm of W. F. Trench," *J. Assoc. Comp. Mach. 16*, 592-601.

Other references pertaining to Toeplitz matrix inversion include

W. F. Trench (1974). "Inversion of Toeplitz Band Matrices," *Math. Comp. 28*, 1089-95.

G. A. Watson (1973). "An Algorithm for the Inversion of Block Matrices of Toeplitz Form," *J. Assoc. Comp. Mach. 20*, 409-15.

The error bounds that we cited were taken from the roundoff analysis in

G. Cybenko (1978). "Error Analysis of Some Signal Processing Algorithms," Ph.D. thesis, Princeton University.

G. Cybenko (1980). "The Numerical Stability of the Levinson-Durbin Algorithm for Toeplitz Systems of Equations," *SIAM J. Sci. & Stat. Comp. 1*, 303-10.

J. Markel and A. Gray (1974). "Fixed-Point Truncation Arithmetic Implementation of a Linear Predication Autocorrelation Vocoder," *IEEE Trans. ASSP 22*, 273-81.

$O(n^2)$ triangular factorization methods for Toeplitz systems also exist:

J. L. Phillips (1971). "The Triangular Decomposition of Hankel Matrices," *Math. Comp. 25*, 599–602.

J. Rissanen (1973). "Algorithms for Triangular Decomposition of Block Hankel and Toeplitz Matrices with Application to Factoring Positive Matrix Polynomials," *Math. Comp. 27*, 147–54.

Finally, we mention the following references which describe some important Toeplitz matrix applications:

J. Makhoul (1975). "Linear Predication: A Tutorial Review," *Proc. IEEE 63 (4)*, 561–80.

J. Markel and A. Gray (1976). *Linear Prediction of Speech*, Springer-Verlag, Berlin and New York.

A. V. Oppenheim (1978). *Applications of Digital Signal Processing*, Prentice-Hall, Englewood Cliffs.

Orthogonalization and Least Squares Methods

This chapter is primarily concerned with the least squares solution of overdetermined systems of equations, i.e., the minimization of $\| Ax - b \|_2$ where $A \in \mathbb{R}^{m \times n}$ ($m > n$) and $b \in \mathbb{R}^m$. The mathematical background for this problem is provided in §6.1. One tactic for solution is to convert the original least squares problem into an equivalent, easy-to-solve problem using orthogonal transformations. Algorithms of this type based on Householder and Givens transformations are described in §6.2 and §6.3. The basic idea is to use these transformations to compute the factorization $A = QR$, where Q is orthogonal and R is upper triangular. This is mathematically equivalent to applying Gram-Schmidt to the columns of A.

In the next two sections we consider methods for handling the difficult situation when A is rank deficient (or nearly so). QR with column pivoting and the SVD algorithm are featured.

In §6.6 we discuss several steps that can be taken to improve the quality of a computed least squares solution. Some remarks about underdetermined systems are offered in §6.7.

Reading Path

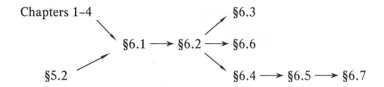

Sec. 6.1. Mathematical Properties of the Least Squares Problem

Consider the problem of finding a vector $x \in \mathbb{R}^n$ such that $Ax = b$ where $A \in \mathbb{R}^{m \times n}$ and $b \in \mathbb{R}^m$ are given and $m > n$. When there are more equations than unknowns, we say that the system $Ax = b$ is *overdetermined*. Usually an overdetermined system has no exact solution, since b must be an element of $R(A)$, a proper subspace of $\mathbb{R}^m$.

This suggests that we strive to minimize $\|Ax - b\|_p$ for some suitable choice of p. Different norms render different optimum solutions. For example, if $A = (1, 1, 1)^T$ and $b = (b_1, b_2, b_3)^T$ and $b_1 \geqslant b_2 \geqslant b_3 \geqslant 0$, then it can be verified that

$$p = 1 \Rightarrow x_{\text{opt}} = b_2$$

$$p = 2 \Rightarrow x_{\text{opt}} = (b_1 + b_2 + b_3)/3$$

$$p = \infty \Rightarrow x_{\text{opt}} = (b_1 + b_3)/2.$$

Minimization in the 1-norm and ∞-norm is complicated by the fact that the function $f(x) = \|Ax - b\|_p$ is not differentiable for those values of p. However, much progress has been made in this area, and there are several good techniques available for 1-norm and ∞-norm minimization. See Bartels, Conn, and Sinclair (1978) and Bartels, Conn, and Charalambous (1978).

On the other hand, $\|Ax - b\|_2$ is a continuously differentiable function of x. The *least squares* (*LS*) problem

$$\min_x \|Ax - b\|_2 \qquad\qquad A \in \mathbb{R}^{m \times n}, b \in \mathbb{R}^m$$

has the added attraction that it can be converted to an equivalent problem

$$\min_x \|(Q^T A)x - (Q^T b)\|_2 \qquad\qquad Q^T Q = I_m$$

by premultiplying both A and b by an orthogonal matrix Q. In subsequent sections we show how to choose Q so that $Q^T A$ has "canonical" form.

In this section we concentrate on the analytical properties of the LS problem beginning with an examination of the set of minimizers **X** defined by

$$\mathbf{X} = \{x \in \mathbb{R}^n \mid \|Ax - b\|_2 = \min\}.$$

This set has a number of easily verified properties:

(6.1-1) $x \in \mathbf{X} \Leftrightarrow A^{\mathrm{T}}(b - Ax) = 0.$

(6.1-2) $\mathbf{X}$ is convex.

(6.1-3) $\mathbf{X}$ has a unique element x_{LS} having minimal 2-norm.

(6.1-4) $\mathbf{X} = \{x_{\mathrm{LS}}\} \Leftrightarrow \mathrm{rank}(A) = n.$

If $x \in \mathbb{R}^n$ then we refer to $r = b - Ax$ as its *residual*. Note that the first property asserts that the residual of an LS solution is orthogonal to the columns of A. The equations $A^{\mathrm{T}}(b - Ax) = 0$ are referred to as the *normal equations*. Properties (6.1-2) through (6.1-4) combine to say that if A is rank deficient, then the LS problem has an infinite number of solutions, but there is exactly one possessing minimal 2-norm. We denote this unique element by x_{LS} and the minimum sum of squares by ρ_{LS}^2:

$$\rho_{\mathrm{LS}}^2 = \|Ax_{\mathrm{LS}} - b\|_2^2.$$

Further insight into the LS problem is furnished by the SVD:

THEOREM 6.1-1.   Let

$$A = \sum_{i=1}^{r=\mathrm{rank}(A)} \sigma_i u_i v_i^{\mathrm{T}} \qquad U = [u_1, \ldots, u_m], \ V = [v_1, \ldots, v_n]$$

be the SVD of $A \in \mathbb{R}^{m \times n}$ $(m \geqslant n)$. If $b \in \mathbb{R}^m$ then

(6.1-5) $$x_{\mathrm{LS}} = \sum_{i=1}^{r} (u_i^{\mathrm{T}} b / \sigma_i) v_i,$$

(6.1-6) $$\rho_{\mathrm{LS}}^2 = \sum_{i=r+1}^{m} (u_i^{\mathrm{T}} b)^2.$$

Proof.   For any $x \in \mathbb{R}^n$ we have

$$\|Ax - b\|_2^2 = \|U^{\mathrm{T}} AV(V^{\mathrm{T}} x) - U^{\mathrm{T}} b\|_2^2$$

$$= \sum_{i=1}^{r} (\sigma_i \alpha_i - u_i^{\mathrm{T}} b)^2 + \sum_{i=r+1}^{m} (u_i^{\mathrm{T}} b)^2,$$

where $\alpha = V^{\mathrm{T}} x$. Clearly, if x solves the LS problem, then $\alpha_i = (u_i^{\mathrm{T}} b / \sigma_i)$ for $i = 1, \ldots, r$. If we set $\alpha_{r+1} = \cdots = \alpha_n = 0$, then $x = x_{\mathrm{LS}}$. $\square$

Figure 6.1-1 helps to clarify the relationship between the various quantities that arise in the LS problem. The vector Ax_{LS} "predicts" b and is the orthogonal projection of b onto the range of A. We define the angle θ between b and $R(A)$ by

(6.1-7)
$$\sin(\theta) = \frac{\rho_{LS}}{\|b\|_2}.$$

When this angle is zero we have a *zero residual* problem.

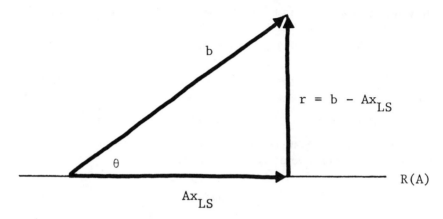

Figure 6.1-1.

Returning to Theorem 6.1-1, note that if we define the matrix A^+ by

$$A^+ = V\Sigma^+ U^{\mathrm{T}},$$

where

$$\Sigma^+ = \mathrm{diag}(\sigma_1^{-1}, \ldots, \sigma_r^{-1}, 0, \ldots, 0) \in \mathbb{R}^{n\times m},$$

then

$$x_{LS} = A^+b \quad \text{and} \quad \rho_{LS} = \|(I - AA^+)b\|_2.$$

A^+ is referred to as the *pseudo-inverse* of A. It is the unique minimal F-norm solution to

(6.1-8)
$$\min_{X\in\mathbb{R}^{n\times m}} \|AX - I_m\|_F.$$

If $\mathrm{rank}(A) = n$, then $A^+ = (A^{\mathrm{T}}A)^{-1}A^{\mathrm{T}}$, while if $m = n = \mathrm{rank}(A)$, then $A^+ = A^{-1}$. Typically, A^+ is defined to be the unique matrix $X \in \mathbb{R}^{n\times m}$ that satisfies the four *Moore-Penrose conditions*:

      (i) $AXA = A$ (iii) $(AX)^{\mathrm{T}} = AX$
      (ii) $XAX = X$ (iv) $(XA)^{\mathrm{T}} = XA$

These conditions amount to the requirement that AA^+ and A^+A be orthogonal projections onto $R(A)$ and $R(A^{\mathrm{T}})$, respectively.

We now examine the sensitivity of the LS problem, quoting first a result concerned with pseudo-inverse perturbation.

THEOREM 6.1-2.   If A and δA are in $\mathbb{R}^{m \times n}$, then

$$(6.1\text{-}9) \quad \|(A + \delta A)^+ - A^+\|_F \leqslant 2\|\delta A\|_F \max\{\|A^+\|_2^2, \|(A + \delta A)^+\|_2^2\}.$$

Proof.   See Wedin (1973) or Stewart (1975a). $\square$

Inequality (6.1-9) is a generalization of the Banach Lemma for nonsingular matrices:

$$\|(A + \delta A)^{-1} - A^{-1}\| \leqslant \|\delta A\|\,\|A^{-1}\|\,\|(A + \delta A)^{-1}\|.$$

However, unlike the square nonsingular case, the upper bound in (6.1-9) does not necessarily tend to zero as δA tends to zero.

EXAMPLE 6.1-1.   If

$$A = \begin{bmatrix} 1 & 0 \\ 0 & 0 \\ 0 & 0 \end{bmatrix} \quad \text{and} \quad \delta A = \begin{bmatrix} 0 & 0 \\ 0 & \epsilon \\ 0 & 0 \end{bmatrix},$$

then

$$A^+ = \begin{bmatrix} 1 & 0 & 0 \\ 0 & 0 & 0 \end{bmatrix} \quad \text{and} \quad (A + \delta A)^+ = \begin{bmatrix} 1 & 0 & 0 \\ 0 & 1/\epsilon & 0 \end{bmatrix}$$

and $\|A^+ - (A + \delta A)^+\|_2 = 1/\epsilon$.

The discontinuous nature of the pseudo-inverse complicates the perturbation analysis of x_{LS}. In the following theorem we simplify matters considerably by assuming that both A and its perturbation $A + \delta A$ have full column rank. Analogous to the linear equation problem, we show that if A's columns are nearly dependent then small changes in its elements can induce large changes in x_{LS}. To make this precise, we need to extend the notion of condition (cf. [2.5-3]) to general rectangular matrices. We do this as follows:

$$\kappa(A) = \|A\|\,\|A^+\|. \qquad\qquad A \in \mathbb{R}^{m \times n}$$

In the important case of 2-norm condition, this gives

$$\kappa_2(A) = \sigma_1(A)/\sigma_r(A)$$

where $r = \text{rank}(A)$.

THEOREM 6.1-3.   Suppose x, r, $\hat{x}$, and $\hat{r}$ satisfy

$$\|Ax - b\|_2 = \min \qquad\qquad r = b - Ax$$

$$\|(A + \delta A)\hat{x} - (b + \delta b)\|_2 = \min \qquad \hat{r} = (b + \delta b) - (A + \delta A)\hat{x},$$

where A and δA are in $\mathbb{R}^{m \times n}$ $(m \geqslant n)$ and $0 \neq b$ and δb are in $\mathbb{R}^m$. Assume

$$\epsilon = \max\left\{\frac{\|\delta A\|_2}{\|A\|_2}, \frac{\|\delta b\|_2}{\|b\|_2}\right\} < \frac{\sigma_n(A)}{\sigma_1(A)}$$

and that

$$\sin(\theta) = \frac{\rho_{\mathrm{LS}}}{\|b\|_2} \neq 1.$$

Then

(6.1-10)
$$\frac{\|\hat{x} - x\|_2}{\|x\|_2} \leqslant \epsilon\left\{\frac{2\kappa_2(A)}{\cos(\theta)} + \tan(\theta)\kappa_2(A)^2\right\} + O(\epsilon^2)$$

(6.1-11)
$$\frac{\|\hat{r} - r\|_2}{\|b\|_2} \leqslant \epsilon(1 + 2\kappa_2(A))\min\{1, m - n\} + O(\epsilon^2).$$

Proof.   The proof is somewhat cumbersome and is deferred to the end of the section. □

An interesting feature of the upper bound in (6.1-10) is the factor $\tan(\theta)\kappa_2(A)^2$. Thus, in nonzero residual problems it is the square of the condition that measures the sensitivity of x_{LS}. In contrast, residual sensitivity depends only linearly on $\kappa_2(A)$.

EXAMPLE 6.1-2.   If

$$A = \begin{bmatrix} 1 & 0 \\ 0 & 10^{-6} \\ 0 & 0 \end{bmatrix}, \quad \delta A = \begin{bmatrix} 0 & 0 \\ 0 & 0 \\ 0 & 10^{-8} \end{bmatrix}, \quad b = \begin{bmatrix} 1 \\ 0 \\ 1 \end{bmatrix}, \quad \text{and } \delta b = \begin{bmatrix} 0 \\ 0 \\ 0 \end{bmatrix},$$

then

$$x = \begin{bmatrix} 1 \\ 0 \end{bmatrix}, \hat{x} \cong \begin{bmatrix} 1 \\ .9999 * 10^4 \end{bmatrix}, r = \begin{bmatrix} 0 \\ 0 \\ 1 \end{bmatrix}, \text{ and } \hat{r} \cong \begin{bmatrix} 0 \\ -.9999 * 10^{-2} \\ .9999 * 10^0 \end{bmatrix}.$$

Since $\kappa_2(A) = 10^6$ we have

$$\frac{\|\hat{x} - x\|_2}{\|x\|_2} \cong .9999 * 10^4 \leqslant \kappa_2(A)^2 \frac{\|\delta A\|_2}{\|A\|_2} = 10^{12} * 10^{-8}$$

and

$$\frac{\|\hat{r} - r\|_2}{\|b\|_2} \cong .7070 * 10^{-2} \leqslant \kappa_2(A) \frac{\|\delta A\|_2}{\|A\|_2} = 10^6 * 10^{-8}.$$

We now use Theorem 6.1-3 to assess one of the most widely used methods for solving the LS problem, the method of normal equations. This technique involves the solution of the symmetric system $A^T A x = A^T b$. If rank$(A) = n$, then this system is positive definite and has solution $x = x_{LS}$ according to equation (6.1-1).

ALGORITHM 6.1-1: *Normal Equations.*   Given $A \in \mathbb{R}^{m \times n}$ with rank$(A) = n$ and $b \in \mathbb{R}^m$, this algorithm computes the solution to the LS problem $\min \|Ax - b\|_2$.

$C := A^T A$
$d := A^T b$

Compute the Cholesky factorization $C = GG^T$ (cf. Algorithm 5.2-1).
Solve $Gy = d$ and $G^T x = y$ (cf. Algorithms 4.1-1 and 4.1-2).

This algorithm requires $(n^2/2)(m + (n/3))$ flops.

Let us consider the accuracy of the computed normal equations solution $\hat{x}$. For clarity, assume that no roundoff errors occur during the formation of $C = A^T A$ and $d = A^T b$. (Typically, inner products are accumulated during this portion of the computation and so this is not a terribly unfair assumption.) It follows from what we know about the roundoff properties of the Cholesky factorization (cf. §5.2) that

$$(A^T A + E)\hat{x} = A^T b,$$

where $\|E\|_2 \cong \mathbf{u}\|A^T A\|_2$ and thus we can expect

$$\frac{\|\hat{x} - x_{LS}\|_2}{\|x_{LS}\|_2} \cong \mathbf{u}\kappa_2(A^T A) = \mathbf{u}\kappa_2(A)^2.$$

In other words, the accuracy of the computed normal equations solution depends on the square of the condition. In view of Theorem 6.1-3, this is consistent with the mathematical sensitivity of x_{LS} in nonzero residual LS problems. However, if $\rho_{LS} = \|Ax_{LS} - b\|_2 = 0$ (or is small), then the normal equation approach can introduce errors that are greater in magnitude than what we would expect with a stable algorithm.

EXAMPLE 6.1-3. If

$$A = \begin{bmatrix} 1 & 1 \\ 10^{-3} & 0 \\ 0 & 10^{-3} \end{bmatrix} \quad \text{and} \quad b = \begin{bmatrix} 2 \\ 10^{-3} \\ 10^{-3} \end{bmatrix},$$

then $\kappa_2(A) \cong 1.4 * 10^3$, $x_{LS} = (1, 1)^T$, and $\rho_{LS} = 0$. If Algorithm 6.1-1 is executed with $\beta = 10$, $t = 6$, chopped arithmetic, then a divide exception occurs during the solution process, since

$$fl(A^TA) = \begin{bmatrix} 1 & 1 \\ 1 & 1 \end{bmatrix}$$

is exactly singular. On the other hand, if $\beta = 10$, $t = 7$, chopped arithmetic is used, then $\hat{x} = (2.000001, 0)^T$ and $\|\hat{x} - x_{LS}\|_2 / \|x_{LS}\|_2 \cong \mathbf{u}\kappa_2(A)^2$.

Proof of Theorem 6.1-3. Let E and f be defined by $E = \delta A/\epsilon$ and $f = \delta b/\epsilon$. By hypothesis, $\|\delta A\|_2 < \sigma_n(A)$ and so by Corollary 2.3-3 we have rank$(A + tE) = n$ for all $t \in [0, \epsilon]$. It follows that the solution $x(t)$ to

$$(6.1-12) \qquad (A + tE)^T(A + tE)x(t) = (A + tE)^T(b + tf)$$

is continuously differentiable for all $t \in [0, \epsilon]$. Since $x = x(0)$ and $\hat{x} = x(\epsilon)$, we have

$$\hat{x} = x + \epsilon\dot{x}(0) + O(\epsilon^2).$$

The assumptions $b \neq 0$ and $\sin(\theta) \neq 1$ ensure that x is nonzero and so

$$(6.1-13) \qquad \frac{\|\hat{x} - x\|_2}{\|x\|_2} = \epsilon \frac{\|\dot{x}(0)\|_2}{\|x\|_2} + O(\epsilon^2).$$

In order to bound $\|\dot{x}(0)\|_2$, we differentiate (6.1-12) and set $t = 0$ in the result. This gives

$$E^TAx + A^TEx + A^TA\dot{x}(0) = A^Tf + E^Tb$$

i.e.,

$$(6.1-14) \qquad \dot{x}(0) = (A^TA)^{-1}A^T(f - Ex) + (A^TA)^{-1}E^Tr.$$

By substituting this result into (6.1-13), taking norms, and using the easily verified inequalities $\|f\|_2 \leqslant \|b\|_2$ and $\|E\|_2 \leqslant \|A\|_2$, we obtain

$$\frac{\|\hat{x} - x\|_2}{\|x\|_2} \leqslant \epsilon \left\{ \|A\|_2 \|(A^TA)^{-1}A^T\|_2 \left(\frac{\|b\|_2}{\|A\|_2\|x\|_2} + 1 \right) \right.$$

$$\left. + \frac{\rho_{LS}}{\|A\|_2\|x\|_2} \|A\|_2^2 \|(A^TA)^{-1}\|_2 \right\} + O(\epsilon^2).$$

Using the SVD of A (cf. Theorem 6.1-1) we have

$$\|A\|_2\|(A^TA)^{-1}A^T\|_2 = \kappa_2(A),$$

$$\|A\|_2^2\|(A^TA)^{-1}\|_2 = \kappa_2(A)^2,$$

$$\|A\|_2^2\|x\|_2^2 = \sigma_1(A)^2 \sum_{i=1}^{n} [u_i^Tb/\sigma_i(A)]^2 \geqslant \sum_{i=1}^{n} (u_i^Tb)^2 = \|b\|_2^2 - \rho_{LS}^2,$$

and thus

$$\frac{\|\hat{x} - x\|_2}{\|x\|_2} \leqslant \epsilon \left\{ \kappa_2(A)\left[\frac{1}{\cos(\theta)} + 1\right] + \kappa_2(A)^2 \frac{\sin(\theta)}{\cos(\theta)} \right\} + O(\epsilon^2).$$

This establishes (6.1-10).

To prove (6.1-11), we define the differentiable vector function $r(t)$ by

$$r(t) = (b + tf) - (A + tE)x(t)$$

and observe that $r = r(0)$ and $\hat{r} = r(\epsilon)$. Using (6.1-14) it can be shown that

$$\dot{r}(0) = (I - A(A^TA)^{-1}A^T)(f - Ex) - A(A^TA)^{-1}E^Tr,$$

and thus, since $\|\hat{r} - r\|_2 = \epsilon\|\dot{r}(0)\|_2 + O(\epsilon^2)$ we have

$$\frac{\|\hat{r} - r\|_2}{\|b\|_2} = \epsilon\frac{\|\dot{r}(0)\|_2}{\|b\|_2} + O(\epsilon^2)$$

$$\leqslant \epsilon\left\{\|I - A(A^TA)^{-1}A^T\|_2\left(1 + \frac{\|A\|_2\|x\|_2}{\|b\|_2}\right)\right.$$

$$\left. + \|A(A^TA)^{-1}\|_2\|A\|_2\frac{\rho_{LS}}{\|b\|_2}\right\} + O(\epsilon^2)$$

Inequality (6.1-11) now follows because

$$\|A\|_2\|x\|_2 = \|A\|_2\|A^+b\|_2 \leqslant \kappa_2(A)\|b\|_2$$

$$\rho_{LS} = \|(I - A(A^TA)^{-1}A^T)b\|_2 \leqslant \|I - A(A^TA)^{-1}A^T\|_2\|b\|_2$$

and

$$\|(I - A(A^TA)^{-1}A^T\|_2 = \min\{m - n, 1\}. \quad \square$$

Problems

P6.1-1.   Use the identity

$$\|A(x + \alpha w) - b\|_2^2 = \|Ax - b\|_2^2 + 2\alpha w^T A^T(Ax - b) + \alpha^2\|Aw\|^2$$

to show that if $x \in \mathbf{X}$, then $A^TAx = A^Tb$.

P6.1-2. Define the function $\phi : \mathbb{R}^n \to \mathbb{R}$ by $\phi(x) = \frac{1}{2}\|Ax - b\|_2^2$. Show that the gradient $\nabla \phi(x)$ is given by $\nabla \phi(x) = A^T(Ax - b)$.

P6.1-3. Show that if $A_k \to A$ and $A_k^+ \to A^+$, then there exists an integer k_0 such that $\text{rank}(A_k)$ is constant for all $k \geqslant k_0$.

P6.1-4. Verify (6.1-2).

P6.1-5. Show that if $A \in \mathbb{R}^{m \times n}$ has rank n, then so does $A + E$ if $\|E\|_2 \|A^+\|_2 < 1$.

P6.1-6. Let $A \in \mathbb{R}^{m \times n}$, and assume that there exists an $X \in \mathbb{R}^{n \times m}$ such that $x = Xb$ minimizes $\|Ax - b\|_2$ for every $b \in \mathbb{R}^m$. Show $AXA = A$ and $(AX)^T = AX$.

P6.1-7. Assume $A^T Ax = A^T b$, $(A^T A + F)\hat{x} = A^T b$, and $2\|F\|_2 \leqslant \sigma_n(A)^2$. If $r = b - Ax$ and $\hat{r} = b - A\hat{x}$, show that

$$\hat{r} - r = [I - A(A^T A + F)^{-1} A^T]Ax$$

and

$$\|\hat{r} - r\|_2 \leqslant 2\kappa_2(A) \frac{\|F\|_2}{\|A\|_2} \|x\|_2 .$$

P6.1-8. Use the method of normal equations to solve the LS problem where

$$A = \begin{bmatrix} 1 & 2 \\ 3 & 4 \\ 5 & 6 \end{bmatrix} \qquad b = \begin{bmatrix} 1 \\ 1 \\ 1 \end{bmatrix} .$$

P6.1-9. Define $B(\lambda) \in \mathbb{R}^{n \times m}$ by $B(\lambda) = (A^T A + \lambda I)^{-1} A^T$, where $\lambda > 0$. Show

$$\|B(\lambda) - A^+\|_2 = \frac{\lambda}{\sigma_r(A)[\sigma_r(A)^2 + \lambda]} \qquad r = \text{rank}(A)$$

and therefore that $B(\lambda) \to A^+$ as $\lambda \to 0$.

P6.1-10. Let $A \in \mathbb{R}^{m \times n}$ and $y \in \mathbb{R}^m$ and define $\bar{A} = [A, y] \in \mathbb{R}^{m \times (n+1)}$. Show $\kappa_2(A) \leqslant \kappa_2(\bar{A})$.

P6.1-11. Let $A \in \mathbb{R}^{m \times n}$, $w \in \mathbb{R}^n$, and define $B = \begin{bmatrix} A \\ w^T \end{bmatrix}$. Show $\sigma_n(B) \geqslant \sigma_n(A)$ and $\sigma_1(B) \leqslant \sqrt{\|A\|_2^2 + \|w\|_2^2}$.

Notes and References for Sec. 6.1

Our restriction to least squares approximation should not be construed as a vote against minimization in other norms. In particular, there are occasions when it is advisable to minimize $\|Ax - b\|_p$ for $p = 1$ and ∞. Some algorithms for doing this are described in

I. Barrodale and F.D.K. Roberts (1973). "An Improved Algorithm for Discrete L_1 Linear Approximation," *SIAM J. Num. Anal. 10*, 839–48.

I. Barrodale and C. Phillips (1975). "Algorithm 495: Solution of an Overdetermined System of Linear Equations in the Chebychev Norm," *ACM Trans. Math. Soft. 1*, 264-70.

A. K. Cline (1976a). "A Descent Method for the Uniform Solution to Overdetermined Systems of Equations," *SIAM J. Num. Anal. 13*, 293-309.

R. H. Bartels, A. R. Conn, and J. W. Sinclair (1978). "Minimization Techniques for Piecewise Differentiable Functions: The L_1 Solution to an Overdetermined Linear System," *SIAM J. Num. Anal. 15*, 224-41.

R. H. Bartels, A. R. Conn, and C. Charalambous (1978). "On Cline's Direct Method for Solving Overdetermined Linear Systems in the L_∞ Sense," *SIAM J. Num. Anal. 15*, 255-70.

The pseudo-inverse literature is vast, as evidenced by the 1,775 references in

M. Z. Nashed (1976). *Generalized Inverses and Applications*, Academic Press, New York.

The differentiation of the pseudo-inverse is further discussed in

C. L. Lawson and R. J. Hanson (1969). "Extensions and Applications of the Householder Algorithm for Solving Linear Least Squares Problems," *Math Comp. 23*, 787-812.

G. H. Golub and V. Pereyra (1973). "The Differentiation of Pseudo-Inverses and Nonlinear Least Squares Problems Whose Variables Separate," *SIAM J. Num. Anal. 10*, 413-32.

Survey treatments of LS perturbation theory may be found in Lawson and Hanson (SLS, chaps. 7-9) as well as in

P. A. Wedin (1973). "Perturbation Theory for Pseudo-Inverses," *BIT 13*, 217-32.

G. W. Stewart (1977a). "On the Perturbation of Pseudo-Inverses, Projections, and Linear Least Squares," *SIAM Review 19*, 634-62.

The causes and effects of the squared condition number in the LS perturbation bound are discussed in

G. H. Golub and J. H. Wilkinson (1966). "Note on the Iterative Refinement of Least Squares Solution," *Numer. Math. 9*, 139-48.

A. van der Sluis (1975). "Stability of the Solutions of Linear Least Squares Problem," *Numer. Math. 23*, 241-54.

Sec. 6.2. Householder and Gram-Schmidt Methods

Let $A \in \mathbb{R}^{m \times n}$ ($m \geqslant n$) and $b \in \mathbb{R}^m$ be given and suppose that an orthogonal matrix $Q \in \mathbb{R}^{m \times m}$ has been computed such that

$$(6.2\text{-}1) \qquad Q^T A = R = \begin{bmatrix} R_1 \\ 0 \end{bmatrix} \begin{matrix} n \\ m-n \end{matrix}$$

is upper triangular. If

$$Q^{\mathrm{T}}b = \begin{bmatrix} c \\ d \end{bmatrix} \begin{matrix} n \\ m-n \end{matrix}$$

then

$$\|Ax - b\|_2^2 = \|Q^{\mathrm{T}}Ax - Q^{\mathrm{T}}b\|_2^2 = \|R_1 x - c\|_2^2 + \|d\|_2^2$$

for any $x \in \mathbb{R}^n$. Clearly, if $\operatorname{rank}(A) = \operatorname{rank}(R_1) = n$, then x_{LS} is defined by the upper triangular system $R_1 x_{LS} = c$. Note that $\rho_{LS}^2 = \|d\|_2^2$.

Thus, the full rank LS problem can be readily solved once we have computed (6.2-1), which we refer to as the *Q-R factorization*. It can be calculated in several ways, as we show in this and the next section.

We begin with a method that utilizes Householder transformations. The essence of the algorithm can be conveyed by a small example. Suppose $m = 6$, $n = 5$, and assume that Householder matrices P_1 and P_2 have been computed so that

$$P_2 P_1 A = \begin{bmatrix} \mathrm{x} & \mathrm{x} & \mathrm{x} & \mathrm{x} & \mathrm{x} \\ 0 & \mathrm{x} & \mathrm{x} & \mathrm{x} & \mathrm{x} \\ 0 & 0 & \boxed{\mathrm{x}} & \mathrm{x} & \mathrm{x} \\ 0 & 0 & \mathrm{x} & \mathrm{x} & \mathrm{x} \\ 0 & 0 & \mathrm{x} & \mathrm{x} & \mathrm{x} \\ 0 & 0 & \mathrm{x} & \mathrm{x} & \mathrm{x} \end{bmatrix}.$$

We now determine a Householder matrix $\tilde{P}_3 \in R^{4 \times 4}$ such that

$$\tilde{P}_3 \begin{bmatrix} \mathrm{x} \\ \mathrm{x} \\ \mathrm{x} \\ \mathrm{x} \end{bmatrix} = \begin{bmatrix} \mathrm{x} \\ 0 \\ 0 \\ 0 \end{bmatrix}$$

and set $P_3 = \operatorname{diag}(I_2, \tilde{P}_3)$. Thus,

$$P_3 P_2 P_1 A = \begin{bmatrix} \mathrm{x} & \mathrm{x} & \mathrm{x} & \mathrm{x} & \mathrm{x} \\ 0 & \mathrm{x} & \mathrm{x} & \mathrm{x} & \mathrm{x} \\ 0 & 0 & \mathrm{x} & \mathrm{x} & \mathrm{x} \\ 0 & 0 & 0 & \mathrm{x} & \mathrm{x} \\ 0 & 0 & 0 & \mathrm{x} & \mathrm{x} \\ 0 & 0 & 0 & \mathrm{x} & \mathrm{x} \end{bmatrix}.$$

Ultimately $P_n P_{n-1} \cdots P_1 A = R$ is upper triangular, and so by setting $Q = P_1 \cdots P_n$ we obtain $A = QR$. Formally, we have

ALGORITHM 6.2-1: *Householder Orthogonalization.*   Given $A \in \mathbb{R}^{m \times n}$ $(m \geq n)$, the following algorithm computes an orthogonal Q such that $Q^T A = R$ is upper triangular. A is overwritten by R.

> For $k = 1, \ldots, n$
> > Determine a Householder $\tilde{P}_k$ of order $m - k + 1$ such that
> > $\tilde{P}_k(a_{kk}, \ldots, a_{mk})^T = (r_{kk}, \ldots, 0)^T$
> $A := \text{diag}(I_{k-1}, \tilde{P}_k)A$

This algorithm requires $n^2(m - n/3)$ flops.

The orthogonal matrix Q in the above can be stored in *factored form*. (Cf. §3.3.) In particular, the information associated with

$$P_k = \text{diag}(I_{k-1}, \tilde{P}_k) \equiv I_m - v^{(k)}[v^{(k)}]^T/\beta_k$$

can be represented as follows:

$$A := \begin{bmatrix} v_1^{(1)} & r_{12} & r_{13} & r_{14} & r_{15} \\ v_2^{(1)} & v_2^{(2)} & r_{23} & r_{24} & r_{25} \\ v_3^{(1)} & v_3^{(2)} & v_3^{(3)} & r_{34} & r_{35} \\ v_4^{(1)} & v_4^{(2)} & v_4^{(3)} & v_4^{(4)} & r_{45} \\ v_5^{(1)} & v_5^{(2)} & v_5^{(3)} & v_5^{(4)} & v_5^{(5)} \\ v_6^{(1)} & v_6^{(2)} & v_6^{(3)} & v_6^{(4)} & v_6^{(5)} \end{bmatrix}, \quad r := \begin{bmatrix} r_{11} \\ r_{22} \\ r_{33} \\ r_{44} \\ r_{55} \\ r_{66} \end{bmatrix}, \quad \beta := \begin{bmatrix} \beta_1 \\ \beta_2 \\ \beta_3 \\ \beta_4 \\ \beta_5 \end{bmatrix}.$$

If the matrix $Q = P_1 \cdots P_n$ is required, it can be accumulated very simply:

> $Q := I$
> For $k = n, \ldots, 1$
> > $Q := P_k Q$

This accumulation requires $2[m^2n - mn^2 + n^3/3]$ flops. See §3.3.

To solve the LS problem, Q does not have to be explicitly formed but merely applied to b:

> For $k = 1, \ldots, n$
> > $b := P_k b.$

This is an $O(mn)$ operation that can be performed either during the reduction or after it has been completed provided Q is stored in factored form. After $b := Q^T b$ has been computed, the solution x follows by back-substitution:

$$x_n := b_n / r_n$$

For $i = n - 1, \ldots, 1$

$$x_i := \left(b_i - \sum_{j=i+1}^{n} a_{ij} x_j \right) \Big/ r_i$$

Here, we assume R is represented as above. Overall, this method for solving the full rank LS problem requires $n^2(m - n/3)$ flops. The computed $\hat{x}$ can be shown to safisfy.

(6.2-2) $$\|(A + \delta A)\hat{x} - (b + \delta b)\|_2 = \min,$$

where

(6.2-3) $$\|\delta A\|_2 \leqslant (6m - 3n + 41)n\mathbf{u}\|A\|_F + O(\mathbf{u}^2)$$

and

(6.2-4) $$\|\delta b\|_2 \leqslant (6m - 3n + 40)n\mathbf{u}\|b\|_2 + O(\mathbf{u}^2).$$

These inequalities are established in Lawson and Hanson (SLE, pp. 90 ff.) and show that $\hat{x}$ satisfies a "nearby" LS problem. The relative error in $\hat{x}$ can be bounded via Theorem 6.1-3.

There is no comparable inverse error analysis for the method of normal equations (Algorithm 6.1-1). The price paid for this stability is not negligible, however, since Householder orthogonalization requires up to twice as many flops.

Of course, the Householder method for solving the LS problem breaks down if $\text{rank}(A) < n$. Numerically, trouble can be expected whenever $\kappa_2(A) \cong \mathbf{u}^{-1}$. This is in contrast to the normal equation approach, where completion of the Cholesky factorization becomes problematical once $\kappa_2(A)$ is in the neighborhood of $\mathbf{u}^{-1/2}$. Hence the claim in Lawson and Hanson (SLS, (pp. 126–27) that for a fixed machine precision, a wider class of LS problems can be solved using Householder orthogonalization.

EXAMPLE 6.2-1. If

$$A = \begin{bmatrix} 1 & 1 \\ 10^{-3} & 0 \\ 0 & 10^{-3} \end{bmatrix} \quad \text{and} \quad b = \begin{bmatrix} 2 \\ 10^{-3} \\ 10^{-3} \end{bmatrix}$$

and the LS problem $\min \|Ax - b\|_2$ is solved using Householder orthogonalization $\beta = 10, t = 6$, chopped arithmetic, then

$$\hat{x}_{LS} = \begin{bmatrix} 1.00000 \\ 1.00000 \end{bmatrix}.$$

Compare with Example 6.1-3. (Note $x_{LS} = (1, 1)^T$.)

Algorithm 6.2-1 can also be used to solve the *orthonormal basis* (OB) problem:

(OB)   Given independent vectors $a_1, \ldots, a_n \in \mathbb{R}^m$, find an orthonormal basis for span$\{a_1, \ldots, a_n\}$.

Indeed, if $A = [a_1, \ldots, a_n] = QR$ and $Q = [q_1, \ldots, q_m]$, then for $k = 1, \ldots, n$

$$(6.2\text{-}5) \qquad\qquad a_k = \sum_{i=1}^{k} r_{ik} q_i,$$

where the orthonormality of the q_i implies

$$r_{ik} = q_i^T a_k. \qquad\qquad i = 1, \ldots, k$$

This says that

$$\text{span}\{a_1, \ldots, a_k\} = \text{span}\{q_1, \ldots, q_k\} \qquad k = 1, \ldots, n$$

and in particular,

$$R(A) = \text{span}\{q_1, \ldots, q_n\}.$$

As a bonus, Householder orthogonalization also provides an orthonormal basis for the orthogonal complement of the range:

$$R(A)^{\perp} = \text{span}\{q_{n+1}, \ldots, q_m\}.$$

The assumption that rank$(A) = n$ implies that r_{kk} is nonzero and so equation (6.2-5) can be solved for q_k:

$$q_k = \frac{1}{r_{kk}} \left(a_k - \sum_{i=1}^{k-1} r_{ik} q_i \right).$$

We can think of q_k as a unit vector in the direction of

$$z_k = a_k - \sum_{i=1}^{k-1} s_{ik} q_i,$$

where to ensure $z_k \in \text{span} \langle q_1, \ldots, q_{k-1} \rangle^{\perp}$ we choose

$$s_{ik} = q_i^T a_k. \qquad\qquad i = 1, \ldots, k-1$$

This leads to the *classical Gram-Schmidt* (CGS) algorithm for solving the OB problem:

For $k = 1, \ldots, n$

$$s_{ik} := q_i^T a_k \quad (i = 1, \ldots, k - 1)$$

$$z_k := a_k - \sum_{i=1}^{k-1} s_{ik} q_i$$

$$r_{kk} := (z_k^T z_k)^{1/2}$$

$$q_k := z_k / r_{kk}$$

$$r_{ik} := s_{ik} / r_{kk} \quad (i = 1, \ldots, k - 1)$$

In the k-th step of CGS, the k-th columns of both Q and R are generated.

Note that CGS does not produce an orthonormal basis for $R(A)^\perp$ as does Householder orthogonalization. However, it does render enough information to solve the LS problem. In particular, CGS computes the factorization

$$A = Q_1 R_1 \quad Q_1 \in \mathbb{R}^{m \times n}, R_1 \in \mathbb{R}^{n \times n}, Q_1^T Q_1 = I_n, R_1 = \nabla.$$

Thus, the normal equations $(A^T A)x = A^T b$ transform to the upper triangular system $R_1 x = Q_1^T b$.

Unfortunately, the CGS method has very poor numerical properties in that there is typically a severe loss of orthogonality among the computed q_i. Interestingly, a rearrangement of the calculation, known as *modified Gram-Schmidt* (MGS), yields a much sounder computational procedure. At the k-th step of MGS, the k-th column of Q (denoted by q_k) and the k-th row of R (denoted by r_k^T) are determined.

To derive the MGS method, define the matrix $A^{(k)}$ by

(6.2-6)
$$\underset{k-1 \quad n-k+1}{[\; 0 \quad , \quad A^{(k)} \;]} = A - \sum_{i=1}^{k-1} q_i r_i^T = \sum_{i=k}^{n} q_i r_i^T.$$

It follows that if

$$A^{(k)} = [z, B] \qquad z \in \mathbb{R}^m, B \in \mathbb{R}^{m \times (n-k)}$$

then

$$r_{kk} = \|z\|_2$$

and

$$q_k = z / r_{kk}$$

$$(r_{k,k+1}, \ldots, r_{kn}) = q_k^T B.$$

We proceed to the next step by setting

$$A^{(k+1)} = B - q_k (r_{k,k+1}, \ldots, r_{kn}).$$

This completely describes the k-th step of MGS.

ALGORITHM 6.2-2: *Modified Gram-Schmidt.* Given $A \in \mathbb{R}^{m \times n}$ with rank$(A) = n$, the following algorithm computes the factorization $A = Q_1 R_1$ where $Q_1 \in \mathbb{R}^{m \times n}$ has orthonormal columns and $R_1 \in \mathbb{R}^{n \times n}$ is upper triangular. A is overwritten by Q_1.

For $k = 1, \ldots, n$
$$r_{kk} := \left(\sum_{i=1}^{m} a_{ik}^2 \right)^{1/2}$$
For $i = 1, \ldots, m$
$$a_{ik} := a_{ik}/r_{kk}$$
For $j = k + 1, \ldots, n$
$$r_{kj} := \sum_{i=1}^{m} a_{ik} a_{ij}$$
For $i = 1, \ldots, m$
$$a_{ij} := a_{ij} - a_{ik} r_{kj}$$

This algorithm requires mn^2 flops.

MGS can be readily adapted to solve the LS problem and is as stable as the Householder approach. (See Bjorck 1967b.) However, the MGS method is slightly more expensive because in it one is always updating m-vectors.

However, if one is interested in the OB problem, then the Householder approach requires $mn^2 - n^3/3$ flops to get Q in factored form and another $mn^2 - n^3/3$ flops to get the first n columns of Q. Therefore, for the problem of finding an orthonormal basis for $R(A)$, MGS is about twice as efficient as Householder orthogonalization. However, Bjorck (1967b) has shown that MGS produces a computed $\hat{Q}_1 = [\hat{q}_1, \ldots, \hat{q}_n]$ that satisfies

$$\hat{Q}_1^T \hat{Q}_1 = I + E_{\text{MGS}}, \qquad \|E_{\text{MGS}}\|_2 \cong \mathbf{u} \kappa_2(A)$$

whereas the corresponding result for the Householder approach is of the form

$$\hat{Q}_1^T \hat{Q}_1 = I + E_H. \qquad \|E_H\|_2 \cong \mathbf{u}$$

Thus, if orthonormality is critical, MGS should be used to solve the OB problem only when the vectors to be orthogonalized are fairly independent.

EXAMPLE 6.2-2. If modified Gram-Schmidt is applied to

$$A = \begin{bmatrix} 1 & 1 \\ 10^{-3} & 0 \\ 0 & 10^{-3} \end{bmatrix} \qquad \kappa_2(A) \cong 1.4 * 10^3$$

with $\beta = 10$, $t = 6$, chopped arithmetic, then

$$[\hat{q}_1, \hat{q}_2] = \begin{bmatrix} 1 & 0 \\ 10^{-3} & -.707107 \\ 0 & .707107 \end{bmatrix}.$$

Problems

P6.2-1. Show that the Q-R decomposition of a full rank matrix is unique if we insist that R have positive diagonal entries.

P6.2-2. Compute the Q-R factorization of $A = \begin{bmatrix} 5 & 9 \\ 12 & 7 \end{bmatrix}$.

P6.2-3. Give an algorithm for solving the LS problem which uses MGS. (No additional two-dimensional arrays are required if a_{ij} is overwritten with r_{ji} ($i \geq j$) and if the quantities $q_i^T b$ are found as the orthonormal q_i are generated.)

P6.2-4. Rearrange Algorithm 6.2-2 so that the decomposition

$$A = US$$

is found where $S \in \mathbb{R}^{n \times n}$ is unit upper triangular and

$$U^T U = \text{diag}(d_1^2, \ldots, d_n^2). \qquad\qquad U \in \mathbb{R}^{m \times n}$$

P6.2-5. Suppose $L \in \mathbb{R}^{m \times n}$ ($m \geq n$) is lower triangular. Show how Householder matrices $P_1, \ldots, P_n$ can be used to determine a lower triangular $L_1 \in \mathbb{R}^{n \times n}$ so that

$$P_n \cdots P_1 L = \begin{bmatrix} L_1 \\ 0 \end{bmatrix} \begin{matrix} n \\ m-n \end{matrix}.$$
$$ n$$

Your algorithm should require $n^2(m - n)$ flops. (Hint: The second step in the 6×3 case involves finding P_2 so that

$$P_2 \begin{bmatrix} x & 0 & 0 \\ x & x & 0 \\ x & x & x \\ x & x & 0 \\ x & x & 0 \\ x & x & 0 \end{bmatrix} = \begin{bmatrix} x & 0 & 0 \\ x & x & 0 \\ x & x & x \\ x & 0 & 0 \\ x & 0 & 0 \\ x & 0 & 0 \end{bmatrix}$$

with the property that rows 1 and 3 are left alone.)

P6.2-6. (Cline 1973)   Suppose that $A \in \mathbb{R}^{m \times n}$ has rank n and that Gaussian elimination with partial pivoting is used to compute the factorization

$PA = LU$, where $L \in \mathbb{R}^{m \times n}$ is unit lower triangular, $U \in \mathbb{R}^{n \times n}$ is upper triangular, and $P \in \mathbb{R}^{m \times m}$ is a permutation. Explain how the decomposition in P6.2-5 can be used to find a vector $z \in \mathbb{R}^n$ such that

$$\| Lz - Pb \|_2 = \min.$$

Show that if $Ux = z$, then $\| Ax - b \|_2 = \min$. Show that this method of solving the LS problem is more efficient than Algorithm 6.2-1 whenever $m \leqslant \frac{5}{3} n$.

P6.2-7. The matrix $C = (A^T A)^{-1}$, where $\text{rank}(A) = n$, arises in many statistical applications and is known as the *variance-covariance matrix*. This problem shows how the decomposition $A = QR$ is useful in computations involving C.

(a) Show $C = (R^T R)^{-1}$.

(b) Give an algorithm for computing $c_{11}, \ldots, c_{nn}$ requiring $n^3/6$ flops.

(c) Show that if $R = \begin{bmatrix} \alpha & v^T \\ 0 & S \end{bmatrix}$ and $C_1 = (S^T S)^{-1}$, then

$$C = (R^T R)^{-1} = \begin{bmatrix} (1 + v^T C_1 v)/\alpha^2 & -v^T C_1/\alpha \\ -C_1 v/\alpha & C_1 \end{bmatrix}.$$

(d) Using (c), give an algorithm which overwrites R with the upper triangular portion of C. Your algorithm should require $\frac{1}{3} n^3$ flops.

Notes and References for Sec. 6.2

The idea of using Householder transformations to solve the LS problem was proposed in

A. S. Householder (1958). "Unitary Triangularization of a Nonsymmetric Matrix," *J. Assoc. Comp. Mach.* 5, 339–42,

and the practical details were worked out in

G. H. Golub (1965). "Numerical Methods for Solving Linear Least Squares Problems," *Numer. Math.* 7, 206–16.

P. Businger and G. H. Golub (1965). "Linear Least Squares Solutions by Householder Transformations," *Numer. Math.* 7, 269–76. See also HACLA, pp. 111–18.

For a discussion of how the Q-R decomposition can be used to solve numerous problems in statistical computation, see

G. H. Golub (1969). "Matrix Decompositions and Statistical Computation," in *Statistical Computation*, ed. R. C. Milton and J. A. Nelder, Academic Press, New York, pp. 365–97.

If a matrix A is perturbed, how does its Q-R decomposition change? This question is analyzed in

G. W. Stewart (1977). "Perturbation Bounds for the QR Factorization of a Matrix," *SIAM J. Num. Anal. 14*, 509–18.

The main result is that the changes in Q and R are bounded by the condition of A times the relative change in A.

The use of Householder matrices to solve sparse LS problems requires careful attention to avoid excessive fill-in. References for this important topic include

J. K. Reid (1967). "A Note on the Least Squares Solution of a Band System of Linear Equations by Householder Reductions," *Comp. J. 10*, 188–89.

I. S. Duff and J. K. Reid (1976). "A Comparison of Some Methods for the Solution of Sparse Over-Determined Systems of Linear Equations," *J. Inst. Math. Applic. 17*, 267–80.

P. E. Gill and W. Murray (1976). "The Orthogonal Factorization of a Large Sparse Matrix," in *Sparse Matrix Computations*, ed. J. R. Bunch and D. J. Rose, Academic Press, New York, pp. 177–200.

L. Kaufman (1979). "Application of Dense Householder Transformations to a Sparse Matrix," *ACM Trans. Math. Soft. 5*, 442–51.

Fortran programs for solving the LS problem via Householder transformations may be found in LINPACK, chap. 9.

The pitfalls of classical Gram-Schmidt are discussed in

J. R. Rice (1966). "Experiments on Gram-Schmidt Orthogonalization," *Math. Comp. 20*, 325–28,

while

A. Bjorck (1967b). "Solving Linear Least Squares Problems by Gram-Schmidt Orthogonalization," *BIT 7*, 1–21,

offers a rigorous roundoff error analysis of the modified Gram-Schmidt process. An ALGOL implementation of the MGS method for solving the LS problem appears in

F. L. Bauer (1965). "Elimination with Weighted Row Combinations for Solving Linear Equations and Least Squares Problems," *Numer. Math 7*, 338–52. See also HACLA, pp. 119–33.

Another way to "repair" the classical algorithm is to reorthogonalize. See

N. N. Abdelmalck (1971). "Roundoff Error Analysis for Gram-Schmidt Method and Solution of Linear Least Squares Problems," *BIT 11*, 345–68.

The use of Gauss transformations to solve the LS problem has attracted some attention because they are cheaper to use than Householder or Givens matrices. See

G. Peters and J. H. Wilkinson (1970). "The Least Squares Problem and Pseudo-Inverses," *Comp. J. 13*, 309–16.

A. K. Cline (1973). "An Elimination Method for the Solution of Linear Least Squares Problems," *SIAM J. Num. Anal. 10*, 283–89.

R. J. Plemmons (1974). "Linear Least Squares by Elimination and MGS," *J. Assoc. Comp. Mach. 21*, 581–85.

Sec. 6.3. Givens and Fast Givens Methods

Givens rotations can also be used to compute the Q-R decomposition. The 4-by-3 case sufficiently illustrates the general idea:

$$\begin{bmatrix} x & x & x \\ x & x & x \\ x & x & x \\ x & x & x \end{bmatrix} \xrightarrow{(1,2)} \begin{bmatrix} x & x & x \\ 0 & x & x \\ x & x & x \\ x & x & x \end{bmatrix} \xrightarrow{(1,3)} \begin{bmatrix} x & x & x \\ 0 & x & x \\ 0 & x & x \\ x & x & x \end{bmatrix} \xrightarrow{(2,3)} \begin{bmatrix} x & x & x \\ 0 & x & x \\ 0 & 0 & x \\ x & x & x \end{bmatrix} \xrightarrow{(1,4)}$$

$$\begin{bmatrix} x & x & x \\ 0 & x & x \\ 0 & 0 & x \\ 0 & x & x \end{bmatrix} \xrightarrow{(2,4)} \begin{bmatrix} x & x & x \\ 0 & x & x \\ 0 & 0 & x \\ 0 & 0 & x \end{bmatrix} \xrightarrow{(3,4)} \begin{bmatrix} x & x & x \\ 0 & x & x \\ 0 & 0 & x \\ 0 & 0 & 0 \end{bmatrix}.$$

Clearly, if J_{pq} denotes the Givens rotation associated with the rotation in the (p, q) plane, then $Q^T A = R$ is upper triangular where

$$Q^T = J_{34} J_{24} J_{14} J_{23} J_{13} J_{12}.$$

For general m and n we have

ALGORITHM 6.3-1: *Givens Orthogonalization.* Given $A \in \mathbb{R}^{m \times n}$, the following algorithm overwrites A with $Q^T A = R$, where R is upper triangular and Q is orthogonal.

> For $q = 2, \ldots, m$
> > For $p = 1, 2, \ldots, \min\{q - 1, n\}$
> > > Find $c = \cos(\theta)$ and $s = \sin(\theta)$ such that
> > >
> > > $$\begin{bmatrix} c & s \\ -s & c \end{bmatrix} \begin{bmatrix} a_{pp} \\ a_{qp} \end{bmatrix} = \begin{bmatrix} x \\ 0 \end{bmatrix}$$
> > >
> > > $A := J(p, q, \theta)A$

This algorithm requires $2n^2(m - n/3)$ flops.

Other sequences of rotations can be used to upper triangularize A. For example, if we replace the "For" statements with

For $p = 1, \ldots, \min\{n, m - 1\}$
  For $q = j + 1, \ldots, m$

then R is computed one column at a time.

By inserting the update $b := J(p, q, \theta)b$ in Algorithm 6.3-1, we can use it to solve the LS problem. The computed solution $\hat{x}$ can be shown to satisfy a nearby problem in the sense of (6.2-2)-(6.2-4). The Givens rotations can be stored in the subdiagonal portion of A. (See §3.4.) Despite all these favorable attributes, Householder orthogonalization is preferred because it is twice as fast.

Nevertheless, the ability to introduce zeros in such a selective fashion makes the Givens approach an important tool in certain structured problems. This has led to the development of "fast Givens" procedures, in which the calculations in Algorithm 6.3-1 are rearranged so that they can be performed with "Householder speed," at least in principle. The idea is to construct a matrix $M \in \mathbb{R}^{m \times m}$ such that

$$MA = S$$

is upper triangular and such that

$$MM^{\mathrm{T}} = D = \mathrm{diag}(d_1, \ldots, d_m). \qquad d_i > 0$$

Since $D^{-1/2}M$ is orthogonal, it follows that

$$A = M^{-1}S = (D^{-1/2}M)^{-1}(D^{-1/2}S) = (M^{\mathrm{T}}D^{-1/2})(D^{-1/2}S)$$

is the Q-R factorization of A. (The matrix $D^{-1/2}S$ is clearly upper triangular.)

As in the description of Givens transformations, the details of the computation can be explained at the 2-by-2 level. Let $x = (x_1, x_2)^{\mathrm{T}}$ and $D = \mathrm{diag}(d_1, d_2)$ be given, where $d_1, d_2 > 0$, and define

$$M_1 = \begin{bmatrix} \beta_1 & 1 \\ 1 & \alpha_1 \end{bmatrix}.$$

Observe that

$$M_1 x = \begin{bmatrix} \beta_1 x_1 + x_2 \\ x_1 + \alpha_1 x_2 \end{bmatrix}$$

and

$$M_1 D M_1^{\mathrm{T}} = \begin{bmatrix} d_2 + \beta_1^2 d_1 & d_1\beta_1 + d_2\alpha_1 \\ d_1\beta_1 + d_2\alpha_1 & d_1 + \alpha_1^2 d_2 \end{bmatrix}.$$

If $x_2 \neq 0$ and $\alpha_1 = -x_1/x_2$ and $\beta_1 = -\alpha_1 d_2/d_1$, then

$$M_1 x = \begin{bmatrix} x_2(1 + \gamma_1) \\ 0 \end{bmatrix}$$

$$M_1 D M_1^T = \begin{bmatrix} d_2(1 + \gamma_1) & 0 \\ 0 & d_1(1 + \gamma_1) \end{bmatrix},$$

where $\gamma_1 = -\alpha_1\beta_1 = (d_2/d_1)(x_1/x_2)^2$.

Analogously, if we assume $x_1 \neq 0$ and define M_2 by

$$M_2 = \begin{bmatrix} 1 & \alpha_2 \\ \beta_2 & 1 \end{bmatrix} \qquad \beta_2 = -x_2/x_1, \qquad \alpha_2 = -(d_1/d_2)\beta_2$$

then

$$M_2 x = \begin{bmatrix} x_1(1 + \gamma_2) \\ 0 \end{bmatrix}$$

and

$$M_2 D M_2^T = \begin{bmatrix} d_1(1 + \gamma_2) & 0 \\ 0 & d_2(1 + \gamma_2) \end{bmatrix},$$

where $\gamma_2 = -\alpha_2\beta_2 = (d_1/d_2)(x_2/x_1)^2$.

It is easy to show that for either $i = 1$ or 2, the matrix $J = D^{-1/2}M_i D^{1/2}$ is orthogonal and that it is designed so that the second component of $J(D^{-1/2}x)$ is zero. Notice that the γ_i satisfy $\gamma_1\gamma_2 = 1$. Thus we can always select M_i in the above so that the "growth factor" $(1 + \gamma_i)$ is bounded by 2. Matrices of the form

$$M_1 = \begin{bmatrix} \beta_1 & 1 \\ 1 & \alpha_1 \end{bmatrix} \qquad M_2 = \begin{bmatrix} 1 & \alpha_2 \\ \beta_2 & 1 \end{bmatrix}$$

satisfying $-1 \leqslant \alpha_i\beta_i \leqslant 0$ are referred to as *fast Givens transformations*. Notice that premultiplication by a fast Givens transformation involves about half the number of multiplies as premultiplication by an "ordinary" Givens transformation. This suggests that the following algorithm may be competitive with the Householder method for solving the LS problem:

ALGORITHM 6.3-2.   Given $A \in \mathbb{R}^{m \times n}$ and $b \in \mathbb{R}^m$, the following algorithm overwrites A with MA and b with Mb where MA is upper triangular and $MM^T = \text{diag}(d_1, \ldots, d_m)$.

$d_i := 1 \qquad (i = 1, \ldots, m)$
For $q = 2, \ldots, m$
    For $p = 1, 2, \ldots, \min\{q - 1, n\}$
        If $a_{qp} \neq 0$
            then

$$\alpha := -a_{pp}/a_{qp}, \qquad \beta := -\alpha d_q/d_p, \qquad \gamma := -\alpha\beta$$

                If $\gamma \leqslant 1$
                    then

$$\begin{bmatrix} a_{pp} & \cdots & a_{pn} & b_p \\ a_{qp} & \cdots & a_{qn} & b_q \end{bmatrix} := \begin{bmatrix} \beta & 1 \\ 1 & \alpha \end{bmatrix} \begin{bmatrix} a_{pp} & \cdots & a_{pn} & b_p \\ a_{qp} & \cdots & a_{qn} & b_q \end{bmatrix}$$

                    Interchange d_p and d_q.

$$d_p := (1 + \gamma)d_p, \qquad d_q := (1 + \gamma)d_q$$

                else
                    Interchange α and β.

$$\alpha := 1/\alpha, \qquad \beta := 1/\beta, \qquad \gamma := 1/\gamma$$

$$\begin{bmatrix} a_{pp} & \cdots & a_{pn} & b_p \\ a_{qp} & \cdots & a_{qn} & b_q \end{bmatrix} := \begin{bmatrix} 1 & \alpha \\ \beta & 1 \end{bmatrix} \begin{bmatrix} a_{pp} & \cdots & a_{pn} & b_p \\ a_{qp} & \cdots & a_{qn} & b_q \end{bmatrix}$$

$$d_p := (1 + \gamma)d_p, \qquad d_q := (1 + \gamma)d_q$$

This algorithm requires $n^2(m - n/3)$ flops.

The solution to the LS problem readily follows from the output of Algorithm 6.3-2. Indeed, if $MM^T = D$ is diagonal, if

$$MA = \begin{bmatrix} S_1 \\ 0 \end{bmatrix} \begin{matrix} n \\ m-n \end{matrix}$$

is upper triangular, and if

$$Mb = \begin{bmatrix} c \\ d \end{bmatrix} \begin{matrix} n \\ m-n \end{matrix},$$

then

$$\|Ax - b\|_2^2 = \|D^{-1/2}MAx - D^{-1/2}Mb\|_2^2 = \left\| D^{-1/2}\left(\begin{bmatrix} S_1 \\ 0 \end{bmatrix} x - \begin{bmatrix} c \\ d \end{bmatrix} \right) \right\|_2^2$$

for any $x \in \mathbb{R}^n$. Clearly, x_{LS} is obtained by solving the nonsingular upper triangular system $S_1x = c$.

The computed solution $\hat{x}$ obtained in this fashion can be shown to solve a nearby LS problem in the sense of (6.2-2)–(6.2-4). This may seem surprising, since large numbers can arise during the calculation. An entry in the scaling matrix D can double in magnitude after a single fast Givens update. However, largeness in D must be exactly compensated for by largeness in M, since $D^{-1/2}M$ is orthogonal at all stages of the computation. It is this phenomenon that enables one to push through a favorable error analysis.

Unfortunately, the possibility of element growth requires continual monitoring of D to avoid overflow. Because of the nontrivial overhead involved, the resulting algorithm is slower than the Householder approach and more complicated to implement.

EXAMPLE 6.3-1.   If Algorithm 6.3-2 is applied to

$$A = \begin{bmatrix} 1 & 4 \\ 2 & 5 \\ 3 & 6 \end{bmatrix} \quad \text{and} \quad b = \begin{bmatrix} 7 \\ 8 \\ 9 \end{bmatrix},$$

then

$$M = \tfrac{1}{24} \begin{bmatrix} 8 & 16 & 24 \\ 40 & 10 & -20 \\ 15 & -30 & 15 \end{bmatrix},$$

$$D = \text{diag}(\tfrac{14}{9} \; \tfrac{175}{48} \; \tfrac{75}{32}),$$

$$M[A, b] = \begin{bmatrix} \tfrac{14}{3} & \tfrac{32}{3} & \tfrac{50}{3} \\ 0 & \tfrac{15}{4} & \tfrac{15}{2} \\ 0 & 0 & 0 \end{bmatrix},$$

and $x_{LS} = (-1, 2)^T$.

Problems

P6.3-1 (Stewart 1976)   Let $\begin{bmatrix} c & s \\ -s & c \end{bmatrix}$ be a Givens rotation and define the scalar ρ by

$$\rho = \begin{cases} 1 & \text{if} \quad c = 0 \\ \text{sign}(c)s/2 & \text{if} \quad |s| \leqslant |c| \\ 2\text{sign}(s)/c & \text{if} \quad |c| < |s| \end{cases}$$

(a) Write an algorithm for determining c and s given ρ. (b) Use this device to store all the information necessary to reconstruct Q in Algorithm 6.3-1. (Overwrite a_{ij} with the ρ_{ij} associated with its zeroing.) (c) Write an algorithm which takes the output from your algorithm in (b) and computes Q. How many flops are required?

P6.3-2. Algorithm 6.3-2 is easily adapted to solve the weighted least squares problem

$$\min_x \| D(Ax - b) \|_2$$

where $A \in \mathbb{R}^{m \times n}$, $b \in \mathbb{R}^m$, and $D = \text{diag}(d_i)$, $d_i > 0$. Indeed, if

$$MD^{-2}M^T = \tilde{D} = \text{diag}(\tilde{d}_i) \qquad\qquad \tilde{d}_i > 0$$

and

$$MA = \begin{bmatrix} S_1 \\ 0 \end{bmatrix} \begin{matrix} n \\ m-n \end{matrix} \qquad Mb = \begin{bmatrix} c \\ d \end{bmatrix} \begin{matrix} n \\ m-n \end{matrix},$$

then the solution to the weighted LS problem is obtained by solving $S_1 x = c$. Show this and indicate what changes in Algorithm 6.3-2 are required to handle this problem.

Notes and References for Sec. 6.3

The idea of using Givens transformations to compute the Q-R decomposition was first suggested in

W. Givens (1958). "Computation of Plane Unitary Rotations Transforming a General Matrix to Triangular Form," *SIAM J. App. Math. 6*, 26–50.

For a detailed analysis of the rounding errors arising in Algorithm 6.3-1 see

M. Gentleman (1973). "Error Analysis of Q-R Decompositions by Givens Transformations," *Lin. Alg. & Its Appl. 10*, 189–97.

An attractive feature of Householder transformations is that a sequence of them can be compactly stored in "factored" form. An analogous technique for compactly representing products of Givens transformations also exists and is described in

G. W. Stewart (1976). "The Economical Storage of Plane Rotations," *Numer. math. 25*, 137–38.

See our comments in §3.4.

Fast Givens transformations are also referred to as "square-root-free" Givens transformations. (Recall that a square root must ordinarily be computed during the formation of Givens transformation.) There are several ways fast Givens calculations can be arranged. See

M. Gentleman (1973). "Least Squares Computations by Givens Transformations without Square Roots," *J. Inst. Math. Appl. 12*, 329–36.
S. Hammarling (1974). "A Note on Modifications to the Givens Plane Rotation," *J. Inst. Math. Appl. 13*, 215–18.
C. F. Van Loan (1973). "Generalized Singular Values with Algorithms and Applications," Ph.D. thesis, University of Michigan, Ann Arbor.

J. H. Wilkinson (1977). "Some Recent Advances in Numerical Linear Algebra," in *The State of the Art in Numerical Analysis*, ed. D.A.H. Jacobs, Academic Press, New York, pp. 1-53.

We wish to repeat that a nontrivial amount of monitoring is necessary in order to successfully implement fast Givens transformations.

Although the computation of the Q-R decomposition is more efficient with Householder reflections, there are some settings where the Givens approach is advantageous. For example, if A is sparse, then the careful application of Givens rotations can minimize fill-in. See

I. S. Duff (1974). "Pivot Selection and Row Ordering in Givens Reduction on Sparse Matrices," *Computing 13*, 239-48.

J. A. George and M. T. Heath (1980). "Solution of Sparse Linear Least Squares Problems Using Givens Rotations," *Lin. Alg. & Its Applic. 34*, 69-83.

Sec. 6.4. Rank Deficiency I: QR with Column Pivoting

If A is rank deficient and $A = QR$ is the Q-R factorization of A, then at least one diagonal entry in R is zero. This means that the back-substitution process used to compute x_{LS} breaks down. On the other hand, we know that if rank$(A) < n$, then there are an infinite number of solutions to the LS problem. This raises three issues:

(1) How can we find *a* solution to the LS problem?

(2) How can we find *the* unique solution having minimal 2-norm?

(3) How can the calculations be performed reliably in the face of A's infinite condition?

Answers to these and other questions are provided in this and the next section.

Let us first examine why the Q-R factorization approach can fail in the case when rank$(A) = r < n$. The "mission" of any orthogonalization method is to compute an orthonormal basis $\{q_1, \ldots, q_r\}$ for $R(A)$. Indeed, if $R(A) = R(Q_1)$ where $Q_1 = [q_1, \ldots, q_r]$ has orthonormal columns, then $A = Q_1 S$ for some $S \in \mathbb{R}^{r \times n}$ and

$$\|Ax - b\|_2 = \|(Q_1 S)x - b\|_2.$$

It follows that any vector x satisfying $Sx = Q_1^T b$ solves the LS problem. Since S has rank r, we are assured that this underdetermined system has a solution.

Unfortunately, if rank $(A) < n$ then the Q-R factorization does not necessarily produce an orthonormal basis for $R(A)$. For example, suppose

$$A = [a_1, a_2, a_3] = [q_1, q_2, q_3] \begin{bmatrix} 1 & 1 & 1 \\ 0 & 0 & 1 \\ 0 & 0 & 1 \end{bmatrix}$$

is the Q-R factorization of A. Note that $\text{rank}(A) = 2$ but that $R(A)$ does not equal any of the subspaces span $\{q_i, q_j\}$, $i \neq j$.

Fortunately, Algorithm 6.2-1 can be modified in a simple way so as to produce an orthonormal basis for A's range. The modified algorithm computes the factorization

(6.4-1) $\qquad A\Pi = QR \qquad R = \begin{bmatrix} R_{11} & R_{12} \\ 0 & 0 \end{bmatrix} \begin{matrix} r \\ m-r \end{matrix}$
$$\qquad\qquad\qquad\qquad\qquad\qquad\quad r \quad\; n-r$$

where $r = \text{rank}(A)$, Q is orthogonal, R_{11} is upper triangular, and Π is a permutation. If $A\Pi = [a_{c_1}, \ldots, a_{c_n}]$ and $Q = [q_1, \ldots, q_m]$, then for $k = 1, \ldots,$ n we have

$$a_{c_k} = \sum_{i=1}^{\min\{r,k\}} r_{ik} q_i \in \text{span}\{q_1, \ldots, q_r\}$$

implying $R(A) = \text{span}[q_1, \ldots, q_r]$.

Once (6.4-1) is computed, the LS problem can be readily solved. Indeed, for any $x \in \mathbb{R}^n$ we have

$$\|Ax - b\|_2^2 = \|Q^T A\Pi)(\Pi^T x) - (Q^T b)\|_2^2 = \|R_{11}y - (c - R_{12}z)\|_2^2 + \|d\|_2^2,$$

where

$$\Pi^T x = \begin{bmatrix} y \\ z \end{bmatrix} \begin{matrix} r \\ n-r \end{matrix}$$

and

$$Q^T b = \begin{bmatrix} c \\ d \end{bmatrix} \begin{matrix} r \\ m-r \end{matrix}.$$

Thus, if $\|Ax - b\|_2 = \min$, then we must have

$$x = \Pi^T \begin{bmatrix} R_{11}^{-1}(c - R_{12}z) \\ z \end{bmatrix}.$$

If z is set to zero in this expression, then we obtain the *basic solution*

$$x_B = \Pi^T \begin{bmatrix} R_{11}^{-1}c \\ 0 \end{bmatrix}.$$

The basic solution is not the minimal 2-norm solution unless the submatrix R_{12} is zero, since

(6.4-2) $\qquad \|x_{LS}\|_2 = \min_{z \in \mathbb{R}^{n-r}} \left\| x_B - \begin{bmatrix} R_{11}^{-1}R_{12} \\ -I_{n-r} \end{bmatrix} z \right\|_2.$

Indeed, this characterization of $\|x_{LS}\|_2$ can be used to show

(6.4-3)
$$1 \leqslant \frac{\|x_B\|_2}{\|x_{LS}\|_2} \leqslant \sqrt{1 + \|R_{11}^{-1}R_{12}\|_2^2} \ .$$

See Golub and Pereyra (1976) for details.

The matrices Q and Π in the decomposition (6.4-1) are determined as products of Householder matrices and elementary permutation matrices respectively:

$$Q = Q_1 Q_2 \cdots Q_r$$

$$\Pi = \Pi_1 \Pi_2 \cdots \Pi_r \ .$$

To see how this is accomplished, assume for some k that we have computed

(6.4-4) $\quad (Q_{k-1} \cdots Q_1)A(\Pi_1 \cdots \Pi_{k-1}) = R^{(k-1)}$

$$= \begin{bmatrix} R_{11}^{(k-1)} & R_{12}^{(k-1)} \\ 0 & R_{22}^{(k-1)} \end{bmatrix} \begin{matrix} k-1 \\ m-k+1 \end{matrix} \ ,$$
$$ \ \ k-1 \quad n-k+1$$

where $R_{11}^{(k-1)}$ is nonsingular and upper triangular. Let

$$R_{22}^{(k-1)} = [v_k^{(k-1)}, \ \ldots, \ v_n^{(k-1)}]$$

be a column partitioning and define the index p $(k \leqslant p \leqslant n)$ by

(6.4-5) $\qquad \|v_p^{(k-1)}\|_2 = \max \{ \|v_k^{(k-1)}\|_2, \ \ldots, \ \|v_n^{(k-1)}\|_2 \}.$

Note that if $k = \mathrm{rank}(A)+1$, then this maximum is zero and we are finished. Otherwise, let Π_k be the identity with columns p and k interchanged and determine a Householder matrix Q_k such that if

$$R^{(k)} = Q_k R^{(k-1)}\Pi_k ,$$

then $r_{k+1,k}^{(k)} = \cdots = r_{mk}^{(k)} = 0$. In other words, Π_k moves the largest column in $R_{22}^{(k-1)}$ to the lead position and Q_k zeros all of its subdiagonal components.

The column norms do not have to be recomputed at each stage if we exploit the property

$$Z^{\mathrm{T}}v = \begin{bmatrix} \alpha \\ w \end{bmatrix} \begin{matrix} 1 \\ s-1 \end{matrix} \quad \Rightarrow \quad \|w\|_2^2 = \|v\|_2^2 - \alpha^2,$$

which holds for any orthogonal matrix $Z \in \mathbb{R}^{s \times s}$. This reduces the overhead associated with column pivoting from $O(mn^2)$ flops to $O(mn)$ flops.

Combining all of the above we obtain the following algorithm established by Businger and Golub (1965):

ALGORITHM 6.4-1: *QR with Column Pivoting.* Given $A \in \mathbb{R}^{m \times n}$, the following algorithm computes the factorization $A \Pi = QR$ defined by (6.4-1). The permutation $\Pi = [e_{c_1}, \ldots, e_{c_n}]$ is determined according to the strategy (6.4-5). The element a_{ij} is overwritten by r_{ij} $(i \leqslant j)$.

$c_j := j \qquad (j = 1, \ldots, n)$

$\gamma_j := \sum\limits_{i=1}^{m} a_{ij}^2 \qquad (j = 1, \ldots, n)$

For $k = 1, \ldots, n$
    Determine p $(k \leqslant p \leqslant n)$ so $\gamma_p = \max\limits_{k \leqslant j \leqslant n} \gamma_j$.

    If $\gamma_p = 0$
        then quit
        else
            Interchange c_k and c_p, γ_k and γ_p, and a_{ik} and a_{ip} for $i = 1, \ldots, m$.

            Determine a Householder $\tilde{Q}_k$ such that

$$\tilde{Q}_k \begin{bmatrix} a_{kk} \\ \vdots \\ a_{mk} \end{bmatrix} = \begin{bmatrix} * \\ 0 \\ \vdots \\ 0 \end{bmatrix}.$$

$$A := \mathrm{diag}(I_{k-1}, \tilde{Q}_k) A$$
$$\gamma_j := \gamma_j - a_{kj}^2 \qquad (j = k + 1, \ldots, n)$$

This algorithm requires $2mnr - r^2(m + n) + 2r^3/3$ flops where $r = \mathrm{rank}(A)$. As is done in Algorithm 6.2-1, it is possible to store the orthogonal matrix Q in factored form in the subdiagonal portion of A.

EXAMPLE 6.4-1.   If Algorithm 6.4-1 is applied to

$$A = \begin{bmatrix} 1 & 2 & 3 \\ 1 & 5 & 6 \\ 1 & 8 & 9 \\ 1 & 11 & 12 \end{bmatrix},$$

then $\Pi = [e_3, e_2, e_1]$ and to three significant digits we obtain

$$\hat{R} = \begin{bmatrix} -16.4 & -14.6 & -1.82 \\ 0 & .816 & -.816 \\ 0 & 0 & 0 \end{bmatrix}$$

$$\hat{Q} = \begin{bmatrix} -.182 & -.816 & .514 & .191 \\ -.365 & .408 & -.827 & .129 \\ .548 & 0 & .113 & -.829 \\ -.730 & .408 & .200 & .510 \end{bmatrix}$$

If Algorithm 6.4-1 is used to compute x_B, then great care must be exercised in the determination of rank(A). In order to appreciate the difficulty of this, suppose

$$fl(Q_k \ldots Q_1 A \Pi_1 \ldots \Pi_k) \equiv \hat{R}^{(k)} = \begin{bmatrix} \hat{R}_{11}^{(k)} & \hat{R}_{12}^{(k)} \\ 0 & \hat{R}_{22}^{(k)} \end{bmatrix}$$

is the matrix computed after k steps of the algorithm have been executed in floating point. Suppose rank$(A) = k$. Because of roundoff error, $\hat{R}_{22}^{(k)}$ will not be exactly zero. However, if $\hat{R}_{22}^{(k)}$ is suitably small in norm then it is reasonable to terminate the reduction and declare A to have rank k. A typical criteria for termination might be

(6.4-6) $$\|\hat{R}_{22}^{(k)}\|_2 \leqslant \epsilon_1 \|A\|_2$$

for some small machine-dependent parameter ϵ_1. In view of the roundoff properties associated with orthogonal matrix computation (cf. §3.4), we know that $\hat{R}^{(k)}$ is orthogonally equivalent to a matrix $A + E_k$, where

$$\|E_k\|_2 \leqslant \epsilon_2 \|A\|_2. \qquad\qquad \epsilon_2 = O(\mathbf{u})$$

Using Corollary 2.3-3 we have

$$\sigma_{k+1}(A + E_k) = \sigma_{k+1}(\hat{R}^{(k)}) \leqslant \|\hat{R}_{22}^{(k)}\|_2,$$

and since $\sigma_{k+1}(A) \leqslant \sigma_{k+1}(A + E_k) + \|E_k\|_2$, it follows that

$$\sigma_{k+1}(A) \leqslant (\epsilon_1 + \epsilon_2) \|A\|_2.$$

In other words, a relative perturbation of $O(\epsilon_1 + \epsilon_2)$ in A can yield a rank-k matrix. With this termination criteria we conclude that Algorithm 6.4-1 "discovers" rank degeneracy *if* in the course of the reduction, $\hat{R}_{22}^{(k)}$ is small for some $k < n$.

Unfortunately, this is not always the case. A matrix can be nearly rank deficient without a single $\hat{R}_{22}^{(k)}$ being particularly small.

EXAMPLE 6.4-2. Let $T_n(c)$ be the matrix

$$T_n(c) = \text{diag}(1, s, \ldots, s^{n-1}) \begin{bmatrix} 1 & -c & -c & \ldots & -c \\ & 1 & -c & \ldots & -c \\ & & & \ddots & \vdots \\ 0 & & & & 1 \end{bmatrix}$$

$$c^2 + s^2 = 1, c, s, > 0.$$

(See Lawson and Hanson [SLS, p.31].) These matrices are unaltered by Algorithm 6.4-1 if we assume no swapping is performed when $\gamma_k = \max [\gamma_k, \ldots, \gamma_n]$. Thus, for $k = 1, \ldots, n - 1$ we have

$$\| R_{22}^{(k)} \|_2 \geqslant s^{n-1}.$$

This inequality implies (for example) that the matrix $T_{100}(.2)$ has no particularly small trailing principal submatrix since $s^{99} \cong .13$. However, it can be shown that σ_n has order 10^{-8}.

From this example we conclude that QR with column pivoting is not entirely reliable as a method for detecting near rank deficiency. However, the "degree of unreliability" is somewhat like that for Gaussian elimination with partial pivoting, a method that works very well in practice.

Problems

P6.4-1. Show that if

$$A = \begin{bmatrix} T & S \\ 0 & 0 \end{bmatrix} \begin{matrix} r \\ m-r \end{matrix} \qquad r = \text{rank}(A)$$
$$ \begin{matrix} r & n-r \end{matrix}$$

and T is nonsingular, then

$$X = \begin{bmatrix} T^{-1} & 0 \\ 0 & 0 \end{bmatrix} \begin{matrix} r \\ n-r \end{matrix}$$
$$ \begin{matrix} r & m-r \end{matrix}$$

satisfies $AXA = A$ and $(AX)^T = (AX)$. In this case we say that X is a (1, 3) pseudo-inverse of A. Show that for general A, $x_B = Xb$ where X is the (1, 3) pseudo-inverse of A.

P6.4-2. Prove (6.4-2) and (6.4-3).

P6.4-3. Show that if

$$R = \begin{bmatrix} R_{11} & R_{12} \\ 0 & R_{22} \end{bmatrix} \begin{matrix} k \\ m-k \end{matrix}$$
$$ \begin{matrix} k & n-k \end{matrix}$$

then $\sigma_{k+1}(A) \leqslant \|R_{22}\|_2$

P6.4-4. Let $A \in \mathbb{R}^{m \times n}$ and assume that $0 \neq v$ satisfies $\|Av\|_2 = \sigma_n(A) \|v\|_2$. Let Π be a permutation such that if $\Pi^T v = w$, then $|w_n| = \|w\|_\infty$. Show that if $A\Pi = QR$ is the Q-R factorization of $A\Pi$, then

$$|r_{nn}| \leqslant \sqrt{n}\, \sigma_n(A).$$

Thus, there always exists a permutation Π such that the decomposition (6.4-1) "displays" near rank deficiency.

P6.4-5. Suppose the decomposition $QR = A\Pi = [a_{c_1}, \ldots, a_{c_n}]$ has been computed via Algorithm 6.4-1. For $k = 1, \ldots, n$ define

$$A_k = [a_{c_1}, \ldots, a_{c_k}]$$

and show that

$$\min_{y \in \mathbb{R}^{k-1}} \|A_{k-1}y - a_{c_k}\|_2 \geqslant \min_{y \in \mathbb{R}^{k-1}} \|A_{k-1}y - a_{c_i}\|_2,$$

for $i = k + 1, \ldots, n$. In other words, at the k-th step in Algorithm 6.1-4 the column furthest away from $R(A_{k-1})$ is pivoted into the k-th position.

P6.4-6. Show that if

$$A = \begin{bmatrix} R & w \\ 0 & v \end{bmatrix} \begin{matrix} k \\ m-k \end{matrix} \qquad b = \begin{bmatrix} c \\ d \end{bmatrix} \begin{matrix} k \\ m-k \end{matrix}$$

and A has full rank, then

$$\min \|Ax - b\|_2^2 = \|d\|_2^2 - (v^T d / \|v\|_2)^2.$$

P6.4-7. Let $x, y \in \mathbb{R}^m$ and $Q \in \mathbb{R}^m$ be given, Q orthogonal. Show that if

$$Q^T x = \begin{bmatrix} \alpha \\ u \end{bmatrix} \begin{matrix} 1 \\ m-1 \end{matrix} \qquad Q^T y = \begin{bmatrix} \beta \\ v \end{bmatrix} \begin{matrix} 1 \\ m-1 \end{matrix}$$

then $u^T v = x^T y - \alpha\beta$.

P6.4-8. Let $A = [a_1, \ldots, a_n] \in \mathbb{R}^{m \times n}$ and $b \in \mathbb{R}^m$ be given. For any subset of A's columns $\{a_{c_1}, \ldots, a_{c_k}\}$ define

$$\text{res}[a_{c_1}, \ldots, a_{c_k}] = \min_{x \in \mathbb{R}^k} \|[a_{c_1}, \ldots, a_{c_k}]x - b\|_2.$$

Describe a pivot selection procedure for Algorithm 6.4-1 such that if

$$QR = A\Pi = [a_{c_q}, \ldots, a_{c_n}]$$

is the final factorization, then

$$\text{res}[a_{c_1}, \ldots, a_{c_k}] = \min_{i \geqslant k} \text{res}[a_{c_1}, \ldots, a_{c_{k-1}}, a_{c_i}]$$

for $k = 1, \ldots, n$.

Notes and References for Sec. 6.4

QR with column pivoting was first discussed in

P. A. Businger and G. H. Golub (1965). "Linear Least Squares Solutions by House-holder Transformations," *Numer. Math. 7*, 269-76. See also HACLA, pp. 111-18.

Even for full rank problems, column pivoting seems to produce more accurate solutions. The error analysis in the following paper attempts to explain why.

L. S. Jennings and M. R. Osborne (1974). "A Direct Error Analysis for Least Squares," *Numer. Math. 22*, 322-32.

Knowing when to stop in Algorithm 6.4-1 is difficult but important. In questions of rank deficiency, it is helpful to obtain information about the smallest singular value of the upper triangular matrix R. This can be done using the techniques of §4.5 or those that are discussed in

I. Karasalo (1974). "A Criterion for Truncation of the Q-R Decomposition Algorithm for the Singular Linear Least Squares Problem," *BIT 14*, 156-66.
N. Anderson and I. Karasalo (1975). "On Computing Bounds for the Least Singular Value of a Triangular Matrix," *BIT 15*, 1-4.

See also Lawson and Hanson (SLS, chap. 6).

Sec. 6.5. Rank Deficiency II: The Singular Value Decomposition

Recall that when QR with column pivoting is used to solve the rank deficient LS problem, it finds x_B and not x_{LS}. In order to find the minimal 2-norm LS solution, it is necessary to zero the submatrix R_{12} in

$$Q^T A \Pi = \begin{bmatrix} R_{11} & R_{12} \\ 0 & 0 \end{bmatrix} \begin{matrix} r \\ m-r \end{matrix} \qquad r = \text{rank}(A)$$
$$\phantom{Q^T A \Pi = \begin{bmatrix} \end{bmatrix}} \begin{matrix} r \quad\; n-r \end{matrix}$$

This can be done by computing Householder matrices $Z_1, \ldots, Z_r \in \mathbb{R}^n$ such that

$$Z_r \ldots Z_1 \begin{bmatrix} R_{11}^T \\ R_{12}^T \end{bmatrix} = \begin{bmatrix} T^T \\ 0 \end{bmatrix} \begin{matrix} r \\ n-r \end{matrix},$$

where T is upper triangular. Using the technique described in P6.2-5, this can be accomplished in $r^2 (n - r)$ flops. It follows that if

$$Z = \Pi Z_1 \ldots Z_r,$$

then

$$Q^T A Z = \begin{bmatrix} T & 0 \\ 0 & 0 \end{bmatrix} \begin{matrix} r \\ m-r \end{matrix}.$$
$$\qquad\qquad\; r \quad n-r$$

Once this decomposition is found, x_{LS} readily follows. Indeed, for any $x \in \mathbb{R}^n$ we have

$$\| Ax - b \|_2^2 = \| Q^T A Z Z^T x - Q^T b \|_2^2 = \| Tw - c \|_2^2 + \| d \|_2^2,$$

where

$$x = Z \begin{bmatrix} w \\ y \end{bmatrix} \begin{matrix} r \\ n-r \end{matrix}$$

and

$$Q^T b = \begin{bmatrix} c \\ d \end{bmatrix} \begin{matrix} r \\ m-r \end{matrix}.$$

Clearly, if x is to minimize the sum of squares then we must have $w = T^{-1}c$. For x to have minimal 2-norm, y must be zero, and thus,

$$x_{LS} = Z \begin{bmatrix} T^{-1}c \\ 0 \end{bmatrix}.$$

In general, if Q and Z are orthogonal and $(Q^T A Z)_{ij} = 0$ for all $i > \text{rank}(A)$ and $j > \text{rank}(A)$, then we say that $B = Q^T A Z$ is a *complete orthogonal decomposition*. To calculate such a decomposition, it is necessary to estimate $\text{rank}(A)$. As we discussed in the previous section, Algorithm 6.4-1 can fail to detect rank in certain examples. Thus, the above algorithm may not be a satisfactory means for computing a complete orthogonal decomposition.

It turns out that the only fully reliable way to treat rank deficiency is to compute the singular value decomposition (SVD). (Note that the SVD is an example of a complete orthogonal factorization). In this section we discuss some of the details associated with using the SVD to solve the rank deficient LS problem.

The particular SVD algorithm that we shall consider is due to Golub and Reinsch (1970). It begins by reducing A to upper bidiagonal form using

Householder matrices. Because this reduction is important in its own right, we describe this portion of the calculation in detail.

The basic idea is to determine products of Householder matrices

$$U_B = U_1 \ldots U_n \in \mathbb{R}^{m \times m}$$

and

$$V_B = V_1 \ldots V_{n-2} \in \mathbb{R}^{n \times n}$$

such that

(6.5-1)
$$U_B^T A V_B = B = \begin{bmatrix} d_1 & f_2 & & & \mathbf{0} \\ & d_2 & \cdot & & \\ & & \cdot & \cdot & \\ \mathbf{0} & & & \cdot & f_n \\ & & & & d_n \\ \hline & & \mathbf{0} & & \end{bmatrix}.$$

A 5-by-4 example illustrates the roles of the individual U_i and V_i:

$$\begin{bmatrix} x & x & x & x \\ x & x & x & x \\ x & x & x & x \\ x & x & x & x \\ x & x & x & x \end{bmatrix} \xrightarrow{U_1} \begin{bmatrix} x & x & x & x \\ 0 & x & x & x \\ 0 & x & x & x \\ 0 & x & x & x \\ 0 & x & x & x \end{bmatrix}$$

$$\xrightarrow{V_1} \begin{bmatrix} x & x & 0 & 0 \\ 0 & x & x & x \\ 0 & x & x & x \\ 0 & x & x & x \\ 0 & x & x & x \end{bmatrix} \xrightarrow{U_2} \begin{bmatrix} x & x & 0 & 0 \\ 0 & x & x & x \\ 0 & 0 & x & x \\ 0 & 0 & x & x \\ 0 & 0 & x & x \end{bmatrix} \xrightarrow{V_2} \begin{bmatrix} x & x & 0 & 0 \\ 0 & x & x & 0 \\ 0 & 0 & x & x \\ 0 & 0 & x & x \\ 0 & 0 & x & x \end{bmatrix}$$

$$\xrightarrow{U_3} \begin{bmatrix} x & x & 0 & 0 \\ 0 & x & x & 0 \\ 0 & 0 & x & x \\ 0 & 0 & 0 & x \\ 0 & 0 & 0 & x \end{bmatrix} \xrightarrow{U_4} \begin{bmatrix} x & x & 0 & 0 \\ 0 & x & x & 0 \\ 0 & 0 & x & x \\ 0 & 0 & 0 & x \\ 0 & 0 & 0 & 0 \end{bmatrix}.$$

In general, U_k introduces zeros into the k-th column, while V_k zeros the appropriate entries in row k. Overall we have:

ALGORITHM 6.5-1: *Householder Bidiagonalization.* Given $A \in \mathbb{R}^{m \times n}$ ($m \geqslant n$), the following algorithm overwrites A with $U_B^T A V_B = B$ where B is upper bidiagonal and U_B and V_B are orthogonal.

> For $k = 1, \ldots, n$
>> Determine a Householder matrix $\tilde{U}_k$ of order $m - k + 1$ such that
>>
>> $$\tilde{U}_k \begin{bmatrix} a_{kk} \\ \vdots \\ a_{mk} \end{bmatrix} = \begin{bmatrix} x \\ 0 \\ \vdots \\ 0 \end{bmatrix}$$
>
> $A := \text{diag}(I_{k-1}, \tilde{U}_k) A$
> If $k \leqslant n - 2$
>> *then*
>>> Determine a Householder matrix $\tilde{V}_k$ of order $n - k + 1$ such that
>>>
>>> $$[a_{k,k+1}, \ldots, a_{kn}] \tilde{V}_k = (x, 0, \ldots, 0)$$
>>> $$A := A \, \text{diag}(I_k, \tilde{V}_k)$$

This algorithm requires $2mn^2 - \frac{2}{3} n^3$ flops. The orthogonal matrices U_B and V_B can be stored in factored form as in Algorithm 6.2-1. In particular, the Householder vectors associated with the $U_k = \text{diag}(I_{k-1}, \tilde{U}_k)$ can be stored in the lower triangular portion of A, while the Householder vectors affiliated the $V_k = \text{diag}(I_k, \tilde{V}_k)$ can be stored in the upper triangular portion of A. Such a technique is used in Golub and Kahan (1965), where Algorithm 6.5-1 is first described.

EXAMPLE 6.5-1. If Algorithm 6.5-1 is applied to

$$A = \begin{bmatrix} 1 & 2 & 3 \\ 4 & 5 & 6 \\ 7 & 8 & 9 \\ 10 & 11 & 12 \end{bmatrix},$$

then to three significant digits we obtain

$$\hat{B} = \begin{bmatrix} 12.8 & 21.8 & 0 \\ 0 & 2.24 & -.613 \\ 0 & 0 & 0 \\ 0 & 0 & 0 \end{bmatrix} \qquad \hat{V}_B = \begin{bmatrix} 1.00 & 0 & 0 \\ 0 & -.667 & -.745 \\ 0 & -.745 & .667 \end{bmatrix}$$

$$\hat{U}_B = \begin{bmatrix} -.0776 & -.833 & .392 & -.383 \\ -.311 & -.451 & -.238 & .802 \\ -.543 & -.069 & .701 & -.457 \\ -.776 & .312 & .547 & .037 \end{bmatrix}.$$

Once the bidiagonalization is complete and U_B and V_B are stored in factored form, then it is possible to solve the full-rank LS problem:

$$b := U_n \ldots U_1 b$$
$$x_n := b_n / d_n$$
$$x_k := (b_k - f_{k+1} x_{k+1}) / d_k \qquad (k = n - 1, \ldots, 1)$$
$$x := V_{n-2} \cdots V_1 x$$

These calculations require $2mn$ flops.

If the matrices U_B and V_B are explicitly desired, then they can be accumulated in $2m^2 n - \frac{4}{3} n^3$ and $\frac{2}{3} n^3$ flops, respectively (cf. §3.4).

A faster method of bidiagonalizing when $m \gg n$ results if we upper triangularize A first before applying Algorithm 6.5-1. In particular, if we compute an orthogonal $Q_1 \in \mathbb{R}^{m \times m}$ such that

$$Q_1^T A = \begin{bmatrix} R_1 \\ 0 \end{bmatrix} \begin{matrix} n \\ m-n \end{matrix}$$

is upper triangular and then apply Algorithm 6.5-1 to R_1 we obtain

$$Q_2^T R_1 V_B = B_1,$$

where Q_2, $V_B \in \mathbb{R}^{n \times n}$ are orthogonal and $B_1 \in \mathbb{R}^{n \times n}$ is upper bidiagonal. Defining $U_B = Q_1 \, \text{diag}(Q_2, I_{m-n})$ we see that

$$U_B^T A V_B = \begin{bmatrix} B_1 \\ 0 \end{bmatrix} \equiv B$$

is a bidiagonalization of A.

The idea of computing the bidiagonalization in this manner is mentioned in Lawson and Hanson (SLS, p. 119) and more fully analyzed in Chan (1982a). We refer to this method as R-bidiagonalization. By comparing its flop count $(mn^2 + n^3)$ with that for Algorithm 6.5-1 $(2mn^2 - \frac{2}{3}n^3)$ we see that it involves fewer computations (approximately) whenever $m \geqslant \frac{5}{3}n$.

Once the bidiagonalization of A has been achieved, the next step in the Golub-Reinsch SVD algorithm is to zero the superdiagonal elements in B. This is an iterative process and is accomplished by an algorithm due to Golub and Kahan (1965). Unfortunately, we must defer our discussion of this iteration until Chapter 8 since it requires an understanding of the symmetric QR algorithm for eigenvalues. Suffice it to say here that it essentially computes orthogonal matrices U_Σ and V_Σ such that

$$U_\Sigma^T B V_\Sigma = \Sigma = \mathrm{diag}(\sigma_1, \ldots, \sigma_n).$$

By defining $U = U_B U_\Sigma$ and $V = V_B V_\Sigma$ we see that $U^T A V = \Sigma$ is the SVD of A. The flop counts associated with this portion of the algorithm depends upon "how much" of the SVD is required. For example, when solving the LS problem, U^T need never be explicitly formed but merely applied to b as it is developed. In other applications, only the matrix U_1 is required where

$$U = [\,U_1, \quad U_2 \ ].$$
$$\quad\quad n \quad m-n$$

Altogether there are six possibilities, and the total amount of work required by the SVD algorithm in each case is summarized in Table 6.5-1. Because of the two possible bidiagonalization schemes, there are two columns of flop counts. If the bidiagonalization is achieved via Algorithm 6.5-1, the Golub-Reinsch (1970) SVD algorithm results, while if R-bidiagonalization is invoked we obtain the Chan (1982b) algorithm. By comparing the entries in this table (which are meant only as approximate estimates of work) we conclude that Chan's approach is more efficient whenever $m \geqslant \frac{5}{3}$.

We now focus our attention on the ability of the SVD to handle the difficult rank-deficient LS problem, even in the presence of roundoff. Our claims apply to either the Golub-Reinsch or the Chan version. Recall that if $A = U\Sigma V^T$ is the SVD of A, then

$$x_{LS} = \sum_{i=1}^{r} (u_i^T b \,/\, \sigma_i)v_i$$

where $r = \mathrm{rank}(A)$. Denote the computed versions of U, V, and $\Sigma = \mathrm{diag}(\sigma_i)$ by $\hat{U}$, $\hat{V}$, and $\hat{\Sigma} = \mathrm{diag}(\hat{\sigma}_i)$. Assume that both sequences of singular values range from largest to smallest. It will be shown in Chapter 8 that

(6.5-2) $\qquad \hat{U} = W + \Delta U \qquad W^T W = I_m \qquad \|\Delta U\|_2 \leqslant \epsilon$

(6.5-3) $\qquad \hat{V} = Z + \Delta V \qquad Z^T Z = I_n \qquad \|\Delta D\|_2 \leqslant \epsilon$

(6.5-4) $\qquad \hat{\Sigma} = W^T(A + \Delta A)Z \qquad \|\Delta A\|_2 \leqslant \epsilon\|A\|_2$

Table 6.5-1. SVD Flopcounts ($m \geq n$)

Required	Golub-Reinsch SVD	Chan SVD
Σ	$2mn^2 - \frac{2}{3}n^3$	$mn^2 + n^3$
Σ, V	$2mn^2 + 4n^3$	$mn^2 + \frac{17}{3}n^3$
Σ, U	$2m^2n + 4mn^2$	$2m^2n + \frac{19}{3}n^3$
Σ, U_1	$7mn^2 - n^3$	$3mn^2 + \frac{16}{3}n^3$
Σ, U, V	$2m^2n + 4mn^2 + \frac{14}{3}n^3$	$2m^2n + 11n^3$
Σ, U_1, V	$7mn^2 + \frac{11}{3}n^3$	$3mn^2 + 10n^3$

where ϵ is a small multiple of **u**, the machine precision. In plain English, the SVD algorithm computes the singular values of the "nearby" matrix $A + \Delta A$.

Note that $\hat{U}$ and $\hat{V}$ are not necessarily close to their exact counterparts. However, we can show that $\hat{\sigma}_k$ is close to σ_k. Using (6.5-4) and Corollary 2.3-3 we have

$$\sigma_k = \min_{\text{rank}(B)=k-1} \| A - B \|_2$$

$$= \min_{\text{rank}(B)=k-1} \| (\hat{\Sigma} - B) - W^T(\Delta A)Z \|_2.$$

Since

$$\| W^T(\Delta A)Z \|_2 \leqslant \epsilon \| A \|_2 = \epsilon \sigma_1$$

and

$$\min_{\text{rank}(B)=k-1} \| \hat{\Sigma} - B \|_2 = \hat{\sigma}_k,$$

it follows that

$$| \sigma_k - \hat{\sigma}_k | \leqslant \epsilon \sigma_1. \qquad\qquad k = 1, 2, \ldots, n$$

This implies that near rank deficiency in A cannot escape detection when the SVD of A is computed.

EXAMPLE 6.5-2.   If the LINPACK SVD algorithm is applied with machine precision $\mathbf{u} \cong 10^{-17}$ to

$$T_n(c) = \text{diag}(1, s, \ldots, s^{n-1}) \begin{bmatrix} 1 & -c & -c & \cdots & -c \\ & 1 & -c & \cdots & -c \\ & & \ddots & \ddots & \vdots \\ \mathbf{0} & & & \ddots & \vdots \\ & & & & 1 \end{bmatrix} \in \mathbb{R}^{n \times n}$$

where $n = 100$, $c = .2$, and $s = \sqrt{1 - c^2}$, then

$$\hat{\sigma}_n = .367805646308792467 * 10^{-8}.$$

See Example 6.4-2.

However, the mere observation of small singular values does not solve the ill-conditioned LS problem, for we must still decide upon a value for rank(A). One approach to this difficult problem is to have a parameter $\delta > 0$ and a convention that A has "numerical rank" $\hat{r}$ if the $\hat{\sigma}_i$ satisfy

$$\hat{\sigma}_1 \geqslant \ldots \geqslant \hat{\sigma}_{\hat{r}} > \delta \geqslant \hat{\sigma}_{\hat{r}+1} \geqslant \ldots \geqslant \hat{\sigma}_n.$$

When this is the case, we can regard

$$x_{\hat{r}} = \sum_{i=1}^{\hat{r}} \frac{\hat{u}_i^T b}{\hat{\sigma}_i} \hat{v}_i$$

as an approximation to x_{LS}.

The parameter δ should be consistent with the machine precision, e.g., $\delta = \mathbf{u} \|A\|_\infty$. However, if the general level of relative error in the data is larger than $\mathbf{u}$, then δ should be correspondingly bigger, e.g., $\delta = 10^{-2} \|A\|_\infty$. Since $\|x_{\hat{r}}\|_2 \cong 1/\hat{\sigma}_r, \leqslant 1/\delta$, δ may also be chosen with the intention of producing an approximate LS solution with suitably small norm. (In §12.3, we discuss more complicated methods for doing this.)

If $\hat{\sigma}_{\hat{r}} \gg \delta$, then we have reason to be comfortable with $x_{\hat{r}}$ because A can then be unambiguously regarded as a rank $\hat{r}$ matrix (modulo δ). Yet $\{\hat{\sigma}_1, \ldots, \hat{\sigma}_n\}$ might not clearly split into subsets of small and large singular values, making the determination of $\hat{r}$ by this means somewhat arbitrary. This leads to more complicated methods for estimating rank.

For example, suppose $r = n$ and assume for the moment that $\Delta A = 0$ in (6.5-4). Thus, $\sigma_i = \hat{\sigma}_i$ for $i = 1, \ldots, n$. Denote the i-th columns of the matrices $\hat{U}$, W, $\hat{V}$, and Z by u_i, w_i, v_i, and z_i, respectively. Subtracting $x_{\hat{r}}$ from x_{LS} and taking norms we obtain

$$\|x_{\hat{r}} - x_{LS}\| \leqslant \sum_{i=1}^{\hat{r}} \frac{1}{\sigma_i} \|(w_i^T b) z_i - (u_i^T b) v_i\|_2 + \sqrt{\sum_{i=\hat{r}+1}^{n} \left(\frac{(w_i^T b)}{\sigma_i} \right)^2}$$

From (6.5-2) and (6.5-3) it is easy to verify that

$$(6.5\text{-}5) \qquad \|(w_i^T b) z_i - (u_i^T b) v_i\|_2 \leqslant 2(1 + \epsilon)\epsilon \|b\|_2,$$

and therefore

$$\|x_{\hat{r}} - x_{LS}\|_2 \leqslant \frac{\hat{r}}{\sigma_{\hat{r}}} 2(1 + \epsilon)\epsilon \|b\|_2 + \sqrt{\sum_{i=\hat{r}+1}^{n} (w_i^T b / \sigma_i)^2}.$$

The parameter $\hat{r}$ can be determined as that integer which minimizes the upper bound. Notice that the first term in the bound increases with $\hat{r}$, while the second decreases.

On occasions when minimizing the residual is more important than accuracy in the solution, we can determine $\hat{r}$ on the basis of how close we surmise $\|b - Ax_{\hat{r}}\|_2$ is to the true minimum. Paralleling the above analysis, it can be shown that

$$\|b - Ax_{\hat{r}}\|_2 - \|b - Ax_{LS}\|_2 \leqslant (n - \hat{r})\|b\|_2 + \epsilon\|b\|_2\left[\hat{r} + \frac{\sigma_1}{\sigma_{\hat{r}}}(1 + \epsilon)\right].$$

Again $\hat{r}$ could be chosen to minimize the upper bound.

These techniques can be incorporated in a computer program by approximating σ_i and w_i by $\hat{\sigma}_i$ and $\hat{u}_i$, respectively. The analysis can also be extended to the case when $\Delta A \neq 0$. See Varah (1973) for details.

APPENDIX

As we mentioned, when solving the LS problem via the SVD, only Σ and V have to be computed. Table 6.5-2 compares the efficiency of this approach with the other algorithms that we have presented.

Table 6.5-2. Solving the LS Problem ($m \geq n$)

Algorithm	Flop Count
Normal Equations	$\dfrac{mn^2}{2} + \dfrac{n^3}{6}$
Householder Orthogonalization	$mn^2 - \dfrac{n^3}{3}$
Modified Gram Schmidt	mn^2
Givens Orthogonalization	$2mn^2 - \frac{2}{3}n^3$
Householder Bidiagonalization	$2mn^2 - \frac{2}{3}n^3$
R-Bidiagonalization	$mn^2 + n^3$
Golub-Reinsch SVD	$2mn^2 + 4n^3$
Chan SVD	$mn^2 + \frac{17}{3}n^3$

Problems

P6.5-1. Givens transformations can be used to reduce an upper triangular matrix R to upper bidiagonal form. The form of the algorithm is as follows:

For $j = n, \ldots, 3$
    For $i = 1, \ldots, j - 2$
        $R := UR$ (U rotates rows i and $i + 1$ and zeros r_{ij})
        $R := RV$ (V rotates columns i and $i + 1$ and zeros
            $r_{i+1,i}$)

Fill in the details of this algorithm and given an operation count.

P6.5-2. Suppose $A \in \mathbb{R}^{m \times n}$ with $m < n$. Give an algorithm for computing the factorization

$$U^T A V = [B, 0],$$

where B is an n-by-m upper bidiagonal matrix. (Hint: Obtain the form

$$\begin{bmatrix} x & x & 0 & 0 & 0 & 0 \\ 0 & x & x & 0 & 0 & 0 \\ 0 & 0 & x & x & 0 & 0 \\ 0 & 0 & 0 & x & x & 0 \end{bmatrix}$$

using Householder matrices and then "chase" the $(n, n + 1)$ entry up the $n + 1$-st column by applying Givens rotations on the right.)

P6.5-3. Verify (6.5-5).

P6.5-4. Compute the SVD of $A = \begin{bmatrix} -105 & 92 \\ -76 & -18 \end{bmatrix}$ and use it to solve $Ax = \begin{bmatrix} 1 \\ 2 \end{bmatrix}$.

Notes and References for Sec. 6.5

Aspects of the complete orthogonal decomposition are discussed in

R. J. Hanson and C. L. Lawson (1969). "Extensions and Applications of the House-holder Algorithm for Solving Linear Least Squares Problems," *Math. Comp. 23*, 787-812.

P. A. Wedin (1973). "On the Almost Rank-Deficient Case of the Least Squares Problem," *BIT 13*, 344-54.

G. H. Golub and V. Pereyra (1976). "Differentiation of Pseudo-Inverses, Separable Nonlinear Least Squares Problems and Other Tales," in *Generalized Inverses and Applications*, ed. M. Z. Nashed, Academic Press, New York, pp. 303-24.

Of course, the most important complete orthogonal decomposition is the SVD. The standard method for computing this decomposition was originally detailed in the papers

G. H. Golub and W. Kahan (1965). "Calculating the Singular Values and Pseudo-Inverse of a Matrix," *SIAM J. Num. Anal. 2*, 205-24.

P. A. Businger and G. H. Golub. (1969). "Algorithm 358: Singular Value Decomposition of a Complex Matrix," *Comm. Assoc. Comp. Mach. 12*, 564-65.

G. H. Golub and C. Reinsch (1970). "Singular Value Decomposition and Least Squares Solutions," *Numer. Math. 14*, 403-20. See also HACLA, pp. 134-51.

Exactly how the SVD algorithm works is the subject of §8.3.

The R-bidiagonalization is described in

T. F. Chan (1982a). "An Improved Algorithm for Computing the Singular Value Decomposition," *ACM Trans. Math. Soft. 8*, 72–83.

T. F. Chan (1982b). "Algorithm 581: An Improved Algorithm for Computing the Singular Value Decomposition," *ACM Trans. Math. Soft. 8*, 84–88.

A FORTRAN code for solving the LS problem via the SVD may be found in LIN-PACK, chap. 11.

The computation of matrix rank is a very difficult problem. Some suggestions are offered in

J. M. Varah (1973). "On the Numerical Solution of Ill-Conditioned Linear Systems with Applications to Ill-Posed Problems," *SIAM J. Num. Anal. 10*, 257–67.

The issue of rank deficiency is discussed further in §§12.1 and 12.2.

Sec. 6.6. Weighting and Iterative Improvement

In §4.5 the concepts of scaling and iterative improvement were introduced in the context of square linear systems. Generalizations of these ideas that are applicable to the least squares problem are now offered. Much of what is known in this area is heuristic and controversial. Those wishing to join in the fray are encouraged to consult the references.

Column Weighting

Suppose $G \in \mathbb{R}^{n \times n}$ is nonsingular. A solution to the LS problem

$$(6.6\text{-}1) \qquad \min \| Ax - b \|_2 \qquad\qquad A \in \mathbb{R}^{m \times n}, b \in \mathbb{R}^m$$

can be obtained by finding the minimum 2-norm solution y_{LS} to

$$(6.6\text{-}2) \qquad \min \| (AG)y - b \|_2$$

and then setting $x_G = G y_{LS}$. If $\text{rank}(A) = n$, then $x_G = x_{LS}$. Otherwise, x_G is the minimum G-norm solution to (6.6-1), where the G-norm is defined by $\| z \|_G = \| G^{-1} z \|_2$.

The choice of G remains to be discussed. Sometimes its selection can be based on a priori knowledge of the uncertainties in x_{LS}. See Lawson and Hanson (SLS, pp. 185–88). On other occasions it may be desirable to normalize the columns of A by setting

$$G = G_0 \equiv \text{diag}(1/ \| a_i \|_2),$$

where $a_i \in \mathbb{R}^m$ is the i-th column of A. Van der Sluis (1969) has shown that if

$$\mu = \min_{G \text{ diagonal}} \kappa_2(AG)$$

then

$$\mu \leqslant \kappa_2(AG_0) \leqslant \sqrt{n}\,\mu.$$

Since the computed accuracy of y_{LS} depends on $\kappa_2(AG)$, a case can be made for setting $G = G_0$.

We remark that column weighting affects singular values. Consequently, a scheme for determining numerical rank may not return the same estimates when applied to A and AG.

Row Weighting

Let $D = \text{diag}(d_1, \ldots, d_m)$ be nonsingular and consider the *weighted least squares problem*:

(6.6-3)   minimize $\| D(Ax - b) \|_2$. $A \in \mathbb{R}^{m \times n}, b \in \mathbb{R}^m$

Assume $\text{rank}(A) = n$ and that x_D solves (6.6-3). It follows that the solution x_{LS} to (6.6-1) satisfies

(6.6-4) $x_D - x_{LS} = (A^T D^2 A)^{-1} A^T (D^2 - I)(b - Ax_{LS})$.

This shows that row weighting in the LS problem affects the solution. (An important exception occurs when $b \in R(A)$, for then $x_D = x_{LS}$.)

One way of determining D is to let d_k be some measure of the uncertainty in b_k, e.g., the reciprocal of b_k's standard deviation. The tendency is for $r_k = e_k^T$ $(b - Ax_D)$ to be small whenever d_k is large. The precise effect of d_k on r_k can be clarified as follows. Define

$$D(\delta) = \text{diag}(d_1, \ldots, d_{k-1}, d_k \sqrt{1 + \delta}, d_{k+1}, \ldots, d_m),$$

where $\delta > -1$. If $x(\delta)$ minimizes $\| D(\delta)(Ax - b) \|_2$, and $r_k(\delta)$ is the k-th component of $b - Ax(\delta)$, then it can be shown that

(6.6-5) $r_k(\delta) = \dfrac{r_k}{1 + \delta d_k^2 e_k^T A (A^T D^2 A)^{-1} A^T e_k}.$

This explicit expression shows that $r_k(\delta)$ is a monotone decreasing function of δ. Of course, how r_k changes when all the weights are varied is much more complicated.

EXAMPLE 6.6-1.   Suppose

$$A = \begin{bmatrix} 1 & 2 \\ 3 & 4 \\ 5 & 6 \\ 7 & 8 \end{bmatrix} \qquad b = \begin{bmatrix} 1 \\ 0 \\ 0 \\ 0 \end{bmatrix}.$$

If $D = I_4$ then $x_D = (-1, .85)^T$ and $r = b - Ax_D = (.3, -.4, -.1, .2)^T$. On the other hand, if $D = \text{diag}(1000, 1, 1, 1)$ then $x_D \cong (-1.43, 1.21)$ and

$$r = b - Ax_D = \begin{bmatrix} .000428 \\ -.571428 \\ -.142853 \\ .285714 \end{bmatrix}.$$

Further discussion of row weighting may be found in Lawson and Hanson (SLS, pp. 183–85).

Generalized Least Squares

In many estimation problems, the vector of observations b is related to x through the equation

(6.6-6) $$b = Ax + w,$$

where the *noise vector* w has zero mean and a symmetric positive definite *variance-covariance* matrix $\sigma^2 W$. Assume that W is known and that $W = BB^T$ for some $B \in \mathbb{R}^{m \times m}$. B might be given or it might be W's Cholesky triangle. In order that all the equations in (6.6-6) contribute equally to the determination of x, statisticians frequently solve the LS problem

(6.6-7) $$\min \| B^{-1}(Ax - b) \|_2.$$

An obvious computational approach to this problem is to form $\bar{A} = B^{-1}A$ and $\bar{b} = B^{-1}b$ and then apply any of our previous techniques to $\min \| \bar{A}x - \bar{b} \|_2$. Unfortunately, x will be poorly determined by this procedure if B is ill conditioned.

A much more stable way of solving (6.6-7) using orthogonal transformations has been suggested by Paige (1979a, 1979b). It is based on the idea that (6.6-7) is equivalent to the *generalized least squares* problem,

(6.6-8) $$\text{minimize } v^T v \quad \text{subject to } b = Ax + Bv.$$

Notice that this problem is defined even if A and B are rank deficient. Although Paige's technique can be applied when this is the case, we shall describe it under the assumption that both these matrices have full rank.

The first step is to compute the Q-R decomposition of A,

$$Q^T A = \begin{bmatrix} R_1 \\ 0 \end{bmatrix} \qquad Q = [\underset{n}{Q_1}, \underset{m-n}{Q_2}].$$

An orthogonal matrix $P \in \mathbb{R}^{m \times m}$ is then determined so that

$$Q_2^T BP = [\underset{n}{0}, \underset{m-n}{S}] \qquad P = [\underset{n}{P_1}, \underset{m-n}{P_2}].$$

where S is upper triangular. With the use of these orthogonal matrices the constraint in (6.6-8) transforms to

$$\begin{bmatrix} Q_1^T b \\ Q_2^T b \end{bmatrix} = \begin{bmatrix} R_1 \\ 0 \end{bmatrix} x + \begin{bmatrix} Q_1^T B P_1 & Q_1^T B P_2 \\ 0 & S \end{bmatrix} \begin{bmatrix} P_1^T v \\ P_2^T v \end{bmatrix}.$$

Notice that the "bottom half" of this equation determines v,

(6.6-9) $$Su = Q_2^T b \qquad v = P_2 u,$$

while the "top half" prescribes x:

(6.6-10) $\quad R_1 x = Q_1^T b - (Q_1^T B P_1 P_1^T + Q_1^T B P_2 P_2^T) v = Q_1^T b - Q_1^T B P_2 u.$

The attractiveness of this method is that all potential ill-conditioning is concentrated in triangular systems (6.6-9) and (6.6-10). Moreover, Paige (1979b) has shown that the above procedure is numerically stable, something that is not true of any method that explicitly forms $B^{-1}A$.

Iterative Improvement

A technique for refining an approximate LS solution has been analyzed by Bjorck (1967, 1968). It is based on the idea that if

(6.6-11) $$\begin{bmatrix} I_m & A \\ A^T & 0 \end{bmatrix} \begin{bmatrix} r \\ x \end{bmatrix} = \begin{bmatrix} b \\ 0 \end{bmatrix}, \qquad A \in \mathbb{R}^{m \times n}, b \in \mathbb{R}^m$$

then $\| b - Ax \|_2 = \min$. This follows because $r + Ax = b$ and $A^T r = 0$ imply $A^T A x = A^T b$. The above augmented system is nonsingular if rank $(A) = n$, which we hereafter assume.

By casting the LS problem in the form of a square linear system, the iterative improvement scheme of §4.5 can be applied:

$r^{(0)} = 0$
$x^{(0)} = 0$
For $k = 0, 1, \ldots$

$$\begin{bmatrix} f^{(k)} \\ g^{(k)} \end{bmatrix} = \begin{bmatrix} b \\ 0 \end{bmatrix} - \begin{bmatrix} I & A \\ A^T & 0 \end{bmatrix} \begin{bmatrix} r^{(k)} \\ x^{(k)} \end{bmatrix}$$

$$\begin{bmatrix} I & A \\ A^T & 0 \end{bmatrix} \begin{bmatrix} p^{(k)} \\ z^{(k)} \end{bmatrix} = \begin{bmatrix} f^{(k)} \\ g^{(k)} \end{bmatrix}$$

$$\begin{bmatrix} r^{(k+1)} \\ x^{(k+1)} \end{bmatrix} = \begin{bmatrix} r^{(k)} \\ x^{(k)} \end{bmatrix} + \begin{bmatrix} p^{(k)} \\ z^{(k)} \end{bmatrix}$$

The residuals $f^{(k)}$ and $g^{(k)}$ must be computed in higher precision, and an original copy of A must be around for this purpose.

If the Q-R decomposition of A is available, the solution of the augmented system is readily obtained. In particular, if $A = Q\begin{bmatrix} R_1 \\ 0 \end{bmatrix}$ then a system of the form

$$\begin{bmatrix} I & A \\ A^T & 0 \end{bmatrix} \begin{bmatrix} p \\ z \end{bmatrix} = \begin{bmatrix} f \\ g \end{bmatrix}$$

transforms to

$$\begin{bmatrix} I_n & 0 & R_1 \\ 0 & I_{m-n} & 0 \\ R_1^T & 0 & 0 \end{bmatrix} \begin{bmatrix} h \\ f_2 \\ z \end{bmatrix} = \begin{bmatrix} f_1 \\ f_2 \\ g \end{bmatrix} \quad \begin{matrix} n \\ m-n \\ n \end{matrix} ,$$

where $Q^T f = [f_1^T, f_2^T]^T$ and $Q^T p = [h^T, f_2^T]^T$. Thus, p and z can be determined by solving the triangular systems $R_1^T h = g$ and $R_1 z = f_1 - h$ and setting $p = Q [h^T, f_2^T]^T$. Assuming that Q is stored in factored form, each iteration requires $4mn - n^2$ flops.

The key to the iteration's success is that *both* the LS residual and solution are updated—not just the solution. Bjorck (1968) shows that if $\kappa_2(A) \cong \beta^q$ and t-digit arithmetic used, then $x^{(k)}$ has approximately $k(t - q)$ correct base β digits, provided the residuals are computed in double precision. Notice that it is $\kappa_2(A)$, not $\kappa_2(A)^2$, that appears in this heuristic.

Problems

P6.6-1. Verify (6.6-4).

P6.6-2. Let $A \in \mathbb{R}^{m \times n}$ have full rank and define the diagonal matrix $\Delta = \text{diag}(1, \ldots, \sqrt{(1 + \delta)}, \ldots 1)$, for $\delta > -1$. Denote the LS solution to $\min \| \Delta(Ax - b_k) \|_2$ by $x(\delta)$ and its residual by $r(\delta) = b - Ax(\delta)$.
(a) Show

$$r(\delta) = \left\{ I - \delta \frac{A(A^T A)^{-1} A^T e_k e_k^T}{1 + \delta e_k^T A (A^T A)^{-1} A^T e_k} \right\} r(0).$$

(b) Letting $r_k(\delta)$ stand for the k-th component of $r(\delta)$, show

$$r_k(\delta) = \frac{r_k(0)}{1 + \delta e_k^T A (A^T A)^{-1} A^T e_k} .$$

Use (b) to verify (6.6-5).

P6.6-3. Show how the SVD can be used to solve the generalized LS problem when the matrices A and B in (6.6-8) are rank deficient.

P6.6-4. Let $A \in \mathbb{R}^{m \times n}$ have rank n and for $\alpha \geq 0$ define

$$M(\alpha) = \begin{bmatrix} \alpha I_m & A \\ A^T & 0 \end{bmatrix}.$$

Show that

$$\sigma_{m+n}\,[M(\alpha)] = \min\left[\alpha,\; -\frac{\alpha}{2} + \sqrt{\sigma_n\,(A)^2 + \left(\frac{\alpha}{2}\right)^2}\,\right].$$

What value of α minimizes $\kappa_2\,[M(\alpha)]$?

P6.6-5. Another iterative improvement method for LS problems is the following

$$x^{(0)} = 0$$
For $k = 0, 1, \ldots$
$\qquad r^{(k)} = b - Ax^{(k)}$ (double precision)
$\qquad \|Az^{(k)} - r^{(k)}\|_2 = \min$
$\qquad x^{(k+1)} = x^{(k)} + z^{(k)}$

(a) Assuming that the Q-R decomposition of A is available, how many flops per iteration are required? (b) Show that the above iteration results by setting $g^{(k)} = 0$ in the iterative improvement scheme given in the text.

Notes and References for Sec. 6.6

Row and column weighting in the LS problem is discussed in Lawson and Hanson (SLS, pp. 180–88). The effect of column normalization on the condition of A is analyzed in

A. van der Sluis (1969). "Condition Numbers and Equilibration of Matrices," *Numer. Math. 14*, 14–23.

The theoretical and computational aspects of the generalized least squares problem appear in

C. C. Paige (1979a). "Computer Solution and Perturbation Analysis of Generalized Least Squares Problems," *Math. Comp. 33*, 171–84.
C. C. Paige (1979b). "Fast Numerically Stable Computations for Generalized Linear Least Squares Problems," *SIAM J. Num. Anal. 16*, 165–71.
S. Kourouklis and C. C. Paige (1981). "A Constrained Squares Approach to the General Gauss-Markov Linear Model," *J. Amer. Stat. Assoc. 76*, 620–25.

Iterative improvement in the LS problem is discussed in

G. H. Golub and J. H. Wilkinson (1966). "Note on Iterative Refinement of Least Squares Solutions," *Numer. Math. 9*, 139–48.
A. Bjorck (1967a). "Iterative Refinement of Linear Least Squares Solution I," *BIT 7*, 257–78.
A. Bjorck (1968). "Iterative Refinement of Linear Least Squares Solution II," *BIT 8*, 8–30.
A. Bjorck and G. H. Golub (1967). "Iterative Refinement of Linear Least Squares Solutions by Householder Transformation," *BIT 7*, 322–37.

Sec. 6.7. A Note on Square and Underdetermined Systems

The orthogonalization methods that have been developed in this chapter can obviously be applied to square systems. For reference purposes, we summarize in Table 6.7-1 the work requirements when these alternatives to Gaussian elimination are used. (The flop count for the SVD algorithm is $\frac{32}{3} n^3$ if b is not available at the time of the decomposition, for it then becomes necessary to accumulate U.)

Table 6.7-1. Solving Square Systems

Method	Flops required to solve $Ax = b$
Gaussian Elimination	$\frac{1}{3} n^3$
Householder Orthogonalization	$\frac{2}{3} n^3$
Modified Gram-Schmidt	n^3
Bidiagonalization	$\frac{4}{3} n^3$
Singular Value Decomposition	$6 n^3$

As the table suggests, Gaussian elimination is the most economical way to solve $Ax = b$. Nevertheless there are three reasons why orthogonalization methods might be considered:

(1) The flop counts tend to exaggerate the Gaussian elimination advantage.

(2) The orthogonalization methods have guaranteed stability; there is no "growth factor" ρ to worry about as in Gaussian elimination.

(3) In cases of ill-conditioning, the reliability of the SVD method is unsurpassed,

We are not expressing a strong preference for orthogonalization methods but merely suggesting viable alternatives to Gaussian elimination.

We now consider methods for solving the underdetermined system

(6.7-1) $Ax = b.$ $A \in \mathbb{R}^{m \times n}, b \in \mathbb{R}^m, m < n$

Notice that (6.7-1) either has no solution or has an infinity of solutions. In the second case it is important to distinguish between algorithms that find the minimum 2-norm solution and those that do not. The first algorithm we present is in the latter category.

ALGORITHM 6.7-1.   Given $A \in \mathbb{R}^{m \times n}$ and $b \in \mathbb{R}^m$ ($m < n$), the following algorithm finds an $x \in \mathbb{R}^n$ such that $Ax = b$ or else indicates that no such x exists.

Use Gaussian elimination with complete pivoting (Algorithm 4.4-1) to obtain the decomposition

$$PAQ = \begin{bmatrix} L_{11} \\ L_{21} \end{bmatrix} \begin{matrix} \\ r \end{matrix} \begin{bmatrix} U_{11}, & U_{12} \end{bmatrix} \begin{matrix} r \\ r \end{matrix} \qquad r = \text{rank}(A)$$

Solve the triangular systems

$$\begin{bmatrix} L_{11} \\ L_{21} \end{bmatrix} y = Pb \qquad \text{(quit if no solution)}$$

$$U_{11}z = y.$$

Set $x = Q \begin{bmatrix} z \\ 0 \end{bmatrix} \begin{matrix} r \\ n-r \end{matrix}$.

This algorithm requires at most $\dfrac{m^2 n}{2} - \dfrac{n^3}{6}$ flops.

If QR with column pivoting is used, then the minimum 2-norm solution can be obtained:

ALGORITHM 6.7-2. Given $A \in \mathbb{R}^{m \times n}$ and $b \in \mathbb{R}^n$ ($m < n$), the following algorithm finds the minimum 2-norm solution to $Ax = b$ or else indicates that no such x exists.

Use Householder's method with column interchanges (Algorithm 6.4-1) to compute the decomposition

$$A^T \Pi = \begin{bmatrix} Q_1 & Q_2 \end{bmatrix} \begin{matrix} \\ r \;\; n-r \end{matrix} \begin{bmatrix} R_{11} & R_{12} \\ 0 & 0 \end{bmatrix} \begin{matrix} r \\ n-r \end{matrix} .$$
$$\begin{matrix} \quad\quad\quad\quad\quad\quad\quad\quad\quad r \quad\;\; m-r \end{matrix}$$

Solve the lower triangular system

$$\begin{bmatrix} R_{11}^T \\ R_{12}^T \end{bmatrix} z = \Pi^T b \qquad \text{(quit if no solution)}$$

$$x := Q_1 z$$

This algorithm requires at most $m^2 n - \dfrac{m^3}{3}$ flops.

The SVD can also be used to compute the minimal norm solution of an underdetermined $Ax = b$ problem:

$$A = \sum_{i=1}^{r} \sigma_i u_i v_i^{\mathrm{T}} \qquad x = \sum_{i=1}^{r} \frac{u_i^{\mathrm{T}} b}{\sigma_i} v_i. \qquad r = \mathrm{rank}(A)$$

As in the least squares problem, the SVD approach is desirable whenever A is nearly rank deficient.

We conclude with a perturbation result for full-rank underdetermined systems.

THEOREM 6.7-1.   Suppose $m \leqslant n$ and that $A \in \mathbb{R}^{m \times n}$, $\delta A \in \mathbb{R}^{m \times n}$, $0 \neq b \in \mathbb{R}^m$, and $\delta b \in \mathbb{R}^m$ satisfy

$$\epsilon = \max \{e_A, e_b\} < \sigma_m (A),$$

where

$$e_A = \|\delta A\|_2 / \|A\|_2 \qquad and \qquad e_b = \|\delta b\|_2 / \|b\|_2.$$

If x and $\hat{x}$ satisfy

$$Ax = b$$

and

$$(A + \delta A)\hat{x} = b + \delta b,$$

then

$$\frac{\|\hat{x} - x\|_2}{\|x\|_2} \leqslant \kappa_2(A) \, [e_A \min\{2, n - m + 1\} + e_b] + O(\epsilon^2).$$

Proof. Let E and f be defined by $\delta A / \epsilon$ and $\delta b / \epsilon$. Note that rank $(A + tE) = m$ for all $0 < t < \epsilon$ and that

$$x(t) = (A + tE)^{\mathrm{T}} [(A + tE)(A + tE)^{\mathrm{T}}]^{-1} (b + tf)$$

satisfies $(A + tE)x(t) = b + tf$. By differentiating this expression for $x(t)$ and setting $t = 0$ in the result we obtain

$$(6.7\text{-}2) \qquad \dot{x}(0) = [I - A^{\mathrm{T}}(AA^{\mathrm{T}})^{-1}A]E^{\mathrm{T}}(AA^{\mathrm{T}})^{-1}b$$
$$+ A^{\mathrm{T}}(AA^{\mathrm{T}})^{-1}[f - Ex].$$

Since

$$\|x\|_2 = \|A^{\mathrm{T}}(AA^{\mathrm{T}})^{-1}b\|_2 \geqslant \sigma_m(A) \| (AA^{\mathrm{T}})^{-1}b\|_2,$$
$$\|I - A^{\mathrm{T}}(AA^{\mathrm{T}})^{-1}A\|_2 = \min \{1, n - m\},$$

and

$$\frac{\|f\|_2}{\|x\|_2} \leqslant \frac{\|f\|_2 \|A\|_2}{\|b\|_2},$$

we have

$$\frac{\|\hat{x} - x\|_2}{\|x\|_2} = \frac{\|x(\epsilon) - x(0)\|_2}{\|x(0)\|_2} = \epsilon \frac{\|\dot{x}(0)\|_2}{\|x\|_2} + O(\epsilon^2)$$

$$\leqslant \epsilon \left\{ \min\{1, n - m\} \frac{\|E\|_2}{\|A\|_2} + \frac{\|f\|_2}{\|b\|_2} + \frac{\|E\|_2}{\|A\|_2} \right\} \kappa_2(A) + O(\epsilon^2),$$

from which the theorem follows. □

Note that there is no $\kappa_2(A)^2$ factor as in the case of overdetermined systems.

Problems

P6.7-1. Show how the Chan-SVD (cf. §6.5) can be used to reduce the flop count in Algorithm 6.7-4 when $n \geq \frac{5}{3}m$.

P6.7-2. Derive (6.7-2).

P6.7-3. Find the minimal norm solution to the system $Ax = b$ where $A = (1, 2, 3)$ and $b = 1$.

Notes and References for Sec. 6.7

A complete survey of methods for undetermined systems is given in

R. E. Cline and R. J. Plemmons (1976). "l_2-Solutions to Undetermined Linear Systems," *SIAM Review 18*, 92–106.

Among the more interesting techniques presented is a mixed Gaussian elimination/orthogonalization method very suitable for slightly underdetermined systems. (See P6.2-6.)

The Unsymmetric Eigenvalue Problem

Having discussed linear equations and least squares, we now direct our attention to the third major problem area in matrix computations, the algebraic eigenvalue problem. The unsymmetric problem is considered in this chapter and the more agreeable symmetric case in the next.

Our first task is to present the decompositions of Schur and Jordan along with the basic properties of eigenvalues and invariant subspaces. The contrasting behavior of these two decompositions sets the stage in §7.2, where we investigate how the eigenvalues and invariant subspaces of a matrix are effected by perturbation. Condition numbers are developed that permit estimation of the errors that can be expected to arise because of roundoff.

The key algorithm of the chapter is the justly famous QR algorithm. This procedure is the most complex algorithm presented in this book, and its development is spread over three sections. We derive the basic QR iteration in §7.3 as a natural generalization of the simple power method. The next two sections are devoted to making this basic iteration computationally feasible. This involves the introduction of the Hessenberg decomposition in §7.4 and the notion of origin shifts in §7.5.

The QR algorithm computes the real Schur form of a matrix, a canonical form that displays eigenvalues but not eigenvectors. Consequently, additional computations must be performed if information regarding invariant subspaces is desired. Section 7.6, which could be subtitled "What to Do after the Real Schur Form Is Calculated," discusses various invariant subspace calculations that can follow the QR algorithm.

Finally, in the last section we consider the generalized eigenvalue problem $Ax = \lambda Bx$ and a variant of the QR algorithm that has been devised to solve it. This algorithm, called the QZ algorithm, underscores the importance of orthogonal matrices in the eigenproblem, a central theme of the chapter.

It is appropriate at this time to make a remark about complex versus real arithmetic. In this book we focus on the development of real arithmetic algorithms for real matrix problems. This chapter is no exception, even though a real unsymmetric matrix can have complex eigenvalues. However, in the derivation of the (real) practical QR algorithm and in the mathematical analysis of the eigenproblem itself, it is convenient to work in the complex field. Thus, the reader will find that we have switched to complex notation in §§7.1, 7.2, and 7.3. In these sections we use without comment the complex Q-R factorization ($A = QR$, Q unitary, R upper triangular) and the complex SVD ($A = U\Sigma V^{\mathrm{H}}$, U and V unitary, Σ diagonal).

Reading Path

$$\text{Chapters 1–4} \rightarrow \S7.1 \rightarrow \S7.2 \rightarrow \S7.3 \rightarrow \S7.4 \rightarrow \S7.5 \rightarrow \S7.6 \rightarrow \S7.7$$
$$\qquad\qquad\qquad \nearrow \qquad\qquad\qquad \nearrow$$
$$\qquad\quad \S6.2 \qquad\qquad\quad \S5.3$$

Sec. 7.1. Properties and Decompositions

The *eigenvalues* of a matrix $A \in \mathbb{C}^{n \times n}$ are the n roots of its *characteristic polynomial* $p(z) = \det(zI - A)$. The set of these roots is called the *spectrum* and is denoted by $\lambda(A)$. If $\lambda(A) = \{\lambda_1, \ldots, \lambda_n\}$, then it follows that

$$\det(A) = \lambda_1 \lambda_2 \cdots \lambda_n.$$

Moreover, if we define the *trace* of A by

$$\text{trace}(A) = \sum_{i=1}^{n} a_{ii}, \qquad\qquad A \in \mathbb{C}^{n \times n}$$

then $\text{trace}(A) = \sum_{i=1}^{n} \lambda_i$.

If $\lambda \in \lambda(A)$ then the nonzero vectors $x \in \mathbb{C}^n$ that satisfy

$$Ax = \lambda x$$

are referred to as *eigenvectors*. An eigenvector defines a one-dimensional subspace that is invariant with respect to premultiplication by A. More generally, a subspace $S \subset \mathbb{C}^n$ with the property that

$$x \in S \Rightarrow Ax \in S$$

is said to be *invariant* (for A). Note that if

$$AX = XB, \qquad\qquad B \in \mathbb{C}^{k \times k}, \; X \in \mathbb{C}^{n \times k}$$

then $R(X)$ is invariant and

$$By = \lambda y \Rightarrow A(Xy) = \lambda(Xy).$$

Thus, if X has full column rank, then $AX = XB$ implies that $\lambda(B) \subset \lambda(A)$. If X is square and nonsingular, then $\lambda(A) = \lambda(B)$ and we say that A and $B = X^{-1}AX$ are *similar*. In this context, X is called a *similarity transformation*.

By using similarity transformations, it is possible to reduce a given matrix to any of several canonical forms. The canonical forms differ in how they display the eigenvalues and in the kind of invariant subspace information that they provide. Because of their numerical stability, we begin by discussing the reductions that can be achieved with unitary similarity.

LEMMA 7.1-1. If $A \in \mathbb{C}^{n \times n}$, $B \in \mathbb{C}^{p \times p}$, and $X \in \mathbb{C}^{n \times p}$ satisfy

(7.1-1) $$\qquad\qquad AX = XB, \qquad\qquad \text{rank}(X) = p$$

then there exists a unitary $Q \in \mathbb{C}^{n \times n}$ such that

(7.1-2) $$Q^H A Q = T = \begin{bmatrix} T_{11} & T_{12} \\ 0 & T_{22} \end{bmatrix} \begin{matrix} p \\ n-p \end{matrix},$$
$$\qquad\qquad\quad p \quad\; n-p$$

where $\lambda(T_{11}) = \lambda(A) \cap \lambda(B)$.

Proof. Let

$$X = Q \begin{bmatrix} R \\ 0 \end{bmatrix} \qquad\qquad Q \in \mathbb{C}^{n \times n}, \; R \in \mathbb{C}^{p \times p}$$

be the Q-R factorization of X. By substituting this into (7.1-1) and rearranging we have

$$\begin{bmatrix} T_{11} & T_{12} \\ T_{21} & T_{22} \end{bmatrix} \begin{bmatrix} R \\ 0 \end{bmatrix} = \begin{bmatrix} R \\ 0 \end{bmatrix} B,$$

where

$$Q^H A Q = \begin{bmatrix} T_{11} & T_{12} \\ T_{21} & T_{22} \end{bmatrix} \begin{matrix} p \\ n-p \end{matrix}.$$
$$\qquad\qquad\quad p \quad\; n-p$$

By using the nonsingularity of R and the equations $T_{21}R = 0 \cdot B$ and $T_{11}R = RB$, we can conclude that $T_{21} = 0$ and $\lambda(T_{11}) = \lambda(B)$. The lemma now follows because $\lambda(A) = \lambda(T) = \lambda(T_{11}) \cup \lambda(T_{22})$. $\square$

EXAMPLE 7.1-1. If

$$A = \begin{bmatrix} 67.00 & 177.60 & -63.20 \\ -20.40 & 95.88 & -87.16 \\ 22.80 & 67.84 & 12.12 \end{bmatrix},$$

$X = (20, -9, -12)^T$, and $B = (25)$, then $AX = XB$. Moreover, if the orthogonal matrix Q is defined by

$$Q = \begin{bmatrix} -.800 & .360 & .480 \\ .360 & .928 & -.096 \\ .480 & -.096 & .872 \end{bmatrix}$$

then $Q^T X = (-25, 0, 0)^T$ and

$$Q^T A Q = T = \begin{bmatrix} 25 & -90 & 5 \\ 0 & 147 & -104 \\ 0 & 146 & 3 \end{bmatrix}.$$

Note: $\lambda(A) = \{25, 75 + 100i, 75 - 100i\}$.

Thus, a matrix can be reduced to block triangular form using unitary similarity transformations if we know one of its invariant subspaces. By inducting on this result, we can readily establish the decomposition of Schur (1909).

THEOREM 7.1-2: *Schur Decomposition.* If $A \in \mathbb{C}^{n \times n}$ then there exists a unitary $Q \in \mathbb{C}^{n \times n}$ such that

$$(7.1-3) \qquad\qquad Q^H A Q = T = D + N$$

where $D = \text{diag}(\lambda_1, \ldots, \lambda_n)$ and $N \in \mathbb{C}^{n \times n}$ is strictly upper triangular. Furthermore, Q can be chosen so that the eigenvalues λ_i appear in any order along the diagonal.

Proof. The theorem obviously holds when $n = 1$. Suppose it holds for all matrices of order $n - 1$ or less. If $Ax = \lambda x$, where $x \neq 0$, then by Lemma 7.1-1 (with $B = (\lambda)$) there exists a unitary U such that

$$U^HAU = \begin{array}{c} \\ \left[\begin{array}{cc} \lambda & w^H \\ 0 & C \end{array}\right] \\ \begin{array}{cc} 1 & n-1 \end{array} \end{array} \begin{array}{c} 1 \\ n-1 \end{array} .$$

By induction there is a unitary $\tilde{U}$ such that $\tilde{U}^H C\tilde{U}$ is upper triangular. Thus, if $Q = U \operatorname{diag}(1, \tilde{U})$, then Q^HAQ is upper triangular. $\square$

EXAMPLE 7.1-2.   If

$$A = \left[\begin{array}{cc} 75 & 200 \\ -50 & 75 \end{array}\right],$$

then

$$Q = \left[\begin{array}{cc} .8944i & .4472 \\ -.4472 & -.8944i \end{array}\right]$$

is unitary and

$$Q^HAQ = T = \left[\begin{array}{cc} 75 + 100i & -150 \\ 0 & 75 - 100i \end{array}\right].$$

If $Q = [q_1, \ldots, q_n]$ is a column partition of the unitary matrix Q in (7.1-3), then the q_i are referred to as *Schur vectors*. By equating columns in the equation $AQ = QT$ we see that the Schur vectors satisfy

$$(7.1\text{-}4) \qquad Aq_k = \lambda_k q_k + \sum_{i=1}^{k-1} n_{ik}q_i. \qquad\qquad k = 1, \ldots, n$$

From this we conclude that the subspaces

$$S_k = \operatorname{span}\{q_1, \ldots, q_k\} \qquad\qquad k = 1, \ldots, n$$

are invariant. Moreover, if $Q_k = [q_1, \ldots, q_k]$, then $\lambda(Q_k^H A Q_k) = \{\lambda_1, \ldots, \lambda_k\}$. Since the eigenvalues in (7.1-3) can be arbitrarily ordered, it follows that there is at least one k-dimensional invariant subspace associated with each subset of k eigenvalues.

Another conclusion to be drawn from (7.1-4) is that the Schur vector q_k is an eigenvector if and only if the k-th column of N is zero. This turns out to be the case for $k = 1, \ldots, n$ whenever $A^HA = AA^H$. Matrices that satisfy this commutivity relation are called *normal*.

COROLLARY 7.1-3. $A \in \mathbb{C}^{n \times n}$ is normal if and only if there exists a unitary $Q \in \mathbb{C}^{n \times n}$ such that $Q^HAQ = \operatorname{diag}(\lambda_1, \ldots, \lambda_n)$.

Proof. It is easy to show that if A is unitarily similar to a diagonal matrix, then A is normal. On the other hand, if A is normal and $Q^H A Q = T$ is its Schur decomposition, then T is also normal. The corrollary follows by showing that a normal, upper triangular matrix is diagonal. $\square$

Note that if $Q^H A Q = T = \text{diag}(\lambda_i) + N$ is a Schur decomposition of a general n-by-n matrix A, then $\|N\|_F$ is independent of the choice of Q:

$$\|N\|_F^2 = \|A\|_F^2 - \sum_{i=1}^{n} |\lambda_i|^2 \equiv \Delta^2(A).$$

This quantity is referred to as A's *departure from normality*.

Thus, to make T "more diagonal," it is necessary to rely on nonunitary similarity transformations. To see what is involved in such a reduction, we examine the block diagonalization of a 2-by-2 block triangular matrix.

LEMMA 7.1-4. Let $T \in \mathbb{C}^{n \times n}$ be partitioned as follows

$$T = \begin{bmatrix} T_{11} & T_{12} \\ 0 & T_{22} \end{bmatrix} \begin{matrix} p \\ q \end{matrix}$$
$$\quad\quad p \quad\ q$$

and define the linear transformation $\phi: \mathbb{C}^{p \times q} \to \mathbb{C}^{p \times q}$ by

$$\phi(X) = T_{11}X - XT_{22}. \qquad\qquad X \in \mathbb{C}^{p \times q}$$

Then ϕ is nonsingular if and only if $\lambda(T_{11}) \cap \lambda(T_{22}) = \emptyset$. If ϕ is nonsingular and Y is defined by

$$Y = \begin{bmatrix} I_p & Z \\ 0 & I_q \end{bmatrix}, \qquad\qquad \phi(Z) = -T_{12}$$

then $Y^{-1}TY = \text{diag}(T_{11}, T_{22})$.

Proof. Suppose $\phi(X) = 0$ for $X \neq 0$ and that

$$U^H X V = \begin{bmatrix} \Sigma_r & 0 \\ 0 & 0 \end{bmatrix} \begin{matrix} r \\ p-r \end{matrix} \qquad \Sigma_r = \text{diag}(\sigma_i), \ r = \text{rank}(X)$$
$$\quad\quad r \quad\ q-r$$

is the SVD of X. Substituting this into the equation $T_{11}X = XT_{22}$ gives

$$\begin{bmatrix} A_{11} & A_{12} \\ A_{21} & A_{22} \end{bmatrix} \begin{bmatrix} \Sigma_r & 0 \\ 0 & 0 \end{bmatrix} = \begin{bmatrix} \Sigma_r & 0 \\ 0 & 0 \end{bmatrix} \begin{bmatrix} B_{11} & B_{12} \\ B_{21} & B_{22} \end{bmatrix},$$

where

$$U^H T_{11} U = (A_{ij}) \qquad \text{and} \qquad V^H T_{22} V = (B_{ij}).$$

By comparing blocks we see that $A_{21} = 0$, $B_{12} = 0$, and $\lambda(A_{11}) = \lambda(B_{11})$. Consequently,

$$\emptyset \neq \lambda(A_{11}) = \lambda(B_{11}) \subset \lambda(T_{11}) \cap \lambda(T_{22}).$$

On the other hand, if $\lambda \in \lambda(T_{11}) \cap \lambda(T_{22})$ then we have nonzero vectors x and y so $T_{11}x = \lambda x$ and $y^H T_{22} = \lambda y^H$. A calculation shows that $\phi(xy^H) = 0$. Finally, if ϕ is nonsingular then the matrix Z above exists and

$$
Y^{-1}TY = \begin{bmatrix} I & -Z \\ 0 & I \end{bmatrix} \begin{bmatrix} T_{11} & T_{12} \\ 0 & T_{22} \end{bmatrix} \begin{bmatrix} I & Z \\ 0 & I \end{bmatrix}
$$

$$
= \begin{bmatrix} T_{11} & T_{11}Z - ZT_{22} + T_{12} \\ 0 & T_{22} \end{bmatrix} = \begin{bmatrix} T_{11} & 0 \\ 0 & T_{22} \end{bmatrix}. \quad \square
$$

EXAMPLE 7.1-3.   If

$$
T = \begin{bmatrix} 1 & 2 & 3 \\ 0 & 3 & 8 \\ 0 & -2 & 3 \end{bmatrix}
$$

and

$$
Y = \begin{bmatrix} 1.0 & 0.5 & -0.5 \\ 0.0 & 1.0 & 0.0 \\ 0.0 & 0.0 & 1.0 \end{bmatrix},
$$

then

$$
Y^{-1}TY = \begin{bmatrix} 1 & 0 & 0 \\ 0 & 3 & 8 \\ 0 & -2 & 3 \end{bmatrix}.
$$

By repeatedly applying Lemma 7.1-4, we can establish the following more general result:

THEOREM 7.1-5: *Block Diagonal Decomposition.*   Suppose

$$(7.1\text{-}5) \qquad Q^H A Q = T = \begin{bmatrix} T_{11} & T_{12} & \cdots & T_{1q} \\ 0 & T_{22} & \cdots & T_{2q} \\ \vdots & \vdots & & \vdots \\ 0 & 0 & \cdots & T_{qq} \end{bmatrix}$$

is a Schur decomposition of $A \in \mathbb{C}^{n \times n}$. If $\lambda(T_{ii}) \cap \lambda(T_{jj}) = \emptyset$ whenever $i \neq j$, then there exists a nonsingular matrix $Y \in \mathbb{C}^{n \times n}$ such that

$$(7.1\text{-}6) \qquad (QY)^{-1}A(QY) = \text{diag}(T_{11}, \ldots, T_{qq}).$$

Proof. We leave the proof of this theorem to the reader. $\square$

If each diagonal block T_{ii} is associated with a distinct eigenvalue, then we obtain

COROLLARY 7.1-6. If $A \in \mathbb{C}^{n \times n}$ then there exists a nonsingular X such that

$$(7.1\text{-}7) \qquad X^{-1}AX = \text{diag}(\lambda_1 I + N_1, \ldots, \lambda_q I + N_q), \qquad N_i \in \mathbb{C}^{n_i \times n_i}$$

where $\lambda_1, \ldots, \lambda_q$ are distinct, the integers $n_1, \ldots, n_q$ satisfy $n_1 + \cdots + n_q = n$, and each N_i is strictly upper triangular.

A number of important terms are connected with decomposition (7.1-7). The integer n_i is referred to as the *algebraic multiplicity* of λ_i. If $n_i = 1$, then λ_i is said to be *simple*. The *geometric multiplicity* of λ_i equals the dimension of $N(N_i)$, i.e., the number of linearly independent eigenvectors associated with λ_i. If the algebraic multiplicity of λ_i exceeds its geometric multiplicity, then it is said to be a *defective eigenvalue*. A matrix with a defective eigenvalue is referred to as a *defective matrix*. Nondefective matrices are also said to be *diagonalizable* in light of the following result:

COROLLARY 7.1-7: *Diagonal Form.* $A \in \mathbb{C}^{n \times n}$ is nondefective if and only if there exists a nonsingular $X \in \mathbb{C}^{n \times n}$ such that

$$(7.1\text{-}8) \qquad X^{-1}AX = \text{diag}(\lambda_1, \ldots, \lambda_n).$$

Proof. A is nondefective if and only if there exist independent vectors $x_1, \ldots, x_n$ in $\mathbb{C}^n$ and scalars $\lambda_1, \ldots, \lambda_n$ such that $Ax_i = \lambda_i x_i$ for $i = 1, \ldots, n$. This is equivalent to the existence of a nonsingular $X = [x_1, \ldots, x_n] \in \mathbb{C}^{n \times n}$ such that $AX = XD$ where $D = \text{diag}(\lambda_1, \ldots, \lambda_n)$. $\square$

Note that if y_i^H is the i-th row of X^{-1}, then $y_i^H A = \lambda_i y_i^H$. Nonzero vectors z that satisfy $z^H A = \lambda z^H$ for $\lambda \in \lambda(A)$ are called *left eigenvectors*. (We sometimes refer to a nonzero vector x satisfying $Ax = \lambda x$ as a *right eigenvector*.)

EXAMPLE 7.1-4. If

$$A = \begin{bmatrix} 5 & -1 \\ -2 & 6 \end{bmatrix} \qquad \text{and} \qquad X = \begin{bmatrix} 1 & 1 \\ 1 & -2 \end{bmatrix}$$

then

$$X^{-1}AX = \text{diag}(4, 7).$$

If we partition the matrix X in (7.1-7),

$$X = [\underset{n_1}{X_1}, \ldots, \underset{n_q}{X_q}]$$

then

$$\mathbb{C}^n = R(X_1) \oplus \cdots \oplus R(X_q),$$

a direct sum of invariant subspaces. If the bases for these subspaces are chosen in a special way, then it is possible to introduce even more zeroes into the upper triangular portion of $X^{-1}AX$:

THEOREM 7.1-8: *Jordan Decomposition.*   If $A \in \mathbb{C}^{n \times n}$, then there exists a nonsingular $X \in \mathbb{C}^{n \times n}$ such that

$$X^{-1}AX = \text{diag}(J_1, \ldots, J_t),$$

where

$$J_i = \begin{bmatrix} \lambda_i & 1 & & & & \\ & \lambda_i & 1 & & \mathbf{0} & \\ & & \ddots & \ddots & & \\ & \mathbf{0} & & & \lambda_i & 1 \\ & & & & & \lambda_i \end{bmatrix}$$

is m_i-by-m_i and $m_1 + \cdots + m_t = n$.

Proof.   See Halmos (1958, pp. 112 *ff.*) $\square$

The J_i are referred to as *Jordan blocks*. The number and dimensions of the Jordan blocks associated with each distinct eigenvalue are unique, although their ordering along the diagonal is not.

Unfortunately, the Jordan block structure of a defective matrix is very difficult to determine numerically. The set of n-by-n diagonalizable matrices is dense in $\mathbb{C}^{n \times n}$, and thus, small changes in a defective matrix can radically alter its Jordan form.

A related difficulty that arises in the eigenvalue problem is that a nearly defective matrix can have a poorly conditioned matrix of eigenvectors. For example, any matrix X that diagonalizes

(7.1-9) $$A = \begin{bmatrix} 1 + \epsilon & 1 \\ 0 & 1 - \epsilon \end{bmatrix} \qquad 0 < \epsilon \ll 1$$

has a 2-norm condition of order $1/\epsilon$.

These observations serve to highlight the difficulties associated with ill-conditioned similarity transformations. Since

(7.1-10) $$fl(X^{-1}AX) = X^{-1}AX + E,$$

where

(7.1-11) $$\|E\|_2 \cong \mathbf{u}\kappa_2(X)\|A\|_2,$$

it is clear that large errors can be introduced into an eigenvalue calculation when we depart from unitary similarity.

EXAMPLE 7.1-5.   If

$$A = \begin{bmatrix} 5 & -1 \\ 2 & 2 \end{bmatrix}, \qquad \lambda(A) = \{3, 4\}$$

$$Q = \begin{bmatrix} .6 & .8 \\ -.8 & .6 \end{bmatrix}, \qquad \kappa_2(Q) = 1$$

and

$$X = \begin{bmatrix} 1. & 1.01 \\ 2. & 2.00 \end{bmatrix}, \qquad \kappa_2(X) \cong 167$$

then with $\beta = 10$, $t = 3$ rounded arithmetic we find

$$\|fl(X^{-1}AX) - X^{-1}AX\|_2 \cong 1.9$$

$$\|fl(Q^{T}AQ) - Q^{T}AQ\|_2 \cong 10^{-3}.$$

Problems

P7.1-1. Show that if $T \in \mathbb{C}^{n \times n}$ is upper triangular and normal, then T is diagonal.

P7.1-2. Verify that if X diagonalizes the 2-by-2 matrix in (7.1-9) and $\epsilon \leqslant \frac{1}{2}$ then $\kappa_1(A) \geqslant \frac{1}{\epsilon}$.

P7.1-3. Show that if $T = \begin{bmatrix} P & R \\ 0 & S \end{bmatrix}$ with P and S square, then $\lambda(T) = \lambda(P) \cup \lambda(S)$.

P7.1-4. Show that if A and B^{H} are in $\mathbb{C}^{m \times n}$ with $m \geqslant n$, then

$$\lambda(AB) = \lambda(BA) \cup \underbrace{\{0, \ldots, 0\}}_{m-n}.$$

P7.1-5. Given $A \in \mathbb{C}^{n \times n}$, show that for every $\epsilon > 0$ there exists a diagonalizable matrix B such that $\|A - B\|_2 \leqslant \epsilon$. (Hint: use the Schur decomposition.) This shows that the set of diagonalizable matrices is dense in $\mathbb{C}^{n \times n}$ and that the Jordan canonical form is not a continuous matrix decomposition.

P7.1-6. Suppose $A_k \to A$ and that $Q_k^H A_k Q_k = T_k$ is a Schur decomposition of A_k. Show that $\{Q_k\}$ has a converging subsequence

$$Q_{k_i} \to Q$$

and that $Q^H A Q = T$ is upper triangular. This shows that the eigenvalues of a matrix are continuous functions of its entries.

P7.1-7. Justify (7.1-10) and (7.1-11).

P7.1-8. Suppose $A \in \mathbb{C}^{n \times n}$ has distinct eigenvalues. Show that if $Q^H A Q = T$ is its Schur decomposition and $AB = BA$, then $Q^H B Q$ is upper triangular.

Notes and References for Sec. 7.1

The mathematical properties of the algebraic eigenvalue problem are elegantly covered in Wilkinson (AEP, chap. 1) and Stewart (IMC, chap. 6). For those who need further review we also recommend

R. Bellman (1970). *Introduction to Matrix Analysis*, 2nd ed., McGraw-Hill, New York.

M. Marcus and H. Minc (1964). *A Survey of Matrix Theory and Matrix Inequalities*, Allyn and Bacon, Boston.

L. Mirsky (1963). *An Introduction to Linear Algebra*, Oxford University Press, Oxford.

The Schur decomposition originally appeared in

I. Schur (1909) "On the Characteristic Roots of a Linear Substitution with an Application to the Theory of Integral Equations," *Math. Ann.* **66**, 488–510 (German).

A proof very similar to ours is given on page 105 of

H. W. Turnbull and A. C. Aitken (1961). *An Introduction to the Theory of Canonical Forms*, Dover, New York.

The Jordan decomposition is proved in many linear algebra texts, e.g.,

P. Halmos (1958). *Finite Dimensional Vector Spaces*, Van Nostrand, New York.

Sec. 7.2. Perturbation Theory

None of the decompositions in the preceding section can be calculated exactly because of roundoff error and because eigenvalue algorithms are iterative and must be terminated after a finite number of steps. Therefore, it is critical that we develop a useful perturbation theory to guide our thinking in subsequent sections where various methods for the eigenproblem are considered.

For example, several eigenvalue routines produce a sequence of similarity transformations X_k with the property that the matrices $X_k^{-1} A X_k$ are progressively "more diagonal." The question naturally arises, how well do the diagonal elements of a matrix approximate its eigenvalues?

THEOREM 7.2-1: *Gershgorin Circle Theorem.* If $X^{-1}AX = \text{diag}(d_1, \ldots, d_n) + F$ and F has zero diagonal entries, then $\lambda(A) \subset \cup_{i=1}^n D_i$ where

$$D_i = \{z \in \mathbb{C} \mid |z - d_i| \leqslant \sum_{i=1}^n |f_{ij}|\}.$$

Proof.   Suppose $\lambda \in \lambda(A)$ and assume without loss of generality that $\lambda \neq d_i$ for $i = 1, \ldots, n$. Since $(D - \lambda I) + F$ is singular, it follows that

$$1 \leqslant \|(D - \lambda I)^{-1}F\|_\infty = \sum_{j=1}^n |f_{kj}| / |d_k - \lambda|$$

for some k, $1 \leqslant k \leqslant n$. But this implies that $\lambda \in D_k$. $\square$

It can also be shown that if the Gershgorin disk D_i is isolated from the other disks, then it contains precisely one of A's eigenvalues. See Wilkinson (AEP, pp. 71 *ff.*).

EXAMPLE 7.2-1.   If

$$A = \begin{bmatrix} 10 & 2 & 3 \\ -1 & 0 & 2 \\ 1 & -2 & 1 \end{bmatrix}$$

then $\lambda(A) \cong \{10.226, .3870 + 2.2216i, .3870 - 2.2216i\}$ and the Gershgorin disks are

$$D_1 = \{|z| \mid |z - 10| \leqslant 5\}$$
$$D_2 = \{|z| \mid |z| \leqslant 3\}$$
$$D_3 = \{|z| \mid |z - 1| \leqslant 3\}.$$

For some very important eigenvalue routines it is possible to show that the computed eigenvalues $\lambda_1, \ldots, \lambda_n$ are the exact eigenvalues of a matrix $A + E$ where E is small in norm. Consequently, we must understand how the eigenvalues of a matrix can be effected by small perturbations. A sample result that sheds light on this issue is the following.

THEOREM 7.2-2: *(Bauer-Fike).* If μ is an eigenvalue of $A + E \in \mathbb{C}^{n \times n}$ and $X^{-1}AX = D = \text{diag}(\lambda_1, \ldots, \lambda_n)$, then

$$\min_{\lambda \in \lambda(A)} |\lambda - \mu| \leqslant \kappa_p(X)\|E\|_p$$

where $\| \cdot \|_p$ denotes any of the Holder norms.

Proof.   We need only consider the case $\mu \notin \lambda(A)$. If $X^{-1}(A + E - \mu I)X$ is singular, then so is $I + (D - \mu I)^{-1}(X^{-1}EX)$. Thus,

$$1 \leqslant \|(D - \mu I)^{-1}(X^{-1}EX)\|_p \leqslant \frac{1}{\min_{\lambda \in \lambda(A)} |\lambda - \mu|} \|X\|_p \|E\|_p \|X^{-1}\|_p,$$

from which the theorem follows. □

An analogous result can be obtained via the Schur decomposition:

THEOREM 7.2-3.   Let $Q^H A Q = D + N$ be a Schur decomposition of $A \in$ $\mathbb{C}^{n \times n}$ as in (7.1-3). If $\mu \in \lambda(A + E)$ and p is the smallest positive integer such that $N^p = 0$, then

$$\min_{\lambda \in \lambda(A)} |\lambda - \mu| \leqslant \max\{\theta, \theta^{1/p}\}$$

where

$$\theta = \|E\|_2 \sum_{k=0}^{p-1} \|N\|_2^k.$$

Proof.   Define $\delta = \min_{\lambda \in \lambda(A)} |\lambda - \mu|$. The theorem is clearly true if $\delta = 0$. If $\delta > 0$ then $I - (\mu I - A)^{-1}E$ is singular and we have

$$1 \leqslant \|(\mu I - A)^{-1}E\|_2 \leqslant \|(\mu I - A)^{-1}\|_2 \|E\|_2$$
$$= \|[(\mu I - D) - N]^{-1}\|_2 \|E\|_2.$$

Since $(\mu I - D)$ is diagonal it follows that $[(\mu I - D)^{-1}N]^p = 0$ and therefore

$$[(\mu I - D) - N]^{-1} = \sum_{k=0}^{p-1} [(\mu I - D)^{-1}N]^k (\mu I - D)^{-1}.$$

Taking norms in this expression and substituting into the above inequality gives

$$1 \leqslant \frac{\|E\|_2}{\delta} \max\left\{1, \frac{1}{\delta^{p-1}}\right\} \sum_{k=0}^{p-1} \|N\|_2^k,$$

from which the theorem readily follows. □

EXAMPLE 7.2-2.   If

$$A = \begin{bmatrix} 1 & 2 & 3 \\ 0 & 4 & 5 \\ 0 & 0 & 4.001 \end{bmatrix} \quad \text{and} \quad E = \begin{bmatrix} 0 & 0 & 0 \\ 0 & 0 & 0 \\ .001 & 0 & 0 \end{bmatrix}$$

then $\lambda(A + E) \cong \{1.0001, 4.0582, 3.9427\}$ and A's matrix of eigenvectors satisfies $\kappa_2(X) \cong 10^7$. The Bauer-Fike bound in Theorem 7.2-2 has order 10^4, while the Schur bound in Theorem 7.2-3 has order 10^0.

Theorems 7.2-2 and 7.2-3 each indicate potential eigenvalue sensitivity if A is non-normal. Specifically, if $\kappa_2(X)$ or $\|N\|_2^{p-1}$ is large, then small changes in A can induce large changes in the eigenvalues.

EXAMPLE 7.2-3.   If

$$A = \begin{bmatrix} 0 & I_9 \\ 0 & 0 \end{bmatrix} \quad \text{and} \quad E = \begin{bmatrix} 0 & 0 \\ 10^{-10} & 0 \end{bmatrix}$$

then for all $\lambda \in \lambda(A)$ and $\mu \in \lambda(A + E)$, $|\lambda - \mu| = 10^{-1}$. In this example a change of order 10^{-10} in A results in a change of order 10^{-1} in its eigenvalues.

The extreme eigenvalue sensitivity of Example 7.2-3 cannot occur if A is normal. On the other hand, non-normality does not necessarily imply eigenvalue sensitivity. Indeed, a non-normal matrix can have a mixture of well-conditioned and ill-conditioned eigenvalues. For this reason it is beneficial to refine our perturbation theory so that it is applicable to individual eigenvalues and not the spectrum as a whole.

To this end, suppose that λ is a simple eigenvalue of $A \in \mathbb{C}^{n \times n}$ and that x and y satisfy $Ax = \lambda x$ and $y^H A = \lambda y^H$ with $\|x\|_2 = \|y\|_2 = 1$. Using classical results from function theory, it can be shown that in a neighborhood of the origin there exist differentiable $x(\epsilon)$ and $\lambda(\epsilon)$ such that

$$(A + \epsilon F)x(\epsilon) = \lambda(\epsilon)x(\epsilon) \qquad\qquad \|x(\epsilon)\|_2 \equiv 1$$

and such that $\lambda(0) = \lambda$ and $x(0) = x$. By differentiating this equation with respect to ϵ and setting $\epsilon = 0$ in the result, we obtain

$$A\dot{x}(0) + Fx = \dot{\lambda}(0)x + \lambda\dot{x}(0).$$

Applying y^H to both sides of this equation, dividing by $y^H x$, and taking absolute values gives

$$|\dot{\lambda}(0)| = \frac{|y^H Fx|}{|y^H x|} \leqslant \frac{1}{|y^H x|}.$$

The upper bound is attained if $F = yx^H$. For this reason we refer to the reciprocal of

$$s(\lambda) \equiv |y^H x|$$

as the *condition of the eigenvalue* λ.

Roughly speaking, the above analysis shows that if order ϵ perturbations are made in A, then an eigenvalue λ may be perturbed by an amount $\epsilon/s(\lambda)$. Thus, if $s(\lambda)$ is small, then λ is appropriately regarded as ill-conditioned. Note that $s(\lambda)$ is the cosine of the angle between the left and right eigenvectors associated with λ and is unique only if λ is simple.

A small $s(\lambda)$ implies that A is near a matrix having a multiple eigenvalue. In particular, if λ is distinct and $s(\lambda) < 1$, then there exists an E such that λ is a repeated eigenvalue of $A + E$ and

$$\|E\|_2 \leqslant \frac{s(\lambda)}{\sqrt{1 - s(\lambda)^2}} .$$

This result is proved in Wilkinson (1972).

EXAMPLE 7.2-4. If

$$A = \begin{bmatrix} 1 & 2 & 3 \\ 0 & 4 & 5 \\ 0 & 0 & 4.001 \end{bmatrix} \quad \text{and} \quad E = \begin{bmatrix} 0 & 0 & 0 \\ 0 & 0 & 0 \\ .001 & 0 & 0 \end{bmatrix}$$

then $\lambda(A + E) \cong \{1.0001, 4.0582, 3.9427\}$ and

$$s(1) \cong .79 \times 10^0$$
$$s(4) \cong .16 \times 10^{-3}$$
$$s(4.001) \cong .16 \times 10^{-3}.$$

Observe that $\|E\|_2/s(\lambda)$ is a good estimate of the perturbation that each eigenvalue undergoes.

If λ is a repeated eigenvalue, then the eigenvalue sensitivity question is more complicated. For example, if

$$A = \begin{bmatrix} 1 & a \\ 0 & 1 \end{bmatrix} \quad \text{and} \quad F = \begin{bmatrix} 0 & 0 \\ 1 & 0 \end{bmatrix}$$

then $\lambda(A + \epsilon F) = \{1 \pm \sqrt{\epsilon a}\}$. Note that if $a \neq 0$, then the eigenvalues of $A + \epsilon F$ are not differentiable at zero; their rate of change at the origin is infinite. In general, if λ is a defective eigenvalue of A, then $O(\epsilon)$ perturbations in A result in $O(\epsilon^{1/p})$ perturbations in λ where $p \geqslant 2$. See Wilkinson (AEP, pp. 77 ff.) for a more detailed discussion.

We now consider how small changes in a matrix affect its invariant subspaces. The case of eigenvectors is dealt with first. Assume $A \in \mathbb{C}^{n \times n}$ has distinct eigenvalues $\lambda_1, \ldots, \lambda_n$ and that $F \in \mathbb{C}^{n \times n}$ satisfies $\|F\|_2 = 1$. A continuity argument ensures that for all ϵ in some neighborhood of the origin we have

$$(A + \epsilon F)x_k(\epsilon) = \lambda_k(\epsilon)x_k(\epsilon) \qquad\qquad \|x_k(\epsilon)\|_2 \equiv 1$$

$$y_k(\epsilon)^{\mathrm{H}}(A + \epsilon F) = \lambda_k(\epsilon)y_k(\epsilon)^{\mathrm{H}} \qquad\qquad \|y_k(\epsilon)\|_2 \equiv 1$$

for $k = 1, \ldots, n$, where each $\lambda_k(\epsilon)$, $x_k(\epsilon)$, and $y_k(\epsilon)$ is differentiable. If we differentiate the first of these equations with respect to ϵ and set $\epsilon = 0$ in the result, we obtain

$$A\dot{x}_k(0) + Fx_k = \dot{\lambda}_k(0)x_k + \lambda_k\dot{x}_k(0),$$

where $\lambda_k = \lambda_k(0)$ and $x_k = x_k(0)$. Since A's eigenvectors form a basis we can write

$$\dot{x}_k(0) = \sum_{i=1}^{n} a_i x_i,$$

and so we have

$$\sum_{\substack{i=1 \\ i \neq k}}^{n} a_i(\lambda_i - \lambda_k)x_i + Fx_k = \dot{\lambda}_k(0)x_k.$$

To obtain expressions for the a_i, note that $y_i(0)^{\mathrm{H}}x_k \equiv y_i^{\mathrm{H}}x_k = 0$ whenever $i \neq k$ and thus,

$$a_i = y_i^{\mathrm{H}}Fx_k/[(\lambda_k - \lambda_i)y_i^{\mathrm{H}}x_i]. \qquad\qquad i \neq k$$

Therefore, the Taylor expansion for $x_k(\epsilon)$ has the following form:

$$x_k(\epsilon) \cong x_k + \epsilon \sum_{\substack{i=1 \\ i \neq k}}^{n} \left\{ \frac{y_i^{\mathrm{H}}Fx_k}{(\lambda_k - \lambda_i)y_i^{\mathrm{H}}x_i} \right\} x_i + O(\epsilon^2).$$

Thus, the sensitivity of x_k depends upon eigenvalue sensitivity *and* the separation of λ_k from the other eigenvalues.

That the separation of the eigenvalues should have a bearing upon eigenvector sensitivity should come as no surprise. Indeed, if λ is a nondefective, repeated eigenvalue, then there are an infinite number of possible eigenvector bases for the associated invariant subspace. The preceding analysis merely indicates that this indeterminancy begins to be felt as the eigenvalues coalesce. In other words: eigenvectors associated with nearby eigenvalues are "wobbly."

EXAMPLE 7.2-5. If

$$A = \begin{bmatrix} 1.01 & .01 \\ 0.00 & .99 \end{bmatrix}$$

then the eigenvalue $\lambda = .99$ has condition $1/s(.99) \cong 1.118$ and associated eigenvector $x = (.4472, -.8944)^{\mathrm{T}}$. On the other hand, the eigenvalue $\hat{\lambda} = 1.00$ of the "nearby" matrix

$$A + E = \begin{bmatrix} 1.01 & 0.01 \\ 0.00 & 1.00 \end{bmatrix}$$

has an eigenvector $\hat{x} = (.7071, -.7071)^T$.

A collection of sensitive eigenvectors can define an insensitive invariant subspace provided the corresponding cluster of eigenvalues is isolated. To be precise, suppose

(7.2-1)
$$Q^H A Q = \begin{matrix} \begin{bmatrix} T_{11} & T_{12} \\ 0 & T_{22} \end{bmatrix} & \begin{matrix} p \\ n-p \end{matrix} \\ \begin{matrix} p & n-p \end{matrix} & \end{matrix}$$

is a Schur decomposition of A with

(7.2-2)
$$Q = \begin{matrix} [\, Q_1, & Q_2 \,]. \\ p & n-p \end{matrix}$$

It is clear from our discussion of eigenvector perturbation that the sensitivity of the invariant subspace $R(Q_1)$ depends on the distance between $\lambda(T_{11})$ and $\lambda(T_{22})$. The proper measure of this distance turns out to be the smallest singular value of the linear transformation $X \rightarrow T_{11}X - XT_{22}$. In particular, if we define the *separation* between the matrices T_{11} and T_{22} by

$$\text{sep}(T_{11}, T_{22}) = \min_{X \neq 0} \frac{\| T_{11}X - XT_{22} \|_F}{\| X \|_F}$$

then we have the following general result:

THEOREM 7.2-4. Suppose that (7.2-1) and (7.2-2) hold and that for $E \in \mathbb{C}^{n \times n}$ we partition $Q^H E Q$ as follows:

$$Q^H E Q = \begin{matrix} \begin{bmatrix} E_{11} & E_{12} \\ E_{21} & E_{22} \end{bmatrix} & \begin{matrix} p \\ n-p \end{matrix} \\ \begin{matrix} p & n-p \end{matrix} & \end{matrix}.$$

If

$$\delta = \text{sep}(T_{11}, T_{22}) - \| E_{11} \|_2 - \| E_{22} \|_2 > 0$$

and

$$\| E_{21} \|_2 (\| T_{12} \|_2 + \| E_{12} \|_2) \leqslant \delta^2/4,$$

then there exists a matrix P in $\mathbb{C}^{(n-k) \times k}$ such that

$$\| P \|_2 \leqslant 2 \| E_{21} \|_2 / \delta$$

and such that the columns of $\hat{Q}_1 = (Q_1 + Q_2 P)(I + P^H P)^{-1/2}$ form an orthonormal basis for a subspace that is invariant for $A + E$.

Proof. See Stewart (1973b, Theorem 4.11). $\square$

Since $Q_1^H \hat{Q}_1 = (I + P^H P)^{-1/2}$, we see that the singular values of P are the tangents of the principle angles between $R(\hat{Q}_1)$ and $R(Q_1)$. (See §2.4.) The theorem shows that these tangents are essentially bounded by $\|E\|_2 / \mathrm{sep}(T_{11}, T_{22})$. Thus the reciprocal of $\mathrm{sep}(T_{11}, T_{22})$ can be thought of as a condition number that measures the sensitivity of $R(Q_1)$ as an invariant subspace.

EXAMPLE 7.2-6. Suppose

$$T_{11} = \begin{bmatrix} 3 & 10 \\ 0 & 1 \end{bmatrix}, \quad T_{22} = \begin{bmatrix} 0 & -20 \\ 0 & 3.01 \end{bmatrix}, \quad \text{and } T_{12} = \begin{bmatrix} 1 & -1 \\ -1 & 1 \end{bmatrix},$$

and that

$$A = T = \begin{bmatrix} T_{11} & T_{12} \\ 0 & T_{22} \end{bmatrix}.$$

Observe that $AQ_1 = Q_1 T_{11}$ where $Q_1 = [e_1, e_2] \in \mathbb{R}^{4 \times 2}$. A calculation shows that $\mathrm{sep}(T_{11}, T_{22}) \cong .0003$. If

$$E_{21} = 10^{-6} \begin{bmatrix} 1 & 1 \\ 1 & 1 \end{bmatrix}$$

and we examine the Schur decomposition of

$$A + E = \begin{bmatrix} T_{11} & T_{12} \\ E_{21} & T_{22} \end{bmatrix},$$

we find that Q_1 gets perturbed to

$$\hat{Q}_1 = \begin{bmatrix} -.9999 & -.0003 \\ .0003 & -.9999 \\ -.0005 & -.0026 \\ .0000 & .0003 \end{bmatrix}.$$

Thus we have $\mathrm{dist}(\hat{Q}_1, Q_1) \cong .0027 \cong 10^{-6}/\mathrm{sep}(T_{11}, T_{22})$.

Problems

P7.2-1. Suppose $Q^H A Q = \mathrm{diag}(\lambda_i) + N$ is a Schur decomposition of $A \in \mathbb{C}^{n \times n}$ and define

$$\nu(A) = \|A^H A - AA^H\|_F.$$

The upper and lower bounds in

$$\frac{\nu(A)^2}{6\|A\|_F^2} \leqslant \|N\|_F^2 \leqslant \sqrt{\frac{n^3 - n}{12}}\,\nu(A)$$

were established by Henrici (1962) and Eberlein (1965), respectively. Verify these results for the case $n = 2$.

P7.2-2. Suppose $A \in \mathbb{C}^{n \times n}$ has distinct eigenvalues and that $X^{-1}AX = \text{diag}(\lambda_1, \ldots, \lambda_n)$. Show that if the columns of X have unit 2-norm, then

$$\kappa_F(X) = \sqrt{n}\left(\sum_{i=1}^{n} \frac{1}{s(\lambda_i)^2}\right)^{1/2}.$$

P7.2-3. (Loizou 1969) Suppose $Q^H A Q = \text{diag}(\lambda_i) + N$ is a Schur decomposition of A and that $X^{-1}AX = \text{diag}(\lambda_i)$. Show

$$\kappa_2(X) \geqslant \left(1 + \frac{\|N\|_F^2}{\|A\|_F^2}\right)^{1/2}.$$

P7.2-4. (Ruhe 1975) If $X^{-1}AX = \text{diag}(\lambda_i)$ and $|\lambda_1| \geqslant \cdots \geqslant |\lambda_n|$, then

$$\frac{\sigma_i(A)}{\kappa_2(A)} \leqslant |\lambda_i| \leqslant \kappa_2(A)\sigma_i(A).$$

Prove this result for the $n = 2$ case.

P7.2-5. Show that if $A = \begin{bmatrix} a & c \\ 0 & b \end{bmatrix}$ then $s(a) = s(b) = (1 + |c/(a - b)|^2)^{-1}$.

P7.2-6. Show that the condition number of an eigenvalue is preserved under unitary similarity transformations.

Notes and References for Sec. 7.2

Many of the results presented in this section may be found in Wilkinson (AEP, chap. 2) as well as in

A. S. Householder (1964). *The Theory of Matrices in Numerical Analysis*. Blaisdell, New York.

F. L. Bauer and C. T. Fike (1960). "Norms and Exclusion Theorems," *Numer. Math. 2*, 137-44.

The following papers are concerned with the effect of perturbations on the eigenvalues of a general matrix:

A. Ruhe (1970). "Perturbation Bounds for Means of Eigenvalues and Invariant Subspaces," *BIT 10*, 343-54.

A. Ruhe (1970). "Properties of a Matrix with a Very Ill-Conditioned Eigenproblem," *Numer. Math. 15*, 57-60.

J. H. Wilkinson (1972). "Note on Matrices with a Very Ill-Conditioned Eigenproblem," *Numer. Math. 19*, 176-78.

The relationship between the eigenvalue condition number, the departure from normality, and the condition of the eigenvector matrix is discussed in:

P. Eberlein (1965). "On Measures of Non-normality for Matrices," *Amer. Math. Soc. Monthly 72*, 995-96.

P. Henrici (1962). "Bounds for Iterates, Inverses, Spectral Variation and Fields of Values of Non-normal Matrices," *Numer. Math. 4*, 24-40.

G. Loizou (1969). "Nonnormality and Jordan Condition Numbers of Matrices," *J. Assoc. Comp. Mach. 16*, 580-84.

R. A. Smith (1967). "The Condition Numbers of the Matrix Eigenvalue Problem," *Numer. Math. 10*, 232-40.

A. van der Sluis (1975). "Perturbations of Eigenvalues of Non-normal Matrices," *Comm. Assoc. Comp. Mach. 18*, 30-36.

The paper by Henrici also contains a result similar to Theorem 7.2-3.
Penetrating treatments of invariant subspace perturbation include

C. Davis and W. M. Kahan (1970). "The Rotation of Eigenvectors by a Perturbation, III," *SIAM J. Num. Anal. 7*, 1-46.

G. W. Stewart (1971). "Error Bounds for Approximate Invariant Subspaces of Closed Linear Operators," *SIAM J. Num. Anal. 8*, 796-808.

G. W. Stewart (1973*b*). "Error and Perturbation Bounds for Subspaces Associated with Certain Eigenvalues Problems," *SIAM Review 15*, 727-64.

A detailed analysis of the function sep(.,.) is given in

J. Varah (1979). "On the Separation of Two Matrices," *SIAM J. Num. Anal. 16*, 216-22.

Gershgorin's Theorem can be used to derive a comprehensive perturbation theory. See Wilkinson (AEP, chap. 2). The theorem itself can be generalized and extended in various ways; see

R. L. Johnston (1971). "Gershgorin Theorems for Partitioned Matrices," *Lin. Alg. & Its Applic. 4*, 205-20.

R. S. Varga (1970). "Minimal Gershgorin Sets for Partitioned Matrices," *SIAM J. Num. Anal. 7*, 493-507.

Finally, we mention the classic reference

T. Kato (1966). *Perturbation Theory for Linear Operators*, Springer-Verlag, New York.

Chapter 2 of this work is a comprehensive treatment of the finite dimensional case.

Sec. 7.3. Power Iterations

Suppose that we are given $A \in \mathbb{C}^{n \times n}$ and a unitary $U_0 \in \mathbb{C}^{n \times n}$. Assume that Householder orthogonalization (Algorithm 6.2-1) can be extended to complex matrices (it can) and consider the following iteration:

$$T_0 = U_0^H A U_0$$

(7.3-1) For $k = 1, 2, \ldots$

Let $T_{k-1} = U_k R_k$ be the Q-R factorization of T_{k-1}.

Set $T_k = R_k U_k$.

Since

(7.3-2) $$T_k = (U_0 U_1 \cdots U_k)^H A (U_0 U_1 \cdots U_k),$$

it is obvious that each T_k is unitarily similar to A. What is *not* obvious and what is the central theme of this section is that the T_k almost always converge to upper triangular form. That is, (7.3-2) almost always "converges" to a Schur decomposition of A.

Iteration (7.3-1) is called the *QR iteration*, and it forms the backbone of the most effective algorithm for computing the Schur decomposition. In order to motivate the method and to derive its convergence properties, two other eigenvalue iterations that are important in their own right are presented first: the power method and the method of orthogonal iteration.

The Power Method

Suppose $A \in \mathbb{C}^{n \times n}$ is diagonalizable and that $X^{-1} A X = \text{diag}(\lambda_1, \ldots, \lambda_n)$ with $X = [x_1, \ldots, x_n]$ and $|\lambda_1| > |\lambda_2| \geqslant \cdots \geqslant |\lambda_n|$. Given $v^{(0)} \in \mathbb{C}^n$, the *power method* produces a sequence of vectors $v^{(k)}$ as follows:

For $k = 1, 2, \ldots$

(7.3-3) $$z^{(k)} = A v^{(k-1)}$$
$$\lambda^{(k)} = z_i^{(k)} \text{ where } |z_i^{(k)}| = \|z^{(k)}\|_\infty \qquad (1 \leqslant i \leqslant n)$$
$$v^{(k)} = z^{(k)} / \lambda^{(k)}$$

If $v^{(0)} = a_1 x_1 + a_2 x_2 + \cdots a_n x_n$ and $a_1 \neq 0$, then it follows that

$$A^k v^{(0)} = a_1 \lambda_1^k \left[x_1 + \sum_{j=2}^n \frac{a_j}{a_1} \left(\frac{\lambda_j}{\lambda_1} \right)^k x_j \right].$$

Since $v^{(k)} \in \text{span}\{A^k v^{(0)}\}$ we conclude that

$$\text{dist}(v^{(k)}, \text{span}\{x_1\}) = O\left[\left(\frac{|\lambda_2|}{|\lambda_1|} \right)^k \right]$$

and moreover,

$$|\lambda_1 - \lambda^{(k)}| = O\left[\left(\frac{|\lambda_2|}{|\lambda_1|} \right)^k \right].$$

If $|\lambda_1| > |\lambda_2| \geqslant \cdots \geqslant |\lambda_n|$ then we say that λ_1 is a *dominant eigenvalue*. Thus, the power method converges if λ_1 is dominant and if $v^{(0)}$ has a component in the direction of the corresponding *dominant eigenvector* x_1.

The behavior of the iteration without these assumptions is discussed in Wilkinson (AEP, pp. 570 ff.) and Parlett and Poole (1973).

EXAMPLE 7.3-1.   If

$$A = \begin{bmatrix} -261 & 209 & -49 \\ -530 & 422 & -98 \\ -800 & 631 & -144 \end{bmatrix} ,$$

then $\lambda(A) = \{10, 4, 3\}$. Applying (7.3-3) with $v^{(0)} = (1, 0, 0)^T$ we find

k	$\lambda^{(k)}$
1	994.49
2	13.0606
3	10.7191
4	10.2073
5	10.0633
6	10.0198
7	10.0063
8	10.0020
9	10.0007
10	10.0002

In practice, the usefulness of the power method depends upon $|\lambda_2|/|\lambda_1|$, since this ratio dictates the rate of convergence. The danger that $v^{(0)}$ is deficient in x_1 is a less worrisome matter because rounding errors sustained during the iteration typically ensure that the subsequent $v^{(k)}$ have a component in this direction. Moreover, it is frequently the case in applications where the dominant eigenvalue and eigenvector are desired that an a priori estimate of x_1 is known. Normally, by setting $v^{(0)}$ to be this estimate, the dangers of a small a_1 are minimized.

Estimates for the error $|\lambda^{(k)} - \lambda_1|$ can be obtained by applying the perturbation theory developed in the previous section. Define $r^{(k)} = Av^{(k)} - \lambda^{(k)}v^{(k)}$ and observe that

$$(A + E^{(k)})v^{(k)} = \lambda^{(k)}v^{(k)}$$

where $E^{(k)} = -r^{(k)}[v^{(k)}]^H/\|v^{(k)}\|_2^2$. This says that $\lambda^{(k)}$ is an eigenvalue of $A + E^{(k)}$ and thus,

$$|\lambda^{(k)} - \lambda_1| \cong \frac{\|E^{(k)}\|_2}{s(\lambda_1)} = \frac{\|r^{(k)}\|_2}{\|v^{(k)}\|_2}s(\lambda_1)^{-1}.$$

By using the power method to generate an approximate dominant left eigenvector, it is possible to obtain an estimate of $s(\lambda_1)$. In particular, if $w^{(k)}$ is a multiple of $(A^H)^k w^{(0)}$, then

$$s(\lambda_1) \cong \frac{|[w^{(k)}]^H v^{(k)}|}{\|w^{(k)}\|_2 \|v^{(k)}\|_2}.$$

Note that the only thing required to implement the power method is a subroutine capable of computing matrix-vector products of the form Av. It is not necessary to store A in an n-by-n array. For this reason, the algorithm can be of interest when A is large and sparse and when there is a sufficient gap between $|\lambda_1|$ and $|\lambda_2|$.

Orthogonal Iteration

A straightforward generalization of the power method can be used to compute higher-dimensional invariant subspaces. Let p be a chosen integer satisfying $1 \leqslant p < n$. Given an n-by-p matrix Q_0 with orthonormal columns, the method of *orthogonal iteration* generates a sequence of matrices $\{Q_k\} \subset \mathbb{C}^{n \times p}$ as follows:

For $k = 1, 2, \ldots$

(7.3-4)
$$Z_k = AQ_{k-1}$$
$$Q_k R_k = Z_k \qquad \text{(Q-R factorization)}$$

Note that if $p = 1$ then this is just the power method.

In order to analyze the behavior of this iteration, suppose that

(7.3-5) $\quad Q^H A Q = T = \text{diag}(\lambda_i) + N \qquad |\lambda_1| \geqslant |\lambda_2| \geqslant \cdots \geqslant |\lambda_n|$

is a Schur decomposition of $A \in \mathbb{C}^{n \times n}$ and partition Q, T, and N as follows:

(7.3-6)
$$Q = [Q_\alpha, Q_\beta] \qquad T = \begin{array}{c} \begin{bmatrix} T_{11} & T_{12} \\ 0 & T_{22} \end{bmatrix} \\ \begin{matrix} p & n-p \end{matrix} \end{array} \begin{matrix} p \\ n-p \end{matrix}$$
$$\begin{matrix} p & n-p \end{matrix}$$

$$N = \begin{array}{c} \begin{bmatrix} N_{11} & N_{12} \\ 0 & N_{22} \end{bmatrix} \\ \begin{matrix} p & n-p \end{matrix} \end{array} \begin{matrix} p \\ n-p \end{matrix}.$$

If $|\lambda_p| > |\lambda_{p+1}|$, then we say that the subspace $D_p(A)$ defined by

$$D_p(A) = R(Q_\alpha)$$

is a *dominant* invariant subspace. It is the unique invariant subspace associated with the eigenvalues $\lambda_1, \ldots, \lambda_p$. The following theorem shows that with reasonable assumptions, the subspaces $R(Q_k)$ generated by (7.3-4) converge to $D_p(A)$ at a rate proportional to $|\lambda_{p+1}/\lambda_p|^k$.

THEOREM 7.3-1.   Let the Schur decomposition of $A \in \mathbb{C}^{n \times n}$ be given by (7.3-5) and (7.3-6). Assume that $|\lambda_p| > |\lambda_{p+1}|$ and that $\Theta \geqslant 0$ satisfies

$$(1 + \Theta)|\lambda_p| > \|N\|_F.$$

If $Q_0 \in \mathbb{C}^{n \times p}$ has orthonormal columns and

$$d = \text{dist}[D_p(A^H), R(Q_0)] < 1,$$

then the matrices Q_k generated by 7.3-4 satisfy

$$\text{dist}[D_p(A), R(Q_k)] \leqslant$$

$$\frac{(1 + \Theta)^{n-2}}{\sqrt{1 - d^2}} \left[1 + \frac{\|T_{12}\|_F}{\text{sep}(T_{11}, T_{22})} \right] \left[\frac{|\lambda_{p+1}| + \|N\|_F / (1 + \Theta)}{|\lambda_p| - \|N\|_F / (1 + \Theta)} \right]^k.$$

Proof. The proof is long and tedious and appears at the end of the section. Note: $d < 1 \Leftrightarrow D_p(A^H)^\perp \cap R(Q_0) = \{0\}$. $\square$

When Θ is chosen large enough, the theorem essentially shows that

$$\text{dist}[D_p(A), R(Q_k)] \leqslant c \, |\lambda_{p+1}/\lambda_p|^k,$$

where c depends on $\text{sep}(T_{11}, T_{22})$ and A's departure from normality. Needless to say, the convergence can be very slow if the gap between $|\lambda_p|$ and $|\lambda_{p+1}|$ is not sufficiently wide.

EXAMPLE 7.3-2.   If (7.3-4) is applied to the matrix A in Example 7.3-1, with $Q_0 = [e_1, e_2]$, we find

k	$\text{dist}[D_2(A), Q_k]$
1	.0052
2	.0047
3	.0039
4	.0030
5	.0023
6	.0017
7	.0013

Clearly, the error is tending to zero with rate $(\lambda_3/\lambda_2)^k = (3/4)^k$.

It is possible to accelerate the convergence in orthogonal iteration using a technique described in Stewart (1976d). (See also §8.5.) In the accelerated scheme, the approximate eigenvalue $\lambda_i^{(k)}$ satisfies

$$|\lambda_i^{(k)} - \lambda_i| \cong |\lambda_{p+1}/\lambda_i|^k. \qquad\qquad i = 1, \ldots, p$$

(Without the acceleration, the right-hand side is $|\lambda_{i+1}/\lambda_i|^k$.) Stewart's algorithm can be very useful in situations where A is large and sparse and a few of its largest eigenvalues are required.

The QR Iteration

We now can "derive" the QR iteration (7.3-1) and examine its convergence. Suppose $p = n$ in (7.3-4) and that the eigenvalues of A satisfy

$$|\lambda_1| > |\lambda_2| > \cdots > |\lambda_n|.$$

Partition the matrix Q in (7.3-5) and Q_k in (7.3-4) as follows:

$$Q = [q_1, \ldots, q_n] \qquad Q_k = [q_1^{(k)}, \ldots, q_n^{(k)}].$$

If

(7.3-7) $\text{dist}[D_i(A^H), \text{span}\{q_1^{(0)}, \ldots, q_i^{(0)}\}] < 1, \qquad i = 1, \ldots, n$

then it follows from Theorem 7.3-1 that

$$\text{dist}[\text{span}\{q_1^{(k)}, \ldots, q_i^{(k)}\}, \text{span}\{q_1, \ldots, q_i\}] \to 0$$

for $i = 1, \ldots, n$. This implies that the matrices T_k defined by

$$T_k = Q_k^H A Q_k$$

are converging to upper triangular form. Thus, it can be said that the method of orthogonal iteration computes the Schur decomposition provided the original iterate $Q_0 \in \mathbb{C}^{n \times n}$ is not deficient.

The QR iteration arises naturally by considering how to compute the matrix T_k directly from its predecessor T_{k-1}. On the one hand, we have from (7.3-4) and the definition of T_{k-1} that

$$T_{k-1} = Q_{k-1}^H A Q_{k-1} = Q_{k-1}^H (A Q_{k-1}) = Q_{k-1}^H Q_k R_k.$$

On the other hand,

$$T_k = Q_k^H A Q_k = (Q_k^H A Q_{k-1})(Q_{k-1}^H Q_k) = R_k(Q_{k-1}^H Q_k).$$

Thus, T_k is determined by computing the QR factorization

$$T_{k-1} = (Q_{k-1}^H Q_k) R_k$$

of T_{k-1} and then multiplying the factors together in the reverse order. This, of course, is precisely what is done in (7.3-1).

EXAMPLE 7.3-3.   If the iteration

For $k = 1, 2, \ldots$

$\quad A = QR \qquad$ (QR factorization)

$\quad A := RQ$

is applied to the matrix of Example 7.3-1, then the subdiagonal elements diminish as shown in Table 7.3-1.

Table 7.3-1

k	$O(|a_{21}|)$	$O(|a_{31}|)$	$O(|a_{32}|)$
1	10^{-1}	10^{-1}	10^{-2}
2	10^{-2}	10^{-2}	10^{-3}
3	10^{-2}	10^{-3}	10^{-3}
4	10^{-3}	10^{-3}	10^{-3}
5	10^{-3}	10^{-4}	10^{-3}
6	10^{-4}	10^{-5}	10^{-3}
7	10^{-4}	10^{-5}	10^{-3}
8	10^{-5}	10^{-6}	10^{-4}
9	10^{-5}	10^{-7}	10^{-4}
10	10^{-6}	10^{-8}	10^{-4}

Note that a single QR iteration is an $O(n^3)$ calculation. Moreover, since convergence is only linear (when it exists), it is clear that the method is a prohibitively expensive way to compute Schur decompositions. Fortunately these practical difficulties can be overcome, as we show in §§7.4 and 7.5.

We conclude with some remarks about power iterations that rely on the L-U rather than the Q-R factorization. Let $G_0 \in \mathbb{C}^{n \times p}$ have rank p. Corresponding to (7.3-4) we have the following iteration:

(7.3-8)
$$\text{For } k = 1, 2, \ldots$$
$$Z_k = AG_{k-1}$$
$$G_k R_k = Z_k \quad \text{(L-U factorization)}$$

Suppose $p = n$ and that we define the matrices T_k by

(7.3-9)
$$T_k = G_k^{-1} A G_k.$$

It can be shown that if we set $L_0 = G_0$, then the T_k can be generated as follows:

(7.3-10)
$$T_0 = L_0^{-1} A L_0$$
$$\text{For } k = 1, 2, \ldots$$
$$T_{k-1} = L_k R_k \quad \text{(L-U factorization)}$$
$$T_k = R_k L_k$$

Iterations (7.3-8) and (7.3-10) are known as *treppeniteration* and the *LR iteration*, respectively. To successfully implement either method, it is necessary to pivot. See Wilkinson (AEP, chap. 9).

Appendix

In order to establish Theorem 7.3-1 we need the following lemma concerned with bounding the powers of a matrix and its inverse.

LEMMA 7.3-2. Let $Q^H A Q = T = D + N$ be a Schur decomposition of $A \in \mathbb{C}^{n \times n}$ where D is diagonal and N strictly upper triangular. Let λ and μ denote the largest and smallest eigenvalues of A in absolute value. If $\Theta \geqslant 0$ then

(7.3-11) $$\| A^k \|_2 \leqslant (1 + \Theta)^{n-1} \left[|\lambda| + \frac{\| N \|_F}{1 + \Theta} \right]^k . \qquad k \geqslant 0$$

If A is nonsingular and $\Theta \geqslant 0$ satisfies $(1 + \Theta) |\mu| > \| N \|_F$, then

(7.3-12) $$\| A^{-k} \|_2 \leqslant (1 + \Theta)^{n-1} \left[\frac{1}{|\mu| - \| N \|_F / (1 + \Theta)} \right]^k . \qquad k \geqslant 0$$

Proof. For $\Theta \geqslant 0$, define the diagonal matrix Δ by

$$\Delta = \mathrm{diag}(1, 1 + \Theta, (1 + \Theta)^2, \ldots, (1 + \Theta)^{n-1})$$

and note that $\kappa_2(\Delta) = (1 + \Theta)^{n-1}$. Since N is strictly upper triangular, it is easy to verify that $\| \Delta N \Delta^{-1} \|_F \leqslant \| N \|_F / (1 + \Theta)$. Thus,

$$\| A^k \|_2 = \| T^k \|_2 = \| \Delta^{-1} (D + \Delta N \Delta^{-1})^k \Delta \|_2$$
$$\leqslant \kappa_2(\Delta) [\| D \|_2 + \| \Delta N \Delta^{-1} \|_2]^k$$
$$\leqslant (1 + \Theta)^{n-1} \left[|\lambda| + \frac{\| N \|_F}{(1 + \Theta)} \right]^k .$$

On the other hand, if A is nonsingular and $(1 + \Theta) |\mu| > \| N \|_F$, then $\| \Delta D^{-1} N \Delta^{-1} \|_2 < 1$ and thus,

$$\| A^{-k} \|_2 = \| T^{-k} \|_2 = \| \Delta^{-1} [(I + \Delta D^{-1} N \Delta^{-1})^{-1} D^{-1}]^k \Delta \|_2$$
$$\leqslant \kappa_2(\Delta) [\| D^{-1} \|_2 / [1 - \| \Delta D^{-1} N \Delta^{-1} \|_2]^k$$
$$\leqslant (1 + \Theta)^{n-1} [1 / [|\mu| - \| N \|_F / (1 + \Theta)]]^k . \quad \square$$

Proof of Theorem 7.3-1. It is easy to show by induction that $A^k Q_0 = Q_k (R_k \cdots R_1)$. By substituting (7.3-5) and (7.3-6) into this equality we obtain

$$T^k \begin{bmatrix} V_0 \\ W_0 \end{bmatrix} = \begin{bmatrix} V_k \\ W_k \end{bmatrix} (R_k \cdots R_1),$$

where $V_k = Q_\alpha^H Q_k$ and $W_k = Q_\beta^H Q_k$. Using Lemma 7.1-4 we know that a matrix $X \in \mathbb{C}^{p \times (n-p)}$ exists such that

$$\begin{bmatrix} I_p & X \\ 0 & I_{n-p} \end{bmatrix}^{-1} \begin{bmatrix} T_{11} & T_{12} \\ 0 & T_{22} \end{bmatrix} \begin{bmatrix} I_p & X \\ 0 & I_{n-p} \end{bmatrix} = \begin{bmatrix} T_{11} & 0 \\ 0 & T_{22} \end{bmatrix} .$$

Moreover, since $\mathrm{sep}(T_{11}, T_{22})$ is the smallest singular value of $\phi(X) = T_{11} X - X T_{22}$ we have from the equation $\phi(X) = -T_{12}$ that

(7.3-13)
$$\|X\|_F \leq \frac{\|T_{12}\|_F}{\text{sep}(T_{11}, T_{22})}.$$

Thus we have

$$\begin{bmatrix} T_{11}^k & 0 \\ 0 & T_{22}^k \end{bmatrix} \begin{bmatrix} V_0 - XW_0 \\ W_0 \end{bmatrix} = \begin{bmatrix} V_k - XW_k \\ W_k \end{bmatrix} (R_k \cdots R_1).$$

If we assume that $V_0 - XW_0$ is nonsingular, then this equation can be solved for W_k:

$$W_k = T_{22}^k W_0 (V_0 - XW_0)^{-1} T_{11}^{-k} (V_k - XW_k).$$

By taking norms in this expression and using the fact that

$$\text{dist}[D_p(A), R(Q_k)] = \|Q_\beta^H Q_k\|_2 = \|W_k\|_2,$$

we get

(7.3-14) $\text{dist}(D_p(A), R(Q_k))$

$$\leq \|T_{22}^k\|_2 \|(V_0 - XW_0)^{-1}\|_2 \|T_{11}^{-k}\|_2 [1 + \|X\|_F].$$

In order for this inequality to be valid, we must show that $V_0 - XW_0$ is nonsingular.

From the equation $A^H Q = Q T^H$ it follows that

$$A^H(Q_\alpha - Q_\beta X^H) = (Q_\alpha - Q_\beta X^H) T_{11}^H,$$

which implies that the orthonormal columns of $Z = (Q_\alpha - Q_\beta X^H) \cdot (I + XX^H)^{-1/2}$ are a basis for $D_p(A^H)$. It can be shown that

$$V_0 - XW_0 = (I + XX^H)^{1/2} Z^H Q_0$$

and therefore

$$\sigma_p(V_0 - XW_0) \geq \sigma_p(Z^H Q_0) =$$
$$\sigma_p(V_0 - XW_0) \geq \sigma_p(Z^H Q_0) = \sqrt{1 - d^2} > 0.$$

This shows that $V_0 - XW_0$ is indeed invertible and that

(7.3-15) $\|(V_0 - XW_0)^{-1}\|_2 \leq 1/\sqrt{1 - d^2}.$

Finally, by using Lemma 7.3-2 it can be shown that

$$\|T_{22}^k\|_2 \leq (1 + \Theta)^{n-p-1}[|\lambda_{p+1}| + \|N\|_F/(1 + \Theta)]^k$$

and

$$\|T_{11}^{-k}\|_2 \leq (1 + \Theta)^{p-1}/[|\lambda_p| - \|N\|_F/(1 + \Theta)]^k.$$

The theorem follows by substituting these inequalities along with (7.3-13) and (7.3-15) into (7.3-14). $\square$

Problems

P7.3-1. Show that if $X \in \mathbb{C}^{n \times n}$ is nonsingular, then

$$\|A\|_X = \|X^{-1}AX\|_2$$

defines a matrix norm with the property that $\|AB\|_X \leqslant \|A\|_X \|B\|_X$.

P7.3-2. Let $A \in \mathbb{C}^{n \times n}$ and set $\rho = \max |\lambda_i|$. Show that for any $\epsilon > 0$ there exists a nonsingular $X \in \mathbb{C}^{n \times n}$ such that

$$\|A\|_X = \|X^{-1}AX\|_2 \leqslant \rho + \epsilon.$$

Conclude that there is a constant M such that $\|A^k\|_2 \leqslant M(\rho + \epsilon)^k$ for all non-negative integers k. (Hint: set $X = Q \operatorname{diag}(1, a, \ldots, a^{n-1})$ where $Q^H A Q = D + N$ is A's Schur decomposition.)

P7.3-3. Verify that (7.3-10) calculates the matrices T_k defined by (7.3-9).

P7.3-4. Suppose the power method (7.3-3) is applied to $A = \begin{bmatrix} 1 & 1 \\ 0 & 2 \end{bmatrix}$ with $v^{(0)} = \begin{bmatrix} 1 \\ 1 \end{bmatrix}$. What is $\lambda^{(20)}$?

P7.3-5. (*Inverse orthogonal iteration*) Suppose $A \in \mathbb{C}^{n \times n}$ is nonsingular and that $Q_0 \in \mathbb{C}^{n \times p}$ has orthonormal columns. Consider the following iteration:

> For $k = 1, 2, \ldots$
> $\quad$ Solve $AZ_k = Q_{k-1}$ for $Z_k \in \mathbb{C}^{n \times p}$.
> $\quad Q_k R_k = Z_k \quad$ (Q-R factorization)

Explain why this iteration can usually be used to compute the p smallest eigenvalues of A. Note that to implement this iteration it is necessary to be able to solve linear systems that involve A. When $p = 1$, the method is referred to as the *inverse power method*.

Notes and References for Sec. 7.3

A detailed practical discussion of the power method is given in Wilkinson (AEP, chap. 10). Methods are discussed for accelerating the basic iteration, for calculating nondominant eigenvalues, and for handling complex conjugate eigenvalue pairs. The connections among the various power iterations are discussed in

B. N. Parlett and W. G. Poole (1973). "A Geometric Theory for the QR, LU, and Power Iterations," *SIAM J. Num. Anal. 10*, 389–412.

The QR iteration was concurrently developed in

J.G.F. Francis (1961). "The QR Transformation: A Unitary Analogue to the LR Transformation," *Comp. J. 4*, 265–71, 332–34.

V. N. Kublanovskaya (1961). "On Some Algorithms for the Solution of the Complete Eigenvalue Problem," *USSR Comp. Math. Phys. 3*, 637–57.

As can be deduced from the title of the first paper, the LR iteration predates the QR iteration. The former very fundamental algorithm was proposed by

H. Rutishauser (1958). "Solution of Eigenvalue Problems with the LR Transformation," *Nat. Bur. Stand. App. Math. Ser. 49*, 47-81.

Numerous papers on the convergence of the QR iteration have appeared. Several of these are

B. N. Parlett (1965). "Convergence of the Q-R Algorithm," *Numer. Math. 7*, 187-93. (Correction in *Numer. Math. 10*, 163-64.)
B. N. Parlett (1966). "Singular and Invariant Matrices under the QR Algorithm," *Math. Comp. 20*, 611-15.
B. N. Parlett (1968). "Global Convergence of the Basic QR Algorithm on Hessenberg Matrices," *Math. Comp. 22*, 803-17.
J. H. Wilkinson (1965). "Convergence of the LR, QR, and Related Algorithms," *Comp. J. 8*, 77-84.

Wilkinson (AEP, chap. 9) also discusses the convergence theory for this important algorithm.

The following papers are concerned with various practical and theoretical aspects of simultaneous iteration:

M. Clint and A. Jennings (1971). "A Simultaneous Iteration Method for the Unsymmetric Eigenvalue Problem," *J. Inst. Math. Applic. 8*, 111-21.
A. Jennings and D.R.L. Orr (1971). "Application of the Simultaneous Iteration Method to Undamped Vibration Problems," *Inst. J. Numer. Meth. Eng. 3*, 13-24.
A. Jennings and W. J. Stewart (1975). "Simultaneous Iteration for the Partial Eigensolution of Real Matrices," *J. Inst. Math. Applic. 15*, 351-62.
H. Rutishauser (1969). "Computational Aspects of F. L. Bauer's Simultaneous Iteration Method," *Numer. Math. 13*, 4-13.
H. Rutishauser (1970). "Simultaneous Iteration Method for Symmetric Matrices," *Numer. Math. 16*, 205-23. See also HACLA, pp. 284-302.
G. W. Stewart (1975c). "Methods of Simultaneous Iteration for Calculating Eigenvectors of Matrices," in *Topics in Numerical Analysis II*, ed. John J. H. Miller, Academic Press, New York, pp. 185-96.
G. W. Stewart (1976d). "Simultaneous Iteration for Computing Invariant Subspaces of Non-Hermitian Matrices," *Numer. Math. 25*, 123-36.

See also chapter 10 of

A. Jennings (1977b). *Matrix Computation for Engineers and Scientists*, John Wiley & Sons, New York.

Simultaneous iteration and the Lanczos algorithm (cf. Chapter 9) are the principal methods for finding a few eigenvalues of a general sparse matrix.

Sec. 7.4. The Hessenberg and Real Schur Decompositions

In this and the next section we show how to make the QR iteration (7.3-1) a fast, effective method for computing Schur decompositions. Because the vast majority of eigenvalue/invariant subspace problems involve real data, we will concentrate on developing the real analog of (7.3-1):

$$H_0 = U_0^T A U_0$$
For $k = 1, 2, \ldots$
(7.4-1) Compute the Q-R factorization $H_{k-1} = U_k R_k$.
Set $H_k = R_k U_k$

Here, $A \in \mathbb{R}^{n \times n}$, each $U_i \in \mathbb{R}^{n \times n}$ is orthogonal, and each $R_i \in \mathbb{R}^{n \times n}$ is upper triangular.

Usually in numerical linear algebra there is no loss in generality by focusing attention on real matrix problems: most real arithmetic algorithms have obvious complex arithmetic analogs. The QR iteration is no exception to this, but there is an additional difficulty for the real iteration (7.4-1) in the event that $A \in \mathbb{R}^{n \times n}$ has complex eigenvalues. When this is the case, the iterates H_k cannot possibly converge to upper triangular form. For this reason we must lower our expectations and be content with the calculation of the following decomposition:

THEOREM 7.4-1: *Real Schur Decomposition*. If $A \in \mathbb{R}^{n \times n}$ then there exists an orthogonal $Q \in \mathbb{R}^{n \times n}$ such that

$$(7.4\text{-}2) \qquad Q^T A Q = \begin{bmatrix} R_{11} & R_{12} & \cdots & R_{1m} \\ 0 & R_{21} & \cdots & R_{2m} \\ \vdots & \vdots & \ddots & \vdots \\ 0 & 0 & \cdots & R_{mm} \end{bmatrix}$$

where each R_{ii} is either a 1-by-1 matrix or a 2-by-2 matrix having complex conjugate eigenvalues.

Proof. The complex eigenvalues of A must come in conjugate pairs, since the characteristic polynomial $\det(zI - A)$ has real coefficients. Let k be the number of complex conjugate pairs in $\lambda(A)$. We prove the theorem by induction on k.

Observe first that Lemma 7.1-1 and Theorem 7.1-2 have obvious real analogs. Thus, the theorem holds if $k = 0$. Now suppose that $k \geq 1$. If $\lambda = \gamma + i\mu \in \lambda(A)$ and $\mu \neq 0$, then there exist vectors y and z in $\mathbb{R}^n$ ($z \neq 0$) such that

$$A(y + iz) = (\gamma + i\mu)(y + iz),$$

i.e.,

$$A[y, z] = [y, z] \begin{bmatrix} \gamma & \mu \\ -\mu & \gamma \end{bmatrix}$$

The assumption that $\mu \neq 0$ implies that y and z span a two-dimensional, real invariant subspace for A. It then follows from Lemma 7.1-1 that an orthogonal $U \in \mathbb{R}^{n \times n}$ exists such that

$$U^T A U = \begin{bmatrix} T_{11} & T_{12} \\ 0 & T_{22} \end{bmatrix} \begin{matrix} 2 \\ n-2 \end{matrix} \qquad \lambda(T_{11}) = \{\lambda, \bar{\lambda}\}$$
$$\qquad \quad 2 \quad\ n-2$$

By induction, there exists an orthogonal $\tilde{U}$ so $\tilde{U}^T T_{22} \tilde{U}$ has the requisite structure. The theorem follows by setting $Q = U \, \mathrm{diag}(I_2, \tilde{U})$. $\square$

The theorem shows that any real matrix is orthogonally similar to an upper quasi-triangular matrix. It is clear that the real and imaginary parts of the complex eigenvalues can be easily obtained from the 2-by-2 diagonal blocks.

We thus turn our attention to the speedy calculation of the real Schur form. In this regard, the most glaring shortcoming associated with (7.4-1) is that each step requires a full Q-R factorization costing $O(n^3)$ flops. Fortunately, the amount of work per iteration can be reduced by an order of magnitude if the orthogonal matrix U_0 is judiciously chosen. In particular, if U_0 is such that $U_0^T A U_0 = H_0 = (h_{ij})$ is upper Hessenberg ($h_{ij} = 0, i > j + 1$), then each subsequent H_k requires only $O(n^2)$ flops to calculate. This follows by examining in detail the following algorithm.

ALGORITHM 7.4-1: *Hessenberg Q-R Step.*   Given the upper Hessenberg matrix $H \in \mathbb{R}^{n \times n}$, the following algorithm computes its Q-R factorization $H = UR$ and then overwrites H with $\bar{H} = RU$.

> For $k = 1, \ldots, n - 1$
>> Determine $c_k = \cos(\theta_k)$ and $s_k = \sin(\theta_k)$ such that
>>
>> $$\begin{bmatrix} c_k & s_k \\ -s_k & c_k \end{bmatrix} \begin{bmatrix} h_{kk} \\ h_{k+1,k} \end{bmatrix} = \begin{bmatrix} * \\ 0 \end{bmatrix}$$
>>
>> For $j = k, \ldots, n$
>>
>> $$\begin{bmatrix} h_{kj} \\ h_{k+1,j} \end{bmatrix} := \begin{bmatrix} c_k & s_k \\ -s_k & c_k \end{bmatrix} \begin{bmatrix} h_{kj} \\ h_{k+1,j} \end{bmatrix}$$
>
> For $k = 1, \ldots, n - 1$
>> For $i = 1, \ldots, k + 1$

$$[h_{ik}, h_{i,k+1}] := [h_{ik}, h_{i,k+1}] \begin{bmatrix} c_k & -s_k \\ s_k & c_k \end{bmatrix}$$

This algorithm requires $4n^2$ flops. Moreover, since

$$U^T = J(n-1, n, \theta_{n-1}) \cdots J(1, 2, \theta_1)$$

is lower Hessenberg, $\bar{H} = RU$ is upper Hessenberg. Thus, the QR iteration preserves Hessenberg structure.

EXAMPLE 7.4-1.   If Algorithm 7.4-1 is applied to

$$H = \begin{bmatrix} 3 & 1 & 2 \\ 4 & 2 & 3 \\ 0 & .01 & 1 \end{bmatrix},$$

then

$$J_1 = \begin{bmatrix} .6 & .8 & 0 \\ -.8 & .6 & 0 \\ 0 & 0 & 1 \end{bmatrix} \text{ and } J_2 = \begin{bmatrix} 1 & 0 & 0 \\ 0 & .9996 & .0249 \\ 0 & -.0249 & .9996 \end{bmatrix}$$

and

$$\bar{H} = \begin{bmatrix} 4.7600 & -2.5442 & 5.4653 \\ .3200 & .1856 & -2.1796 \\ .0000 & .0263 & 1.0540 \end{bmatrix}.$$

It remains for us to show how the *Hessenberg decomposition*

(7.4-3) $$\qquad U_0^T A U_0 = H = \diagdown \qquad\qquad U_0^T U_0 = I$$

can be computed. Fortunately, the calculation is not difficult and involves computing U_0 as a product of Householder matrices $P_1, \ldots, P_{n-2}$. The "role" of P_k is to zero the k-th column below the subdiagonal. In the $n = 6$ case we have

$$A = \begin{bmatrix} \times & \times & \times & \times & \times & \times \\ \times & \times & \times & \times & \times & \times \\ \times & \times & \times & \times & \times & \times \\ \times & \times & \times & \times & \times & \times \\ \times & \times & \times & \times & \times & \times \\ \times & \times & \times & \times & \times & \times \end{bmatrix} \xrightarrow{P_1} \begin{bmatrix} \times & \times & \times & \times & \times & \times \\ \times & \times & \times & \times & \times & \times \\ 0 & \times & \times & \times & \times & \times \\ 0 & \times & \times & \times & \times & \times \\ 0 & \times & \times & \times & \times & \times \\ 0 & \times & \times & \times & \times & \times \end{bmatrix} \xrightarrow{P_2}$$

$$P_2 \longrightarrow \begin{bmatrix} x & x & x & x & x & x \\ x & x & x & x & x & x \\ 0 & x & x & x & x & x \\ 0 & 0 & x & x & x & x \\ 0 & 0 & x & x & x & x \\ 0 & 0 & x & x & x & x \end{bmatrix} \quad P_3 \longrightarrow \begin{bmatrix} x & x & x & x & x & x \\ x & x & x & x & x & x \\ 0 & x & x & x & x & x \\ 0 & 0 & x & x & x & x \\ 0 & 0 & 0 & x & x & x \\ 0 & 0 & 0 & x & x & x \end{bmatrix}$$

$$P_4 \longrightarrow \begin{bmatrix} x & x & x & x & x & x \\ x & x & x & x & x & x \\ 0 & x & x & x & x & x \\ 0 & 0 & x & x & x & x \\ 0 & 0 & 0 & x & x & x \\ 0 & 0 & 0 & 0 & x & x \end{bmatrix}$$

In general, after $k - 1$ steps we have computed Householder matrices $P_1, \ldots, P_{k-1}$ such that

$$(P_1 \cdots P_{k-1})^T A (P_1 \cdots P_{k-1}) = \begin{bmatrix} B & \vdots & C \\ \hline 0 & b & D \end{bmatrix} \begin{matrix} k \\ \\ n-k \end{matrix},$$
$$\phantom{(P_1 \cdots P_{k-1})^T A =} \begin{matrix} k-1 & 1 & n-k \end{matrix}$$

where B is upper Hessenberg. If $\bar{P}_k$ is an order $n - k$ Householder matrix for which $\bar{P}_k b$ is a multiple of $e_1^{(n-k)}$ and if $P_k = \operatorname{diag}(I_k, \bar{P}_k)$, then

$$(P_1 \cdots P_k)^T A (P_1 \cdots P_k) = \begin{bmatrix} B & \vdots & C\bar{P}_k \\ \hline 0 & \bar{P}_k b & \bar{P}_k D \bar{P}_k \end{bmatrix}$$

is a matrix that is upper Hessenberg in its first k columns. Overall we have

ALGORITHM 7.4-2: *Householder Reduction to Hessenberg Form.* Given $A \in \mathbb{R}^{n \times n}$, the following algorithm overwrites A with $H = U_0^T A U_0$, where H is upper Hessenberg and $U_0 = P_1 \cdots P_{n-2}$ is a product of Householder matrices.

For $k = 1, \ldots, n - 2$
    Determine a Householder matrix $\bar{P}_k$ of order $n - k$ such that

$$\bar{P}_k \begin{bmatrix} a_{k+1,k} \\ \vdots \\ a_{nk} \end{bmatrix} = \begin{bmatrix} x \\ 0 \\ \vdots \\ 0 \end{bmatrix}$$

$$A := P_k^T A P_k, \quad P_k = \operatorname{diag}(I_k, \bar{P}_k)$$

This algorithm requires $\frac{5}{3}n^3$ flops. U_0 can be stored in factored form below the subdiagonal of A. See Martin and Wilkinson (1968d). If U_0 is explicitly formed, an additional $\frac{2}{3}n^3$ flops are required.

The roundoff properties of this method for reducing A to Hessenberg form are very desirable. Wilkinson (AEP, p. 351) states that the computed Hessenberg matrix $\hat{H}$ satisfies $\hat{H} = Q^T(A + E)Q$, where $Q^TQ = I$,

$$\|E\|_F \leqslant cn^2 \mathbf{u} \|A\|_F,$$

and c is a small constant.

EXAMPLE 7.4-2. If

$$A = \begin{bmatrix} 1 & 5 & 7 \\ 3 & 0 & 6 \\ 4 & 3 & 1 \end{bmatrix} \quad \text{and} \quad Q = \begin{bmatrix} 1 & 0 & 0 \\ 0 & .6 & .8 \\ 0 & .8 & -.6 \end{bmatrix},$$

then

$$Q^TAQ = H = \begin{bmatrix} 1.00 & 8.60 & -0.20 \\ 5.00 & 4.96 & -0.72 \\ 0.00 & 2.28 & -3.96 \end{bmatrix}.$$

The Hessenberg decomposition is not unique. If Z is any orthogonal matrix in $\mathbb{R}^{n \times n}$ and we apply Algorithm 7.4-2 to Z^TAZ, then $Q^TAQ = H$ is upper Hessenberg where $Q = ZU_0$. However, $Qe_1 = Z(U_0 e_1) = Z e_1$ suggesting that H is unique once the first column of Q is specified. This is essentially the case provided H has no zero subdiagonal entries. Hessenberg matrices with this property are said to be *unreduced*.

THEOREM 7.4-2: *Implicit Q Theorem.* Suppose $Q = [q_1, \ldots, q_n]$ and $V = [v_1, \ldots, v_n]$ are orthogonal matrices with the property that both $Q^TAQ = H$ and $V^TAV = G$ are upper Hessenberg. Let k denote the smallest positive integer for which $h_{k+1,k} = 0$ with the convention that $k = n$ if H is unreduced. If $v_1 = q_1$ then $v_i = \pm q_i$ and $|h_{i,i-1}| = |g_{i,i-1}|$ for $i = 2, \ldots, k$. Moreover, if $k < n$ then $g_{k+1,k} = 0$.

Proof. Define the orthogonal matrix $W = [w_1, \ldots, w_n]$ by $W = V^TQ$ and observe that $GW = WH$. Thus, for $i = 2, \ldots, k$ we have

$$h_{i,i-1}w_i = Gw_{i-1} - \sum_{j=1}^{i-1} h_{j,i-1}w_j.$$

Since $w_1 = e_1$ it follows that $[w_1, \ldots, w_k]$ is upper triangular and thus, $w_i = \pm e_i$ for $i = 2, \ldots, k$. Since $w_i = V^T q_i$ and $h_{i,i-1} = w_i^T G w_{i-1}$ it follows that $v_i = \pm q_i$ and $|h_{i,i-1}| = |g_{i,i-1}|$ for $i = 2, \ldots, k$. If $h_{k+1,k} = 0$ then ignoring signs we have

$$g_{k+1,k} = e_{k+1}^T G e_k = e_{k+1}^T G W e_k = (e_{k+1}^T W)(H e_k)$$

$$= e_{k+1}^T \sum_{i=1}^{k} h_{ik} W e_i = \sum_{i=1}^{k} h_{ik} e_{k+1}^T e_i = 0 \ \square$$

The gist of the implicit Q theorem is that if $Q^T A Q = H$ and $Z^T A Z = G$ are each unreduced upper Hessenberg matrices and Q and Z have the same first column, then G and H are "essentially equal" in the sense that $G = D^{-1} H D$ where $D = \text{diag}(\pm 1, \ldots, \pm 1)$.

Another way to characterize the non-uniqueness of the Hessenberg decomposition is to consider the *Krylov matrices* $K(x, A, j)$ defined by

$$K(x, A, j) = [x, Ax, A^2 x, \ldots, A^{j-1} x].$$

In particular, if $Q^T A Q = H$ is a Hessenberg decomposition of A, then

$$K(q_1, A, n-1) = Q K(e_1, H, n-1) \qquad q_1 = Q e_1$$

is a Q-R factorization of $K(q_1, A, n-1)$. If $R = K(e_1, H, n-1)$, then it follows that $r_{ii} = h_{21} h_{32} \cdots h_{i,i-1}$, and so we see that there is more or less a correspondance between nonsingular Krylov matrices $K(x, A, n-1)$ and orthogonal similarity reductions to unreduced Hessenberg form.

Just as the Schur decomposition has a nonunitary analog in the Jordan decomposition, so does the Hessenberg decomposition have a nonunitary analog in the *companion matrix decomposition*. Suppose that for some $x \in \mathbb{R}^n$ we have $A K(x, A, n-1) = K(x, A, n-1) C$, where C has the form

$$(7.4\text{-}4) \qquad C = \begin{bmatrix} 0 & 0 & \cdots & 0 & -c_0 \\ 1 & 0 & \cdots & 0 & -c_1 \\ 0 & 1 & \cdots & 0 & -c_2 \\ \vdots & \vdots & \ddots & \vdots & \vdots \\ 0 & 0 & \cdots & 1 & -c_{n-1} \end{bmatrix}$$

The matrix C is said to be a *companion matrix*. Since $\det(zI - C) = c_0 + c_1 z + \cdots + c_{n-1} z^{n-1} + z^n$, it follows that if $Y = K(x, A, n-1)$ is nonsingular, then the decomposition $Y^{-1} A Y = C$ displays A's characteristic polynomial. This, coupled with the sparseness of C, has led to "companion matrix methods" in various application areas. These techniques typically involve

(a) Computing the Hessenberg decomposition $U_0^T A U_0 = H$.

(b) Hoping H is unreduced and setting $Y = [e_1, He_1, \ldots, H^{n-1}e_1]$.

(c) Solving $YC = HY$ for C.

Unfortunately, this calculation can be highly unstable. A is similar to an unreduced Hessenberg matrix only if each eigenvalue has unit geometric multiplicity. (H unreduced $\Rightarrow$ rank$[H - \lambda I] \geqslant n - 1$.) Matrices that have this property are called *nonderogatory*. It follows that the matrix Y above can be very poorly conditioned if A is close to a derogatory matrix.

A full discussion of the dangers associated with companion matrix computation can be found in Wilkinson (AEP, pp. 405 ff.).

While we are on the subject of nonorthogonal reduction to Hessenberg form, we should mention that Gauss transformations can be used in lieu of Householder matrices in Algorithm 7.4-2. Suppose permutations $\Pi_1, \ldots,$ Π_{k-1} and Gauss transformations $M_1, \ldots, M_{k-1}$ have been determined such that

$$(M_{k-1}\Pi_{k-1} \cdots M_1\Pi_1)A(M_{k-1}\Pi_{k-1} \cdots M_1\Pi_1)^{-1}$$

$$= \begin{array}{c} \\ \left[\begin{array}{c|c|c} B & C \\ \hline 0 & b & D \end{array} \right] \\ \begin{array}{ccc} k-1 & 1 & n-k \end{array} \end{array} \begin{array}{c} k \\ \\ n-k \end{array},$$

where B is upper Hessenberg. A permutation $\bar{\Pi}_k$ of order $n - k$ is then determined such that the first element of $\bar{\Pi}_k b$ is maximal in absolute value. This makes it possible to determine a stable Gauss transformation $\bar{M}_k = I - z_k e_1^T$, also of order $n - k$, such that all but the first component of $\bar{M}_k(\bar{\Pi}_k b)$ is zero. Defining $\Pi_k = \text{diag}(I_k, \bar{\Pi}_k)$ and $M_k = \text{diag}(I_k, \bar{M}_k)$ we see that

$$(M_k \Pi_k) \begin{bmatrix} B & C \\ \hline 0 & b & D \end{bmatrix} (M_k \Pi_k)^{-1} = \begin{bmatrix} B & D\bar{\Pi}_k^T \bar{M}_k^{-1} \\ \hline 0 & \bar{M}_k\bar{\Pi}_k b & \bar{M}_k\bar{\Pi}_k D\bar{\Pi}_k^T \bar{M}_k^{-1} \end{bmatrix}$$

is upper Hessenberg in its first k columns. Note that $\bar{M}_k^{-1} = I + z_k e_1^T$.

A careful operation count reveals that the Gauss reduction to Hessenberg form requires only half the number of flops of the Householder method. However, as in the case of Gaussian elimination with partial pivoting, there is a (fairly remote) chance of 2^n growth. See Businger (1969). Another difficulty associated with the Gauss approach is that the eigenvalue condition numbers—the $s(\lambda)^{-1}$—are not preserved with nonorthogonal similarity transformations. This somewhat complicates the error estimation schemes that are presented in §7.6.

Problems

P7.4-1. Show that if $A \in \mathbb{R}^{n \times n}$ and $z \in \mathbb{R}^n$, then there exists an orthogonal Q such that $Q^T A Q$ is upper Hessenberg and $Q^T z$ is a multiple of e_1. (Hint: Reduce z first and then apply Algorithm 7.4-2.)

P7.4-2. Fill in the details of the Gauss reduction to Hessenberg form and verify that it only requires $\frac{5}{6} n^3$ flops.

P7.4-3. In some situations it is necessary to solve the linear system $(A + zI)x = b$ for many different values of $z \in \mathbb{R}$ and $b \in \mathbb{R}^n$. Show how this problem can be efficiently and stably solved through application of Algorithms 5.3-4 and 7.4-2.

P7.4-4. Show that if an upper Hessenberg matrix has a repeated nondefective eigenvalue, then it must have a zero subdiagonal entry.

P7.4-5. Suppose $H \in \mathbb{R}^{n \times n}$ is an unreduced upper Hessenberg matrix. Show that there exists a diagonal matrix D such that each subdiagonal element of $D^{-1} H D$ is equal to one. What is $\kappa_2(D)$?

P7.4-6. Suppose $W, Y \in \mathbb{R}^{n \times n}$ and define the matrices C and B by

$$C = W + iY, \qquad B = \begin{bmatrix} W & -Y \\ Y & W \end{bmatrix}.$$

Show that if $\lambda \in \lambda(C)$ is real, then $\lambda \in \lambda(B)$. Relate the corresponding eigenvectors.

P7.4-7. Suppose $A = \begin{bmatrix} w & x \\ y & z \end{bmatrix}$ is a real matrix having eigenvalues $\lambda \pm i\mu$, where μ is nonzero. Give an algorithm that stably determines $c = \cos(\theta)$ and $s = \sin(\theta)$ such that

$$\begin{bmatrix} c & s \\ -s & c \end{bmatrix}^T \begin{bmatrix} w & x \\ y & z \end{bmatrix} \begin{bmatrix} c & s \\ -s & c \end{bmatrix} = \begin{bmatrix} \lambda & \beta \\ \alpha & \lambda \end{bmatrix}$$

where $\alpha\beta = -\mu^2$.

P7.4-8. Suppose λ and x are a known eigenvalue-eigenvector pair for the upper Hessenberg matrix $H \in \mathbb{R}^{n \times n}$. Give an algorithm for computing an orthogonal matrix P such that

$$P^T H P = \begin{bmatrix} \lambda & w^T \\ 0 & H_1 \end{bmatrix},$$

where $H_1 \in \mathbb{R}^{(n-1) \times (n-1)}$ is upper Hessenberg. (Hint: P is a product of Jacobi rotations.)

Notes and References for Sec. 7.4

The real Schur decomposition was originally presented in

F. D. Murnaghan and A. Wintner (1931). "A Canonical Form for Real Matrices under Orthogonal Transformations," *Proc. Nat. Acad. Sci. 17*, 417-20.

A thorough treatment of the reduction to Hessenberg form is given in Wilkinson (AEP, chap. 6), and ALGOL procedures for both the Householder and Gauss methods appear in

R. S. Martin and J. H. Wilkinson (1968d). "Similarity Reduction of a General Matrix to Hessenberg Form," *Numer. Math. 12*, 349-68. See also HACLA, pp. 339-58.

FORTRAN versions of the ALGOL procedures in the last reference are in EISPACK. The possibility of exponential growth in the Gauss transformation approach was first pointed out in

P. Businger (1969). "Reducing a Matrix to Hessenberg Form," *Math. Comp. 23*, 819-21.

However, the algorithm should be regarded in the same light as Gaussian elimination with partial pivoting—stable for all practical purposes. See EISPACK, pp. 56-58.

Aspects of the Hessenberg decomposition for sparse matrices are discussed in

I. S. Duff and J. K. Reid (1975). "On the Reduction of Sparse Matrices to Condensed Forms by Similarity Transformations," *J. Inst. Math. Applic. 15*, 217-24.

Once an eigenvalue of an unreduced upper Hessenberg matrix is known, it is possible to zero the last subdiagonal entry using a sequence of Givens similarity transformations. See

P. A. Businger (1971). "Numerically Stable Deflation of Hessenberg and Symmetric Tridiagonal Matrices," *BIT 11*, 262-70.

Some interesting mathematical properties of the Hessenberg form may be found in

Y. Ikebe (1979). "On Inverses of Hessenberg Matrices," *Lin. Alg. & Its Applic. 24*, 93-97.
B. N. Parlett (1967). "Canonical Decomposition of Hessenberg Matrices," *Math. Comp. 21*, 223-27.

Although the Hessenberg decomposition is largely appreciated as a "front end" decomposition for the QR iteration, it is increasingly popular as a cheap alternative to the more expensive Schur decomposition in certain problems. For a sampling of applications where it has proven to be very useful, consult

G. H. Golub, S. Nash, and C. Van Loan (1979). "A Hessenberg-Schur Method for the Problem $AX + XB = C$," *IEEE Trans. Auto. Cont. AC-24*, 909-13.
A. Laub (1981). "Efficient Multivariable Frequency Response Computations," *IEEE Trans. Auto. Cont. AC-26*, 407-8.
W. Enwright (1979). "On the Efficient and Reliable Numerical Solution of Large Linear Systems of O.D.E.'s," *IEEE Trans. Auto. Cont. AC-24*, 905-8.
C. C. Paige (1981). "Properties of Numerical Algorithms Related to Computing Controllability," *IEEE Trans. Auto. Cont. AC-26*, 130-38.
C. Van Loan (1982b). "Using the Hessenberg Decomposition in Control Theory," in

Algorithms and Theory in Filtering and Control, ed. D. C. Sorenson and R. J. Wets, Mathematical Programming Study No. 18, North Holland, Amsterdam, pp. 102–11.

Sec. 7.5. The Practical QR Algorithm

We return to the Hessenberg QR iteration which we write as follows:

$$H = U_0^T A U_0$$
$$\text{For } k = 1, 2, \ldots$$
(7.5-1) $$H = UR \qquad \text{(Q-R factorization)}$$
$$H := RU$$

Without loss of generality we may assume that each Hessenberg matrix produced by this iteration is unreduced. If not, then at some stage we have

$$H = \begin{bmatrix} H_{11} & H_{12} \\ 0 & H_{22} \end{bmatrix} \begin{matrix} p \\ n-p \end{matrix} \qquad 1 \leqslant p < n$$
$$\quad\; p \quad\;\; n-p$$

and the problem "decouples" into two smaller problems involving H_{11} and H_{22}. The term *deflation* is also used in this context, usually when $p = n - 1$ or $n - 2$.

In practice, decoupling occurs whenever a subdiagonal entry in H is suitably small. For example, in EISPACK if

(7.5-2) $$|h_{p+1,p}| \leqslant \mathbf{u}(|h_{pp}| + |h_{p+1,p+1}|)$$

then $h_{p+1,p}$ is "declared" to be zero. This is justified since rounding errors of order $\mathbf{u} \, \|H\|$ are already present throughout the matrix.

Our aim in this section is to show how the convergence of (7.5-1) can be accelerated by incorporating "shifts." Let $\mu \in \mathbb{R}$ and consider the iteration

$$H = U_0^T A U_0 \qquad \text{(unreduced upper Hessenberg)}$$
$$\text{For } k = 1, 2, \ldots$$
(7.5-3) $$H - \mu I = UR \qquad \text{(Q-R factorization)}$$
$$H := RU + \mu I$$

The scalar μ is referred to as a *shift*. Each matrix H generated in (7.5-3) is similar to A, since $RU + \mu I = U^T(UR + \mu I)U = U^T H U$.

If we order the eigenvalues λ_i of A so that

$$|\lambda_1 - \mu| \geqslant \ldots \geqslant |\lambda_n - \mu| \, ,$$

then the theory of §7.3 says that the p-th subdiagonal entry in H converges to zero with rate

$$\left| \frac{\lambda_{p+1} - \mu}{\lambda_p - \mu} \right|^k .$$

Of course, if $\lambda_p = \lambda_{p+1}$, then there is no convergence at all. But if μ is much closer to λ_n than to the other eigenvalues, then convergence is rapid. In the extreme case we have the following:

THEOREM 7.5-1. Let μ be an eigenvalue of an n-by-n unreduced Hessenberg matrix H. If $\bar{H} = RU + \mu I$, where $(H - \mu I) = UR$ is the Q-R factorization of $H - \mu I$, then $\bar{h}_{n,n-1} = 0$ and $\bar{h}_{nn} = \mu$.

Proof. If H is unreduced, then so is the upper Hessenberg matrix $H - \mu I$. Since $U^T(H - \mu I) = R$ is singular and since it can be shown that

(7.5-4) $|r_{ii}| \geq |h_{i+1,i}|$, $i = 1, 2, \ldots, n - 1$

it follows that $r_{nn} = 0$. Consequently, the bottom row of $\bar{H}$ is equal to $(0, \ldots, 0, \mu)$. $\square$

Thus, if we shift by an eigenvalue, we can decouple in one step.

EXAMPLE 7.5-1. If

$$H = \begin{bmatrix} 9 & -1 & -2 \\ 2 & 6 & -2 \\ 0 & 1 & 5 \end{bmatrix} ,$$

then $6 \in \lambda(H)$. If $UR = H - 6I$ is the Q-R factorization, then $\bar{H} = RQ + 6I$ is given by

$$\bar{H} \cong \begin{bmatrix} 8.5384 & -3.7313 & -1.0090 \\ 0.6343 & 5.4615 & 1.3867 \\ 0.0000 & 0.0000 & 6.0000 \end{bmatrix} .$$

(Unfortunately, there are examples where $\bar{h}_{nn}$ is non-negligible due to round-off.)

This suggests that we vary μ during the iteration as new information about $\lambda(A)$ emerges. Heuristically, h_{nn} is usually the best approximate eigenvalue to be found along the diagonal. If we shift by this quantity during each iteration, we obtain the *single-shift QR iteration*:

(7.5-5)
$$U^T A U = H = \text{\reflectbox{$\mathbb{N}$}} , \qquad U^T U = I$$
For $k = 1, 2, \ldots$
$$H - h_{nn}I = UR \qquad \text{(Q-R factorization)}$$
$$H := RU + h_{nn}I$$

If the $(n, n - 1)$ entry converges to zero, it is likely to do so at a quadratic rate. To see this, we borrow an example from Stewart (IMC, p. 366). Suppose H is an unreduced upper Hessenberg matrix of the form

$$H = \begin{bmatrix} x & x & x & x & x \\ x & x & x & x & x \\ & x & x & x & x \\ & & x & x & x \\ & & & \epsilon & h_{nn} \end{bmatrix}$$

and that we perform one step of the single-shift QR algorithm: $UR = H - h_{nn}I, \bar{H} = RU + h_{nn}I$. After $n - 2$ steps in the reduction of $H - h_{nn}I$ to upper triangular we obtain a matrix with the following structure:

$$\begin{bmatrix} x & x & x & x & x \\ & x & x & x & x \\ & & x & x & x \\ & & & a & b \\ & & & \epsilon & 0 \end{bmatrix}.$$

It is not hard to show that the $(n, n - 1)$ entry in $\bar{H}$ is given by

$$\bar{h}_{n,n-1} = \frac{\epsilon^2 b}{\epsilon^2 + a^2}.$$

If we assume that $\epsilon \ll a$, then it is clear that the new $(n, n - 1)$ entry has order ϵ^2, precisely what we would expect of a quadratically converging algorithm.

EXAMPLE 7.5-2.   If

$$H = \begin{bmatrix} 1 & 2 & 3 \\ 4 & 5 & 6 \\ 0 & .001 & 7 \end{bmatrix}.$$

and $UR = H - 7I$ is the Q-R factorization, then $\bar{H} + RU + 7I$ is given by

$$\bar{H} \cong \begin{bmatrix} -0.5384 & -1.6908 & 0.8351 \\ 0.3076 & 6.5264 & -6.6555 \\ 0.0000 & 2 \times 10^{-5} & 7.0119 \end{bmatrix}.$$

Unfortunately, difficulties with (7.5-5) can be expected if at some stage the eigenvalues a_1 and a_2 of

$$(7.5\text{-}6) \qquad G = \begin{bmatrix} h_{mm} & h_{mn} \\ h_{nm} & h_{nn} \end{bmatrix} \qquad\qquad m = n - 1$$

are complex, for then h_{nn} would tend to be a poor approximate eigenvalue.

A way around this difficulty is to perform two single-shift QR steps in succession, using a_1 and a_2 as shifts:

$$(7.5\text{-}7) \qquad \begin{aligned} H - a_1 I &= U_1 R_1 \\ H_1 &= R_1 U_1 + a_1 I \\ H_1 - a_2 I &= U_2 R_2 \\ H_2 &= R_2 U_2 + a_2 I. \end{aligned}$$

A calculation shows

$$(7.5\text{-}8) \qquad (U_1 U_2)(R_2 R_1) = M,$$

where M is defined by

$$(7.5\text{-}9) \qquad M = (H - a_1 I)(H - a_2 I).$$

Note that M is a real matrix even if G's eigenvalues are complex, since

$$M = H^2 - sH + tI$$

where

$$s = a_1 + a_2 = h_{mm} + h_{nn} = \text{trace}(G) \in \mathbb{R}$$

and

$$t = a_1 a_2 = h_{mm} h_{nn} - h_{mn} h_{nm} = \det(G) \in \mathbb{R}.$$

Thus, (7.5-8) is the Q-R factorization of a real matrix and we may choose U_1 and U_2 so that $Z = U_1 U_2$ is real orthogonal. It then follows that

$$H_2 = U_2^H H_1 U_2 = U_2^H (U_1^H H U_1) U_2 = (U_1 U_2)^H H (U_1 U_2) = Z^T H Z$$

is real.

Unfortunately, roundoff error will almost always prevent this fortuitous return to the real field. A real H_2 could be guaranteed if we

(a) explicitly form the real matrix $M = H^2 - sH + tI$,

(b) compute the real Q-R factorization $M = ZR$, and

(c) set $H_2 = Z^T H Z$.

But since (a) requires $O(n^3)$ flops, this is not a practical course of action.

In light of the Implicit Q Theorem, however it is possible to effect the transition from H to H_2 in $O(n^2)$ flops if we (a') compute Me_1, the first column of M;

(b') determine a Householder matrix P_0 such that $P_0 (Me_1)$ is a multiple of e_1;

(c') compute Householder matrices $P_1, \ldots, P_{n-2}$ such that if $Z_1 = P_0 P_1 \cdots P_{n-2}$, then $Z_1^T H Z_1$ is upper Hessenberg *and* the first columns of Z and Z_1 are the same.

Under these circumstances, the theorem permits us to conclude that if $Z^T H Z$ and $Z_1^T H Z_1$ are both unreduced upper Hessenberg matrices, then they are essentially equal. Note that if these Hessenberg matrices are not unreduced, then we can effect a decoupling and proceed with smaller subproblems.

Considering (a')-(c') in detail, observe first that P_0 can be determined in $O(1)$ flops, since $Me_1 = (x, y, z, 0, \ldots, 0)^T$ where

$$
\begin{aligned}
x &= h_{11}^2 + h_{12} h_{21} - s h_{11} + t \\
Y &= h_{21}(h_{11} + h_{22} - s) \\
z &= h_{21} h_{32}.
\end{aligned}
$$

Since a similarity transformation with P_0 only changes rows and columns 1, 2, and 3, we see that

$$
P_0 H P_0 = \begin{bmatrix}
x & x & x & x & x & x \\
x & x & x & x & x & x \\
x & x & x & x & x & x \\
x & x & x & x & x & x \\
0 & 0 & 0 & x & x & x \\
0 & 0 & 0 & 0 & x & x
\end{bmatrix}.
$$

Now the mission of the Householder matrices in (c') is to restore this matrix to upper Hessenberg form, a calculation that proceeds as follows:

$$
\begin{bmatrix}
x & x & x & x & x & x \\
x & x & x & x & x & x \\
x & x & x & x & x & x \\
x & x & x & x & x & x \\
0 & 0 & 0 & x & x & x \\
0 & 0 & 0 & 0 & x & x
\end{bmatrix}
\xrightarrow{P_1}
\begin{bmatrix}
x & x & x & x & x & x \\
x & x & x & x & x & x \\
0 & x & x & x & x & x \\
0 & x & x & x & x & x \\
0 & x & x & x & x & x \\
0 & 0 & 0 & 0 & x & x
\end{bmatrix}
$$

$$P_2 \longrightarrow \begin{bmatrix} x & x & x & x & x & x \\ x & x & x & x & x & x \\ 0 & x & x & x & x & x \\ 0 & 0 & x & x & x & x \\ 0 & 0 & x & x & x & x \\ 0 & 0 & x & x & x & x \end{bmatrix} \qquad P_3 \longrightarrow \begin{bmatrix} x & x & x & x & x & x \\ x & x & x & x & x & x \\ 0 & x & x & x & x & x \\ 0 & 0 & x & x & x & x \\ 0 & 0 & 0 & x & x & x \\ 0 & 0 & 0 & x & x & x \end{bmatrix}$$

$$P_4 \longrightarrow \begin{bmatrix} x & x & x & x & x & x \\ x & x & x & x & x & x \\ 0 & x & x & x & x & x \\ 0 & 0 & x & x & x & x \\ 0 & 0 & 0 & x & x & x \\ 0 & 0 & 0 & 0 & x & x \end{bmatrix}$$

Clearly, the general P_k has the form $P_k = \text{diag}(I_k, \bar{P}_k, I_{n-k-3})$ where $\bar{P}_k$ is a 3-by-3 Householder matrix. For example,

$$P_2 = \begin{bmatrix} 1 & 0 & 0 & 0 & 0 & 0 \\ 0 & 1 & 0 & 0 & 0 & 0 \\ 0 & 0 & x & x & x & 0 \\ 0 & 0 & x & x & x & 0 \\ 0 & 0 & x & x & x & 0 \\ 0 & 0 & 0 & 0 & 0 & 1 \end{bmatrix}$$

Note that P_{n-2} is an exception to this, having the form $P_{n-2} = \text{diag}(I_{n-2}, \bar{P}_{n-2})$.

The applicability of Theorem 7.4-3 follows from the observation that

$$P_k e_1 = e_1 \qquad\qquad k = 1, \ldots, n-2$$

and that P_0 and Z have the same first column. Hence $Z_1 e_1 = Z e_1$, and we can assert that Z_1 essentially equals Z provided that the upper Hessenberg matrices $Z^T H Z$ and $Z_1^T H Z_1$ are each unreduced. (If they are not, then the problem decouples, and we can proceed with a smaller subproblem.)

The implicit determination of H_2 from H was first described by Francis (1961) and we refer to (a$'$)–(c$'$) as a *Francis QR step*. The overall process is summarized as follows:

ALGORITHM 7.5-1: *Francis QR Step.* Given the unreduced upper Hessenberg matrix $H \in \mathbb{R}^{n \times n}$ whose trailing 2-by-2 principal submatrix has eigenvalues a_1 and a_2, the following algorithm overwrites H with $Z^T H Z$, where $Z = P_1 \cdots P_{n-2}$ is a product of Householder matrices and $Z^T(H - a_1 I) \cdot (H - a_2 I)$ is upper triangular.

$m := n - 1$
$s := h_{mm} + h_{nn}$
$t := h_{mm}h_{nn} - h_{mn}h_{nm}$
$x := h_{11}^2 + h_{12}h_{21} - s h_{11} + t$
$y := h_{21}(h_{11} + h_{22} - s)$
$z := h_{21}h_{32}$
For $k = 0, \ldots, n - 2$
$\quad$ If $k < n - 2$
$\qquad$ *then*
$\qquad\qquad$ Determine a Householder matrix $\bar{P}_k \in \mathbb{R}^{3 \times 3}$ such that

$$\bar{P}_k \begin{bmatrix} x \\ y \\ z \end{bmatrix} = \begin{bmatrix} * \\ 0 \\ 0 \end{bmatrix}.$$

$$H := P_k H P_k^T, \quad P_k = \mathrm{diag}(I_k, \bar{P}_k, I_{n-k-3})$$

$\qquad$ *else*
$\qquad\qquad$ Determine a Householder matrix $\bar{P}_{n-2} \in \mathbb{R}^{2 \times 2}$ such that

$$\bar{P}_{n-2} \begin{bmatrix} x \\ y \end{bmatrix} = \begin{bmatrix} * \\ 0 \end{bmatrix}.$$

$$H := P_{n-2} H P_{n-2}^T, \quad P_{n-2} = \mathrm{diag}(I_{n-2}, \bar{P}_{n-2})$$

$\quad x := h_{k+2,k+1}$
$\quad y := h_{k+3,k+1}$
$\quad$ If $k < n - 3$ then $z := h_{k+4,k+1}$

If modified Householder matrices (cf. §3.3) are used, then this algorithm requires $6n^2$ flops. If Z is accumulated into a given orthogonal matrix, an additional $6n^2$ flops are necessary.

Reducing A to Hessenberg form using Algorithm 7.4-1 and then iterating with Algorithm 7.5-1 to produce the real Schur form is the standard means by which the dense unsymmetric eigenproblem is solved. During the iteration it is necessary to monitor the subdiagonal elements in H in order to spot any possible decoupling. How this is done is illustrated in the following algorithm:

ALGORITHM 7.5-2: *QR Algorithm.*   Given $A \in \mathbb{R}^{n \times n}$ and a tolerance ϵ greater than unit roundoff, this algorithm computes the real Schur decomposition $Q^T A Q = T$. A is overwritten with the Hessenberg decomposition. If Q and T are desired, then T is stored in H. If only eigenvalues are desired, then diagonal blocks in T are stored in the corresponding positions in H.

> Use Algorithm 7.4-2 to compute the Hessenberg decomposition $U_0^T A U_0 = H$, where $U_0 = P_1 \cdots P_{n-2}$. Store the Hessenberg matrix in H.
> If Q is desired then
> > $Q := I$
> > For $k = n - 2, \ldots, 1$
> > > $Q := P_k Q$
>
> *Repeat*:
> > Set to zero all subdiagonal elements that satisfy
> >
> > $$|h_{i,i-1}| \leqslant \epsilon \, (\, |h_{ii}| + |h_{i-1,i-1}| \,).$$
> >
> > Find the largest non-negative q and the smallest non-negative p such that
> >
> > $$H = \begin{bmatrix} H_{11} & H_{12} & H_{13} \\ 0 & H_{22} & H_{23} \\ 0 & 0 & H_{33} \end{bmatrix} \begin{matrix} p \\ n-p-q \\ q \end{matrix}$$
> > $$\begin{matrix} p & n-p-q & q \end{matrix}$$
> >
> > where H_{33} is upper quasi-triangular and H_{22} is unreduced. (Note: either p or q may be zero.)
> > If $q = n$ then upper triangularize all 2-by-2 diagonal blocks in H that have real eigenvalues, accumulate the orthogonal transformations if necessary, and quit.
> > Apply a Francis QR step to H_{22}:
> > > $H_{22} := Z^T H_{22} Z$
> > > If Q and T are desired
> > > > *then*
> > > > > $Q := Q \, \text{diag}(I_p, Z, I_q)$
> > > > > $H_{12} := H_{12} Z$
> > > > > $H_{23} := Z^T H_{23}$
>
> > Go to *Repeat*.

This algorithm requires $15n^3$ flops if Q and T are computed. If only the eigenvalues are desired, then $8n^3$ flops are necessary. These flop counts are very approximate and are based on the empirical observation that on average only two Francis iterations are required before the lower 1-by-1 or 2-by-2 decouples. A more detailed look at the time required to execute the QR algorithm is given in EISPACK, p. 119.

EXAMPLE 7.5-3. If the QR algorithm (Algorithm 7.5-2) is applied to

$$A = H = \begin{bmatrix} 2 & 3 & 4 & 5 & 6 \\ 4 & 4 & 5 & 6 & 7 \\ 0 & 3 & 6 & 7 & 8 \\ 0 & 0 & 2 & 8 & 9 \\ 0 & 0 & 0 & 1 & 10 \end{bmatrix}$$

then the subdiagonal entries converge as shown in Table 7.5-1.

Table 7.5-1.

Iteration	$O(|h_{21}|)$	$O(|h_{32}|)$	$O(|h_{43}|)$	$O(|h_{54}|)$
1	10^0	10^0	10^0	10^0
2	10^0	10^0	10^0	10^0
3	10^0	10^0	10^{-1}	10^0
4	10^0	10^0	10^{-3}	10^{-3}
5	10^0	10^0	10^{-6}	10^{-5}
6	10^{-1}	10^0	10^{-13}	10^{-13}
7	10^{-1}	10^0	10^{-28}	10^{-13}
8	10^{-4}	10^0	converg.	converg.
9	10^{-8}	10^0		
10	10^{-8}	10^0		
11	10^{-16}	10^0		
12	10^{-32}	10^0		
	converg.	converg.		

The roundoff properties of the QR algorithm are what one would expect of any orthogonal matrix technique. The computed real Schur form $\hat{T}$ is orthogonally similar to a matrix near to A, i.e.,

$$Q^T(A + E)Q = \hat{T},$$

where $Q^T Q = I$ and $\|E\|_2 \cong \mathbf{u} \|A\|_2$. The computed $\hat{Q}$ is almost orthogonal in the sense that $\hat{Q}^T \hat{Q} = I + F$ where $\|F\|_2 \cong \mathbf{u}$.

The order of the eigenvalues along T is somewhat arbitrary. But as we discuss in §7.6, any ordering can be achieved by using a simple procedure for swapping two adjacent diagonal entries.

Finally, we mention that if the elements of A have widely varying magnitudes, then A should be *balanced* before applying the QR algorithm. This is an $O(n^2)$ calculation in which a diagonal matrix D of the form

$$D = \text{diag}(b^{i_1}, \ldots, b^{i_n}) \qquad\qquad b = \text{machine base}$$

is determined such that if

$$D^{-1}AD = [c_1, \ldots, c_n] = \begin{bmatrix} r_1^T \\ \vdots \\ r_n^T \end{bmatrix}$$

then $\|r_i\|_\infty$ and $\|c_i\|_\infty$ are approximately equal, $i = 1, \ldots, n$. Note that $D^{-1}AD$ can be calculated without roundoff. When A is balanced, the computed eigenvalues are often more accurate. See Parlett and Reinsch (1969).

Problems

P7.5-1. Show that if $\bar{H} = Q^T H Q$ is obtained by performing a single-shift Q-R step with $H = \begin{bmatrix} w & x \\ y & z \end{bmatrix}$, then

$$|h_{21}| \leqslant |y^2 x| / [(w - z)^2 + y^2]$$

P7.5-2. Give a formula for the 2-by-2 diagonal matrix D that minimizes

$$\|D^{-1}AD\|_F$$

where $A = \begin{bmatrix} w & x \\ y & z \end{bmatrix}$.

P7.5-3. Explain how the single-shift QR step $H - \mu I = UR, \bar{H} = RU + \mu I$ can be carried out implicitly. That is, show how the transition from H to $\bar{H}$ can be carried out without subtracting the shift μ from the diagonal of H.

P7.5-4. Suppose H is upper Hessenberg and that we compute the factorization $PH = LU$ via Gaussian elimination with partial pivoting. (See Algorithm 5.3-4.) Show that $\bar{H} = U(P^T L)$ is upper Hessenberg and similar to H. (This is the basis of the *modified LR algorithm*.)

P7.5-5. Show that if $H = H_0$ is given and we generate the matrices H_k via $H_k - \mu_k I = U_k R_k, H_{k+1} = R_k U_k + \mu_k I$, then

$$(U_1 \cdots U_j)(R_j \cdots R_1) = (H - \mu_1 I) \ldots (H - \mu_j I).$$

Notes and References for Sec. 7.5

The development of the practical QR algorithm began with the important paper

H. Rutishauser (1958). "Solution of Eigenvalue Problems with the LR Transformation," *Nat. Bur. Stand. App. Math. Ser. 49*, 47-81.

The algorithm described here was then "orthogonalized" in

J.G.F. Francis (1961). "The QR Transformation: A Unitary Analogue to the LR Transformation, Parts I and II," *Comp. J. 4*, 265-72, 332-45.

Descriptions of the practical QR algorithm may be found in Wilkinson (AEP, chap. 8) and Stewart (IMC, chap. 7). ALGOL procedures for LR and QR methods are given in

R. S. Martin, G. Peters, and J. H. Wilkinson (1970). "The QR algorithm for Real Hessenberg Matrices," *Numer. Math. 14*, 219-31. See also HACLA, pp. 359-71.
R. S. Martin and J. H. Wilkinson (1968). "The Modified LR Algorithm for Complex Hessenberg Matrices," *Numer. Math. 12*, 369-76. See also HACLA, pp. 396-403.

Their FORTRAN equivalents are in EISPACK.

The problem of minimizing $\| D^{-1}AD \|$ over all diagonal matrices is investigated in

E. E. Osborne (1960). "On Preconditioning of Matrices," *JACM 7*, 338-45.

A practical scheme for balancing that incorporates some of Osborne's ideas is given in

B. N. Parlett and C. Reinsch (1969). "Balancing a Matrix for Calculation of Eigenvalues and Eigenvectors," *Numer. Math. 13*, 292-304. See also HACLA, pp. 315-26.

Sec. 7.6. Computing Eigenvectors and Invariant Subspaces

Several important invariant subspace problems can be solved once the real Schur decomposition $Q^{T}AQ = T$ has been computed. In this section we discuss how to

(a) compute the eigenvectors associated with some subset of $\lambda(A)$,

(b) compute an orthonormal basis for a given invariant subspace,

(c) block-diagonalize A using well-conditioned similarity transformations,

(d) compute a basis of eigenvectors regardless of their condition, and

(e) attempt to compute the Jordan canonical form of A.

Eigenvector/invariant subspace computation for sparse matrices is discussed elsewhere in the book. See §7.3 as well as portions of Chapters 8 and 9.

Computing Selected Eigenvectors via Inverse Iteration

Let $z^{(0)} \in \mathbb{R}^n$ be a given nonzero vector and assume that $A - \mu I \in \mathbb{R}^{n \times n}$ is nonsingular. The following is referred to as *inverse iteration*:

For $k = 1, 2, \ldots$

(7.6-1)   Solve: $(A - \mu I)v^{(k)} = z^{(k-1)}$

Normalize: $z^{(k)} = \alpha_k v^{(k)}$ (e.g., $\alpha_k^{-1} = \| v^{(k)} \|_\infty$)

Inverse iteration is just the power method applied with $(A - \mu I)^{-1}$.

To analyze the behavior of (7.6-1), assume that A has a basis of eigenvectors $\{x_1, \ldots, x_n\}$ and that $Ax_i = \lambda_i x_i$ for $i = 1, \ldots, n$. If

$$z^{(0)} = \sum_{i=1}^{n} \beta_i x_i$$

then

$$z^{(k)} = \alpha_1 \cdots \alpha_k \sum_{i=1}^{n} \frac{\beta_i}{(\lambda_i - \mu)^k} x_i.$$

Clearly, if μ is much closer to an eigenvalue λ_j than to the other eigenvalues, then $z^{(k)}$ becomes increasingly rich in the direction of x_j, provided $\beta_j \neq 0$.

A sample stopping criteria for (7.6-1) might be to quit as soon as the residual

$$r^{(k)} = (A - \mu I)z^{(k)}$$

satisfies

(7.6-2) $$\|r^{(k)}\|_{\infty} \leqslant c\mathbf{u}\|A\|_{\infty}\|z^{(k)}\|_{\infty},$$

where c is a constant of order unity. Since

$$(A + E_k)z^{(k)} = \mu z^{(k)}$$

with $E_k = -r^{(k)}[z^{(k)}]^{\mathrm{T}}/[z^{(k)}]^{\mathrm{T}}z^{(k)}$, it follows that (7.6-2) forces μ and $z^{(k)}$ to be an exact eigenpair for a nearby matrix.

Inverse iteration is used in conjunction with the QR algorithm as follows:

(a) Compute the Hessenberg decomposition $U_0^{\mathrm{T}}AU_0 = H$.

(b) Apply the double Francis iteration to H *without* accumulating transformations.

(c) For each computed eigenvalue $\hat{\lambda}$ whose corresponding eigenvector x is sought, apply (7.6-1) with $A = H$ and $\mu = \hat{\lambda}$ to produce a vector z such that $Hz \cong \mu z$.

(d) Set $x = U_0 z$.

Inverse iteration with H is very economical because (1) we do not have to accumulate transformations during the double Francis iteration, (2) we can factor matrices of the form $H - \hat{\lambda}I$ in $O(n^2)$ flops, and (3) only one inverse iteration is typically required to produce an adequate approximate eigenvector.

This last point is perhaps the most interesting aspect of inverse iteration and requires some justification, since $\hat{\lambda}$ can be comparatively inaccurate. Assume for simplicity that $\hat{\lambda}$ is real and let

$$H - \hat{\lambda}I = \sum_{i=1}^{n} \sigma_i u_i v_i^{\mathrm{T}} = U\Sigma V^{\mathrm{T}}$$

be the SVD of $H - \hat{\lambda}I$. From what we said about the roundoff properties of the QR algorithm in §7.5, there exists a matrix $E \in \mathbb{R}^{n \times n}$ such that $H + E - \hat{\lambda}I$ is singular and

$$\|E\|_2 \cong \mathbf{u} \|H\|_2.$$

It follows that $\sigma_n \cong \mathbf{u}\sigma_1$ and $\|(H - \hat{\lambda}I)v_n\|_2 \cong \mathbf{u}\sigma_1$, i.e., v_n is a good approximate eigenvector. Clearly if

$$z^{(0)} = \sum_{i=1}^{n} \gamma_i u_i,$$

then

$$z^{(1)} = \sum_{i=1}^{n} (\gamma_i / \sigma_i) v_i$$

is "rich" in the direction v_n. Note that if $s(\lambda) \cong |u_n^T v_n|$ is small, then $z^{(1)}$ will be rather deficient in the direction u_n. This explains (heuristically) why another step of inverse iteration is not likely to produce an improved eigenvector approximate, especially if λ is ill-conditioned. For more details, see Peters and Wilkinson (1979).

EXAMPLE 7.6-1.   The matrix

$$A = \begin{bmatrix} 1 & 1 \\ 10^{-10} & 1 \end{bmatrix}$$

has eigenvalues $\lambda_1 = .99999$ and $\lambda_2 = 1.00001$ and corresponding eigenvectors $x_1 = (1, -10^{-5})^T$ and $x_2 = (1, 10^{-5})^T$. The condition of both eigenvalues is of order 10^5. The approximate eigenvalue $\mu = 1$ is an exact eigenvalue of $A + E$ where

$$E = \begin{bmatrix} 0 & 0 \\ -10^{-10} & 0 \end{bmatrix}.$$

Thus, the quality of μ is typical of the quality of an eigenvalue produced by the QR algorithm when executed in 10-digit floating point.

If (7.6-1) is applied with starting vector $z^{(0)} = (0, 1)^T$, then $z^{(1)} = (1, 0)^T$ and $\|Az^{(1)} - \mu z^{(1)}\|_2 = 10^{-10}$. However, one more step produces $z^{(2)} = (0, 1)^T$ for which $\|Az^{(2)} - \mu z^{(2)}\|_2 = 1$.

This example is discussed in Peters and Wilkinson (1979).

Ordering Eigenvalues in the Real Schur Form

Recall that the real Schur decomposition provides information about invariant subspaces. If

$$Q^T A Q = T = \begin{bmatrix} T_{11} & T_{12} \\ 0 & T_{22} \end{bmatrix} \begin{matrix} p \\ q \end{matrix}$$

and $\lambda(T_{11}) \cap \lambda(T_{22}) = \emptyset$, then the first p columns of Q span the unique invariant subspace associated with $\lambda(T_{11})$. (See Lemma 7.1-1.) Unfortunately, the Francis iteration supplies us with a real Schur decomposition $Q_F^T A Q_F = T_F$ in which the eigenvalues appear somewhat randomly along the diagonal of T_F. This poses a problem if we want an orthonormal basis for an invariant subspace whose associated eigenvalues are not at the top of T_F's diagonal. Clearly, we need a method for computing an orthogonal matrix Q_D such that $Q_D^T T_F Q_D$ is upper quasi-triangular with appropriate eigenvalue ordering.

A look at the 2-by-2 case suggests how this can be accomplished. Suppose

$$Q_F^T A Q_F = T_F = \begin{bmatrix} \lambda_1 & t_{12} \\ 0 & \lambda_2 \end{bmatrix} \qquad \lambda_1 \neq \lambda_2$$

and we wish to reverse the order of the eigenvalues. Note that $T_F x = \lambda_2 x$ where

$$x = \begin{bmatrix} t_{12} \\ \lambda_2 - \lambda_1 \end{bmatrix}.$$

Let Q_D be a Givens rotation such that the second component of $Q_D^T x$ is zero. If $Q = Q_F Q_D$ then

$$(Q^T A Q)e_1 = Q_D^T T_F (Q_D e_1) = \lambda_2 Q_D^T (Q_D e_1) = \lambda_2 e_1$$

and so $Q^T A Q$ must have the form

$$Q^T A Q = \begin{bmatrix} \lambda_2 & \pm t_{12} \\ 0 & \lambda_1 \end{bmatrix}.$$

By systematically interchanging adjacent pairs of eigenvalues using this technique, we can move any subset of $\lambda(A)$ to the top of T's diagonal:

ALGORITHM 7.6-1. Given an orthogonal matrix $Q \in \mathbb{R}^{n \times n}$, an upper triangular matrix $T = Q^T A Q$, and a subset $\Delta = \{\lambda_1, \ldots, \lambda_p\}$ of $\lambda(A)$, the following algorithm computes an orthogonal matrix Q_D such that $Q_D^T T Q_D = S$ is upper triangular and $\{s_{11}, \ldots, s_{pp}\} = \Delta$. The matrices Q and T are overwritten by QQ_D and S respectively.

> Do until $\{t_{11}, \ldots, t_{pp}\} = \Delta$
>   For $k = 1, \ldots, n - 1$
>   If $t_{kk} \notin \Delta$ and $t_{k+1,k+1} \in \Delta$ then
>   Determine $c = \cos(\theta)$ and $s = \sin(\theta)$ such that
>
> $$\begin{bmatrix} c & s \\ -s & c \end{bmatrix} \begin{bmatrix} t_{k,k+1} \\ t_{kk} - t_{k+1,k+1} \end{bmatrix} = \begin{bmatrix} * \\ 0 \end{bmatrix}$$
>
> $T := J(k, k + 1, \theta) T J(k, k + 1, \theta)^T$
> $Q := Q J(k, k + 1, \theta)^T$

This algorithm requires $k(8n)$ flops, where k is the total number of required swaps. The integer k is never greater than $(n - p)p$.

The swapping gets a little more complicated when T has 2-by-2 blocks along its diagonal. See Ruhe (1970a) and Stewart (1976a) for details. Of course, these interchanging techniques can be used to *sort* the eigenvalues, say from maximum to minimum modulus.

Computing invariant subspaces by manipulating the real Schur decomposition is extremely stable. If $\hat{Q} = [\hat{q}_1, \ldots, \hat{q}_n]$ denotes the computed orthogonal matrix Q, then $\|\hat{Q}^T\hat{Q} - I\|_2 \cong \mathbf{u}$ and there exists a matrix E satisfying $\|E\|_2 \cong \mathbf{u}\|A\|_2$ such that

$$(A + E)\hat{q}_i \in \mathrm{span}\{\hat{q}_1, \ldots, \hat{q}_p\}. \qquad\qquad i = 1, \ldots, p$$

Block Diagonalization

Let

(7.6-3)
$$T = \begin{bmatrix} T_{11} & T_{12} & \cdots & T_{1q} \\ & T_{22} & \cdots & T_{2q} \\ & & \ddots & \vdots \\ \mathbf{0} & & & T_{qq} \end{bmatrix} \begin{matrix} \} n_1 \\ \} n_2 \\ \vdots \\ \} n_q \end{matrix}$$

be a partitioning of some real Schur form $Q^TAQ = T \in \mathbb{R}^{n \times n}$ such that $\lambda(T_{11}), \ldots, \lambda(T_{qq})$ are disjoint. By Theorem 7.1-5 there exists a matrix Y such that $Y^{-1}TY = \mathrm{diag}(T_{11}, \ldots, T_{qq})$. A practical procedure for determining Y is now given together with an analysis of Y's sensitivity as a function of the above partitioning.

Partition $I_n = [E_1, \ldots, E_q]$ conformably with T and define the matrix $Y_{ij} \in \mathbb{R}^{n \times n}$ as follows:

$$Y_{ij} = I_n + E_i Z_{ij} E_j^T. \qquad\qquad i < j, Z_{ij} \in \mathbb{R}^{n_i \times n_j}$$

In other words, Y_{ij} looks just like the identity except that Z_{ij} occupies the (i, j) block position. It follows that if $Y_{ij}^{-1}TY_{ij} = \bar{T} = (\bar{T}_{ij})$ then T and $\bar{T}$ are identical except that

$$\begin{aligned} \bar{T}_{ij} &= T_{ii}Z_{ij} - Z_{ij}T_{jj} + T_{ij} \\ \bar{T}_{ik} &= T_{ik} - Z_{ij}T_{jk} \quad (k = j + 1, \ldots, q) \\ \bar{T}_{kj} &= T_{ki}Z_{ij} + T_{kj} \quad (k = 1, \ldots, i - 1) \end{aligned}$$

Thus, T_{ij} can be zeroed provided we have an algorithm for solving the *Sylvester equation*

(7.6-4)
$$FZ - ZG = C,$$

where $F \in \mathbb{R}^{p \times p}$ and $G \in \mathbb{R}^{r \times r}$ are given upper quasi-triangular matrices and $C \in \mathbb{R}^{p \times r}$.

Bartels and Stewart (1972) have devised a method for doing this. Let $C = [c_1, \ldots, c_r]$ and $Z = [z_1, \ldots, z_r]$ be column partitionings. If $g_{k+1,k} = 0$, then by comparing columns in (7.6-4) we find

$$Fz_k - \sum_{i=1}^{k} g_{ik} z_i = c_k.$$

Thus, once we know $z_1, \ldots, z_{k-1}$ then we can solve the quasi-triangular system

$$(F - g_{kk}I)z_k = c_k + \sum_{i=1}^{k-1} g_{ik} z_i$$

for z_k. If $g_{k+1,k} \neq 0$, then z_k and z_{k+1} can be simultaneously found by solving the $2p$-by-$2p$ system

$$(7.6\text{-}5) \quad \begin{bmatrix} F - g_{kk}I & -g_{mk}I \\ -g_{km}I & F - g_{mm}I \end{bmatrix} \begin{bmatrix} z_k \\ z_m \end{bmatrix} = \begin{bmatrix} c_k \\ c_m \end{bmatrix} + \sum_{i=1}^{k-1} \begin{bmatrix} g_{ik} z_i \\ g_{im} z_i \end{bmatrix}.$$

$$(m = k + 1)$$

By reordering the equations according to the permutation $(1, p + 1, 2, p + 2, \ldots, p, 2p)$, a banded system is obtained that can be solved in $O(p^2)$ flops. The details may be found in Bartels and Stewart (1972). We summarize the overall process for the simple case when F and G are both triangular:

ALGORITHM 7.6-2: *Bartels-Stewart Algorithm.*   Given $C \in \mathbb{R}^{p \times r}$ and upper triangular matrices $F \in \mathbb{R}^{p \times p}$ and $G \in \mathbb{R}^{r \times r}$ satisfying $\lambda(F) \cap \lambda(G) = \emptyset$, the following algorithm overwrites C with the solution to the equation $FZ - ZG = C$.

> For $k = 1, \ldots, r$
> > For $s = 1, \ldots, p$
> > $$b_s := c_{sk} + \sum_{i=1}^{k-1} g_{ik} c_{si}$$
> > Solve $(F - g_{kk}I)z = b$
> > $c_{ik} := z_i \quad (i = 1, \ldots, p)$

This algorithm requires $pr(p + r)/2$ flops.

Clearly, by zeroing the superdiagonal blocks in T in the appropriate order, we can reduce the entire matrix to block diagonal form:

ALGORITHM 7.6-3. Given an orthogonal matrix $Q \in \mathbb{R}^{n \times n}$, an upper quasi-triangular matrix $T = Q^T A Q$, and the partitioning (7.6-3), the following algorithm overwrites Q with QY where $Y^{-1}TY = \text{diag}(T_{11}, \ldots, T_{qq})$.

For $j = 2, \ldots, q$
    For $i = 1, \ldots, j - 1$
        Solve $T_{ii}Z - ZT_{jj} = -T_{ij}$ for Z using Algorithm 7.6-2.
        For $k = j + 1, \ldots, q$
            $T_{ik} := T_{ik} - ZT_{jk}$
        For $k = 1, \ldots, q$
            $Q_{kj} := Q_{ki}Z + Q_{kj}$

This algorithm requires at most $O(n^3)$ flops.

The choice of the real Schur form T and its partitioning in (7.6-3) determines the sensitivity of the Sylvester equations that must be solved in Algorithm 7.6-3. This in turn affects the condition of the matrix Y and the overall usefulness of the block diagonalization. The reason for these dependencies is that the relative error of the computed solution $\hat{Z}$ to

$$(7.6-6) \qquad\qquad T_{ii}Z - ZT_{jj} = -T_{ij}$$

satisfies

$$\frac{\|\hat{Z} - Z\|_F}{\|Z\|_F} \cong \mathbf{u} \, \frac{\|T\|_F}{\text{sep}(T_{ii}, T_{jj})}$$

For details, see Golub, Nash, and Van Loan (1979). Since

$$\text{sep}(T_{ii}, T_{jj}) = \min_{X \neq 0} \frac{\|T_{ii}X - XT_{jj}\|_F}{\|X\|_F} \leqslant \min_{\substack{\lambda \in \lambda(F) \\ \mu \in \lambda(G)}} |\lambda - \mu|,$$

there can be a substantial loss of accuracy whenever the subsets $\lambda(T_{ii})$ are insufficiently separated. Moreover, if Z satisfies (7.6-6) then

$$\|Z\|_F \leqslant \frac{\|T_{ij}\|_F}{\text{sep}(T_{ii}, T_{jj})}.$$

Thus, large-norm solutions can be expected if $\text{sep}(T_{ii}, T_{jj})$ is small. This tends to make the matrix Y in Algorithm 7.6-3 ill-conditioned since it is the product of the matrices

$$Y_{ij} = \begin{bmatrix} I & Z \\ 0 & I \end{bmatrix}.$$

Note: $\kappa_F(Y_{ij}) = 2n + \|Z\|_F^2$.

Confronted with these difficulties, Bavely and Stewart (1979) have developed an algorithm for block diagonalizing that dynamically determines the eigenvalue ordering and partitioning in (7.6-3) so that all the "Z matrices" in Algorithm 7.6-3 are bounded in norm by some user-supplied tolerance. They find that the condition of Y can be controlled by controlling the condition of the Y_{ij}.

Eigenvector Bases

If the blocks in the partitioning (7.6-3) are all 1-by-1, then Algorithm 7.6-3 produces a basis of eigenvectors. As with the method of inverse iteration, the computed eigenvalue-eigenvector pairs are exact for some "nearby" matrix. A widely followed rule of thumb for deciding upon a suitable eigenvector method is to use inverse iteration whenever fewer than 25% of the eigenvectors are desired.

We point out, however, that the real Schur form can be used to determine selected eigenvectors. Suppose

$$Q^T A Q = \begin{bmatrix} T_{11} & u & T_{13} \\ 0 & \lambda & v^T \\ 0 & 0 & T_{33} \end{bmatrix} \begin{matrix} k-1 \\ 1 \\ n-k \end{matrix}$$

is upper quasi-triangular and that $\lambda \notin \lambda(T_{11}) \cup \lambda(T_{33})$. It follows that if we solve the linear systems

$$(T_{11} - \lambda I)w = -u$$
$$(T_{33} - \lambda I)^T z = -v$$

then

$$x = Q \begin{bmatrix} w \\ 1 \\ 0 \end{bmatrix} \quad \text{and} \quad y = Q \begin{bmatrix} 0 \\ 1 \\ z \end{bmatrix}$$

are the associated right and left eigenvectors, respectively. Note that

$$s(\lambda) = 1/\sqrt{(1 + w^T w)(1 + z^T z)} \ .$$

Ascertaining Jordan Block Structures

Suppose that we have computed the real Schur decomposition $A = QTQ^T$, identified clusters of "equal" eigenvalues, and calculated the corresponding block diagonalization $T = Y \operatorname{diag}(T_{11}, \ldots, T_{qq})Y^{-1}$. As we have seen, this can be a formidable task. However, even greater numerical problems confront us if we attempt to ascertain the Jordan block structure of each T_{ii}. A brief examination of these difficulties will serve to highlight the limitations of the Jordan decomposition in numerical analysis.

Assume for clarity that $\lambda(T_{ii})$ is real. The reduction of T_{ii} to Jordan form begins by replacing it with a matrix C that has the form

$$C = \lambda I + N,$$

where N is the strictly upper triangular portion of T_{ii} and where λ, say, is the mean of its eigenvalues.

Recall that the dimension of a Jordan block $J(\lambda)$ is the smallest non-negative integer k for which $[J(\lambda) - \lambda I]^k = 0$. Thus, if $p_i = \dim[(N(N^i)]$, for $i = 0, 1, \ldots, n$, then $p_i - p_{i-1}$ equals the number of blocks in C's Jordan form that have dimension i or greater. A concrete example helps to make this assertion clear and to illustrate the role of the SVD in Jordan form computations.

Assume that C is 7-by-7. Suppose we compute the SVD $U_1^T N V_1 = \Sigma_1$ and "discover" that N has rank 3. If we order the singular values from small to large then it follows that the matrix $N_1 = V_1^T N V_1$ has the form

$$N_1 = \begin{bmatrix} 0 & K \\ 0 & L \end{bmatrix} \begin{matrix} 4 \\ 3 \end{matrix}.$$
$$ \begin{matrix} 4 & 3 \end{matrix}$$

At this point we know that the geometric multiplicity of λ is 4—i.e., C's Jordan form has 4 blocks ($p_1 - p_0 = 4 - 0 = 4$).

Now suppose $\tilde{U}_2^T L \tilde{V}_2 = \Sigma_2$ is the SVD of L and that we find $\text{rank}(L) = 1$. If we again order the singular values from small to large, then $L_2 = \tilde{V}_2^T L \tilde{V}_2$ clearly has the following structure:

$$L_2 = \begin{bmatrix} 0 & 0 & a \\ 0 & 0 & b \\ 0 & 0 & c \end{bmatrix}.$$

However, $\lambda(L_2) = \lambda(L) = \{0, 0, 0\}$ and so $c = 0$. Thus, if

$$V_2 = \text{diag}(I_4, \tilde{V}_2)$$

then $N_2 = V_2^T N_1 V_2$ has the following form:

$$N_2 = \begin{bmatrix} 0 & 0 & 0 & 0 & x & x & x \\ 0 & 0 & 0 & 0 & x & x & x \\ 0 & 0 & 0 & 0 & x & x & x \\ 0 & 0 & 0 & 0 & x & x & x \\ 0 & 0 & 0 & 0 & 0 & 0 & a \\ 0 & 0 & 0 & 0 & 0 & 0 & b \\ 0 & 0 & 0 & 0 & 0 & 0 & 0 \end{bmatrix}.$$

Besides allowing us to introduce more zeros into the upper triangle, the SVD of L also enables us to deduce the dimension of the null space of N^2. Since

$$N_1^2 = \begin{bmatrix} 0 & KL \\ 0 & L^2 \end{bmatrix} = \begin{bmatrix} 0 & K \\ 0 & L \end{bmatrix} \begin{bmatrix} 0 & K \\ 0 & L \end{bmatrix}$$

and $\begin{bmatrix} K \\ L \end{bmatrix}$ has full column rank,

$$p_2 = \dim[N(N^2)] = \dim[N(N_1^2)] = 4 + \dim[N(L)] = p_1 + 2.$$

Hence, we can conclude at this stage that the Jordan form of C has at least two blocks of dimension 2 or greater.

Finally, it is easy to see that $N_1^3 = 0$, from which we conclude that there is $p_3 - p_2 = 7 - 6 = 1$ block of dimension 3 or larger. If we define $V = V_1 V_2$ then it follows that the decomposition

$$V^T C V = \begin{bmatrix} \lambda & 0 & 0 & 0 & x & x & x \\ 0 & \lambda & 0 & 0 & x & x & x \\ 0 & 0 & \lambda & 0 & x & x & x \\ 0 & 0 & 0 & \lambda & x & x & x \\ 0 & 0 & 0 & 0 & \lambda & 0 & a \\ 0 & 0 & 0 & 0 & 0 & \lambda & b \\ 0 & 0 & 0 & 0 & 0 & 0 & \lambda \end{bmatrix} \begin{matrix} \left. \begin{matrix} \\ \\ \\ \\ \end{matrix} \right\} \text{4 blocks of order 1 or larger} \\ \\ \left. \begin{matrix} \\ \\ \end{matrix} \right\} \text{2 blocks of order 2 or larger} \\ \\ \left. \begin{matrix} \\ \end{matrix} \right\} \text{1 block of order 3 or larger} \end{matrix}$$

"displays" C's Jordan block structure: 2 blocks of order 1, 1 block of order 2, and 1 block of order 3.

To actually compute the Jordan decomposition it is necessary to resort to nonorthogonal transformations. We refer the reader to either Golub and Wilkinson (1976) or Kagstrom and Ruhe (1980a, 1980b) for how to proceed with this phase of the reduction.

The above calculations with the SVD amply illustrate that difficult rank decisions must be made at each stage and that the final computed block structure depends critically on those decisions. Fortunately, the stable Schur decomposition can almost always be used in lieu of the Jordan decomposition in practical applications.

EXAMPLE 7.6-2. An example from Kagstrom and Ruhe (1980a) serves to illustrate some of the reductions discussed in this section. Let

$$A = \begin{bmatrix} 1 & 1 & 1 & -2 & 1 & -1 & 2 & -2 & 4 & -3 \\ -1 & 2 & 3 & -4 & 2 & -2 & 4 & -4 & 8 & -6 \\ -1 & 0 & 5 & -5 & 3 & -3 & 6 & -6 & 12 & -9 \\ -1 & 0 & 3 & -4 & 4 & -4 & 8 & -8 & 16 & -12 \\ -1 & 0 & 3 & -6 & 5 & -4 & 10 & -10 & 20 & -15 \\ -1 & 0 & 3 & -6 & 2 & -2 & 12 & -12 & 24 & -18 \\ -1 & 0 & 3 & -6 & 2 & -5 & 15 & -13 & 28 & -21 \\ -1 & 0 & 3 & -6 & 2 & -5 & 12 & -11 & 32 & -24 \\ -1 & 0 & 3 & -6 & 2 & -5 & 12 & -14 & 37 & -26 \\ -1 & 0 & 3 & -6 & 2 & -5 & 12 & -14 & 36 & -25 \end{bmatrix}.$$

After computing the Schur decomposition, sorting the eigenvalues, and zero-ing upper triangular entries via Algorithm 7.6-3 we obtain a block diagonal matrix $D = \text{diag}(D_{11}, D_{22}, D_{33})$ where

$$D_{11} \cong \begin{bmatrix} 3.00 & 12.31 & 30.98 & -34.65 \\ 0.00 & 3.00 & 0.00 & -2.68 \\ 0.00 & 0.00 & 3.00 & 1.06 \\ 0.00 & 0.00 & 0.00 & 3.00 \end{bmatrix},$$

$$D_{22} \cong \begin{bmatrix} 2.00 & 0.00 & 4.55 & 13.17 & 13.94 \\ 0.00 & 2.00 & -0.20 & 3.22 & 3.14 \\ 0.00 & 0.00 & 2.00 & 0.00 & 1.44 \\ 0.00 & 0.00 & 0.00 & 2.00 & .09 \\ 0.00 & 0.00 & 0.00 & 0.00 & 2.00 \end{bmatrix},$$

and

$$D_{33} \cong [1.00].$$

Using the SVD (and appropriately defined user-specified tolerances) it is found that $\text{rank}(D_{11} - 3I) = 2$ and $\text{rank}(D_{22} - 2I) = 3$. These two blocks are then transformed into an unnormalized Jordan form:

$$Z_1^{-1}D_{11}Z_1 = \begin{bmatrix} 3.00 & 48.12 & 0.00 & 0.00 \\ 0.00 & 3.00 & 0.00 & 0.00 \\ 0.00 & 0.00 & 3.00 & 1.98 \\ 0.00 & 0.00 & 0.00 & 3.00 \end{bmatrix}$$

and

$$Z_2^{-1}D_{22}Z_2 = \begin{bmatrix} 2.00 & 0.54 & 0.00 & 0.00 & 0.00 \\ 0.00 & 2.00 & 14.36 & 0.00 & 0.00 \\ 0.00 & 0.00 & 2.00 & 0.00 & 0.00 \\ 0.00 & 0.00 & 0.00 & 2.00 & 13.56 \\ 0.00 & 0.00 & 0.00 & 0.00 & 2.00 \end{bmatrix}$$

and the calculation is complete.

Problems

P7.6-1. Give a complete algorithm for solving a real, n-by-n, upper quasi-triangular system $Tx = b$.

P7.6-2. Suppose $U^{-1}AU = \text{diag}(\alpha_1, \ldots, \alpha_m)$ and $V^{-1}BV = \text{diag}(\beta_1, \ldots, \beta_n)$. Show that if $\phi(X) = AX + XB$, then

$$\lambda(\phi) = \{\alpha_i + \beta_j \mid i = 1, \ldots, m \quad j = 1, \ldots, n\}.$$

What are the corresponding eigenvectors? How can these decompositions be used to solve $AX + XB = C$?

P7.6-3. Show that if $Y = \begin{bmatrix} I & Z \\ 0 & I \end{bmatrix}$, then $\kappa_2(Y) = [2 + \sigma^2 + \sqrt{4\sigma^2 + \sigma^4}]/2$, where $\sigma = \|Z\|_2$.

P7.6-4. Derive the system (7.6-5).

P7.6-5. Suppose the matrix T_{22} in

$$T = \begin{bmatrix} T_{11} & T_{12} & T_{13} \\ 0 & T_{22} & T_{23} \\ 0 & 0 & T_{33} \end{bmatrix} \qquad T \in \mathbb{R}^{n \times n}$$

is 2-by-2 with complex eigenvalues that are disjoint from $\lambda(T_{11})$ and $\lambda(T_{33})$. Give an algorithm for computing the 2-dimensional real invariant subspace associated with T_{22}'s eigenvalues.

Notes and References for Sec. 7.6

Much of the material discussed in this section may be found in the survey paper

G. H. Golub and J. H. Wilkinson (1976). "Ill-Conditioned Eigensystems and the Computation of the Jordan Canonical Form," *SIAM Review 18*, 578-619.

Papers that specifically analyze the method of inverse iteration for computing eigenvectors include

J. Varah (1968a). "The Calculation of the Eigenvectors of a General Complex Matrix by Inverse Iteration," *Math. Comp. 22*, 785-91.
J. Varah (1968b). "Rigorous Machine Bounds for the Eigensystem of a General Complex Matrix," *Math. Comp. 22*, 793-801.
J. Varah (1970). "Computing Invariant Subspaces of a General Matrix When the Eigensystem Is Poorly Determined," *Math. Comp. 24*, 137-49.
G. Peters and J. H. Wilkinson (1979). "Inverse Iteration, Ill-Conditioned Equations, and Newton's Method," *SIAM Review 21*, 339-60.

The ALGOL version of the EISPACK inverse iteration subroutine is given in

G. Peters and J. H. Wilkinson (1971). "The Calculation of Specified Eigenvectors by Inverse Iteration," in *HACLA*, pp. 418-39.

The problem of ordering the eigenvalues in the real Schur form is the subject of the following papers:

A. Ruhe (1970a). "An Algorithm for Numerical Determination of the Structure of a General Matrix," *BIT 10*, 196-216.
G. W. Stewart (1976a). "Algorithm 406: HQR3 and EXCHNG: FORTRAN Subroutines for Calculating and Ordering the Eigenvalues of a Real Upper Hessenberg Matrix," *ACM Trans. Math. Soft. 2*, 275-80.

FORTRAN programs for computing block diagonalizations and Jordan forms are described in

C. Bavely and G. W. Stewart (1979). "An Algorithm for Computing Reducing Subspaces by Block Diagonalization," *SIAM J. Num. Anal. 16*, 359-67.
B. Kagstrom and A. Ruhe (1980a). "An Algorithm for Numerical Computation of the Jordan Normal Form of a Complex Matrix," *ACM Trans. Math. Soft. 6*, 398-419.
B. Kagstrom and A. Ruhe (1980b). "Algorithm 560 JNF: An Algorithm for Numerical Computation of the Jordan Normal Form of a Complex Matrix," *ACM Trans. Math. Soft. 6*, 437-43.

Papers that are concerned with estimating the roundoff error in a computed eigenvalue and/or eigenvector include

S. P. Chan and B. N. Parlett (1977). "Algorithm 517: A Program for Computing the Condition Numbers of Matrix Eigenvalues without Computing Eigenvectors," *ACM Trans. Math. Soft. 3*, 186-203.

H. J. Symm and J. H. Wilkinson (1980). "Realistic Error Bounds for a Simple Eigenvalue and Its Associated Eigenvector," *Numer. Math. 35*, 113–26.

As we have seen, the sep(.,.) function is of great importance in the assessment of a computed invariant subspace. Aspects of this quantity are discussed in

J. Varah (1979). "On the Separation of Two Matrices," *SIAM J. Num. Anal. 16*, 212–22.

Numerous algorithms have been proposed for the Sylvester equation $FX - XG = C$, but those described in

R. H. Bartels and G. W. Stewart (1972). "Solution of the Equation $AX + XB = C$," *Comm. Assoc. Comp. Mach. 15*, 820–26.
G. H. Golub, S. Nash, and C. Van Loan (1979). "A Hessenberg-Schur Method for the Matrix Problem $AX + XB = C$," *IEEE Trans. Auto. Cont. AC-24*, 909–13.

are among the more reliable in that they rely on orthogonal transformations. The Lyapunov problem $FX + XF^T = -C$ where C is non-negative definite has a very important role to play in control theory. See

S. Barnett and C. Storey (1968). "Some Applications of the Lyapunov Matrix Equation," *J. Inst. Math. Applic. 4*, 33–42.

Finally, several authors have considered generalizations of the Sylvester equation, i.e., $\Sigma F_i X G_i = C$. These include

P. Lancaster (1970). "Explicit Solution of Linear Matrix Equations," *SIAM Review 12*, 544–66.
W. J. Vetter (1975). "Vector Structures and Solutions of Linear Matrix Equations," *Lin. Alg. & Its Applic. 10*, 181–88.
H. Wimmer and A. D. Ziebur (1972). "Solving the Matrix Equations $\Sigma f_p(A) \cdot X g_p(A)$," *SIAM Review 14*, 318–23.

Sec. 7.7. The QZ Algorithm and the $Ax = \lambda Bx$ Problem

Let A and B be two n-by-n matrices. The set of all matrices of the form $A - \lambda B$ with $\lambda \in \mathbb{C}$ is said to be a *pencil*. The eigenvalues of the pencil are elements of the set $\lambda(A, B)$ defined by

$$\lambda(A, B) = \{z \in \mathbb{C} \mid \det(A - zB) = 0\}.$$

If $\lambda \in \lambda(A, B)$ and

(7.7-1) $$Ax = \lambda Bx, \qquad\qquad x \neq 0$$

then x is referred to as an eigenvector of $A - \lambda B$.

In this section we briefly survey some of the mathematical properties of the generalized eigenproblem (7.7-1) and present a stable method for its solution.

The important case when A and B are symmetric with the latter positive definite is discussed in §8.6.

The first thing to observe about the generalized eigenvalue problem is that there are n eigenvalues if and only if rank$(B) = n$. If B is rank deficient then $\lambda(A, B)$ may be finite, empty, or infinite.

EXAMPLE 7.7-1.

$$\text{(i)}\quad A = \begin{bmatrix} 1 & 2 \\ 0 & 3 \end{bmatrix} \quad B = \begin{bmatrix} 1 & 0 \\ 0 & 0 \end{bmatrix} \Rightarrow \lambda(A, B) = \{1\}$$

$$\text{(ii)}\quad A = \begin{bmatrix} 1 & 2 \\ 0 & 3 \end{bmatrix} \quad B = \begin{bmatrix} 0 & 1 \\ 0 & 0 \end{bmatrix} \Rightarrow \lambda(A, B) = \emptyset$$

$$\text{(iii)}\quad A = \begin{bmatrix} 1 & 2 \\ 0 & 0 \end{bmatrix} \quad B = \begin{bmatrix} 1 & 0 \\ 0 & 0 \end{bmatrix} \Rightarrow \lambda(A, B) = \mathbb{C}$$

Note that if $0 \neq \lambda \in \lambda(A, B)$ then $(1/\lambda) \in \lambda(B, A)$. Moreover, if B is nonsingular then $\lambda(A, B) = \lambda(B^{-1}A, I) = \lambda(B^{-1}A)$.

This last observation suggests one method for solving the $A - \lambda B$ problem when B is nonsingular:

(a) Apply Gaussian elimination with pivoting to $B : PB = LU$,

(b) Solve $BC = A$ for C.

(c) Use the QR algorithm to compute the eigenvalues of C.

Note that C will be contaminated by roundoff errors of order $\mathbf{u}\|A\|_2\|B^{-1}\|_2$. If B is ill-conditioned, then this can rule out the possibility of computing any generalized eigenvalues accurately—even those eigenvalues that may be regarded as well-conditioned.

EXAMPLE 7.7-2.   If

$$A = \begin{bmatrix} 1.746 & .940 \\ 1.246 & 1.898 \end{bmatrix} \quad \text{and} \quad B = \begin{bmatrix} .780 & .563 \\ .913 & .659 \end{bmatrix},$$

then $\lambda(A, B) = \{2, 1.07 \times 10^6\}$. Using 7-digit floating point arithmetic, we find $\lambda[fl(AB^{-1})] = \{1.562539, 1.01 \times 10^6\}$. The poor quality of the small eigenvalue is because $\kappa_2(B) \cong 2 \times 10^6$. On the other hand we find that $\lambda(I, fl(A^{-1}B)) \cong \{2.000001, 1.06 \times 10^6\}$. The accuracy of the small eigenvalue is improved because $\kappa_2(A) \cong 4$.

The example suggests we seek an alternative approach to the $A - \lambda B$ problem. One idea is to choose nonsingular Q and Z such that the matrices

$$A_1 = Q^{-1}AZ$$

(7.7-2)

$$B_1 = Q^{-1}BZ$$

are each in canonical form. Note that $\lambda(A, B) = \lambda(A_1, B_1)$ since

$$Ax = \lambda Bx \iff A_1 y = \lambda B_1 y, \quad x = Zy$$

We say that the pencils $A - \lambda B$ and $A_1 - \lambda B_1$ are *equivalent* if (7.7-2) holds with nonsingular Q and Z.

As in the standard eigenproblem $A - \lambda I$ there is a choice between canonical forms. Analogous to the Jordan form is a decomposition of Kronecker's in which both A_1 and B_1 are block diagonal. The blocks are similar to Jordan blocks. The Kronecker canonical form poses the same numerical difficulties as the Jordan form. However, this decomposition does provide insight into the mathematical properties of the pencil $A - \lambda B$. See Wilkinson (1978) for details.

More attractive from the numerical point of view is the following decomposition described in Moler and Stewart (1973).

THEOREM 7.7-1: *Generalized Schur Decomposition.* If A and B are in $\mathbb{C}^{n \times n}$, then there exist unitary Q and Z such that $Q^H AZ = T$ and $Q^H BZ = S$ are upper triangular. If for some k, t_{kk} and s_{kk} are both zero, then $\lambda(A, B) = \mathbb{C}$. Otherwise,

$$\lambda(A, B) = \{t_{ii}/s_{ii} \mid s_{ii} \neq 0\}.$$

Proof. Let B_k be a sequence of nonsingular matrices that converge to B. For each k, let $Q_k^H (AB_k^{-1})Q_k = R_k$ be a Schur decomposition of AB_k^{-1}. Let Z_k be unitary such that $Z_k^H (B_k^{-1} Q_k) \equiv S_k^{-1}$ is upper triangular. It follows that both $Q_k^H AZ_k = R_k S_k$ and $Q_k^H B_k Z_k = S_k$ are also upper triangular.

Using the Bolzano-Weierstrass theorem, we know that the bounded sequence $\{(Q_k, Z_k)\}$ has a converging subsequence, $\lim(Q_{k_i}, Z_{k_i}) = (Q, Z)$. It is easy to show that Q and Z are unitary and that $Q^H AZ$ and $Q^H BZ$ are upper triangular.

The assertions about $\lambda(A, B)$ follow from the identity

$$\det(A - \lambda B) = \det(QZ^H) \prod_{i=1}^{n} (t_{ii} - \lambda s_{ii}). \quad \square$$

This decomposition illuminates the issue of eigenvalue sensitivity for the $A - \lambda B$ problem. Clearly, small changes in A and B can induce large changes in the eigenvalue $\lambda_i = t_{ii}/s_{ii}$ if s_{ii} is small. However, as Stewart (1978) argues, it may not be appropriate to regard such an eigenvalue as "ill-conditioned." The reason is that the reciprocal $\mu_i = s_{ii}/t_{ii}$ might be a very well behaved eigenvalue for the pencil $\mu A - B$. In the Stewart analysis, A and B are treated symmetrically and the eigenvalues are regarded more as ordered pairs (t_{ii}, s_{ii}) than as quotients. With this point of view it becomes appropriate to measure eigenvalue perturbations in the *chordal metric* chord(a, b) defined by

$$\text{chord}(a, b) = \frac{|a - b|}{\sqrt{1 + a^2}\ \sqrt{1 + b^2}}.$$

Stewart shows that if λ is a distinct eigenvalue of $A - \lambda B$ and λ_ϵ is the corresponding eigenvalue of the ϵ-perturbed pencil $\tilde{A} - \lambda \tilde{B}$, then

$$\text{chord}(\lambda, \lambda_\epsilon) \leqslant \frac{\epsilon}{(y^H A x)^2 + (y^H B x)^2} + O(\epsilon^2)$$

where x and y have unit 2-norm and satisfy $Ax = \lambda Bx$ and $y^H A = \lambda y^H B$. Note that the denominator in the upper bound is symmetric in A and B. The "truly" ill-conditioned eigenvalues are those for which this denominator is small.

The extreme case when $t_{kk} = s_{kk} = 0$ for some k has been studied by Wilkinson (1979). He makes the interesting observation that when this occurs, the remaining quotients t_{ii}/s_{ii} can assume arbitrary values.

Turning to algorithmic developments, we now describe a generalization of the QR algorithm that can be used to solve the $A - \lambda B$ problem. As in the standard eigenproblem $A - \lambda I$, we restrict the discussion to real arithmetic algorithms for real pencils. The decomposition of interest is the following:

THEOREM 7.7-2: *Generalized Real Schur Decomposition.* If A and B are in $\mathbb{R}^{n \times n}$ then there exist orthogonal matrices Q and Z such that $Q^T A Z$ is upper quasi-triangular and $Q^T B Z$ is upper triangular.

Proof. See Stewart (1972). $\square$

The method begins by reducing A to upper Hessenberg and B to upper triangular via orthogonal transformations. We first determine an orthogonal U such that $U^T B$ is upper triangular. Of course, to preserve eigenvalues, we must also update A in exactly the same way:

$$A := U^T A = \begin{bmatrix} x & x & x & x & x \\ x & x & x & x & x \\ x & x & x & x & x \\ x & x & x & x & x \\ x & x & x & x & x \end{bmatrix}, \qquad B := U^T B = \begin{bmatrix} x & x & x & x & x \\ 0 & x & x & x & x \\ 0 & 0 & x & x & x \\ 0 & 0 & 0 & x & x \\ 0 & 0 & 0 & 0 & x \end{bmatrix}.$$

$$n = 5$$

Next, we reduce A to upper Hessenberg form, preserving all the while B's upper triangular form. First, a Jacobi rotation $Q_{45} = J(4, 5, \theta)$ is determined to zero a_{51}:

$$A := Q_{45}A = \begin{bmatrix} x & x & x & x & x \\ x & x & x & x & x \\ x & x & x & x & x \\ x & x & x & x & x \\ 0 & x & x & x & x \end{bmatrix}, \quad B := Q_{45}B = \begin{bmatrix} x & x & x & x & x \\ 0 & x & x & x & x \\ 0 & 0 & x & x & x \\ 0 & 0 & 0 & x & x \\ 0 & 0 & 0 & x & x \end{bmatrix}.$$

The nonzero entry arising in b_{54} can be zeroed by postmultiplying with an appropriate Givens rotation $Z_{45} = J(4, 5, \theta)$:

$$A := AZ_{45} = \begin{bmatrix} x & x & x & x & x \\ x & x & x & x & x \\ x & x & x & x & x \\ x & x & x & x & x \\ 0 & x & x & x & x \end{bmatrix}, \quad B := BZ_{45} = \begin{bmatrix} x & x & x & x & x \\ 0 & x & x & x & x \\ 0 & 0 & x & x & x \\ 0 & 0 & 0 & x & x \\ 0 & 0 & 0 & 0 & x \end{bmatrix}.$$

Zeros are similarly introduced into the $(4, 1)$ and $(3, 1)$ positions in A:

$$A := Q_{34}A = \begin{bmatrix} x & x & x & x & x \\ x & x & x & x & x \\ x & x & x & x & x \\ 0 & x & x & x & x \\ 0 & x & x & x & x \end{bmatrix} \quad B := Q_{34}B = \begin{bmatrix} x & x & x & x & x \\ 0 & x & x & x & x \\ 0 & 0 & x & x & x \\ 0 & 0 & x & x & x \\ 0 & 0 & 0 & 0 & x \end{bmatrix}$$

$$A := AZ_{34} = \begin{bmatrix} x & x & x & x & x \\ x & x & x & x & x \\ x & x & x & x & x \\ 0 & x & x & x & x \\ 0 & x & x & x & x \end{bmatrix} \quad B := BZ_{34} = \begin{bmatrix} x & x & x & x & x \\ 0 & x & x & x & x \\ 0 & 0 & x & x & x \\ 0 & 0 & 0 & x & x \\ 0 & 0 & 0 & 0 & x \end{bmatrix}$$

$$
A := Q_{23}A = \begin{bmatrix} x & x & x & x & x \\ x & x & x & x & x \\ 0 & x & x & x & x \\ 0 & x & x & x & x \\ 0 & x & x & x & x \end{bmatrix}
\qquad
B := Q_{23}B = \begin{bmatrix} x & x & x & x & x \\ 0 & x & x & x & x \\ 0 & x & x & x & x \\ 0 & 0 & 0 & x & x \\ 0 & 0 & 0 & 0 & x \end{bmatrix}
$$

$$
A := AZ_{23} = \begin{bmatrix} x & x & x & x & x \\ x & x & x & x & x \\ 0 & x & x & x & x \\ 0 & x & x & x & x \\ 0 & x & x & x & x \end{bmatrix}
\qquad
B := BZ_{23} = \begin{bmatrix} x & x & x & x & x \\ 0 & x & x & x & x \\ 0 & 0 & x & x & x \\ 0 & 0 & 0 & x & x \\ 0 & 0 & 0 & 0 & x \end{bmatrix}.
$$

A is now upper Hessenberg through its first column. The reduction is completed by zeroing a_{52}, a_{42}, and a_{53}. As is evident above, two orthogonal transformations are required for each a_{ij} that is zeroed—one to do the zeroing and the other to restore B's triangularity. Either Givens rotations or 2-by-2 modified Householder transformations can be used, the latter being a little more efficient. Overall we have:

ALGORITHM 7.7-1: *Hessenberg-Triangular Reduction.* Given A and B in $\mathbb{R}^{n \times n}$, the following algorithm overwrites A with an upper Hessenberg matrix $Q^T A Z$ and B with an upper triangular matrix $Q^T B Z$ where both Q and Z are orthogonal.

> Using Algorithm 6.2-1, compute Householder matrices $P_1, \ldots, P_{n-1}$
> such that if $U = P_1 \cdots P_{n-1}$, then $U^T B$ is upper triangular.
> For $i = 1, \ldots, n - 1$
> $\quad A := P_i A$
> $\quad B := P_i B$
> For $j = 1, \ldots, n - 2$
> $\quad$ For $i = n, n - 1, \ldots, j + 2$
> $\quad\quad$ Determine a Householder matrix $\bar{P} \in \mathbb{R}^{2 \times 2}$ such that
> $$\bar{P} \begin{bmatrix} a_{i-1,j} \\ a_{ij} \end{bmatrix} = \begin{bmatrix} * \\ 0 \end{bmatrix}$$
> $\quad\quad A := PA$, $B := PB$, where $P = \mathrm{diag}(I_{i-2}, \bar{P}, I_{n-i})$
> $\quad\quad$ Determine a Householder matrix $\bar{P} \in \mathbb{R}^{2 \times 2}$ such that
> $$[b_{i,i-1}, b_{ii}]\bar{P} = [0 \; *]$$
> $\quad\quad A := AP$, $B := BP$, where $P = \mathrm{diag}(I_{i-2}, \bar{P}, I_{n-i})$

This algorithm requires about $5n^3$ flops. The accumulation of Q and Z requires $\frac{7}{3}n^3$ and $\frac{3}{2}n^3$ flops, respectively.

EXAMPLE 7.7-3.   If

$$A = \begin{bmatrix} 10 & 1 & 2 \\ 1 & 3 & -1 \\ 1 & 1 & 2 \end{bmatrix} \quad \text{and} \quad B = \begin{bmatrix} 1 & 2 & 3 \\ 4 & 5 & 6 \\ 7 & 8 & 9 \end{bmatrix}$$

and orthogonal matrices Q and Z are defined by

$$Q = \begin{bmatrix} -.1231 & -.9917 & .0378 \\ -.4924 & .0279 & -.8699 \\ -.8616 & .1257 & .4917 \end{bmatrix}$$

and

$$Z = \begin{bmatrix} 1.0000 & .0000 & .0000 \\ 0.0000 & -.8944 & -.4472 \\ 0.0000 & .4472 & -.8944 \end{bmatrix},$$

then

$$Q^T A Z = \begin{bmatrix} -2.5849 & 1.5413 & 2.4221 \\ -9.7631 & .0874 & 1.9239 \\ .0000 & 2.7233 & -.7612 \end{bmatrix}$$

and

$$Q^T B Z = \begin{bmatrix} -8.1240 & 3.6332 & 14.2024 \\ 0.0000 & 0.0000 & 1.8739 \\ 0.0000 & 0.0000 & .7612 \end{bmatrix}.$$

The reduction of $A - \lambda B$ to Hessenberg-triangular form serves as a "front end" decomposition for a generalized QR iteration known as the *QZ iteration*. In describing the QZ iteration we may assume without loss of generality that A is an unreduced upper Hessenberg matrix and that B is a nonsingular upper triangular matrix. The first of these assertions is obvious, for if $a_{k+1,k} = 0$ then

$$A - \lambda B = \begin{bmatrix} A_{11} - \lambda B_{11} & A_{12} - \lambda B_{12} \\ 0 & A_{22} - \lambda B_{22} \end{bmatrix} \begin{matrix} k \\ n-k \end{matrix}$$

and we may proceed to solve the two smaller problems $A_{11} - \lambda B_{11}$ and $A_{22} - \lambda B_{22}$. On the other hand, if $b_{kk} = 0$ for some k, then it is possible introduce a zero in A's $(n, n - 1)$ position and thereby deflate. Illustrating by example, suppose

$$
A = \begin{bmatrix}
x & x & x & x & x \\
x & x & x & x & x \\
0 & x & x & x & x \\
0 & 0 & x & x & x \\
0 & 0 & 0 & x & x
\end{bmatrix}
\qquad
B = \begin{bmatrix}
x & x & x & x & x \\
0 & x & x & x & x \\
0 & 0 & 0 & x & x \\
0 & 0 & 0 & x & x \\
0 & 0 & 0 & 0 & x
\end{bmatrix}
\qquad n = 5, \; k = 3
$$

The zero on B's diagonal can be "pushed down" to the $(5, 5)$ position as follows using Givens rotations (or modified 2-by-2 Householders):

$$
A := Q_{34}A = \begin{bmatrix}
x & x & x & x & x \\
x & x & x & x & x \\
0 & x & x & x & x \\
0 & x & x & x & x \\
0 & 0 & 0 & x & x
\end{bmatrix}
\qquad
B := Q_{34}B = \begin{bmatrix}
x & x & x & x & x \\
0 & x & x & x & x \\
0 & 0 & 0 & x & x \\
0 & 0 & 0 & 0 & x \\
0 & 0 & 0 & 0 & x
\end{bmatrix}
$$

$$
A := AZ_{23} = \begin{bmatrix}
x & x & x & x & x \\
x & x & x & x & x \\
0 & x & x & x & x \\
0 & 0 & x & x & x \\
0 & 0 & 0 & x & x
\end{bmatrix},
\qquad
B := BZ_{23} = \begin{bmatrix}
x & x & x & x & x \\
0 & x & x & x & x \\
0 & 0 & 0 & x & x \\
0 & 0 & 0 & 0 & x \\
0 & 0 & 0 & 0 & x
\end{bmatrix}
$$

$$
A := Q_{45}A = \begin{bmatrix}
x & x & x & x & x \\
x & x & x & x & x \\
0 & x & x & x & x \\
0 & 0 & x & x & x \\
0 & 0 & x & x & x
\end{bmatrix}
\qquad
B := Q_{45}B = \begin{bmatrix}
x & x & x & x & x \\
0 & x & x & x & x \\
0 & 0 & x & x & x \\
0 & 0 & 0 & x & x \\
0 & 0 & 0 & 0 & 0
\end{bmatrix}
$$

$$A := AZ_{34} = \begin{bmatrix} x & x & x & x & x \\ x & x & x & x & x \\ 0 & x & x & x & x \\ 0 & 0 & x & x & x \\ 0 & 0 & 0 & x & x \end{bmatrix} \qquad B := BZ_{34} = \begin{bmatrix} x & x & x & x & x \\ 0 & x & x & x & x \\ 0 & 0 & x & x & x \\ 0 & 0 & 0 & 0 & x \\ 0 & 0 & 0 & 0 & 0 \end{bmatrix}$$

$$A := AZ_{45} = \begin{bmatrix} x & x & x & x & x \\ x & x & x & x & x \\ 0 & x & x & x & x \\ 0 & 0 & x & x & x \\ 0 & 0 & 0 & 0 & x \end{bmatrix} \qquad B := BZ_{45} = \begin{bmatrix} x & x & x & x & x \\ 0 & x & x & x & x \\ 0 & 0 & x & x & x \\ 0 & 0 & 0 & x & x \\ 0 & 0 & 0 & 0 & 0 \end{bmatrix}$$

This zero-chasing technique is perfectly general and can be used to zero $a_{n,n-1}$ regardless of where the zero appears along B's diagonal.

We are now in a position to describe a QZ step. The basic idea is to update A and B as follows

$$(\tilde{A} - \lambda \tilde{B}) = \tilde{Q}^T (A - \lambda B)\tilde{Z},$$

where $\tilde{A}$ is upper Hessenberg, $\tilde{B}$ is upper triangular, $\tilde{Q}$ and $\tilde{Z}$ are each orthogonal, and $\tilde{A}\tilde{B}^{-1}$ is essentially the same matrix that would result if a Francis QR step (Algorithm 7.5-2) were applied to AB^{-1}. This can be done with some clever zero-chasing and an appeal to the implicit Q theorem.

Let $M = AB^{-1}$ and let v be the first column of the matrix $(M - aI)(M - bI)$, where a and b are the eigenvalues of M's lower 2-by-2 submatrix. Note that v can be calculated in $O(1)$ flops. If Q_1 is a Householder matrix such that $Q_1 v$ is a multiple of e_1, then

$$A := Q_1 A = \begin{bmatrix} x & x & x & x & x & x \\ x & x & x & x & x & x \\ x & x & x & x & x & x \\ 0 & 0 & x & x & x & x \\ 0 & 0 & 0 & x & x & x \\ 0 & 0 & 0 & 0 & x & x \end{bmatrix},$$

$$B := Q_1 B = \begin{bmatrix} x & x & x & x & x & x \\ x & x & x & x & x & x \\ x & x & x & x & x & x \\ 0 & 0 & 0 & x & x & x \\ 0 & 0 & 0 & 0 & x & x \\ 0 & 0 & 0 & 0 & 0 & x \end{bmatrix}.$$

The idea now is to restore these matrices to Hessenberg-triangular form by chasing the unwanted nonzero elements down the diagonal.

To this end, we first determine a pair of Householder matrices Z_1 and Z_2 to zero b_{31}, b_{32}, and b_{21}:

$$A := AZ_1Z_2 = \begin{bmatrix} x & x & x & x & x & x \\ x & x & x & x & x & x \\ x & x & x & x & x & x \\ x & x & x & x & x & x \\ 0 & 0 & 0 & x & x & x \\ 0 & 0 & 0 & 0 & x & x \end{bmatrix},$$

$$B := BZ_1Z_2 = \begin{bmatrix} x & x & x & x & x & x \\ 0 & x & x & x & x & x \\ 0 & 0 & x & x & x & x \\ 0 & 0 & 0 & x & x & x \\ 0 & 0 & 0 & 0 & x & x \\ 0 & 0 & 0 & 0 & 0 & x \end{bmatrix}.$$

Then a Householder matrix Q_2 is used to zero a_{31} and a_{41}:

$$A := Q_2A = \begin{bmatrix} x & x & x & x & x & x \\ x & x & x & x & x & x \\ 0 & x & x & x & x & x \\ 0 & 0 & x & x & x & x \\ 0 & 0 & 0 & x & x & x \\ 0 & 0 & 0 & 0 & x & x \end{bmatrix},$$

$$B := Q_2 B = \begin{bmatrix} x & x & x & x & x & x \\ 0 & x & x & x & x & x \\ 0 & x & x & x & x & x \\ 0 & x & x & x & x & x \\ 0 & 0 & 0 & 0 & x & x \\ 0 & 0 & 0 & 0 & 0 & x \end{bmatrix}.$$

Notice that with this step the unwanted nonzero elements have been shifted down and to the right from their original position. This illustrates a typical step in the QZ iteration. Notice that $\tilde{Q} = Q_1 Q_2 \cdots$ has the same first column as Q_1. By the way this initial Householder matrix was determined, we can apply the implicit Q theorem and assert that $\tilde{A}\tilde{B}^{-1} = \tilde{Q}^T(AB^{-1})\tilde{Q}$ is indeed essentially the same matrix that we would obtain by applying the Francis iteration to $M = AB^{-1}$ directly. Overall we have

ALGORITHM 7.7-2: *The QZ Step.* Given an unreduced upper Hessenberg matrix $A \in \mathbb{R}^{n \times n}$ and a nonsingular upper triangular matrix $B \in \mathbb{R}^{n \times n}$, the following algorithm overwrites A with the upper Hessenberg matrix $\tilde{Q}^T A \tilde{Z}$ and B with the upper triangular matrix $\tilde{Q}^T B \tilde{Z}$ where $\tilde{Q}$ and $\tilde{Z}$ are orthogonal and $\tilde{Q}$ has the same first column as the orthogonal similarity transformation in Algorithm 7.5-1 when it is applied to AB^{-1}.

> Let $M = AB^{-1}$ and compute $(M - aI)(M - bI)e_1 = (x, y, z, 0, \ldots, 0)^T$ where a and b are the eigenvalues of M's lower 2-by-2 principal submatrix.
> For $k = 1, 2, \ldots, n - 2$
> > Let Q_k be a Householder matrix such that $Q_k \begin{bmatrix} x \\ y \\ z \end{bmatrix} = \begin{bmatrix} * \\ 0 \\ 0 \end{bmatrix}$.
> > $A := \text{diag}(I_{k-1}, Q_k, I_{n-k-2})A$, $B := \text{diag}(I_{k-1}, Q_k, I_{n-k-2})B$
> > Let Z_{k1} be a Householder matrix such that
> > $$(b_{k+2,k}, b_{k+2,k+1}, b_{k+2,k+2})Z_{k1} = (0, 0, *)$$
> > $A := A \, \text{diag}(I_{k-1}, Z_{k1}, I_{n-k-2})$, $B := B \, \text{diag}(I_{k-1}, Z_{k1}, I_{n-k-2})$
> > Let Z_{k2} be a Householder matrix such that
> > $$(b_{k+1,k}, b_{k+1,k+1})Z_{k2} = (0, *)$$
> > $A := A \, \text{diag}(I_{k-1}, Z_{k2}, I_{n-k-1})$, $B := B \, \text{diag}(I_{k-1}, Z_{k2}, I_{n-k-1})$
> > $x := a_{k+1,k}$
> > $y := a_{k+2,k}$
> > If $k < n - 2$ then $z := a_{k+3,k}$
> Let Q_{n-1} be a Householder matrix such that $Q_{n-1} \begin{bmatrix} x \\ y \end{bmatrix} = \begin{bmatrix} * \\ 0 \end{bmatrix}$.
> $A := \text{diag}(I_{n-2}, Q_{n-1})$; $B := \text{diag}(I_{n-2}, Q_{n-1})B$
> Let Z_{n-1} be a Householder matrix such that $(b_{n,n-1}, b_{nn})Z_{n-1} = (0 \, *)$
> $A := A \, \text{diag}(I_{n-2}, Z_{n-1})$; $B := B \, \text{diag}(I_{n-2}, Z_{n-1})$

This algorithm requires $13n^2$ flops. Q and Z can be accumulated for an additional $5n^2$ *and* $8n^2$ flops, respectively.

By applying a sequence of QZ steps to the Hessenberg-triangular pencil $A - \lambda B$, it is possible to reduce A to quasi-triangular form. In doing this it is necessary to monitor A's subdiagonal and B's diagonal in order to bring about decoupling whenever possible. The complete process, due to Moler and Stewart (1973), is as follows:

ALGORITHM 7.7-3.   Given $A \in \mathbb{R}^{n \times n}$ and $B \in \mathbb{R}^{n \times n}$, the following algorithm computes orthogonal Q and Z such that $Q^{\mathsf{T}}AZ = T$ is upper quasi-triangular and $Q^{\mathsf{T}}BZ = S$ is upper triangular. A is overwritten by T and B by S.

Apply Algorithm 7.7-1 to $A - \lambda B$. Accumulate Q and Z if necessary.
Repeat:

   Set all subdiagonal elements in A to zero that satisfy

$$|a_{i,i-1}| \leqslant \mathbf{u}[\,|a_{i-1,i-1}| + |a_{ii}|\,].$$

   Find the largest nonnegative q and the smallest nonnegative p such
      that if

$$A = \begin{bmatrix} A_{11} & A_{12} & A_{13} \\ 0 & A_{22} & A_{23} \\ 0 & 0 & A_{33} \end{bmatrix} \begin{matrix} p \\ n-p-q \\ q \end{matrix}$$

   then A_{33} is upper quasi-triangular and A_{22} is unreduced.
   If $q = n$ *then* quit.
   Partition B conformably:

$$B = \begin{bmatrix} B_{11} & B_{12} & B_{13} \\ 0 & B_{22} & B_{23} \\ 0 & 0 & B_{33} \end{bmatrix} \begin{matrix} p \\ n-p-q \\ q \end{matrix}$$

   If B_{22} is singular, then zero $a_{n-q,n-q-1}$ and go to *Repeat*.
   Apply Algorithm 7.7-2 to $A_{22} - \lambda B_{22}$:

$$A := \text{diag}(I_p, Q, I_q)^{\mathsf{T}}A\,\text{diag}(I_p, Z, I_q)$$
$$B := \text{diag}(I_p, Q, I_q)^{\mathsf{T}}B\,\text{diag}(I_p, Z, I_q)$$

This algorithm requires $15n^3$ flops. If Q is desired, an additional $8n^3$ are necessary. If Z is required, an additional $10n^3$ are needed.

These estimates of work are based on the experience that about two QZ iterations per eigenvalue are necessary. Thus, the convergence properties of

QZ are the same as for QR. The speed of the QZ algorithm is in no way affected by rank deficiency in B.

The computed S and T can be shown to satisfy

$$Q_0^T(A + E)Z_0 = \hat{T} \qquad Q_0^T(B + F)Z_0 = \hat{S}$$

where Q_0 and Z_0 are exactly orthogonal and $\|E\| \cong \mathbf{u}\|A\|$ and $\|F\| \cong \mathbf{u}\|B\|$.

EXAMPLE 7.7-5. If the QZ algorithm is applied to

$$A = \begin{bmatrix} 2 & 3 & 4 & 5 & 6 \\ 4 & 4 & 5 & 6 & 7 \\ 0 & 3 & 6 & 7 & 8 \\ 0 & 0 & 2 & 8 & 9 \\ 0 & 0 & 0 & 1 & 10 \end{bmatrix} \text{ and } B = \begin{bmatrix} 1 & -1 & -1 & -1 & -1 \\ 0 & 1 & -1 & -1 & -1 \\ 0 & 0 & 1 & -1 & -1 \\ 0 & 0 & 0 & 1 & -1 \\ 0 & 0 & 0 & 0 & 1 \end{bmatrix},$$

then the subdiagonal elements of H converge as indicated in Table 7.7-1.

Table 7.7-1.

Iteration	$O(|h_{21}|)$	$O(|h_{32}|)$	$O(|h_{43}|)$	$O(|h_{54}|)$
1	10^0	10^1	10^0	10^{-1}
2	10^0	10^0	10^0	10^{-1}
3	10^0	10^1	10^{-1}	10^{-3}
4	10^0	10^0	10^{-1}	10^{-8}
5	10^0	10^1	10^{-1}	10^{-16}
6	10^0	10^0	10^{-2}	converg.
7	10^0	10^{-1}	10^{-4}	
8	10^1	10^{-1}	10^{-8}	
9	10^0	10^{-1}	10^{-19}	
10	10^0	10^{-2}	converg.	
11	10^{-1}	10^{-4}		
12	10^{-2}	10^{-11}		
13	10^{-3}	10^{-27}		
14	converg.	converg.		

Many of the invariant subspace computations discussed in §7.6 carry over to the generalized eigenvalue problem. For example, approximate eigenvectors can be found via inverse iteration:

$z^{(0)} \in \mathbb{R}^n$ given.
For $k = 1, 2, \ldots$
    Solve $(A - \mu B)v^{(k)} = Bz^{(k-1)}$
    Normalize: $z^{(k)} = v^{(k)}/\|v^{(k)}\|_\infty$

When B is nonsingular, this is equivalent to applying (7.6-1) with the matrix $B^{-1}A$. Typically, only a single iteration is required if μ is an approximate eigenvalue computed by the QZ algorithm. By inverse iterating with the Hessenberg-triangular pencil, costly accumulation of the Z-transformations during the QZ iteration can be avoided.

Corresponding to the notion of an invariant subspace for a single matrix, we have the notion of a *deflating* subspace for the pencil $A - \lambda B$. In particular, we say that a k-dimensional subspace $S \subset \mathbb{R}^n$ is "deflating" for the pencil $A - \lambda B$ if the subspace $\{Ax + By \mid x, y \in S\}$ has dimension k or less. Note that the columns of the matrix Z in the generalized Schur decomposition define a family of deflating subspaces, since if $Q = [q_1, \ldots, q_n]$ and $Z = [z_1, \ldots, z_n]$ then

$$A \operatorname{span}\{z_1, \ldots, z_k\} \subset \operatorname{span}\{q_1, \ldots, q_k\}$$

and

$$B \operatorname{span}\{z_1, \ldots, z_k\} \subset \operatorname{span}\{q_1, \ldots, q_k\}$$

Properties of deflating subspaces and their behavior under perturbation are described in Stewart (1972).

Problems

P7.7-1. Suppose A and B are in $\mathbb{R}^{n \times n}$ and that

$$U^T B V = \begin{bmatrix} D & 0 \\ 0 & 0 \end{bmatrix} \begin{matrix} r \\ n-r \end{matrix} \qquad U = [U_1, U_2], \ V = [V_1, V_2] \atop \ \ \ \ \ r \ \ n-r \ \ \ \ \ \ r \ \ n-r$$

is the SVD of B, where D is r-by-r and $r = \operatorname{rank}(B)$. Show that if $\lambda(A, B) = \mathbb{C}$ then $U_2^T A V_2$ is singular.

P7.7-2. Define $F : \mathbb{R}^n \rightarrow R$ by

$$F(x) = \tfrac{1}{2} \| Ax - \frac{x^T B^T A x}{x^T B^T B x} Bx \|_2^2,$$

where A and B are in $\mathbb{R}^{m \times n}$. Show that if $\nabla F(x) = 0$, then Ax is a multiple of Bx.

P7.7-3. Suppose A and B are in $\mathbb{R}^{n \times n}$. Give an algorithm for computing orthogonal Q and Z such that $Q^T A Z$ is upper Hessenberg and $Z^T B Q$ is upper triangular.

P7.7-4. Show that if $\mu \notin \lambda(A, B)$ then $A_1 = (A - \mu B)^{-1}A$ and $B_1 = (A - \mu B)^{-1}B$ commute.

Notes and References for Sec. 7.7

Mathematical aspects of the generalized eigenvalue problem are covered in

I. Erdelyi (1967). "On the Matrix Equation $Ax = \lambda Bx$," *J. Math. Anal. and Applic.* *17*, 119-32.
F. Gantmacher (1959). *The Theory of Matrices*, vol. 2, Chelsea, New York.
H. W. Turnbull and A. C. Aitken (1961). *An Introduction to the Theory of Canonical Matrices*, Dover, New York.

References that are particularly concerned with the Kronecker canonical form include

P. Van Dooren (1979). "The Computation of Kronecker's Canonical Form of a Singular Pencil," *Lin. Alg. & Its Applic.* *27*, 103-40.
J. H. Wilkinson (1978). "Linear Differential Equations and Kronecker's Canonical Form," in *Recent Advances in Numerical Analysis*, ed. C. deBoor and G. H. Golub, Academic Press, New York, pp. 231-65.

Stewart deals with questions of eigenvalue sensitivity in

G. W. Stewart (1972). "On the Sensitivity of the Eigenvalue Problem $Ax = \lambda Bx$," *SIAM J. Num. Anal. 9*, 669-86.
G. W. Stewart (1973*b*). "Error and Perturbation Bounds for Subspaces Associated with Certain Eigenvalue Problems," *SIAM Review 15*, 727-64.
G. W. Stewart (1975*b*). "Gershgorin Theory for the Generalized Eigenvalue Problem $AX = \lambda Bx$," *Math. Comp. 29*, 600-606.
G. W. Stewart (1978). "Perturbation Theory for the Generalized Eigenvalue Problem," in *Recent Advances in Numerical Analysis*, ed. C. deBoor and G. H. Golub, Academic Pres, New York.

Rectangular generalized eigenvalue problems arise in certain applications. See

G. L. Thompson and R. L. Weil (1970). "Reducing the Rank of $A - \lambda B$," *Proc. Amer. Math. Sec. 26*, 548-54.
G. L. Thompson and R. L. Weil (1972). "Roots of Matrix Pencils $Ay = \lambda By$: Existence, Calculations, and Relations to Game Theory," *Lin. Alg. & Its Applic. 5*, 207-26

Nonorthogonal computational procedures for the generalized eigenvalue problem are proposed in

G. Peters and J. H. Wilkinson (1970*a*). "$Ax = \lambda Bx$ and the Generalized Eigenproblem," *SIAM J. Num. Anal. 7*, 479-92.
V. N. Kublanovskaja and V. N. Fadeeva (1964). "Computational Methods for the Solution of a Generalized Eigenvalue Problem," *Amer. Math. Soc. Transl. 2*, 271-90.

The QZ algorithm is in EISPACK 2 and was originally presented in

C. B. Moler and G. W. Stewart (1973). "An Algorithm for Generalized Matrix Eigenvalue Problems," *SIAM J. Num. Anal. 10*, 241-56.

Improvements in the original algorithm are discussed in

R. C. Ward (1975). "The Combination Shift QZ Algorithm," *SIAM J. Num. Anal. 12*, 835-53.

L. Kaufman (1977). "Some Thoughts on the QZ Algorithm for Solving the Generalized Eigenvalue Problem," *ACM Trans. Math. Soft. 3*, 65-75.

A method similar to QZ but relying on Gauss transformations is presented in

L. Kaufman (1974). "The LZ Algorithm to Solve the Generalized Eigenvalue Problem," *SIAM J. Num. Anal. 11*, 997-1024,

while

C. F. Van Loan (1975). "A General Matrix Eigenvalue Algorithm," *SIAM J. Num. Anal. 12*, 819-34,

is concerned with a generalization of the QZ algorithm that handles the problem $A^T C x = \lambda B^T D x$ without inversion or formation of $A^T C$ and $B^T D$.

The behavior of the QZ algorithm on pencils $A - \lambda B$ that are always nearly singular is discussed in

J. W. Wilkinson (1979). "Kronecker's Canonical Form and the QZ Algorithm," *Lin. Alg. & Its Applic. 28*, 285-303.

Other approaches to the problem include

H. R. Schwartz (1974). "The Method of Coordinate Relaxation for $(A - \lambda B)x = 0$," *Numer. Math. 23*, 135-52.

G. Rodrigue (1973). "A Gradient Method for the Matrix Eigenvalue Problem $Ax = \lambda Bx$," *Numer. Math 22*, 1-16.

A. Jennings and M. R. Osborne (1977). "Generalized Eigenvalue Problems for Certain Unsymmetric Band Matrices," *Lin. Alg. & Its Applic. 29*, 139-50.

A. Ruhe (1978). "A Note on the Efficient Solution of Matrix Pencil Systems," *BIT 18*, 276-81.

The Symmetric Eigenvalue Problem

The perturbation theory and algorithmic developments in the previous chapter undergo considerable simplification when the matrix A is symmetric. Indeed, the symmetric eigenvalue problem with its rich mathematical structure is one of the most aesthetically pleasing problems in all of numerical algebra.

We begin our presentation with a brief discussion of the mathematical properties underlying the symmetric eigenproblem. In §8.2 we specialize the algorithms of §§7.4 and 7.5 and obtain the elegant symmetric QR algorithm. A variant of this routine capable of computing the singular value decomposition is detailed in §8.3.

One of the earliest matrix algorithms to appear in the literature is Jacobi's method for the symmetric eigenproblem. There is renewed interest in this algorithm because of parallel computation, and for this reason we have devoted a section to its presentation.

Because the eigenvalues of a symmetric matrix can be characterized in several ways, there are a host of alternatives to the QR algorithm. Some of these methods are described in §8.5. In the final section we discuss the $A - \lambda B$ problem for the important case when A is symmetric and B is symmetric positive definite. Although no suitable analog of the QZ algorithm exists for this specially structured problem, there are several successful methods that can be applied. The generalized singular value decomposition is also discussed.

Much of the material in this chapter may be found in the new work by Parlett (SEP).

Reading Path

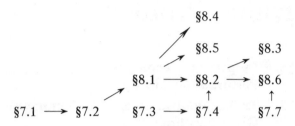

Sec. 8.1. Properties, Decompositions, Perturbation Theory

Symmetry simplifies the real eigenvalue problem $Ax = \lambda x$ in two ways: it ensures that all of A's eigenvalues are real, and it ensures that there is an orthonormal basis of eigenvectors. These properties are a consequence of

THEOREM 8.1-1: *Real Schur Decomposition for Symmetric Matrices.* If $A \in \mathbb{R}^{n \times n}$ is symmetric, then there exists an orthogonal Q such that

$$Q^{T}AQ = \text{diag}(\lambda_1, \ldots, \lambda_n).$$

Proof. Let $Q^{T}AQ = T$ be the real Schur decomposition (Theorem 7.4-1) of A. Since T is also symmetric, it follows that it must be a direct sum of 1-by-1 and 2-by-2 matrices. It is easy to verify, however, that a 2-by-2 symmetric matrix cannot have complex eigenvalues. Consequently, T can have no 2-by-2 blocks along its diagonal. □

EXAMPLE 8.1-1. If

$$A = \begin{bmatrix} 6.8 & 2.4 \\ 2.4 & 8.2 \end{bmatrix} \text{ and } Q = \begin{bmatrix} .6 & -.8 \\ .8 & .6 \end{bmatrix}$$

then Q is orthogonal and $Q^{T}AQ = \text{diag}(10, 5)$.

If $A^{T} = A \in \mathbb{R}^{n \times n}$, then we let $\lambda_i(A)$ denote the i-th largest eigenvalue of A. Thus,

$$\lambda_n(A) \leqslant \lambda_{n-1}(A) \leqslant \cdots \leqslant \lambda_2(A) \leqslant \lambda_1(A).$$

It is easy to show that each eigenvalue of a symmetric matrix is a stationary value of the map $x \to x^{T}Ax/x^{T}x$ where $x \neq 0$ and, moreover, satisfies the following "minimax" characterization:

THEOREM 8.1-2: *Courant-Fischer Minimax Characterization.* If $A \in \mathbb{R}^{n \times n}$ is symmetric, then for $k = 1, \ldots, n$,

$$\lambda_k(A) = \max_{\dim(S)=k} \min_{0 \neq y \in S} y^T A y / y^T y.$$

Proof. Wilkinson (AEP, pp. 100–101). $\square$

Using this result it is possible to establish several very useful corollaries.

COROLLARY 8.1-3. If A and $A + E$ are n-by-n symmetric matrices, then for $k = 1, 2, \ldots, n$

$$\lambda_k(A) + \lambda_n(E) \leq \lambda_k(A + E) \leq \lambda_k(A) + \lambda_1(E).$$

Proof. Wilkinson (AEP, pp. 101–2). $\square$

COROLLARY 8.1-4. If A_r denotes the leading r-by-r principal submatrix of an n-by-n symmetric matrix A, then for $r = 1, 2, \ldots, n - 1$ the following *interlacing property* holds:

$$\lambda_{r+1}(A_{r+1}) \leq \lambda_r(A_r) \leq \lambda_r(A_{r+1}) \leq \cdots \leq \lambda_2(A_{r+1}) \leq \lambda_1(A_r) \leq \lambda_1(A_{r+1}).$$

Proof. Wilkinson (AEP, pp. 103–4). $\square$

EXAMPLE 8.1-2. If

$$A = \begin{bmatrix} 6.8 & 2.4 \\ 2.4 & 8.2 \end{bmatrix} \quad \text{and} \quad E = \begin{bmatrix} 2.0 & 0.0 \\ 0.0 & 1.0 \end{bmatrix}$$

then

$$\lambda(A) = \{5, 10\}$$
$$\lambda(E) = \{1, 2\}$$
$$\lambda(A + E) = \{6.5917, 11.4083\}$$

EXAMPLE 8.1-3. If

$$A = \begin{bmatrix} 1 & 1 & 1 & 1 \\ 1 & 2 & 3 & 4 \\ 1 & 3 & 6 & 10 \\ 1 & 4 & 10 & 20 \end{bmatrix}$$

then

$$\lambda(A_1) = \{1\}$$
$$\lambda(A_2) = \{.3820, 2.6180\}$$
$$\lambda(A_3) = \{.1270, 1.0000, 7.873\}$$
$$\lambda(A_4) = \{.0380, .4538, 7.2034, 26.3047\}$$

COROLLARY 8.1-5.   Suppose $B = A + \tau cc^T$ where $A \in \mathbb{R}^{n \times n}$ is symmetric, $c \in \mathbb{R}^n$ has unit 2-norm, and $\tau, \in \mathbb{R}$. If $\tau \geqslant 0$ then

$$\lambda_i(B) \in [\lambda_i(A), \lambda_{i-1}(A)], \qquad\qquad i = 2, \ldots, n$$

while if $\tau \leqslant 0$ then

$$\lambda_i(B) \in [\lambda_{i+1}(A), \lambda_i(A)]. \qquad\qquad i = 1, \ldots, n - 1$$

In either case,

$$\lambda_i(B) = \lambda_i(A) + m_i \tau, \qquad\qquad i = 1, \ldots, n$$

where $m_1 + \cdots + m_n = 1$ and each $m_i \geqslant 0$.

Proof.   Wilkinson (AEP, pp. 94–97). □

The effect of rank one perturbations on the eigenvalues of a symmetric matrix is of interest in several settings, e.g., quasi-Newton methods for unconstrained optimization.

Our next perturbation result is of a rather different variety. It bounds the difference between $\lambda(A)$ and $\lambda(A + E)$ in terms of $\| E \|_F$.

THEOREM 8.1-6: *Wielandt-Hoffman.*   If A and $A + E$ are n-by-n symmetric matrices, then

$$\sum_{i=1}^{n} [\lambda_i(A + E) - \lambda_i(A)]^2 \leqslant \| E \|_F^2.$$

Proof.   See Wilkinson (AEP, pp. 104–8). □

EXAMPLE 8.1-4.   If

$$A = \begin{bmatrix} 6.8 & 2.4 \\ 2.4 & 8.2 \end{bmatrix} \qquad \text{and} \qquad E = \begin{bmatrix} 2.0 & 0.0 \\ 0.0 & 1.0 \end{bmatrix}$$

then

$$4.5168 = \sum_{i=1}^{2} [\lambda_i(A + E) - \lambda_i(A)]^2 \leqslant \| E \|_F^2 = 5.$$

See Example 8.1-2.

The invariant subspaces of a symmetric matrix are not necessarily well-conditioned; the separation of the eigenvalues is a critical factor just as in the unsymmetric case. However, the bound given in our general perturbation theorem for invariant subspaces (Theorem 7.2-4) undergoes some simplification in the symmetric case.

THEOREM 8.1-7. Suppose A and $A + E$ are n-by-n symmetric matrices and that

$$Q = [\underset{k}{Q_1} , \underset{n-k}{Q_2}]$$

is an orthogonal matrix such that $R(Q_1)$ is an invariant subspace for A. Partition the matrices Q^TAQ and Q^TEQ as follows:

$$Q^TAQ = \begin{bmatrix} A_{11} & 0 \\ 0 & A_{22} \end{bmatrix}, \qquad Q^TEQ = \begin{bmatrix} E_{11} & E_{12} \\ E_{21} & E_{22} \end{bmatrix}.$$

If

$$\delta = \min_{\substack{\lambda \in \lambda(A_{11}) \\ \mu \in \lambda(A_{22})}} |\lambda - \mu| - \|E_{11}\|_2 - \|E_{22}\|_2 > 0$$

and $\|E_{12}\|_2 \leqslant \delta/2$, then there exists a matrix $P \in R^{(n-k)\times k}$ satisfying

$$\|P\|_2 \leqslant 2 \, \|E_{21}\|_2/\delta$$

such that the columns of $Q_1 = (Q_1 + Q_2 P)(I + P^T P)^{-1/2}$ form an orthonormal basis for a subspace that is invariant for $A + E$.

Proof. See Stewart (1973b). $\square$

The theorem essentially says that the sensitivity of an invariant subspace of a symmetric matrix depends solely on the isolation of the associated eigenvalues.

EXAMPLE 8.1-5. If $A = \text{diag}(.999, 1.001, 2.)$, and

$$E = \begin{bmatrix} 0.00 & 0.01 & 0.01 \\ 0.01 & 0.00 & 0.01 \\ 0.01 & 0.01 & 0.00 \end{bmatrix},$$

then $\hat{Q}^T(A + E)\hat{Q} = \text{diag}(.9899, 1.0098, 2.0002)$ where

$$\hat{Q} = \begin{bmatrix} -.7418 & .6706 & .0101 \\ .6708 & .7417 & .0101 \\ .0007 & -.0143 & .9999 \end{bmatrix}$$

is orthogonal. Let $\hat{q}_i = \hat{Q}e_i$, $i = 1, 2, 3$. Thus, $\hat{q}_i$ is the perturbation of A's eigenvector $q_i = e_i$. A calculation shows that

$$\text{dist}[\text{span}\{q_1\}, \text{span}\{\hat{q}_1\}] = .67$$

$$\text{dist}[\text{span}\{q_2\}, \text{span}\{\hat{q}_2\}] = .67$$

$$\text{dist}[\text{span}\{q_3\}, \text{span}\{\hat{q}_3\}] = \text{dist}[\text{span}\{q_1, q_2\}, \text{span}\{\hat{q}_1, \hat{q}_2\}] = .01$$

We next present a collection of results that are concerned with approximate invariant subspaces. These results will be used in subsequent sections and will aid in the interpretation of various algorithms.

THEOREM 8.1-8. Suppose $A \in \mathbb{R}^{n \times n}$ and $S \in \mathbb{R}^{k \times k}$ are symmetric and that

$$AQ_1 - Q_1 S = E_1$$

where $Q_1 \in \mathbb{R}^{n \times k}$ satisfies $Q_1^T Q_1 = I_k$. Then there exist eigenvalues $\mu_1, \ldots, \mu_k \in \lambda(A)$ such that

$$|\mu_i - \lambda_i(S)| \leqslant \sqrt{2}\,\|E_1\|_2. \qquad\qquad i = 1, \ldots, k$$

Proof. Let $Q_2 \in \mathbb{R}^{n \times (n-k)}$ be any matrix such that $Q = [Q_1, Q_2]$ is orthogonal. It follows that

$$Q^T A Q = \begin{bmatrix} S & 0 \\ 0 & Q_2^T A Q_2 \end{bmatrix} + \begin{bmatrix} Q_1^T E_1 & E_1^T Q_2 \\ Q_2^T E_1 & 0 \end{bmatrix} \equiv B + E,$$

and so by using Corollary 8.1-3 we have

$$|\lambda_i(A) - \lambda_i(B)| \leqslant \|E\|_2. \qquad\qquad (i = 1, \ldots, n)$$

Since $\lambda(S) \subset \lambda(B)$, there exist $\mu_1, \ldots, \mu_k \subset \lambda(A)$ such that

$$|\mu_i - \lambda_i(S)| \leqslant \|E\|_2. \qquad\qquad (i = 1, \ldots, k)$$

The theorem follows by noting that for any $x \in \mathbb{R}^k$ and $y \in \mathbb{R}^{n-k}$ we have

$$\left\| E\binom{x}{y} \right\|_2 \leqslant \|E_1 x\|_2 + \|E_1^T Q_2 y\|_2 \leqslant \|E_1\|_2 \|x\|_2 + \|E_1\|_2 \|y\|_2,$$

from which we readily conclude that $\|E\|_2 \leqslant \sqrt{2}\,\|E_1\|_2$. $\square$

EXAMPLE 8.1-6. If

$$A = \begin{bmatrix} 6.8 & 2.4 \\ 2.4 & 8.2 \end{bmatrix}, \qquad Q_1 = \begin{bmatrix} .7994 \\ .6007 \end{bmatrix}, \qquad S = (5.1),$$

then

$$AQ_1 - Q_1S = \begin{bmatrix} -.0828 \\ -.0562 \end{bmatrix} = E_1.$$

The theorem predicts that A has an eigenvalue within $\sqrt{2}\,\|E_1\|_2 \cong .1415$ of 5.1. This is true since $\lambda(A) = \{5, 10\}$.

The usefulness of Theorem 8.1-8 is enhanced if we weaken the assumption that the columns of Q_1 are orthonormal. As can be expected, the bounds deteriorate with loss of orthogonality:

THEOREM 8.1-9. Suppose $A \in \mathbb{R}^{n \times n}$ and $S \in \mathbb{R}^{k \times k}$ are symmetric and that

$$AX_1 - X_1S = F_1,$$

where $X_1 \in \mathbb{R}^{n \times k}$ satisfies $\sigma_k(X_1) > 0$. Then there exist $\mu_1, \ldots, \mu_k \in \lambda(A)$ such that

$$|\mu_i - \lambda_i(S)| \leqslant \sqrt{2}\,\frac{\|F_1\|_2}{\sigma_k(X_1)}.$$

Proof. Let $X_1 = Q_1R_1$ be the Q-R factorization of X_1 ($Q_1^TQ_1 = I_k$, $R_1 \in \mathbb{R}^{k \times k}$). When this is substituted into $AX_1 - X_1 S = F_1$ we get

$$AQ_1 - Q_1S_1 = E_1,$$

where $S_1 = R_1SR_1^{-1}$ and $E_1 = F_1R_1^{-1}$. The theorem follows by applying Theorem 8.1-8 and noting that $\lambda(S) = \lambda(S_1)$ and $\|E_1\|_2 \leqslant \|F_1\|_2/\sigma_k(X_1)$. $\square$

The eigenvalue bounds in Theorem 8.1-8 depend on the size of the residual of the approximate invariant subspace, i.e., upon the size of $\|AQ_1 - Q_1S\|$. Given A and Q_1, the following theorem tells how to choose S so that this quantity is minimized when $\|\cdot\| = \|\cdot\|_F$.

THEOREM 8.1-10. If $A \in \mathbb{R}^{n \times n}$ is symmetric and $Q_1 \in \mathbb{R}^{n \times k}$ satisfies $Q_1^TQ_1 = I_k$, then

$$\min_{S \in \mathbb{R}^{k \times k}} \|AQ_1 - Q_1S\|_F = \|AQ_1 - Q_1(Q_1^TAQ_1)\|_F$$
$$= \|(I - Q_1Q_1^T)AQ_1\|_F$$

Proof. Let $Q_2 \in \mathbb{R}^{n \times (n-k)}$ be such that $Q = [Q_1, Q_2]$ is orthogonal. For any $S \in \mathbb{R}^{k \times k}$ we have

$$\|AQ_1 - Q_1S\|_F^2 = \|Q^TAQ_1 - Q^TQ_1S\|_F^2 = \|Q_1^TAQ_1 - S\|_F^2 + \|Q_2^TAQ_1\|_F^2.$$

Clearly, the minimizing S is given by $S = Q_1^TAQ_1$. $\square$

This result enables us to associate with any k-dimensional subspace $R(Q_1)$, a set of k "optimal" eigenvalue-eigenvector approximates.

THEOREM 8.1-11. Suppose $A \in \mathbb{R}^{n \times n}$ is symmetric and that $Q_1 \in \mathbb{R}^{n \times k}$ satisfies $Q_1^T Q_1 = I_k$. If

$$Z^T(Q_1^T A Q_1)Z = \text{diag}(\theta_1, \ldots, \theta_k) = D$$

is the Schur decomposition of $Q_1^T A Q_1$ and $Q_1 Z = [y_1, \ldots, y_k]$, then

$$\|A y_i - \theta_i y_i\|_2 = \|(I - Q_1 Q_1^T) A Q_1 Z e_i\|_2 \leqslant \|(I - Q_1 Q_1^T) A Q_1\|_2$$

for $i = 1, \ldots, k$. The θ_i are called *Ritz values*, the y_i are called *Ritz vectors*, and the (θ_i, y_i) are called *Ritz pairs*.

Proof.

$$A y_i - \theta_i y_i = A Q_1 Z e_i - Q_1 Z D e_i = (A Q_1 - Q_1(Q_1^T A Q_1)) Z e_i.$$

The theorem follows by taking norms. $\square$

We conclude the section by proving the Sylvester law of inertia. The *inertia* of a symmetric matrix A is a triplet of integers (m, z, p) where $m, z,$ and p are the number of negative, zero, and positive elements of $\lambda(A)$.

THEOREM 8.1-12: *Sylvester Law of Inertia.* If $A \in \mathbb{R}^{n \times n}$ is symmetric and $X \in \mathbb{R}^{n \times n}$ is nonsingular, then A and $X^T A X$ have the same inertia.

Proof. Suppose $\lambda_r(A) > 0$ and define the subspace $S_0 \subset R^n$ by

$$S_0 = \text{span } \{X^{-1}q_1, \ldots, X^{-1}q_r\}, \qquad q_i \neq 0$$

where $A q_i = \lambda_i(A) q_i$ and $i = 1, \ldots, r$. From the minimax characterization of $\lambda_r(X^T A X)$ we have

$$\lambda_r(X^T A X) = \max_{\dim(S)=r} \min_{y \in S} \frac{y^T(X^T A X)y}{y^T y} \geqslant \min_{y \in S_0} \frac{y^T(X^T A X)y}{y^T y}.$$

Now for any $y \in \mathbb{R}^n$ we have

$$\frac{y^T(X^T X)y}{y^T y} \geqslant \sigma_n(X)^2,$$

while for $y \in S_0$ it is clear that

$$\frac{y^T(X^T A X)y}{y^T(X^T X)y} \geqslant \lambda_r(A).$$

Thus,

$$\lambda_r(X^TAX) \geqslant \min_{y \in S_0} \left\{ \frac{y^T(X^TAX)y}{y^T(X^TX)y} \quad \frac{y^T(X^TX)y}{y^Ty} \right\} \geqslant \lambda_r(A)\sigma_n(X)^2.$$

An analogous argument with the roles of A and X^TAX reversed shows that

$$\lambda_r(A) \geqslant \lambda_r(X^TAX)\sigma_n(X^{-1})^2 = \lambda_r(X^TAX)/\sigma_1(X)^2.$$

It follows that A and X^TAX have the same number of positive eigenvalues. If we apply this result to $-A$, we conclude that A and X^TAX have the same number of negative eigenvalues. Obviously, the number of zero eigenvalues possessed by each matrix is also the same. □

EXAMPLE 8.1-7. If $A = \text{diag}(3, 2, -1)$ and

$$X = \begin{bmatrix} 1 & 4 & 5 \\ 0 & 1 & 2 \\ 0 & 0 & 1 \end{bmatrix},$$

then

$$X^TAX = \begin{bmatrix} 3 & 12 & 15 \\ 12 & 50 & 64 \\ 15 & 64 & 82 \end{bmatrix}$$

and $\lambda(X^TAX) = \{134.769, .3555, -.1252\}$.

Problems

P8.1-1. Show that the eigenvalues of a 2-by-2 symmetric matrix must be real.

P8.1-2. Suppose that A and E are n-by-n symmetric matrices. Assume that there exists a continuously differentiable orthogonal matrix $Q(\epsilon)$ on $[0, 1]$ such that

$$Q(\epsilon)^T(A + \epsilon E)Q(\epsilon) = \text{diag}[\lambda_1(\epsilon), \ldots, \lambda_n(\epsilon)] = D(\epsilon).$$

(a) Show that if $Q(\epsilon) = [q_1(\epsilon), \ldots, q_n(\epsilon)]$ then

$$\dot{D} = \text{diag}(q_1^TEq_1, \ldots, q_n^TEq_n).$$

(b) By integrating the above result from 0 to 1 and taking norms, show that $\|D(1) - D(0)\|_2 \leqslant \|E\|_F$, thereby establishing the Wielandt-Hoffman theorem for the case when the family of Schur decompositions above is continuously differentiable.

P8.1-3. Show that if the gradient of the function $f(x) = x^TAx/x^Tx$ is zero for some vector z, then Az must be a multiple of z.

P8.1-4. Compute the Schur decomposition of $A = \begin{bmatrix} 1 & 2 \\ 2 & 3 \end{bmatrix}$.

Notes and References for Sec. 8.1

The perturbation theory for the symmetric eigenvalue problem is surveyed in Wilkinson (AEP, chap. 2) and Parlett (SEP, chaps. 10 and 11). Some representative papers in this well-documented area include

W. Kahan (1975). "Spectra of Nearly Hermitian Matrices," *Proc. Amer. Math. Soc. 48*, 11-17.

W. Kahan (1967). "Inclusion Theorems for Clusters of Eigenvalues of Hermitian Matrices," Computer Science Report, University of Toronto.

C. C. Paige (1974b). "Eigenvalues of Perturbed Hermitian Matrices," *Lin. Alg. & Its Applic. 8*, 1-10.

A. Ruhe (1975). "On the Closeness of Eigenvalues and Singular Values for Almost Normal Matrices," *Lin. Alg. & Its Applic. 11*, 87-94.

A. Schonhage (1979). "Arbitrary Perturbations of Hermitian Matrices," *Lin. Alg. & Its Applic. 24*, 143-49.

G. W. Stewart (1973b). "Error and Perturbation Bounds for Subspaces Associated with Certain Eigenvalue Problems," *SIAM Review 15*, 727-64.

Sec. 8.2. Tridiagonalization and the Symmetric QR Algorithm

We now investigate how the practical QR algorithm developed in Chapter 7 can be specialized when $A \in \mathbb{R}^{n \times n}$ is symmetric. There are three obvious observations to make immediately:

(a) If $U_0^T A U_0 = H$ is upper Hessenberg, then $H = H^T$ must be tridiagonal.
(b) Symmetry and tridiagonal band structure are preserved when a single-shift QR step is performed.
(c) There is no need to consider complex shifts, since $\lambda(A) \subset \mathbb{R}$.

These simplifications together with the nice mathematical properties of the symmetric eigenproblem combine to make the algorithms of this chapter among the most elegant in all of matrix computations.

We begin by deriving the Householder reduction to tridiagonal form. Suppose that Householder matrices $P_1, \ldots, P_{k-1}$ have been determined such that submatrix B in

$$
A_{k-1} = (P_1 \cdots P_{k-1})^T A \, (P_1 \cdots P_{k-1}) =
\begin{array}{c}
\left[
\begin{array}{c|c|c}
B & \begin{matrix} \\ \end{matrix} & 0 \\
& & b^T \\
\hline
0 & b & D
\end{array}
\right] \\
\begin{array}{ccc} k-1 & 1 & n-k \end{array}
\end{array}
\begin{array}{c}
k-1 \\ 1 \\ n-k
\end{array}
$$

is tridiagonal. If $\bar{P}_k$ is an order $n - k$ Householder matrix such that $\bar{P}_k b$ is a multiple of $e_1^{(n-k)}$, and if $P_k = \text{diag}(I_k, \bar{P}_k)$, then the leading k-by-k principal submatrix of

$$A_k = P_k A_{k-1} P_k = \begin{bmatrix} B & \vline & 0 \\ & \vline & \overline{} \\ & \vline & b^T \overline{P}_k \\ \hline 0 & \vline \overline{P}_k b & \vline \overline{P}_k D \overline{P}_k \end{bmatrix}$$

is tridiagonal. Clearly, if $Q = P_1 P_2 \cdots P_{n-2}$, then $Q^T A Q = T$ is tridiagonal.

In the calculation of A_k it is important to exploit symmetry during the formation of the matrix $\overline{P}_k D \overline{P}_k$. To be specific, suppose that $\overline{P}_k$ has the form

$$\overline{P}_k = I - \frac{2}{v^T v} v v^T . \qquad\qquad 0 \neq v \in \mathbb{R}^{n-k}$$

Note that if

$$p = \frac{2}{v^T v} D v \qquad \text{and} \qquad w = p - \frac{p^T v}{v^T v} v,$$

then the update $\overline{P}_k D \overline{P}_k$ has the form

$$\overline{P}_k D \overline{P}_k = D - v w^T - w v^T.$$

Since only the upper triangular portion of this matrix needs to be calculated, we see that the transition from A_{k-1} to A_k can be accomplished in only $2(n-k)^2$ flops.

Overall we have the following tridiagonalization procedure:

ALGORITHM 8.2-1: *Householder Tridiagonalization*. Given an n-by-n symmetric matrix A, the following algorithm overwrites A with $T = U_0^T A U_0$, where T is tridiagonal and $U_0 = P_1 \ldots P_{n-2}$ is the product of Householder transformations.

For $k = 1, \ldots, n - 2$
Determine a Householder matrix $\dot{P}_k \in \mathbb{R}^{n-k}$ such that

$$\overline{P}_k \begin{bmatrix} a_{k+1,k} \\ \vdots \\ a_{nk} \end{bmatrix} = \begin{bmatrix} x \\ 0 \\ \vdots \\ 0 \end{bmatrix} .$$

$$A := P_k A P_k^T, \qquad P_k = \mathrm{diag}(I_k, \overline{P}_k)$$

This algorithmn requires $\frac{2}{3} n^3$ flops. The matrix U_0 can be stored in factored form in the subdiagonal portion of A. If U_0 is explicitly required, it can be formed with an additional $\frac{2}{3} n^3$ flops.

EXAMPLE 8.2-1.   If

$$A = \begin{bmatrix} 1 & 3 & 4 \\ 3 & 2 & 8 \\ 4 & 8 & 3 \end{bmatrix} \quad \text{and} \quad Q = \begin{bmatrix} 1 & 0 & 0 \\ 0 & .6 & .8 \\ 0 & .8 & -.6 \end{bmatrix},$$

then $Q^T Q = I$ and

$$Q^T A Q = \begin{bmatrix} 1 & 5 & 0 \\ 5 & 10.32 & 1.76 \\ 0 & 1.76 & -5.32 \end{bmatrix}.$$

Let $\hat{T}$ denote the computed version of T obtained by Algorithm 8.2-1. It can be shown that $\hat{T} = Q^T(A + E)Q$ where Q is exactly orthogonal and E is a symmetric matrix satisfying

$$\|E\|_F \leqslant c\mathbf{u} \|A\|_F$$

where c is a small constant. See Wilkinson (AEP, p. 297).

We now consider the single-shift QR iteration for symmetric matrices:

(8.2-1)

$$\begin{aligned} &T = U_0^T A U_0 \text{ (tridiagonal)} \\ &\text{For } k = 0, 1, \dots \\ &\quad T - \mu I = UR \quad \text{(Q-R factorization)} \\ &\quad T := RU + \mu I \end{aligned}$$

Since this iteration preserves Hessenberg structure (cf. Algorithm 7.4-1) and symmetry, it preserves tridiagonal form. Moreover, the Q-R factorization of a tridiagonal matrix requires only $O(n)$ flops, since R has upper bandwidth 2. Thus, there is an order of magnitude less work than in the unsymmetric case. (If orthogonal matrices are accumulated in (8.2-1), however, then $O(n^2)$ flops per iteration are necessary.)

Symmetry can also be exploited in conjunction with the calculation of the shift μ. Denote T by

$$T = \begin{bmatrix} a_1 & b_2 & & & \mathbf{0} \\ b_2 & a_2 & b_3 & & \\ & \ddots & \ddots & \ddots & b_n \\ \mathbf{0} & & & b_n & a_n \end{bmatrix}.$$

We can obviously set $\mu = a_n$. But a more effective choice is to shift by the eigenvalue of

$$\begin{bmatrix} a_{n-1} & b_n \\ b_n & a_n \end{bmatrix}$$

that is closer to a_n. This is known as the *Wilkinson shift* and is given by

(8.2-2) $$\mu = a_n + d - \text{sign}(d)\sqrt{d^2 + b_n^2},$$

where

$$d = (a_{n-1} - a_n)/2.$$

Wilkinson (1968b) has shown that (8.2-1) is cubically convergent with either shift strategy, but gives heuristic reasons why (8.2-2) is preferred.

As in the unsymmetric QR iteration, it is possible to shift implicitly in (8.2-1). That is, we can effect the transition from T to $\bar{T} = RU + \mu I = U^T TU$ without explicitly forming the matrix $T - \mu I$. Let $c = \cos(\theta)$ and $s = \sin(\theta)$ be computed such that

$$\begin{bmatrix} c & s \\ -s & c \end{bmatrix}^T \begin{bmatrix} a_2 - \mu \\ b_2 \end{bmatrix} = \begin{bmatrix} * \\ 0 \end{bmatrix}.$$

If we set $J_1 = J(1, 2, \theta)$ then $J_1 e_1 = U e_1$ and

$$J_1^T T J_1 = \begin{bmatrix} x & x & + & 0 & 0 & 0 \\ x & x & x & 0 & 0 & 0 \\ + & x & x & x & 0 & 0 \\ 0 & 0 & x & x & x & 0 \\ 0 & 0 & 0 & x & x & x \\ 0 & 0 & 0 & 0 & x & x \end{bmatrix}. \qquad n = 6$$

We are thus in a position to apply the implicit Q theorem provided we can compute rotations $J_2, \ldots, J_{n-1}$ with the property that if $Z = J_1 J_2 \ldots J_{n-1}$ then $Z e_1 = J_1 e_1 = U e_1$ and $Z^T TZ$ is tridiagonal.

Note that the first column of Z and U are identical provided we take each J_i to be of the form $J_i = J(i, i + 1, \theta_i)$, $i = 2, 3, \ldots, n - 1$. But J_i of this form can be used to chase the unwanted nonzero element "$+$" out of the matrix as follows:

$$T := J_2^T T J_2 = \begin{bmatrix} x & x & 0 & 0 & 0 & 0 \\ x & x & x & + & 0 & 0 \\ 0 & x & x & x & 0 & 0 \\ 0 & + & x & x & x & 0 \\ 0 & 0 & 0 & x & x & x \\ 0 & 0 & 0 & 0 & x & x \end{bmatrix} \qquad T := J_3^T T J_3 = \begin{bmatrix} x & x & 0 & 0 & 0 & 0 \\ x & x & x & 0 & 0 & 0 \\ 0 & x & x & x & + & 0 \\ 0 & 0 & x & x & x & 0 \\ 0 & 0 & + & x & x & x \\ 0 & 0 & 0 & 0 & x & x \end{bmatrix}$$

$$
T := J_4^T T J_4 = \begin{bmatrix} x & x & 0 & 0 & 0 & 0 \\ x & x & x & 0 & 0 & 0 \\ 0 & x & x & x & 0 & 0 \\ 0 & 0 & x & x & x & + \\ 0 & 0 & 0 & x & x & x \\ 0 & 0 & 0 & + & x & x \end{bmatrix} \qquad T := J_5^T T J_5 = \begin{bmatrix} x & x & 0 & 0 & 0 & 0 \\ x & x & x & 0 & 0 & 0 \\ 0 & x & x & x & 0 & 0 \\ 0 & 0 & x & x & x & 0 \\ 0 & 0 & 0 & x & x & x \\ 0 & 0 & 0 & 0 & x & x \end{bmatrix}
$$

Thus, it follows from the implicit Q theorem that the tridiagonal matrix $(Z^T T Z)$ produced by this zero-chasing technique is essentially the same as the tridiagonal matrix T obtained by the explicit method. (We may assume that all tridiagonal matrices in question are unreduced, since otherwise the problem decouples.)

Note that at any stage of the zero-chasing, there is only one nonzero entry outside the tridiagonal band. How this nonzero entry moves down the matrix during the update $T := J_k^T T J_k$ is illustrated in the following:

$$
\begin{bmatrix} a_k & b_p & 0 & 0 \\ b_p & a_p & b_q & z_r \\ 0 & b_q & a_q & b_r \\ 0 & z_r & b_r & a_r \end{bmatrix} :=
$$

$$
\begin{bmatrix} 1 & 0 & 0 & 0 \\ 0 & c & -s & 0 \\ 0 & s & c & 0 \\ 0 & 0 & 0 & 1 \end{bmatrix} \begin{bmatrix} a_k & b_p & z_q & 0 \\ b_p & a_p & b_q & 0 \\ z_q & b_b & a_q & b_r \\ 0 & 0 & b_r & a_r \end{bmatrix} \begin{bmatrix} 1 & 0 & 0 & 0 \\ 0 & c & s & 0 \\ 0 & -s & c & 0 \\ 0 & 0 & 0 & 1 \end{bmatrix} ,
$$

where $(p, q, r) = (k + 1, k + 2, k + 3)$. This update can be performed in only 10 flops once c and s have been determined from the equation $b_p s + z_q c = 0$. Overall we obtain:

ALGORITHM 8.2-2: *Implicit Symmetric Q-R Step with Wilkinson Shift.* Given an unreduced symmetric tridiagonal matrix $T \in \mathbb{R}^{n \times n}$, the following algorithm overwrites T with $\bar{Z}^T T \bar{Z}$, where $\bar{Z} = J_1 \ldots J_{n-1}$ is a product of Givens rotations with the property that $\bar{Z}^T (T - \mu I)$ is upper triangular and μ is that eigenvalue of T's trailing 2-by-2 principal submatrix closer to t_{nn}.

$$
\begin{aligned}
d &:= (t_{n-1,n-1} - t_{nn})/2 \\
\mu &:= t_{nn} - t_{n,n-1}^2/[d + \text{sign}(d) \sqrt{d^2 + t_{n,n-1}^2}] \\
x &:= t_{11} - \mu \\
z &:= t_{21}
\end{aligned}
$$

For $k = 1, \ldots, n - 1$
    Determine $c = \cos(\theta)$ and $s = \sin(\theta)$ such that

$$\begin{bmatrix} c & -s \\ s & c \end{bmatrix} \begin{bmatrix} x \\ z \end{bmatrix} = \begin{bmatrix} * \\ 0 \end{bmatrix}$$

$T := J_k^T T J_k, J_k = J(k, k + 1, \theta)$
If $k < n - 1$
    then
        $x := t_{k+1,k}$
        $z := t_{k+2,k}$

This algorithm requires about $14n$ flops and n square roots. If a given orthogonal matrix Q is overwritten with $QJ_1 \ldots J_{n-1}$, then an additional $4\,n^2$ flops are needed. Of course, in any practical implementation the tridiagonal matrix T would be stored in a pair of n-vectors and not in an n-by-n array.

EXAMPLE 8.2-2. If Algorithm 8.2-2 is applied to

$$T = \begin{bmatrix} 1 & 1 & 0 & 0 \\ 1 & 2 & 1 & 0 \\ 0 & 1 & 3 & .01 \\ 0 & 0 & .01 & 4 \end{bmatrix}$$

then the new tridiagonal matrix $\bar{T}$ is given by

$$\bar{T} = \begin{bmatrix} .5000 & .5916 & 0 & 0 \\ .5916 & 1.785 & .1808 & 0 \\ 0 & .1808 & 3.7140 & .0000044 \\ 0 & 0 & .0000044 & 4.002497 \end{bmatrix}.$$

Algorithm 8.2-2 forms the basis of the symmetric QR algorithm—the standard means for computing the Schur decomposition of a dense symmetric matrix.

ALGORITHM 8.2-3: *Symmetric QR Algorithm.* Given an n-by-n symmetric matrix A and ϵ, a small multiple of the unit roundoff, the following algorithm overwrites A with $Q^T A Q = D + E$ where Q is orthogonal, D is diagonal, and E satisfies $\|E\|_2 \cong \mathbf{u}\, \|A\|_2$.

Using Algorithm 8.2-1, compute the tridiagonalization

$$A := (P_1 \cdots P_{n-2})^T A (P_1 \cdots P_{n-2}) = T.$$

Repeat

Set $a_{i,i+1}$ and $a_{i+1,i}$ to zero if

$$|a_{i+1,i}| = |a_{i,i+1}| \leqslant \epsilon (|a_{ii}| + |a_{i+1,i+1}|)$$

for any $i = 1, \ldots, n - 1$.

Find the largest q and the smallest p such that if

$$A = \begin{bmatrix} A_{11} & 0 & 0 \\ 0 & A_{22} & 0 \\ 0 & 0 & A_{33} \end{bmatrix} \begin{matrix} p \\ n - p - q \\ q \end{matrix}$$

then A_{33} is diagonal and A_{22} has no zero subdiagonal elements.
If $q = n$ *then* quit.
Apply Algorithm 8.2-2 to A_{22},

$$A := \text{diag}(I_p, \bar{Z}, I_q)^T A \ \text{diag}(I_p, \bar{Z}, I_q)$$

Go to *Repeat*.

This algorithm requires about $\frac{2}{3}n^3$ flops if Q is not accumulated and about
$5 n^3$ flops if Q is accumulated.

EXAMPLE 8.2-3. Suppose Algorithm 8.2-3 is applied to the tridiagonal matrix

$$A = \begin{bmatrix} 1 & 2 & 0 & 0 \\ 2 & 3 & 4 & 0 \\ 0 & 4 & 5 & 6 \\ 0 & 0 & 6 & 7 \end{bmatrix}.$$

The subdiagonal entries change as indicated in Table 8.2-1 during the QR iteration. Upon completion we find $\lambda(A) \cong \{-2.4848, .7046, 4.9366, 12.831\}$.

As in the unsymmetric QR algorithm, the computed eigenvalues $\hat{\lambda}_1, \ldots, \hat{\lambda}_n$ obtained via Algorithm 8.2-3 are the exact eigenvalues of a matrix that is near to A:

$$Q_0^T(A + E)Q_0 = \text{diag}(\hat{\lambda}_i) \qquad Q_0^T Q_0 = I, \|E\|_2 \cong \mathbf{u} \|A\|_2$$

However, unlike the unsymmetric case, we can claim (via Corollary 8.1-3) that the absolute error in each $\hat{\lambda}_i$ is small:

$$|\hat{\lambda}_i - \lambda_i| \cong \mathbf{u} \|A\|_2.$$

Table 8.2-1.

Iteration	a_{21}	a_{32}	a_{43}
1	1.6817	3.2344	.8649
2	1.6142	2.5755	.0006
3	1.6245	1.6965	10^{-13}
4	1.6245	1.6965	converg.
5	1.5117	.0150	
6	1.1195	10^{-9}	
7	.7071	converg.	
8	converg.		

If $\hat{Q} = [\hat{q}_1, \ldots, \hat{q}_n]$ is the computed matrix of orthonormal eigenvectors, then the accuracy of $\hat{q}_i$ depends on the separation of λ_i from the remainder of the spectrum. See Theorem 8.1-6.

If all of the eigenvalues and a few of the eigenvectors are desired, then it is advisable not to accumulate Q in Algorithm 8.2-3. Instead, the desired eigenvectors can be found via inverse iteration with T. If only a few eigenvalues and eigenvectors are required, then some of the special techniques in §8.5 are appropriate.

Problems

P8.2-1. Let $A = \begin{bmatrix} w & x \\ x & z \end{bmatrix}$ be real and suppose we perform the following shifted QR step:

$$A - zI = UR$$
$$\bar{A} = RU + zI$$

Show that if $\bar{A} = \begin{bmatrix} \bar{w} & \bar{x} \\ \bar{x} & \bar{z} \end{bmatrix}$ then

$$\bar{w} = w + x^2(w - z)/[(w - z)^2 + x^2]$$
$$\bar{z} = z - x^2(w - z)/[(w - z)^2 + x^2]$$
$$\bar{x} = -x^3/[(w - z)^2 + x^2].$$

P8.2-2. Suppose $A \in \mathbb{R}^{n \times n}$ is symmetric and positive definite and consider the following iteration:

$$
\begin{aligned}
&A_0 = A \\
&\text{For } k = 1, 2, \ldots \\
&\quad A_{k-1} = L_k L_k^T \quad \text{(Cholesky)} \\
&\quad A_k = L_k^T L_k
\end{aligned}
$$

(a) Show that this iteration is defined. (b) show that if

$$A = \begin{bmatrix} a & b \\ b & c \end{bmatrix} \qquad a \geq c$$

has eigenvalues $\lambda_1 \geq \lambda_2 \geq 0$, then the A_k converge to $\text{diag}(\lambda_1, \lambda_2)$.

P8.2-3. Suppose $A \in \mathbb{R}^{n \times n}$ is skew symmetric ($A^T = -A$). Show how to construct Householder matrices $P_1, \ldots, P_{n-2}$ such that $(P_1 \cdots P_{n-2})^T A (P_1 \cdots P_{n-2})$ is tridiagonal. How many flops are required by your algorithm?

P8.2-4. Suppose $A \in \mathbb{C}^{n \times n}$ is Hermitian ($A^H = A$). Show how to construct Householder matrices $P_1, \ldots, P_{n-2} \in \mathbb{C}^{n \times n}$ such that

$$(P_1 \cdots P_{n-2})^H A (P_1 \cdots P_{n-2})$$

is real symmetric tridiagonal.

P8.2-5. Show that if $A = B + iC$ is Hermitian, then $M = \begin{bmatrix} B & -C \\ C & B \end{bmatrix}$ is symmetric. Relate the eigenvalues and eigenvectors of A and M.

P8.2-6. Rewrite Algorithm 8.2-2 for the case when A is stored in two n-vectors. Justify the claim that only $14n$ flops and n square roots are required.

Notes and References for Sec. 8.2

The tridiagonalization of a symmetric matrix is discussed in

R. S. Martin and J. H. Wilkinson (1968a). "Householder's Tridiagonalization of a Symmetric Matrix," *Numer. Math. 11*, 181–95. See also HACLA, pp. 212–26.

H. R. Schwartz (1968). "Tridiagonalization of a Symmetric Band Matrix," *Numer. Math 12*, 231–41. See also HACLA, pp. 273–83.

N. E. Gibbs and W. G. Poole, Jr. (1974). "Tridiagonalization by Permutations," *Comm. Assoc. Comp. Mach. 17*, 20–24.

The first two references contain ALGOL programs.

ALGOL procedures for the explicit and implicit tridiagonal QR algorithm are given in

H. Bowdler, R. S. Martin, C. Reinsch, and J. H. Wilkinson (1968). "The QR and QL Algorithms for Symmetric Matrices," *Numer. Math. 11*, 293–306. See also HACLA, pp. 227–40.

A. Dubrulle, R. S. Martin, and J. H. Wilkinson (1968). "The Implicit QL Algorithm," *Numer. Math. 12*, 377–83. See also HACLA, pp. 241–48.

The "QL" algorithm is identical to the QR algorithm except that at each step the matrix $T - \lambda I$ is factored into a product of an orthogonal matrix and a lower triangular matrix. Other papers concerned with these methods include

G. W. Stewart (1970). "Incorporating Origin Shifts into the QR Algorithm for Symmetric Tridiagonal Matrices," *Comm. Assoc. Comp. Mach. 13*, 365–67.

A. Dubrulle (1970). "A Short Note on the Implicit QL Algorithm for Symmetric Tridiagonal Matrices," *Numer. Math. 15*, 450.

Extensions to Hermitian and skew-symmetric matrices are described in

D. Mueller (1966). "Householder's Method for Complex Matrices and Hermitian Matrices," *Numer. Math. 8*, 72–92.

R. C. Ward and L. J. Gray (1978). "Eigensystem Computation for Skew-Symmetric and a Class of Symmetric Matrices," *ACM Trans. Math. Soft. 4*, 278-85.

The convergence properties of Algorithm 8.2-3 are detailed in Lawson and Hanson (SLS, app. B), as well as in

J. H. Wilkinson (1968b). "Global Convergence of Tridiagonal QR Algorithm With Origin Shifts," *Lin. Alg. & Its Applic. 1*, 409-20.
W. Hoffman and B. N. Parlett (1978). "A New Proof of Global Convergence for the Tridiagonal QL Algorithm," *SIAM J. Num. Anal. 15*, 929-37.
T. J. Dekker and J. F. Traub (1971). "The Shifted QR Algorithm for Hermitian Matrices," *Lin. Alg. & Its Applic. 4*, 137-54.

For an analysis of the method when it is applied to normal matrices see

C. P. Huang (1981). "On the Convergence of the QR Algorithm with Origin Shifts for Normal Matrices," *IMA J. Num. Anal. 1*, 127-33.

Interesting papers concerned with shifting in the tridiagonal QR algorithm include

F. L. Bauer and C. Reinsch (1968). "Rational QR Transformation with Newton Shift for Symmetric Tridiagonal Matrices," *Numer. Math. 11*, 264-72. See also HACLA, pp. 257-65.
G. W. Stewart (1970). "Incorporating Origin Shifts into the QR Algorithm for Symmetric Tridiagonal Matrices," *Comm. Assoc. Comp. Mach. 13*, 365-67.

Sec. 8.3. Once Again: The Singular Value Decomposition

There are important relationships between the singular value decomposition of a matrix A and the Schur decompositions of the symmetric matrices $A^T A$, $A A^T$, and $\begin{bmatrix} 0 & A^T \\ A & 0 \end{bmatrix}$. Indeed, if

$$U^T A V = \text{diag}(\sigma_1, \ldots, \sigma_n)$$

is the SVD of $A \in \mathbb{R}^{m \times n}$ ($m \geq n$) then

(8.3-1) $$V^T(A^T A)V = \text{diag}(\sigma_1^2, \ldots, \sigma_n^2) \in \mathbb{R}^{n \times n}$$

and

(8.3-2) $$U^T(A A^T)U = \text{diag}(\sigma_1^2, \ldots, \sigma_n^2, 0, \ldots, 0) \in \mathbb{R}^{m \times m}.$$

Moreover, if $U = [\underset{n}{U_1}, \underset{m-n}{U_2}]$ and we define the orthogonal matrix Q by

$$Q = \frac{1}{\sqrt{2}} \begin{bmatrix} V & V & 0 \\ U_1 & -U_1 & \sqrt{2}U_2 \end{bmatrix},$$

then

$$(8.3-3) \quad Q^T \begin{bmatrix} 0 & A^T \\ A & 0 \end{bmatrix} Q = \text{diag}(\sigma_1, \ldots, \sigma_n, -\sigma_1, \ldots, -\sigma_n, 0, \ldots, 0).$$

These connections to the symmetric eigenproblem allow us to adapt the mathematical and algorithmic developments of the previous two sections to the singular value problem.

We first establish some perturbation results for the SVD based on the theorems of §8.1. Recall that $\sigma_i(A)$ denotes the i-th largest singular value of A.

Theorem 8.3-1. If $A \in \mathbb{R}^{m \times n}$ then for $k = 1, \ldots, \min\{m, n\}$

$$\sigma_k(A) = \max_{\substack{\dim(S)=k \\ \dim(T)=k}} \min_{\substack{x \in S \\ y \in T}} \frac{y^T A x}{\|x\|_2 \|y\|_2} = \max_{\dim(S)=k} \min_{x \in S} \frac{\|Ax\|_2}{\|x\|_2}.$$

It is clear in this expression that $S \subset \mathbb{R}^n$ and $T \subset \mathbb{R}^m$ are subspaces.

Proof. The right-most characterization follows by applying Theorem 8.1-2 to $A^T A$. The remainder of the proof we leave as an exercise. $\square$

By applying Corollary 8.1-3 to $\begin{bmatrix} 0 & A^T \\ A & 0 \end{bmatrix}$ and $\begin{bmatrix} 0 & (A+E)^T \\ A+E & 0 \end{bmatrix}$ and

Corollary 8.1-4 to $A^T A$ we obtain

Corollary 8.3-2. If A and $A + E$ are in $\mathbb{R}^{m \times n}$ ($m \geqslant n$), then for $k = 1, 2, \ldots, n$

$$|\sigma_k(A + E) - \sigma_k(A)| \leqslant \sigma_1(E) = \|E\|_2.$$

Corollary 8.3-3. Let $A = [a_1, \ldots, a_n]$ be a column partitioning of $A \in \mathbb{R}^{m \times n}$ ($m \geqslant n$). If $A_r = [a_1, \ldots, a_r]$ then for $r = 1, \ldots, n-1$

$$\sigma_1(A_{r+1}) \geqslant \sigma_1(A_r) \geqslant \sigma_2(A_{r+1}) \geqslant \cdots \geqslant \sigma_r(A_{r+1}) \geqslant \sigma_r(A_r) \geqslant \sigma_{r+1}(A_{r+1}).$$

This last result says that by adding a column to a matrix, the largest singular value increases and the smallest singular value is diminished.

Example 8.3-1. If

$$A = \begin{bmatrix} 1 & 4 \\ 2 & 5 \\ 3 & 6 \end{bmatrix} \quad \text{and} \quad A + E = \begin{bmatrix} 1 & 4 \\ 2 & 5 \\ 3 & 6.01 \end{bmatrix},$$

then $\sigma(A) \cong \{9.5080, .7729\}$ and $\sigma(A + E) \cong \{9.5145, .7706\}$. It is clear that $|\sigma_i(A + E) - \sigma_i(A)| \leqslant \|E\|_2 = .01$ for $i = 1, 2$.

EXAMPLE 8.3-2.

$$A = \begin{bmatrix} 1 & 6 & 11 \\ 2 & 7 & 12 \\ 3 & 8 & 13 \\ 4 & 9 & 14 \\ 5 & 10 & 15 \end{bmatrix} \implies \begin{array}{l} \sigma(A_1) = \{7.4162\} \\ \sigma(A_2) = \{19.5377, \, 1.8095\} \\ \sigma(A_3) = \{35.1272, \, 2.4654, \, 0.0000\}. \end{array}$$

The next result amounts to a Wielandt-Hoffman theorem for singular values:

THEOREM 8.3-5.   If A and $A + E$ are in $\mathbb{R}^{m \times n}$ $(m \geq n)$, then

$$\sum_{k=1}^{n} [\sigma_k(A + E) - \sigma_k(A)]^2 \leq \|E\|_F^2.$$

Proof.   Apply Theorem 8.1-5 to $\begin{bmatrix} 0 & A^T \\ A & 0 \end{bmatrix}$ and $\begin{bmatrix} 0 & (A + E)^T \\ A + E & 0 \end{bmatrix}$. $\square$

EXAMPLE 8.3-3.   If

$$A = \begin{bmatrix} 1 & 4 \\ 2 & 5 \\ 3 & 6 \end{bmatrix} \quad \text{and} \quad A + E = \begin{bmatrix} 1 & 4 \\ 2 & 5 \\ 3 & 6.01 \end{bmatrix},$$

then $\displaystyle\sum_{k=1}^{2} |\sigma_k(A + E) - \sigma_k(A)|^2 \cong .472 \times 10^{-4} \leq 10^{-4} = \|E\|_F^2$.
See Example 8.3-1.

For $A \in \mathbb{R}^{m \times n}$ we say that the k-dimensional subspaces $S \subset \mathbb{R}^n$ and $T \subset \mathbb{R}^m$ form a *singular subspace pair* if $x \in S$ and $y \in T$ imply $Ax \in T$ and $A^T y \in S$. The following result is concerned with the perturbation of singular subspace pairs.

THEOREM 8.3-6.   Let $A, E \in \mathbb{R}^{m \times n}$ $(m \geq n)$ be given and suppose that

$$V = \underset{k \quad n-k}{[V_1, \, V_2]} \in \mathbb{R}^{n \times n} \qquad U = \underset{k \quad m-k}{[U_1, \, U_2]} \in \mathbb{R}^{m \times m}$$

are orthogonal with the property that $R(V_1)$ and $R(U_1)$ form a singular subspace pair for A. Let

$$U^H A V = \begin{bmatrix} A_{11} & 0 \\ 0 & A_{22} \end{bmatrix} \begin{matrix} k \\ m-k \end{matrix} \qquad U^H E V = \begin{bmatrix} E_{11} & E_{12} \\ E_{21} & E_{22} \end{bmatrix} \begin{matrix} k \\ m-k \end{matrix},$$
$$\begin{matrix} k & n-k \end{matrix} \qquad\qquad\qquad \begin{matrix} k & n-k \end{matrix}$$

and define

$$\epsilon = \| [E_{21}, E_{12}^T] \|_F$$

and

$$\delta = \min_{\substack{\sigma \in \sigma(A_{11}) \\ \gamma \in \sigma(A_{22})}} |\sigma - \gamma| - \|E_{11}\|_2 - \|E_{22}\|_2.$$

If $\epsilon / \delta \leq \frac{1}{2}$, then there exist matrices $P \in \mathbb{R}^{(n-k) \times k}$ and $Q \in \mathbb{R}^{(m-k) \times k}$ satisfying

$$\left\| \begin{pmatrix} Q \\ P \end{pmatrix} \right\|_F \leq 2 \frac{\epsilon}{\delta}$$

such that $R(V_1 + V_2 Q)$ and $R(U_1 + U_2 P)$ are a pair of singular subspaces for $A + E$.

Proof.   See Stewart (1973), Theorem 6.4. $\square$

Roughly speaking, the theorem says that ϵ changes in A can alter a singular subspace by an amount ϵ / δ, where δ measures the isolation of the relevant singular values.

EXAMPLE 8.3-4.   The matrix $A = \mathrm{diag}(2.000, 1.001, .999) \in \mathbb{R}^{4 \times 3}$ has singular subspace pairs $\langle \mathrm{span}\{v_i\}, \mathrm{span}\{u_i\} \rangle$ for $i = 1, 2, 3$ where $v_i = e_i^{(3)}$ and $u_i = e_i^{(4)}$. Suppose

$$A + E = \begin{bmatrix} 2.000 & .01 & .01 \\ .01 & 1.001 & .01 \\ .01 & .01 & .999 \\ .01 & .01 & .01 \end{bmatrix}.$$

The corresponding columns of the matrices

$$[\hat{u}_1, \hat{u}_2, \hat{u}_3] = \begin{bmatrix} .9999 & -.0144 & .0007 \\ .0101 & .7415 & .6708 \\ .0101 & .6707 & -.7616 \\ .0051 & .0138 & -.0007 \end{bmatrix}$$

$$[\hat{v}_1, \hat{v}_2, \hat{v}_3] = \begin{bmatrix} .9999 & -.0143 & .0007 \\ .0101 & .7416 & .6708 \\ .0101 & .6707 & -.7416 \end{bmatrix}$$

define singular subspace pairs for $A + E$. Note that $\langle \text{span}\{\hat{v}_i\}, \text{span}\{\hat{u}_i\}\rangle$ is close to $\langle \text{span}\{v_i\}, \text{span}\{u_i\}\langle$ for $i = 1$ but not for $i = 2$ or 3. On the other hand, the singular subspace pair $\langle \text{span}\{\hat{v}_2, \hat{v}_3\}, \text{span}\{\hat{u}_2, \hat{u}_3\}\rangle$ is close to $\langle \text{span}\{v_2, v_3\}, \text{span}\{u_2, u_3\}\rangle$.

We now show how a variant of the QR algorithm can be used to compute the SVD of a matrix. At first glance, this appears straightforward. Equation (8.3-1) suggests that we

(a) form $C = A^T A$;
(b) use the symmetric QR algorithm to compute $V_1^T C V_1 = \text{diag}(\sigma_i^2)$;
(c) use QR with column pivoting to upper triangularize $B = AV_1$:

$$U^T(AV_1)\Pi = R.$$

Since R has orthogonal columns, it follows that $U^T A(V_1 \Pi)$ is diagonal. However, as we saw in §6.1, the formation of $A^T A$ can lead to a loss of information. The situation is not quite so bad here, since the original A is used in (c) to compute U.

A preferable method for computing the SVD is described in Golub and Kahan (1965). Their technique finds U and V simultaneously by *implicitly* applying the symmetric QR algorithm to $A^T A$. The first step is to reduce A to upper bidiagonal form using Algorithm 6.4-2:

$$U_B^T A V_B = \begin{bmatrix} B \\ -- \\ 0 \end{bmatrix} = \begin{bmatrix} d_1 & f_2 & & & & \mathbf{0} \\ & d_2 & f_3 & & & \\ & & \ddots & \ddots & & \\ & & & \ddots & f_n & \\ \mathbf{0} & & & & d_n \\ \hline & & & \mathbf{0} & & \end{bmatrix}.$$

The remaining problem is thus to compute the SVD of B. To this end, consider applying an implicit-shift QR step (Algorithm 8.2-2) to the tridiagonal matrix $T = B^T B$:

(a) Compute the eigenvalue λ of

$$\begin{bmatrix} d_m^2 + f_m^2 & d_m f_n \\ d_m f_n & d_n^2 + f_n^2 \end{bmatrix} \qquad m = n - 1$$

that is closer to $d_n^2 + f_n^2$.

(b) Compute $c_1 = \cos(\theta_1)$ and $s_1 = \sin(\theta_1)$ such that

$$\begin{bmatrix} c_1 & -s_1 \\ s_1 & c_1 \end{bmatrix} \begin{bmatrix} d_1^2 - \lambda \\ d_1 f_2 \end{bmatrix} = \begin{bmatrix} * \\ 0 \end{bmatrix}$$

and set $J_1 = J(1, 2, \theta_1)$.

(c) Compute Givens rotations $J_2, \ldots, J_{n-1}$ such that if $Q = J_1 \ldots J_{n-1}$ then $Q^T T Q$ is tridiagonal and $Q e_1 = J_1 e_1$.

Note that these calculations require the explicit formation of $B^T B$, which, as we have seen, is unwise from the numerical standpoint.

Suppose instead that we apply the Givens rotation J_1 above to B directly. This gives

$$B := BJ_1 = \begin{bmatrix} x & x & & & & \\ + & x & x & & & \\ & & x & x & & \\ & & & x & x & \\ & & & & x & x \\ & & & & & x \end{bmatrix} \qquad n = 6$$

We then can determine Givens rotations $U_1, V_2, U_2, \ldots, V_{n-1}$, and U_{n-1} to chase the unwanted nonzero element down the bidiagonal:

$$B := U_1^T B = \begin{bmatrix} x & x & + & & & \\ & x & x & & & \\ & & x & x & & \\ & & & x & x & \\ & & & & x & x \\ & & & & & x \end{bmatrix} \qquad B := BV_2 = \begin{bmatrix} x & x & & & & \\ & x & x & & & \\ & + & x & x & & \\ & & & x & x & \\ & & & & x & x \\ & & & & & x \end{bmatrix}$$

$$B := U_2^T B = \begin{bmatrix} x & x & & & & \\ & x & x & + & & \\ & & x & x & & \\ & & & x & x & \\ & & & & x & x \\ & & & & & x \end{bmatrix} \qquad B := BV_3 = \begin{bmatrix} x & x & & & & \\ & x & x & & & \\ & & x & x & & \\ & & + & x & x & \\ & & & & x & x \\ & & & & & x \end{bmatrix}$$

etc. The process terminates with a new bidiagonal $\bar{B}$ that is related to B as follows:

$$\bar{B} = (U_{n-1}^{T} \cdots U_1^{T})B(J_1 V_2 \cdots V_{n-1}) = \bar{U}^{T}B\bar{V}.$$

Since each V_i has the form $V_i = J(i, i + 1, \theta_i)$, where $i = 2, \ldots, n - 1$, it follows that $Ve_1 = Qe_1$. By the implicit Q theorem we can assert that V and Q are essentially the same. Thus, we can implicitly effect the transition from T to $\bar{T} = \bar{B}^{T}\bar{B}$ by working directly on the bidiagonal matrix B.

Of course, for these claims to hold it is necessary that the underlying tridiagonal matrices be unreduced. Since the subdiagonal entries of $B^{T}B$ are of the form $d_{i-1}f_i$, it is clear that we must search the bidiagonal band for zeros. If $f_{k+1} = 0$ for some k, then

$$B = \begin{bmatrix} B_1 & 0 \\ 0 & B_2 \end{bmatrix} \begin{matrix} k \\ n-k \end{matrix}$$

and the original SVD problem decouples into two smaller problems involving the matrices B_1 and B_2. If $d_k = 0$ for some k, then premultiplication by a sequence of Givens transformations can zero f_{k+1}. For example, suppose

$$B = \begin{bmatrix} x & x & 0 & 0 & 0 & 0 \\ 0 & x & x & 0 & 0 & 0 \\ 0 & 0 & 0 & x & 0 & 0 \\ 0 & 0 & 0 & x & x & 0 \\ 0 & 0 & 0 & 0 & x & x \\ 0 & 0 & 0 & 0 & 0 & x \end{bmatrix}. \qquad n = 6, k = 3$$

By rotating in planes (3, 4), (3, 5), and (3, 6), row 3 can be zeroed:

$$\xrightarrow[\text{(3, 4)}]{} \begin{bmatrix} x & x & 0 & 0 & 0 & 0 \\ 0 & x & x & 0 & 0 & 0 \\ 0 & 0 & 0 & 0 & x & 0 \\ 0 & 0 & 0 & x & x & 0 \\ 0 & 0 & 0 & 0 & x & x \\ 0 & 0 & 0 & 0 & 0 & x \end{bmatrix} \xrightarrow[\text{(3, 5)}]{} \begin{bmatrix} x & x & 0 & 0 & 0 & 0 \\ 0 & x & x & 0 & 0 & 0 \\ 0 & 0 & 0 & 0 & 0 & x \\ 0 & 0 & 0 & x & x & 0 \\ 0 & 0 & 0 & 0 & x & x \\ 0 & 0 & 0 & 0 & 0 & x \end{bmatrix}$$

$$\begin{bmatrix} x & x & 0 & 0 & 0 & 0 \\ 0 & x & x & 0 & 0 & 0 \\ 0 & 0 & 0 & 0 & 0 & 0 \\ 0 & 0 & 0 & x & x & 0 \\ 0 & 0 & 0 & 0 & x & x \\ 0 & 0 & 0 & 0 & 0 & x \end{bmatrix}.$$

$(3, 6)$ $\longrightarrow$

Thus, we can decouple whenever $f_2 \cdots f_n = 0$ or $d_1 \cdots d_{n-1} = 0$.

ALGORITHM 8.3-1: *Golub-Kahan SVD Step.* Given a bidiagonal matrix $B \in \mathbb{R}^{n \times n}$ having no zeros on its diagonal or superdiagonal, the following algorithm overwrites B with the bidiagonal matrix $\bar{B} = \bar{U}^T B \bar{V}$ where $\bar{U}$ and $\bar{V}$ are orthogonal and $\bar{V}$ is essentially the orthogonal matrix that would be obtained by applying Algorithm 8.2-2 to $T = B^T B$.

Let μ be the eigenvalue of the trailing 2-by-2 submatrix of $T = B^T B$ that is closer to t_{nn}.
$y := t_{11} - \mu$
$z := t_{12}$
For $k = 1, \ldots, n - 1$
    Determine $c = \cos(\theta)$ and $s = \sin(\theta)$ such that

$$[y \ \ z] \begin{bmatrix} c & s \\ -s & c \end{bmatrix} = [* \ \ 0]$$

$B := BJ(k, k + 1, \theta)$
$y := b_{kk}$
$z := b_{k+1,k}$
Determine $c = \cos(\theta)$ and $s = \sin(\theta)$ such that

$$\begin{bmatrix} c & -s \\ s & c \end{bmatrix} \begin{bmatrix} y \\ z \end{bmatrix} = \begin{bmatrix} * \\ 0 \end{bmatrix}$$

$B := J(k, k + 1, \theta)^T B$
If $k < n - 1$
    then
        $y := b_{k,k+1}$
        $z := b_{k,k+2}$

An efficient implementation of this algorithm would store B's diagonal and superdiagonal in vectors $(a_1, \ldots, a_n)$ and $(f_2, \ldots, f_n)$ respectively and would require $20n$ flops and $2n$ square roots. Accumulating $\bar{U}$ requires $4mn$ flops. Accumulating $\bar{V}$ requires $4n^2$ flops.

Typically, after a few of the above SVD iterations, the superdiagonal entry f_n becomes negligible. Criteria for smallness within B's band are usually of the form

$$|f_i| \leqslant \epsilon (|d_{i-1}| + |d_i|)$$

$$|d_i| \leqslant \epsilon \|B\|$$

where ϵ is a small multiple of the unit roundoff and $\|\cdot\|$ is some computationally convenient norm.

Combining Algorithm 6.4-1 (bidiagonalization), Algorithm 8.3-1, and the decoupling calculations mentioned earlier gives:

ALGORITHM 8.3-2: *The SVD Algorithm.*   Given $A \in \mathbb{R}^{m \times n}$ $(m \geqslant n)$ and ϵ, a small multiple of the unit roundoff, the following algorithm overwrites A with $U^T A V = D + E$, where $U \in \mathbb{R}^{m \times n}$ is orthogonal, $V \in \mathbb{R}^{n \times n}$ is orthogonal, $D \in \mathbb{R}^{m \times n}$ is diagonal, and E satisfies $\|E\|_2 \cong \mathbf{u} \|A\|_2$.

Use Algorithm 6.4-1 to compute the bidiagonalization

$$A := (U_1 \ \ldots \ U_n)^T A (V_1 \ \ldots \ V_{n-2})$$

Repeat
    Set $a_{i,i+1}$ to zero if $|a_{i,i+1}| \leqslant \epsilon (|a_{ii}| + |a_{i+1,i+1}|)$ for any $i = 1,$
    $\ldots, n - 1$.
    Find the largest q and the smallest p such that if

$$A = \begin{bmatrix} A_{11} & 0 & 0 \\ 0 & A_{22} & 0 \\ 0 & 0 & A_{33} \\ 0 & 0 & 0 \end{bmatrix} \begin{matrix} p \\ n-p-q \\ q \\ m-n \end{matrix}$$

    then A_{33} is diagonal and A_{22} has a nonzero superdiagonal.
    If $q = n$ *then* quit.
    If any diagonal entry in A_{22} is zero, then zero the superdiagonal entry in the same row and go to *Repeat*.
    Apply Algorithm 8.3-1 to A_{22},

$$A := \mathrm{diag}(I_p, \bar{U}, I_{q+m-n})^T A \, \mathrm{diag}(I_p, \bar{V}, I_q)$$

    Go to *Repeat*.

The amount of work required by this algorithm and its numerical properties are discussed in §6.5.

EXAMPLE 8.3-5.   If Algorithm 8.3-2 is applied to

$$A = \begin{bmatrix} 1 & 1 & 0 & 0 \\ 0 & 2 & 1 & 0 \\ 0 & 0 & 3 & 1 \\ 0 & 0 & 0 & 4 \end{bmatrix}$$

then the superdiagonal elements converge to zero as Table 8.3-1 indicates.

Table 8.3-1.

Iteration	$O(a_{12})$	$O(a_{23})$	$O(a_{34})$
1	10^0	10^0	10^0
2	10^0	10^0	10^0
3	10^0	10^0	10^0
4	10^0	10^{-1}	10^{-2}
5	10^0	10^{-1}	10^{-8}
6	10^0	10^{-1}	10^{-27}
7	10^0	10^{-1}	converg.
8	10^0	10^{-4}	
9	10^{-1}	10^{-14}	
10	10^{-1}	converg.	
11	10^{-4}		
12	10^{-12}		
13	converg.		

Problems

P8.3-1. Show that if $B \in \mathbb{R}^{n \times n}$ is an upper bidiagonal matrix having a re-peated singular value, then B must have a zero either on its diagonal or superdiagonal.

P8.3-2. Give formulae for the eigenvectors of $\begin{bmatrix} 0 & A^T \\ A & 0 \end{bmatrix}$ in terms of the singular vectors of $A \in \mathbb{R}^{m \times n}$, where $m \geqslant n$.

P8.3-3. Give an algorithm for reducing a complex matrix A to *real* bi-diagonal form using complex Householder transformations.

P8.3-4. Relate the singular values and vectors of $A = B + iC$ $(B, C \in \mathbb{R}^{m \times n})$ to those of $\begin{bmatrix} B & -C \\ C & B \end{bmatrix}$.

P8.3-5. Justify the flop count for Algorithm 8.3-1.

Notes and References for Sec. 8.3

The SVD and its mathematical properties are discussed in

A. R. Amir-Moez (1965). *Extremal Properties of Linear Transformations and Geometry of Unitary Spaces*, Texas Tech University Mathematics Series, no. 243, Lubbock, Tex.

P. A. Wedin (1972). "Perturbation Bounds in Connection with the Singular Value Decomposition," *BIT 12*, 99-111.

G. W. Stewart (1973b). "Error and Perturbation Bounds for Subspaces Associated with Certain Eigenvalue Problems," *SIAM Review 15*, 727-64.

The last two references are expressly concerned with perturbation properties of the decomposition. Other papers dealing with singular value sensitivity include

A. Ruhe (1975). "On the Closeness of Eigenvalues and Singular Values for Almost Normal Matrices," *Lin. Alg. & Its Applic. 11*, 87-94.

G. W. Stewart (1979). "A Note on the Perturbation of Singular Values," *Lin. Alg. & Its Applic. 28*, 213-16.

The idea of adapting the symmetric QR algorithm to compute the SVD first appeared in

G. H. Golub and W. Kahan (1965). "Calculating the Singular Values and Pseudo-Inverse of a Matrix," *SIAM J. Num. Anal. Ser. B 2*, 205-24,

and the first working program in

P. A. Businger and G. H. Golub (1969). "Algorithm 358: Singular Value Decomposition of a Complex Matrix," *Comm. Assoc. Comp. Mach. 12*, 564-65.

The ALGOL procedure in the following article is the basis for the EISPACK 2 and LINPACK SVD subroutines:

G. H. Golub and C. Reinsch (1970). "Singular Value Decomposition and Least Squares Solutions," *Numer. Math. 14*, 403-20. See also HACLA, pp. 134-51.

Other FORTRAN SVD programs are given in Lawson and Hanson (SLS, chap. 9) and in

G. E. Forsythe, M. Malcolm and C. B. Moler (1977). *Computer Methods for Mathematical Computations*, Prentice-Hall, Englewood Cliffs.

Sec. 8.4. Jacobi Methods

Jacobi (1846) proposed a method for reducing a symmetric matrix $A \in \mathbb{R}^{n \times n}$ to diagonal form using what we have been calling Givens rotations. Although his method has essentially been eclipsed by the symmetric QR algorithm, it is important to understand because of its significant role in parallel computation. Jacobi's method is also useful for finding the eigenvalues of nearly diagonal symmetric matrices.

In honor of the method's inventor, we will refer to "Givens rotations" as "Jacobi rotations" throughout this section.

The idea behind Jacobi's method is to systematically reduce the quantity

$$\text{off}(A) = \sum_{i=1}^{n} \sum_{\substack{j=1 \\ j \neq i}}^{n} a_{ij}^2,$$

i.e., the "norm" of the off-diagonal elements. In a given step, the object is to determine a Jacobi rotation $J(p, q, \theta)$ such that off(B) is minimized where

$$B = J^T AJ, \qquad\qquad J = J(p, q, \theta).$$

The matrix A is then updated, $A := J^T AJ$, and the process repeated.

To see how the rotation J can be determined, observe that the matrix B above agrees with A except in rows and columns p and q. Moreover,

(8.4-1)
$$\begin{bmatrix} b_{pp} & b_{pq} \\ b_{qp} & b_{qq} \end{bmatrix} = \begin{bmatrix} c & s \\ -s & c \end{bmatrix}^T \begin{bmatrix} a_{pp} & a_{pq} \\ a_{qp} & a_{qq} \end{bmatrix} \begin{bmatrix} c & s \\ -s & c \end{bmatrix}$$

and consequently,

$$b_{pp}^2 + b_{qq}^2 + 2b_{pq}^2 = a_{pp}^2 + a_{qq}^2 + 2a_{pq}^2.$$

Since $\|B\|_F^2 = \|A\|_F^2$ it follows that

$$\text{off}(B) = \|B\|_F^2 - \sum_{i=1}^{n} b_{ii}^2 = \|A\|_F^2 - \sum_{\substack{i=1 \\ i \neq p, q}}^{n} a_{ii}^2 - (b_{pp}^2 + b_{qq}^2)$$

$$= \text{off}(A) - 2a_{pq}^2 + 2b_{pq}^2.$$

Thus, for a given index pair (p,q), off(B) is minimized by determining $J(p, q, \theta)$ such that $b_{pq} = b_{qp} = 0$.

Formulae for computing $c = \cos(\theta)$ and $s = \sin(\theta)$ can be derived from (8.4-1). In particular we have the requirement that

$$b_{pq} = a_{pq}(c^2 - s^2) + (a_{pp} - a_{qq})cs = 0$$

and therefore

(8.4-2)
$$\tau = \text{ctn}(2\theta) = \frac{a_{qq} - a_{pp}}{2a_{pq}}. \qquad\qquad a_{pq} \neq 0$$

(If $a_{pq} = 0$, then just set $c = 1$ and $s = 0$.) Using more trigonometry we find that $t = \tan(\theta)$ satisfies

$$t^2 + 2\tau t - 1 = 0$$

whereupon

(8.4-3)
$$c = (1 + t^2)^{-1/2}$$
$$s = tc$$

It is customary to choose t to be the smaller of the two roots to the above quadratic:

(8.4-4)
$$t = \frac{\text{sign}(\tau)}{|\tau| + \sqrt{1 + \tau^2}}$$

This ensures that $|\theta| \leqslant \pi/4$ and has the effect of minimizing the difference between B and A because

(8.4-5)
$$\|B - A\|_F^2 = 4(1 - c) \sum_{\substack{i=1 \\ i \neq p, q}}^{n} (a_{ip}^2 + a_{iq}^2) + 2a_{pq}^2/c^2.$$

These computations for zeroing a given off-diagonal entry can be repeated, making A ever closer to diagonal form. To have a complete algorithm, however, we need a method for determining the rotation index pair (p, q). Noting that off(A) is reduced by an amount $2a_{pq}^2$, it is altogether reasonable to select these indices so that a_{pq} is the largest off-diagonal entry in modulus. This defines the *classical Jacobi iteration*:

$A^{(0)} = A$

For $k = 0, 1, \ldots$

    Let $|a_{pq}^{(k)}| = \max_{i<j} a_{ij}^{(k)}$

    Determine $c_k = \cos(\theta_k)$ and $s_k = \sin(\theta_k)$ via formulae (8.4-2), (8.4-4), and (8.4-3).

    $A^{(k+1)} = J_k^T A^{(k)} J_k, \quad J_k = J(p, q, \theta_k)$

Since

$$2[a_{pq}^{(k)}]^2 \geqslant \text{off}(A^{(k)})/N,$$

where $N = n(n - 1)/2$, it follows that

$$\text{off}(A^{(k+1)}) \leqslant \left(1 - \frac{1}{N}\right) \text{off}(A^{(k)}).$$

By induction we have

$$\text{off}(A^{(k)}) \leqslant \left(1 - \frac{1}{N}\right)^k \text{off}(A^{(0)}),$$

thereby proving that the classical Jacobi iteration converges.

EXAMPLE 8.4-1. Applying the classical Jacobi iteration to

$$A = \begin{bmatrix} 1 & 1 & 1 & 1 \\ 1 & 2 & 3 & 4 \\ 1 & 3 & 6 & 10 \\ 1 & 4 & 10 & 20 \end{bmatrix},$$

we find

k	$O[\text{off}(A^{(k)})]$
0	10^2
N	10^1
$2N$	10^{-2}
$3N$	10^{-11}
$4N$	10^{-17}

These results highlight the fact that for small dimensions, the Jacobi method is very effective.

The example suggests that the asymptotic convergence rate of the method is considerably better than linear. This turns out to be the case. Schonage (1964) and van Kempen (1966) show that for k large enough, there is a constant c such that

$$\text{off}(A^{(k+N)}) \leq c \, \text{off}(A^{(k)})^2,$$

i.e., quadratic convergence. An earlier paper by Henrici (1958) established the same result for the special case when A has distinct eigenvalues.

In the convergence theory for the Jacobi iteration, it is critical that $|\theta| \leq \pi/4$. Among other things this precludes the possibility of "interchanging" nearly converged diagonal entries. This follows from the formulae $b_{pp} = a_{pp} - ta_{pq}$ and $b_{qq} = a_{qq} + ta_{pq}$, which can be derived from (8.4-1).

If symmetry is exploited, then the Jacobi update $A := J^T A J$ requires $4n$ flops. Indeed, if $J = J(p, q, \theta), c = \cos(\theta), s = \sin(\theta)$, and $t = s/c$, then the update is calculated as follows:

$$a_{pp} := a_{pp} - ta_{pq}$$
$$a_{qq} := a_{qq} + ta_{pq}$$

$$(a_{jp}, a_{jq}) := (a_{jp}, a_{jq}) \begin{bmatrix} c & s \\ -s & c \end{bmatrix} \qquad j = 1, \ldots, p-1$$

$$(a_{pj}, a_{jq}) := (a_{pj}, a_{jq}) \begin{bmatrix} c & s \\ -s & c \end{bmatrix} \qquad j = p+1, \ldots, q-1$$

$$(a_{pj}, a_{qj}) := (a_{pj}, a_{qj}) \begin{bmatrix} c & s \\ -s & c \end{bmatrix} \qquad j = q+1, \ldots, n$$

(Recall that t is defined if $a_{pq} \neq 0$.) However, for each update $O(n^2)$ comparisons are required in order to locate the largest off-diagonal element. Thus, much more time is spent searching than updating.

To reduce the excessive amount of searching, one can choose the rotation pairs cyclically, e.g.,

$$(p, q) = (1, 2), (1, 3), \ldots, (1, n), (2, 3), \ldots,$$
$$(2, n), \ldots, (n - 1, n), (1, 2), (1, 3), \ldots$$

The resulting method is an example of a *cyclic Jacobi method*, and it, too, has quadratic convergence. (See Wilkinson (1962) and van Kempen (1966).) However, since it does not require off-diagonal search, it is considerably faster than Jacobi's original algorithm.

In the cyclic method, it is customary to refer to each set of $n(n - 1)/2$ rotations as a *sweep*. Thus, in a given sweep there is one rotation for each superdiagonal entry in the matrix. There are different ways to order the rotations within a sweep. The above row-wise scheme gives rise to the *serial Jacobi method*:

ALGORITHM 8.4-1: *Serial Jacobi.*   Given a symmetric matrix $A \in \mathbb{R}^{n \times n}$ and $\delta \geq \mathbf{u}$, the unit roundoff, the following algorithm overwrites A with $U^T A U = D + E$ where U is orthogonal, D is diagonal, and E has a zero diagonal and satisfies $\|E\|_F \leq \delta \|A\|_F$:

$\delta := \delta \|A\|_F$
Do Until off$(A) \leq \delta^2$
    For $p = 1, 2, \ldots, n - 1$
        For $q = p + 1, \ldots, n$
            Find $J = J(p, q, \theta)$ such that the (p, q) entry of $J^T A J$ is
            zero.
            $A := J^T A J$

This algorithm requires $2n^3$ flops per sweep. An additional $2n^3$ flops are required if the orthogonal matrix U is accumulated.

EXAMPLE 8.4-2.   If the serial Jacobi method is applied to the matrix in Example 8.4-1 we find

Sweep	$O[\text{off}(A)]$
0	10^2
1	10^1
2	10^{-1}
3	10^{-6}
4	10^{-16}

When implementing serial Jacobi, it is sensible to skip the annihilation of a_{pq} if its modulus is less than some small (sweep-dependent) parameter, because the net reduction in off(A) is not worth the cost. This leads to what is called the *threshold Jacobi method*. Details concerning this variant of Jacobi's algorithm may be found in Wilkinson (AEP, p. 277 *ff.*). Using Wilkinson's error analysis it is possible to show that if r sweeps are needed in Algorithm 8.4-1 then the computed $\hat{d}_i$ satisfy

$$\frac{\sqrt{\sum\limits_{i=1}^{n} (\hat{d}_i - \lambda_i)^2}}{\|A\|_F} \leqslant \delta + k_r \mathbf{u}$$

for some ordering of A's eigenvalues $\lambda_1, \ldots, \lambda_n$. The parameter k_r depends mildly on r.

Although the serial Jacobi method converges quadratically, it is not generally competitive with the symmetric QR algorithm. One sweep of Jacobi with accumulation requires as many flops as complete calculation of the Schur decomposition by the symmetric QR algorithm. There are, however, situations where the Jacobi iteration is attractive. For example, A might already be close to the diagonal form, as is the case when a good approximate eigensystem is known. In this situation, the QR algorithm loses its advantage over Jacobi.

Another reason why the Jacobi iteration is of interest is that it is readily adapted to parallel computation. A parallel computer has numerous central processing units which run concurrently. A given computational task, such as a sweep, can be shared among the various CPUs thereby reducing the overall computation time. We look briefly into this interesting and important development.

For concreteness, suppose that A is 8-by-8 and that we have a parallel computer with 4 CPUs. In order to distribute the workload among these CPUs in the course of performing a sweep, we partition the set of 28 rotations associated with a single sweep into 7 *rotation sets*. The following is but one of the several ways that this can be done:

Rotation Set Number	Member Rotations
1	(1, 2) (3, 4) (5, 6) (7, 8)
2	(1, 3) (2, 4) (5, 7) (6, 8)
3	(1, 4) (2, 3) (5, 8) (6, 7)
4	(1, 5) (2, 6) (3, 7) (4, 8)
5	(1, 6) (2, 5) (3, 8) (4, 7)
6	(1, 7) (2, 8) (3, 5) (4, 6)
7	(1, 8) (2, 7) (3, 6) (4, 5)

The idea behind this partitioning is that the four rotations in a given rotation set can be carried out concurrently. Consider rotation set 1. The update $A := J_{12}^T A J_{12}$ does not affect the 2-by-2 submatrices used to determine J_{34}, J_{56}, and J_{78}. Thus, each CPU can independently compute one of the rotations J_{12}, J_{34}, J_{56}, and J_{78}. More importantly, the updates

$$A := AJ_{12} \qquad A := AJ_{34} \qquad A := AJ_{56} \qquad A := AJ_{78}$$

are mutually independent and can be carried out concurrently. Once this is done, the 4 CPUs can simultaneously perform the row updates:

$$A := J_{12}^T A \qquad A := J_{34}^T A \qquad A := J_{56}^T A \qquad A := J_{78}^T A.$$

In this way, the time required for a given sweep in Algorithm 8.4-1 is reduced by a factor of 4.

Problems

P8.4-1. Let

$$C = \begin{bmatrix} w & x \\ y & z \end{bmatrix}$$

be real. (a) Give a stable algorithm for computing c and s with $c^2 + s^2 = 1$ such that

$$\begin{bmatrix} c & s \\ -s & c \end{bmatrix} C$$

is symmetric. (b) Combine (a) with the Jacobi trig calculations in the text to obtain a stable algorithm for computing the SVD of C. (c) Part (b) can be used to develop a Jacobi-like algorithm for computing the SVD of $A \in \mathbb{R}^{n \times n}$. For a given (p, q) with $p < q$, Jacobi transformations $J(p, q, \theta_1)$ and $J(p, q, \theta_2)$ are determined such that if

$$B = J(p, q, \theta_1) A J(p, q, \theta_2),$$

then $b_{pq} = b_{qp} = 0$. Show

$$\text{off}(B) = \text{off}(A) - b_{pq}^2 - b_{qp}^2.$$

How might p and q be determined? How could the algorithm be adapted to handle the case when $A \in \mathbb{R}^{m \times n}$ with $m > n$?

P8.4-2. Let x and y be in $\mathbb{R}^n$ and define the orthogonal matrix Q by

$$Q = \begin{bmatrix} c & s \\ -s & c \end{bmatrix}.$$

Give a stable algorithm for computing c and s such that the columns of $[x, y]Q$ are orthogonal to one another. (Nash (1975) used this idea to develop a "one-sided" Jacobi-like SVD algorithm. It finds an orthogonal V such that the columns of AV are mutually orthogonal.)

P8.4-3. Let the scalar γ be given along with the matrix

$$A = \begin{bmatrix} w & x \\ x & z \end{bmatrix}.$$

It is desired to compute an orthogonal matrix

$$J = \begin{bmatrix} c & s \\ -s & c \end{bmatrix}$$

such that the $(1, 1)$ entry of $J^T A J$ equals γ. Show that this requirement leads to the equation

$$(w - \gamma)^2 - 2x\tau + (z - \gamma) = 0,$$

where $\tau = c/s$. Verify that this quadratic has real roots if γ satisfies $\lambda_2 \leqslant \gamma \leqslant \lambda_1$, where λ_1 and λ_2 are the eigenvalues of A.

P8.4-4. Let $A \in \mathbb{R}^{n \times n}$ be symmetric. Using P8.4-3, give an algorithm that computes the factorization

$$Q^T A Q = \gamma I + F$$

where Q is orthogonal, $\gamma = \text{trace}(A)/n$, and F has zero diagonal entries. (Hint: $Q = J_{12} J_{23} \cdots J_{n-1,n}$.) Discuss the uniqueness of Q.

P8.4-5. Partition the n-by-n real symmetric matrix A as follows:

$$A = \begin{bmatrix} a & v^T \\ v & A_1 \end{bmatrix} \begin{matrix} 1 \\ n-1 \end{matrix}.$$
$$\begin{matrix} 1 & n-1 \end{matrix}$$

Let Q be a Householder matrix such that if $B = Q^T A Q$, then $b_{31} = b_{41} = \cdots = b_{n1} = 0$. Let $J = J(1, 2, \theta)$ be determined such that if $C = J^T B J$, then $c_{12} = 0$ and $c_{11} \geqslant c_{22}$. Show

$$c_{11} \geqslant a + \|v\|_2.$$

(La Budde (1964) formulated an algorithm for the symmetric eigenvalue problem based upon repetition of this Householder-Jacobi computation.)

P8.4-6. Let $A = (A_{ij})$ and $U = (U_{ij})$ be comformably partitioned block matrices with A symmetric and U orthogonal. Assume that U agrees with I in all block positions except (p, p), (p, q), (q, p), and (q, q). Let the submatrices in these positions be such that the matrix

$$\begin{bmatrix} U_{pp} & U_{pq} \\ U_{qp} & U_{qq} \end{bmatrix}^T \begin{bmatrix} A_{pp} & A_{pq} \\ A_{qp} & A_{qq} \end{bmatrix} \begin{bmatrix} U_{pp} & U_{pq} \\ U_{qp} & U_{qq} \end{bmatrix}$$

is diagonal. Show that

$$\text{off}(U^{T}AU) = \text{off}(A) - [2\|A_{pq}\|_F^2 + \text{off}(A_{pp}) + \text{off}(A_{qq})].$$

Notes and References for Sec. 8.4

Jacobi's original paper is one of the earliest references found in the numerical analysis literature:

C.G.J. Jacobi (1846). "Uber ein Leichtes Verfahren Die in der Theorie der Sacularstorungen Vorkommendern Gleichungen Numerisch Aufzulosen," *Crelle's J. 30*, 51-94.

Prior to the QR algorithm, the Jacobi technique was the standard method for solving dense symmetric eigenvalue problems. Early attempts to improve upon it include

C. D. La Budde (1964). "Two Classes of Algorithms for Finding the Eigenvalues and Eigenvectors of Real Symmetric Matrices," *J. Assoc. Comp. Mach. 11*, 53-58.

M. Lotkin (1956). "Characteristic Values of Arbitrary Matrices," *Quart. Appl. Math. 14*, 267-75.

D. A. Pope and C. Tompkins (1957). "Maximizing Functions of Rotations: Experiments Concerning Speed of Diagonalization of Symmetric Matrices Using Jacobi's Method," *J. Assoc. Comp. Mach. 4*, 459-66.

The computational aspects of Algorithm 8.4-1 are described in Wilkinson (AEP, pp. 265 *ff.*), while an ALGOL procedure can be found in

H. Rutishauser (1966). "The Jacobi Method for Real Symmetric Matrices," *Numer. Math. 9*, 1-10. See also HACLA, pp. 202-11.

Although the symmetric QR algorithm is generally much faster than the Jacobi method, there are special settings where the latter technique is of interest. As we illustrated, on a parallel computer it is possible to perform several rotations concurrently, thereby accelerating the reduction of the off-diagonal elements. See

A. Sameh (1971). "On Jacobi and Jacobi-like Algorithms for a Parallel Computer," *Math. Comp. 25*, 579-90.

F. Luk (1980). "Computing the Singular Value Decomposition on the ILLIAC IV," *ACM Trans. Math. Soft. 6*, 524-39.

Jacobi methods are also of interest in minicomputing environments, where their compactness makes them attractive. See

J. C. Nash (1975). "A One-Sided Transformation Method for the Singular Value Decomposition and Algebraic Eigenproblem," *Comp. J. 18*, 74-76.

They are also useful when a nearly diagonal matrix must be diagonalized. See

J. H. Wilkinson (1968). "Almost Diagonal Matrices with Multiple or Close Eigenvalues," *Lin. Alg. & Its Applic. 1*, 1-12.

Establishing the quadratic convergence of the classical and cyclic Jacobi iterations has engrossed several authors:

K. W. Brodlie and M.J.D. Powell (1975). "On the Convergence of Cyclic Jacobi Methods," *J. Inst. Math. Applic. 15*, 279–87.

E. R. Hansen (1962). "On Quasicyclic Jacobi Methods," *ACM J. 9*, 118–35.

E. R. Hansen (1963). "On Cyclic Jacobi Methods," *SIAM J. Applied Math. 11*, 448–59.

P. Henrici (1958). "On the Speed of Convergence of Cyclic and Quasicyclic Jacobi Methods for Computing the Eigenvalues of Hermitian Matrices," *SIAM J. Applied Math. 6*, 144–62.

P. Henrici and K. Zimmermann (1968). "An Estimate for the Norms of Certain Cyclic Jacobi Operators," *Lin. Alg. & Its Applic. 1*, 489–501.

A. Schonhage (1964). "On the Quadratic Convergence of the Jacobi Process," *Numer. Math. 6*, 410–12.

H.P.M. van Kempen (1966). "On Quadratic Convergence of the Special Cyclic Jacobi Method," *Numer. Math. 9*, 19–22.

J. H. Wilkinson (1962). "Note on the Quadratic Convergence of the Cyclic Jacobi Process," *Numer. Math. 4*, 296–300.

Attempts have been made to extend the Jacobi iteration to other classes of matrices and to push through corresponding convergence results. The case of normal matrices is discussed in

H. H. Goldstine and L. P. Horowitz (1959). "A Procedure for the Diagonalization of Normal Matrices," *J. Assoc. Comp. Mach. 6*, 176–95.

A. Ruhe (1967). "On the Quadratic Convergence of the Jacobi Method for Normal Matrices," *BIT 7*, 305–13.

G. Loizou (1972). "On the Quadratic Convergence of the Jacobi Method for Normal Matrices," *Comp. J. 15*, 274–76.

See also

M.H.C. Paardekooper (1971). "An Eigenvalue Algorithm for Skew Symmetric Matrices," *Numer. Math. 17*, 189–202.

Essentially, the analysis and algorithmic developments presented in the text carry over to the normal case with minor modification. The same comment applies to Jacobi-like methods for the SVD. See

G. E. Forsythe and P. Henrici (1960). "The Cyclic Jacobi Method for Computing the Principle Values of a Complex Matrix," *Trans. Amer. Math. Soc. 94*, 1–23.

J. C. Nash (1975). "A One-Sided Transformation Method for the Singular Value Decomposition and Algebraic Eigenproblem," *Comp. J. 18*, 74–76.

For non-normal matrices, the situation is considerably more difficult. Attempts to compute the Schur decomposition by using Jacobi transformations to reduce the norm of the subdiagonal elements have not met with much success. Consult

C. E. Froberg (1965). "On Triangularization of Complex Matrices by Two-Dimensional Unitary Transformations," *BIT 5*, 230–34.

C. P. Huang (1975). "A Jacobi-Type Method for Triangularizing an Arbitrary Matrix," *SIAM J. Num. Anal. 12*, 566–70.

Jacobi-like techniques for general matrices which use a combination of orthogonal transformations and "plane shears" have also been used. See

J. Boothroyd and P. J. Eberlein (1968). "Solution to the Eigenproblem by a Norm-Reducing Jacobi-Type Method (Handbook)," *Numer. Math. 11*, 1–12. See also HACLA, pp. 327–38.

P. J. Eberlein (1970). "Solution to the Complex Eigenproblem by a Norm-Reducing Jacobi-Type Method," *Numer. Math. 14*, 232–45. See also HACLA, pp. 404–17.

A. Ruhe (1968). "On the Quadratic Convergence of a Generalization of the Jacobi Method to Arbitrary Matrices," *BIT 8*, 210–31.

Using nonorthogonal transformations makes it difficult to assess the reduction of the off-diagonal entries. Some background analysis of this problem appears in

A. Ruhe (1969). "The Norm of a Matrix after a Similarity Transformation," *BIT 9*, 53–58.

Jacobi methods for complex symmetric matrices have also been developed. See

P. Anderson and G. Loizou (1973). "On the Quadratic Convergence of an Algorithm which Diagonalizes a Complex Symmetric Matrix," *J. Inst. Math. Applic. 12*, 261–71.

P. Anderson and G. Loizou (1976). "A Jacobi-Type Method for Complex Symmetric Matrices (Handbook)," *Numer. Math. 25*, 347–63.

P. J. Eberlein (1971). "On the Diagonalization of Complex Symmetric Matrices," *J. Inst. Math. Applic. 7*, 377–83.

J. J. Seaton (1969). "Diagonalization of Complex Symmetric Matrices Using a Modified Jacobi Method," *Comp. J. 12*, 156–57.

Sec. 8.5. Some Special Methods

By exploiting the rich mathematical structure of the symmetric eigenproblem, it is possible to devise useful alternatives to the symmetric QR algorithm. Many of these techniques are appropriate when only a few eigenvalues and/or eigenvectors are desired. Three such methods are described in this section: bisection, Rayleigh quotient iteration, and orthogonal iteration with Ritz acceleration.

Bisection

Let T_r denote the leading r-by-r principal submatrix of

$$(8.5\text{-}1) \qquad T = \begin{bmatrix} a_1 & b_2 & & & \mathbf{0} \\ b_2 & a_2 & b_3 & & \\ & \ddots & \ddots & \ddots & \\ & & \ddots & \ddots & b_n \\ \mathbf{0} & & & b_n & a_n \end{bmatrix}$$

and define the polynomials $p_0(x), \ldots, p_n(x)$ by

$$p_0(x) \equiv 1$$

$$p_r(x) = \det(T_r - xI)$$

for $r = 1, 2, \ldots, n$. A simple determinantal expansion can be used to show that

(8.5-2) $\qquad p_r(x) = (a_r - x)p_{r-1}(x) - b_r^2 p_{r-2}(x). \qquad r = 2, \ldots, n$

Because $p_n(x)$ can be evaluated in $O(n)$ flops, it is feasible to find its roots using the method of bisection. For example, if $p(y)p(z) < 0$ and $y < z$, then the iteration

$$
\begin{aligned}
&\text{Do While } [\, |y - z| > \mathbf{u}(|y| + |z|)\,] \\
&\quad x := (y + z)/2 \\
&\quad \text{if } p_n(x)p_n(y) < 0 \\
&\qquad \textit{then } z := x \\
&\qquad \textit{else } y := x
\end{aligned}
$$

is guaranteed to converge to a zero of $p_n(x)$, i.e., to an eigenvalue of T. The iteration converges linearly: error is approximately halved at each step.

Sometimes it is necessary to compute the k-th largest eigenvalue of T for some prescribed value of k. This can be done efficiently by using the bisection idea and the following classical result:

THEOREM 8.5-1: *Sturm Sequence Property.* If the tridiagonal matrix T in (8.5-1) is unreduced, then the eigenvalues of T_{r-1} strictly separate the eigenvalues of T_r:

$$\lambda_r(T_r) < \lambda_{r-1}(T_{r-1}) < \lambda_{r-1}(T_r) < \cdots < \lambda_2(T_r) < \lambda_1(T_{r-1}) < \lambda_1(T_r).$$

Moreover, if $a(\lambda)$ denotes the number of sign changes in the sequence

$$\{\, p_0(\lambda), p_1(\lambda), \ldots, p_n(\lambda)\,\}$$

then $a(\lambda)$ equals the number of T's eigenvalues that are less than λ. (Convention: $p_r(\lambda)$ has the opposite sign from $p_{r-1}(\lambda)$ if $p_r(\lambda) = 0$.)

Proof.   It follows from Corollary 8.1-4 that the eigenvalues of T_{r-1} weakly separate those of T_r. To prove that the separation must be strict, suppose that $p_r(\mu) = p_{r-1}(\mu) = 0$ for some r and μ. It then follows from (8.5-2) and the assumption that T is unreduced that $p_0(\mu) = p_1(\mu) = \cdots = p_r(\mu) = 0$, an obvious contradiction. Thus, we must have strict separation.

The assertion about $a(\lambda)$ is established in Wilkinson (AEP, pp. 300–301). $\square$

EXAMPLE 8.5-1. If

$$T = \begin{bmatrix} 1 & -1 & 0 & 0 \\ -1 & 2 & -1 & 0 \\ 0 & -1 & 3 & -1 \\ 0 & 0 & -1 & 4 \end{bmatrix},$$

then $\lambda(T) \cong \{.254, 1.82, 3.18, 4.74\}$. The sequence

$$\{p_0(2), p_1(2), p_2(2), p_3(2), p_4(2)\} = \{1, -1, -1, 0, 1\}$$

confirms that there are two eigenvalues less than $\lambda = 2$.

Suppose we wish to compute $\lambda_k(T)$. From the Gershgorin circle theorem (Theorem 7.2-1) it follows that $\lambda_k(T) \in [y, z]$ where

$$y = \min_i a_i - |b_i| - |b_{i-1}|$$

$$z = \max_i a_i + |b_i| + |b_{i-1}|$$

$$b_0 = b_{n+1} \equiv 0$$

With these starting values, it is clear from the Sturm sequence property that the iteration

(8.5-3)
$$\begin{aligned} &\text{Do While } [\,|z - y| > \mathbf{u}(|y| + |z|)\,] \\ &\quad x := (y + z)/2 \\ &\quad \text{if } a(x) \geq k \\ &\qquad \text{then } z := x \\ &\qquad \text{else } y := x \end{aligned}$$

produces a sequence of subintervals that are repeatedly halved in length but which always contain $\lambda_k(T)$.

EXAMPLE 8.5-2. If (8.5-3) is applied to the matrix of Example 8.5-1 with $k = 2$ we find the values shown in Table 8.5-1, from which we conclude that $\lambda_3(T) \in [1.7969, 1.9375]$. Note: $\lambda_3(T) \cong 1.82$.

Table 8.5-1.

y	z	x	$a(x)$
0.0000	5.0000	2.5000	2
0.0000	2.5000	1.2500	1
1.2500	2.5000	1.3750	1
1.3750	2.5000	1.9375	2
1.3750	1.9375	1.6563	1
1.6563	1.9375	1.7969	1

During the execution of (8.5-3), information about the location of other eigenvalues is obtained. By systematically keeping track of this information it is possible to devise an efficient scheme for computing "contiguous" subsets of $\lambda(T)$, e.g., $\{\lambda_k(T), \lambda_{k+1}(T), \ldots, \lambda_{k+j}(T)\}$. See Barth, Martin, and Wilkinson (1967).

If selected eigenvalues of a general symmetric matrix A are desired, then it is necessary first to compute the tridiagonalization $T = U_0^T T U_0$ before the above bisection schemes can be applied. This can be done using Algorithm 8.2-1 or by the Lanczos algorithm discussed in the next chapter. In either case, the corresponding eigenvectors can be readily found via inverse iteration (see §7.6), since tridiagonal systems can be solved in $O(n)$ flops.

In those applications where the original matrix A already has tridiagonal form, bisection computes eigenvalues with small relative error, regardless of their magnitude. This is in contrast to the tridiagonal QR iteration, where the computed eigenvalues $\hat{\lambda}_i$ can be guaranteed only to have small absolute error: $|\hat{\lambda}_i - \lambda_i(T)| \cong \mathbf{u}\|T\|_2$.

Finally, it is possible to compute specific eigenvalues of a symmetric matrix by using the L-D-L^T factorization (see §5.2). The idea is to use the Sylvester inertia theorem (Theorem 8.1-12). If

$$A - \mu I = LDL^T \qquad\qquad A = A^T \in \mathbb{R}^{n \times n}$$

is the L-D-L^T factorization of $A - \mu I$ with $D = \operatorname{diag}(d_1, \ldots, d_n)$, then the number of negative d_i equals the number of $\lambda_i(A)$ that are less than μ. See Parlett (SEP, pp. 46 *ff.*) for details.

Rayleigh Quotient Iteration

Suppose $A \in \mathbb{R}^{n \times n}$ is symmetric and that x is a given nonzero n-vector. A simple differentiation reveals that

$$\lambda = r(x) \equiv \frac{x^T A x}{x^T x}$$

minimizes $\|(A - \lambda I)x\|_2$. The scalar $r(x)$ is called the *Rayleigh quotient* of x. Clearly, if x is an approximate eigenvector, then $r(x)$ is a reasonable choice for the corresponding eigenvalue. On the other hand, if λ is an approximate eigenvalue, then inverse iteration theory tells us that the solution to $(A - \lambda I)x = b$ will almost always be a good approximate eigenvector.

Combining these two ideas in the natural way give rise to the *Rayleigh quotient iteration*:

(8.5-4)
$$
\begin{aligned}
&x_0 \text{ given, } \|x_0\|_2 = 1 \\
&\text{For } k = 0, 1, \ldots \\
&\qquad \mu_k = r(x_k) \\
&\qquad \text{Solve } (A - \mu_k I)z_{k+1} = x_k \text{ for } z_{k+1}. \\
&\qquad x_{k+1} = z_{k+1} / \|z_{k+1}\|_2
\end{aligned}
$$

Note that for any k we have

$$(A + E_k)z_{k+1} = \mu_k z_{k+1}$$

where

$$E_k = -x_k z_{k+1}^T / \|z_{k+1}\|_2^2.$$

It follows from Corollary 8.1-3 that $|\mu_k - \lambda| \leqslant 1/\|z_{k+1}\|_2$ for some $\lambda \in \lambda(A)$.

EXAMPLE 8.5-3. If (8.5-4) is applied to

$$A = \begin{bmatrix} 1 & 1 & 1 & 1 & 1 & 1 \\ 1 & 2 & 3 & 4 & 5 & 6 \\ 1 & 3 & 6 & 10 & 15 & 21 \\ 1 & 4 & 10 & 20 & 35 & 56 \\ 1 & 5 & 15 & 35 & 70 & 126 \\ 1 & 6 & 21 & 56 & 126 & 252 \end{bmatrix}$$

with $x_0 = (1, 1, 1, 1, 1, 1)^T/\sqrt{6}$, then

k	μ_k
0	153.8333
1	120.0571
2	49.5011
3	13.8687
4	15.4959
5	15.5534

The iteration is converging to the eigenvalue $\lambda = 15.5534732737$.

Parlett (1974b) has shown that (8.5-4) converges globally and that the convergence is ultimately cubic. We demonstrate this for the case $n = 2$. Without loss of generality, we may assume that $A = \text{diag}(\lambda_1, \lambda_2)$, for $\lambda_1 > \lambda_2$. Denoting x_k by

$$x_k = \begin{bmatrix} c_k \\ s_k \end{bmatrix}, \qquad\qquad c_k^2 + s_k^2 = 1$$

it follows that

$$\mu_k = \lambda_1 c_k^2 + \lambda_2 s_k^2$$

and

$$z_{k+1} = \frac{1}{\lambda_1 - \lambda_2} \begin{bmatrix} c_k/s_k^2 \\ -s_k/c_k^2 \end{bmatrix}.$$

A calculation shows that

$$c_{k+1} = c_k^3/(c_k^6 + s_k^6)^{1/2}$$
$$s_{k+1} = -s_k^3/(c_k^6 + s_k^6)^{1/2}.$$

From these equations it is clear that the x_k converge cubically either to span$\{e_1\}$ or span$\{e_2\}$ provided $|c_k| \neq |s_k|$.

Details associated with the practical implementation of the Rayleigh quotient iteration may be found in Parlett (1974b).

It is interesting to note the connection between Rayleigh quotient iteration and the symmetric QR algorithm. Suppose we apply the latter to the tridiagonal matrix $T \in \mathbb{R}^{n \times n}$ with shift $\sigma = e_n^T T e_n = t_{nn}$. If $T - \sigma I = QR$, then we obtain $\hat{T} = RQ + \sigma I$. From the equation $(T - \sigma I)Q = R^T$ it follows that

$$(T - \sigma I)q_n = r_{nn}e_n,$$

where q_n is the last column of the orthogonal matrix Q. Thus, if we apply (8.5-4) with $x_0 = e_n$, then $x_1 = q_n$.

Orthogonal Iteration with Ritz Acceleration

Recall the method of orthogonal iteration from §7.3:

(8.5-5)
$$\begin{aligned} &Q_0 \in \mathbb{R}^{n \times p}, \, Q_0^T Q_0 = I_p \\ &\text{For } k = 1, 2, \ldots \\ &\quad Z_k = A Q_{k-1} \\ &\quad Q_k R_k = Z_k \quad \text{(Q-R factorization)} \end{aligned}$$

Suppose A is symmetric with Schur decomposition $Q^T A Q = \text{diag}(\lambda_i)$ where

$$Q = [q_1, \ldots, q_n]$$

and

$$|\lambda_1| > |\lambda_2| > \cdots > |\lambda_n|.$$

It follows from Theorem 7.3-2 that if $d = \text{dist}[D_p(A), R(Q_0)] < 1$, then

$$\text{dist}[D_p(A), R(Q_k)] \leq \frac{1}{\sqrt{1 - d^2}} \left| \frac{\lambda_{p+1}}{\lambda_p} \right|^k,$$

where $D_p(A) = \text{span}\{q_1, \ldots, q_p\}$. Moreover, from the analysis in §7.3 we know that if $R_k = [r_{ij}^{(k)}]$, then

$$|r_{ii}^{(k)} - \lambda_i| = O\left(\left|\frac{\lambda_{i+1}}{\lambda_i}\right|^k\right). \qquad i = 1, \ldots, p$$

This can be an unacceptably slow rate of convergence if λ_i and λ_{i+1} are of nearly equal modulus.

This difficulty can be partially surmounted by replacing Q_k with its Ritz vectors at each step:

(8.5-6)
$$\begin{aligned}
&Q_0 \in \mathbb{R}^{n \times p} \text{ given, } Q_0^T Q_0 = I_p \\
&\text{For } k = 1, 2, \ldots \\
&\quad Z_k = A Q_{k-1} \\
&\quad \bar{Q}_k R_k = Z_k \qquad \text{(Q-R factorization)} \\
&\quad S_k = \bar{Q}_k^T A \bar{Q}_k \\
&\quad U_k^T S_k U_k = D_k \qquad \text{(Schur decomposition)} \\
&\quad Q_k = \bar{Q}_k U_k
\end{aligned}$$

it can be shown that if

$$D_k = \text{diag}[\theta_1^{(k)}, \ldots, \theta_p^{(k)}], \qquad |\theta_1^{(k)}| \geqslant \cdots \geqslant |\theta_p^{(k)}|$$

then

$$|\theta_i^{(k)} - \lambda_i(A)| = \left|\frac{\lambda_{p+1}}{\lambda_i}\right|^k. \qquad i = 1, \ldots, p$$

Thus, the Ritz values $\theta_i^{(k)}$ converge at a much more favorable rate than the $r_{ii}^{(k)}$ in (8.5-5). For details, see Stewart (1969).

EXAMPLE 8.5-4. If we apply (8.5-6) to

$$A = \begin{bmatrix} 100 & 1 & 1 & 1 \\ 1 & 99 & 1 & 1 \\ 1 & 1 & 2 & 1 \\ 1 & 1 & 1 & 1 \end{bmatrix}$$

with

$$Q_0 = \begin{bmatrix} 1 & 0 \\ 0 & 1 \\ 0 & 0 \\ 0 & 0 \end{bmatrix},$$

then

k	dist$[D_2(A), Q_k]$
0	$.2 \times 10^{-1}$
1	$.5 \times 10^{-3}$
2	$.1 \times 10^{-4}$
3	$.3 \times 10^{-6}$
4	$.8 \times 10^{-8}$

Clearly, convergence is taking place at the rate $(2/99)^k$.

Problems

P8.5-1. Suppose λ is an eigenvalue of a symmetric tridiagonal matrix T. Show that if λ has algebraic multiplicity k, then at least $k - 1$ of T's subdiagonal elements are zero.

P8.5-2. Suppose A is symmetric and has bandwidth p. Show that if we perform the shifted QR step $A - \mu I = QR$, $\bar{A} = RQ + \mu I$, then $\bar{A}$ has bandwidth p.

P8.5-3. Suppose $B \in \mathbb{R}^{n \times n}$ is upper bidiagonal with diagonal entries $d_1, \ldots,$ d_n and superdiagonal entries $f_2, \ldots, f_n$. State and prove a singular value version of Theorem 8.5-1.

P8.5-4. How many positive eigenvalues does the following tridiagonal matrix have?

$$A = \begin{bmatrix} 2 & 1 & & & \\ 1 & 0 & 3 & & \text{\huge 0} \\ & 3 & 4 & 1 & \\ \text{\huge 0} & & 1 & 1 & 2 \\ & & & 2 & 7 \end{bmatrix}.$$

Notes and References for Sec. 8.5

A bisection subroutine is in EISPACK. The ALGOL program upon which it is based is described in

W. Barth, R. S. Martin, and J. H. Wilkinson (1967). "Calculation of the Eigenvalues of a Symmetric Tridiagonal Matrix by the Method of Bisection," *Numer. Math. 9*, 386-93. See also HACLA, pp. 249-56.

Another way to compute a specified subset of eigenvalues is via the *rational QR* algorithm. In this method the shift is determined using Newton's method. This makes it possible to "steer" the iteration towards desired eigenvalues. See

C. Reinsch and F. L. Bauer (1968). "Rational QR Transformation with Newton's Shift for Symmetric Tridiagonal Matrices," *Numer. Math. 11*, 264-72. See also HACLA, pp. 257-65.

Papers concerned with the symmetric QR algorithm for banded matrices include

R. S. Martin and J. H. Wilkinson (1967). "Solution of Symmetric and Unsymmetric Band Equations and the Calculation of Eigenvectors of Band Matrices," *Numer. Math. 9*, 279-301. See also HACLA, pp. 70-92.

R. S. Martin, C. Reinsch, and J. H. Wilkinson (1970). "The QR Algorithm for Band Symmetric Matrices," *Numer. Math. 16*, 85-92. See also HACLA, pp. 266-72.

The following references are concerned with the method of simultaneous iteration for sparse symmetric matrices:

M. Clint and A. Jennings (1970). "The Evaluation of Eigenvalues and Eigenvectors of Real Symmetric Matrices by Simultaneous Iteration," *Comp. J. 13*, 76-80.

H. Rutishauser (1970). "Simultaneous Iteration Method for Symmetric Matrices," *Numer. Math. 16*, 205-23. See also HACLA, pp. 284-302.

G. W. Stewart (1969). "Accelerating the Orthogonal Iteration for the Eigenvalues of a Hermitian Matrix,," *Numer. Math. 13*, 362-76.

The literature on special methods for the symmetric eigenproblem is vast. Some representative papers include

K. J. Bathe and E. L. Wilson (1973). "Solution Methods for Eigenvalue Problems in Structural Mechanics," *Int. J. Numer. Meth. Eng. 6*, 213-26.

C. F. Bender and I. Shavitt (1970). "An Iterative Procedure for the Calculation of the Lowest Real Eigenvalue and Eigenvector of a Non-Symmetric Matrix," *J. Comp. Physics 6*, 146-49.

J.J.M. Cuppen (1981). "A Divide and Conquer Method for the Symmetric Eigen-problem," *Numer. Math. 36*, 177-95.

K. K. Gupta (1972). "Solution of Eigenvalue Problems by Sturm Sequence Method," *Int. J. Numer. Meth. Eng. 4*, 379-404.

P. S. Jenson (1972). "The Solution of Large Symmetric Eigenproblems by Section-ing," *SIAM J. Num. Anal. 9*, 534-45.

S. F. McCormick (1972). "A General Approach to One-Step Iterative Methods with Application to Eigenvalue Problems," *J. Comput. Sys. Sci. 6*, 354-72.

B. N. Parlett (1974*b*). "The Rayleigh Quotient Iteration and Some Generalizations for Nonnormal Matrices," *Math. Comp. 28*, 679-93.

A. Ruhe (1974). "SOR Methods for the Eigenvalue Problem with Large Sparse Matri-ces," *Math. Comp. 28*, 695-710.

A. Sameh, J. Lermit and K. Noh (1975). "On the Intermediate Eigenvalues of Sym-metric Sparse Matrices," *BIT 12*, 543-54.

G. W. Stewart (1974). "The Numerical Treatment of Large Eigenvalue Problems," *Proc. IFIP Congress 74*, North-Holland, pp. 666-72.

J. Vandergraft (1971). "Generalized Rayleigh Methods with Applications to Finding Eigenvalues of Large Matrices," *Lin. Alg. & Its Applic. 4*, 353-68.

Sec. 8.6. More Generalized Eigenvalue Problems

In the generalized eigenvalue problem $Ax = \lambda Bx$ it is frequently the case that A is symmetric and B is symmetric and positive definite. Pencils of this variety are referred to as *symmetric-definite pencils*. This property is preserved under congruence transformations:

$$\left.\begin{array}{l} A - \lambda B \text{ symmetric-definite} \\[6pt] \qquad X \text{ nonsingular} \end{array}\right\} \iff (X^T A X) - \lambda(X^T B X) \text{ symmetric-definite}.$$

Although the QZ algorithm of §7.7 can be used to solve the symmetric-definite problem, it has the flaw of destroying both symmetry and definiteness. What we seek is a stable efficient algorithm that computes X such that $X^T A X$ and $X^T B X$ are both in "canonical form." The obvious form to aim for is diagonal form.

THEOREM 8.6-1. Suppose A and B are n-by-n symmetric matrices, and define $C(\mu)$ by

$$(8.6\text{-}1) \qquad\qquad C(\mu) = \mu A + (1 - \mu)B.$$

If there exists a $\mu \in [0, 1]$ such that $C(\mu)$ is non-negative definite and $N[C(\mu)] = N(A) \cap N(B)$, then there exists a nonsingular X such that both $X^T A X$ and $X^T B X$ are diagonal.

Proof. Let $\mu \in [0, 1]$ be chosen so that $C(\mu)$ is non-negative definite with the property that $N[C(\mu)] = N(A) \cap N(B)$. Let

$$Q_1^T C(\mu) Q_1 = \begin{bmatrix} D & 0 \\ 0 & 0 \end{bmatrix} \qquad D = \mathrm{diag}(d_1, \ldots, d_k),\ d_i > 0$$

be the Schur decomposition of $C(\mu)$ and define $X_1 = Q_1 \mathrm{diag}(D^{-1/2}, I_{n-k})$. If $A_1 = X_1^T A X_1$, $B_1 = X_1^T B X_1$, and $C_1 = X_1^T C(\mu) X_1$, then

$$C_1 = \begin{bmatrix} I_k & 0 \\ 0 & 0 \end{bmatrix} = \mu A_1 + (1 - \mu)B_1.$$

Since $\mathrm{span}\{e_{k+1}, \ldots, e_n\} = N(C_1) = N(A_1) \cap N(B_1)$ it follows that A_1 and B_1 have the following block structure:

$$A_1 = \begin{bmatrix} A_{11} & 0 \\ 0 & 0 \end{bmatrix} \begin{matrix} k \\ n-k \end{matrix} \qquad B_1 = \begin{bmatrix} B_{11} & 0 \\ 0 & 0 \end{bmatrix} \begin{matrix} k \\ n-k \end{matrix} .$$
$$ \begin{matrix} k & n-k \end{matrix} \qquad\qquad \begin{matrix} k & n-k \end{matrix}$$

Moreover, $I_k = \mu A_{11} + (1 - \mu)B_{11}$.

Suppose $\mu \ne 0$. It then follows that if

$$Z^T B_{11} Z = \mathrm{diag}(b_1, \ldots, b_k)$$

is the Schur decomposition of B_{11} and we set $X = X_1 \mathrm{diag}(Z, I_{n-k})$, then

$$X^T B X = \text{diag}(b_1, \ldots, b_k, 0, \ldots, 0) \equiv D_B$$

and

$$X^T A X = \frac{1}{\mu} X^T [C(\mu) - (1 - \mu)B]X$$

$$= \frac{1}{\mu} \left(\begin{bmatrix} I_k & 0 \\ 0 & 0 \end{bmatrix} - (1 - \mu)D_B \right) \equiv D_A.$$

On the other hand, if $\mu = 0$, then let $Z^T A_{11} Z = \text{diag}(a_1, \ldots, a_k)$ be the Schur decomposition of A_{11} and set $X = X_1 \text{diag}(Z, I_{n-k})$. It is easy to verify that in this case as well, both $X^T A X$ and $X^T B X$ are diagonal. $\square$

COROLLARY 8.6-2. If $A - \lambda B \in \mathbb{R}^{n \times n}$ is symmetric-definite, then there exists a nonsingular $X = [x_1, \ldots, x_n]$ such that

$$X^T A X = \text{diag}(a_1, \ldots, a_n)$$

and

$$X^T B X = \text{diag}(b_1, \ldots, b_n)$$

Moreover, $A x_i = \lambda_i x_i$ for $i = 1, \ldots, n$ where $\lambda_i = a_i / b_i$.

Proof. By setting $\mu = 0$ in Theorem 8.6-1 we see that symmetric-definite pencils can be simultaneously diagonalized. The rest of the corollary is easily verified. $\square$

EXAMPLE 8.6-1. If

$$A = \begin{bmatrix} 229 & 163 \\ 163 & 116 \end{bmatrix} \quad \text{and} \quad B = \begin{bmatrix} 81 & 59 \\ 59 & 43 \end{bmatrix}$$

then $A - \lambda B$ is symmetric-definite and $\lambda(A, B) = \{5, -\frac{1}{2}\}$. If

$$X = \begin{bmatrix} 3 & -5 \\ -4 & 7 \end{bmatrix}$$

then $X^T A X = \text{diag}(5, -1)$ and $X^T B X = \text{diag}(1, 2)$.

Stewart (1979b) has worked out a perturbation theory for symmetric pencils $A - \lambda B$ that satisfy

(8.6-2) $$c(A, B) = \min_{\|x\|_2 = 1} (x^T A x)^2 + (x^T B x)^2 > 0.$$

The scalar $c(A, B)$ is called the *Crawford number* of the pencil $A - \lambda B$.

THEOREM 8.6-3. Suppose $A - \lambda B$ is an n-by-n symmetric-definite pencil with eigenvalues $\lambda_1 \geqslant \lambda_2 \geqslant \cdots \geqslant \lambda_n$. Suppose E_A and E_F are symmetric n-by-n matrices that satisfy

$$\epsilon^2 = \|E_A\|_2^2 + \|E_B\|_2^2 < c(A, B).$$

Then $(A + E_A) - \lambda(B + E_B)$ is symmetric-definite with eigenvalues $\mu_1 \geqslant \cdots \geqslant \mu_n$ that satisfy

$$|\arctan(\lambda_i) - \arctan(\mu_i)| \leqslant \arctan[\epsilon/c(A, B)]$$

for $i = 1, \ldots, n$.

Proof. See Stewart (1979b). □

Turning to algorithmic developments, we first present a method for solving the symmetric-definite problem that utilizes both the Cholesky decomposition and the symmetric QR algorithm.

ALGORITHM 8.6-1. Given $A = A^T \in \mathbb{R}^{n \times n}$ and $B = B^T \in \mathbb{R}^{n \times n}$ with B positive definite, the following algorithm computes a nonsingular X such that $X^T B X = I_n$ and $X^T A X = \text{diag}(a_1, \ldots, a_n)$.

> Compute the Cholesky decomposition $B = GG^T$ using Algorithm 5.2-1.
> Set $C = G^{-1} A G^{-T}$.
> Use the symmetric QR algorithm to compute $Q^T A Q = \text{diag}(a_1, \ldots, a_n)$.
> Set $X = G^{-T} Q$.

This algorithm requires about $7n^3$ flops. In a practical implementation, A can be overwritten by the matrix C. See Martin and Wilkinson (1968c) for details. Note that $\lambda(A, B) = \lambda(A, GG^T) = \lambda(G^{-1}AG^{-T}, I) = \lambda(C) = \{a_1, \ldots, a_n\}$.

If $\hat{a}_i$ is a computed eigenvalue obtained by Algorithm 8.6-1, then it can be shown that

$$\hat{a}_i \in \lambda(G^{-1}AG^{-T} + E_i),$$

where $\|E_i\|_2 \cong \mathbf{u}\|A\|_2\|B^{-1}\|_2$. Thus, if B is ill-conditioned, then $\hat{a}_i$ may be severely contaminated with roundoff error even if a_i is well-conditioned. The problem, of course, is that in this case the matrix C will have some very large entries. This difficulty can sometimes be overcome by replacing the matrix G in Algorithm 8.6-1 with $VD^{-1/2}$ where $V^T B V = D$ is the Schur decomposition of B. If the diagonal entries of D are ordered from smallest to largest, then the large entries in C are concentrated in the upper left-hand corner. The small eigenvalues of C can then be computed without excessive roundoff error contamination (or so the heuristic goes). For further discussion, consult Wilkinson (AEP, pp. 337–38).

EXAMPLE 8.6-2.   If

$$A = \begin{bmatrix} 1 & 2 & 3 \\ 2 & 4 & 5 \\ 3 & 5 & 6 \end{bmatrix} \quad \text{and} \quad G = \begin{bmatrix} .001 & 0 & 0 \\ 1 & .001 & 0 \\ 2 & 1 & .001 \end{bmatrix}$$

and $B = GG^T$, then the two smallest eigenvalues of $A - \lambda B$ are

$$a_1 = -.619402940600584$$

$$a_2 = 1.627440079051887.$$

If 17-digit floating point arithmetic is used, then these eigenvalues are computed to full machine precision when the symmetric QR algorithm is applied to $fl[D^{-1/2}V^TAVD^{-1/2}]$, where $B = VDV^T$ is the Schur decomposition of B. On the other hand, if Algorithm 8.6-1 is applied, then

$$\hat{a}_1 = -.619373517376444$$

$$\hat{a}_2 = 1.627516601905228.$$

The reason for obtaining only 4 correct significant digits is that $\kappa_2(B) \cong 10^{18}$.

The condition of the matrix X in Algorithm 8.6-1 can sometimes be improved by replacing B with a suitable convex combination of A and B. The connection between the eigenvalues of the modified pencil and those of the original are detailed in the proof of Theorem 8.6-1.

Other difficulties concerning Algorithm 8.6-1 revolve around the fact that $G^{-1}AG^{-T}$ is generally full even when A and B are sparse. This is a serious problem, since many of the symmetric-definite problems that arise in practice are large and sparse.

Crawford (1973) has shown how to implement Algorithm 8.6-1 effectively when A and B are banded. Aside from this case, however, the simultaneous diagonalization approach is impractical for the large, sparse symmetric-definite problem.

An alternative idea is to extend the Rayleigh quotient iteration (8.5-3) as follows:

(8.6-3)
$$\begin{aligned} &x_0 \text{ given, } \|x_0\|_2 = 1 \\ &\text{For } k = 0, 1, \ldots \\ &\quad \mu_k = x_k^T A x_k / x_k^T B x_k \\ &\quad \text{Solve } (A - \mu_k B)z_{k+1} = Bx_k \text{ for } z_{k+1}. \\ &\quad x_{k+1} = z_{k+1} / \|z_{k+1}\|_2 \end{aligned}$$

The mathematical basis for this iteration is that

(8.6-4) $$\lambda = r(x) = x^T A x / x^T B x$$

minimizes

(8.6-5) $$f(\lambda) = \|A x - \lambda B x\|_B,$$

where $\|\cdot\|_B$ is defined by $\|z\|_B^2 = z^T B^{-1} z$. The mathematical properties of (8.6-3) are similar to those of (8.5-3). Its applicability depends on whether or not systems of the form $(A - \mu B)z = x$ can be readily solved.

A similar comment pertains to the following generalized orthogonal iteration:

(8.6-6)
$$\begin{aligned} &Q_0 \in \mathbb{R}^{n \times p} \text{ given, with } Q_0^T Q_0 = I_p \\ &\text{For } k = 1, 2, \ldots \\ &\quad \text{Solve } B Z_k = A Q_{k-1} \text{ for } Z_k. \\ &\quad Q_k R_k = Z_k \quad \text{(Q-R factorization)} \end{aligned}$$

This is mathematically equivalent to (7.3-4) with A replaced by $B^{-1}A$. Its practicality depends on how easy it is to solve linear systems of the form $Bz = y$.

Sometimes A and B are so large that neither (8.6-3) nor (8.6-6) can be invoked. In this situation, one can resort to any of a number of gradient and coordinate relaxation algorithms. See Stewart (1976b) for an extensive guide to the literature.

We conclude with some remarks about symmetric pencils that have the form $A^T A - \lambda B^T B$ where $A \in \mathbb{R}^{m \times n}$ and $B \in \mathbb{R}^{p \times n}$. This pencil underlies the *generalized singular value decomposition* (GSVD), a decomposition that is useful in several constrained least squares problems. (Cf. §12.1.) Note that by Theorem 8.6-1 there exists a nonsingular $X \in \mathbb{R}^{n \times n}$ such that $X^T(A^T A)X$ and $X^T(B^T B)X$ are both diagonal. The value of the GSVD is that these diagonalizations can be achieved without forming $A^T A$ and $B^T B$.

THEOREM 8.6-4: GSVD. If $A \in \mathbb{R}^{m \times n}$ $(m \geq n)$ and $B \in \mathbb{R}^{p \times n}$ then there exist orthogonal $U \in \mathbb{R}^{m \times m}$ and $V \in \mathbb{R}^{p \times p}$ and an invertible $X \in \mathbb{R}^{n \times n}$ such that

$$U^T A X = D_A = \mathrm{diag}(\alpha_1, \ldots, \alpha_n) \qquad \alpha_i \geq 0$$

and

$$V^T B X = D_B = \mathrm{diag}(\beta_1, \ldots, \beta_q) \qquad \beta_i \geq 0;\ q = \min\{p, n\}$$

where

$$\beta_1 \geq \cdots \geq \beta_r > \beta_{r+1} = \cdots = \beta_q = 0 \qquad r = \mathrm{rank}(B)$$

Proof. The proof of this decomposition appears in Van Loan (1976). We present a more constructive proof along the lines of Paige and Saunders

(1981). Without loss of generality we assume for clarity that $m = n = p$ and $N(A) \cap N(B) = \{0\}$.

Let

(8.6-7)
$$\begin{bmatrix} A \\ B \end{bmatrix} = \begin{bmatrix} Q_{11} & Q_{12} \\ Q_{21} & Q_{22} \end{bmatrix} \begin{bmatrix} R \\ 0 \end{bmatrix}$$

be the Q-R factorization where

$$Q = \begin{matrix} \begin{bmatrix} Q_{11} & Q_{12} \\ Q_{21} & Q_{22} \end{bmatrix} & \begin{matrix} n \\ n \end{matrix} \\ \begin{matrix} n \quad\; n \end{matrix} & \end{matrix}$$

is orthogonal and $R \in \mathbb{R}^{n \times n}$ is upper triangular. Note that $N(A) \cap N(B) = \{0\}$ implies that R is nonsingular. Let

(8.6-8)
$$\begin{bmatrix} Q_{11} & Q_{12} \\ Q_{21} & Q_{22} \end{bmatrix} = \begin{bmatrix} U & 0 \\ 0 & -V \end{bmatrix} \begin{bmatrix} C & S \\ -S & C \end{bmatrix} \begin{bmatrix} W & 0 \\ 0 & Y \end{bmatrix}^{T}$$

be the C-S decomposition of Q. (See Theorem 2.4-1.) Here U, V, W, and Y are orthogonal, $C = \text{diag}(c_i)$, $c_i \geq 0$, $S = \text{diag}(s_i)$, $s_i \geq 0$ and $C^2 + S^2 = I_n$. By comparing blocks in (8.6-7) and using (8.6-8) we find

$$A = Q_{11}R = UC(W^T R)$$
$$B = Q_{21}R = VS(W^T R).$$

The theorem follows by setting $D_A = C$, $D_B = S$, and $X = (W^T R)^{-1}$. $\square$

The elements of the set $\sigma(A, B) \equiv \{\alpha_1/\beta_1, \ldots, \alpha_r/\beta_r\}$ are referred to as the *generalized singular values* of A and B. Note that $\sigma \in \sigma(A, B)$ implies that $\sigma^2 \in \lambda(A^T A, B^T B)$. The theorem is a generalization of the SVD in that if $B = I_n$, then $\sigma(A, B) = \sigma(A)$.

This proof of the GSVD is constructive since Stewart (1982) has shown how to stably compute the C-S decomposition.

Problems

P8.6-1. Suppose $A \in \mathbb{R}^{n \times n}$ is symmetric and $G \in \mathbb{R}^{n \times n}$ is lower triangular and nonsingular. Give an algorithm for computing $C = G^{-1}AG^{-T}$ in $n^3/3$ flops.

P8.6-2. Suppose $A \in \mathbb{R}^{n \times n}$ is symmetric and $B \in \mathbb{R}^{n \times n}$ is symmetric positive definite. Give an algorithm for computing the eigenvalues of AB that uses the Cholesky decomposition and the symmetric QR algorithm.

P8.6-3. Show that if C is diagonalizable, then there exist symmetric matrices A and B, B nonsingular, such that $C = AB^{-1}$. (This shows that symmetric pencils $A - \lambda B$ are essentially perfectly general.)

P8.6-4. Suppose $A - \lambda B$ is a symmetric definite n-by-n pencil and that $Ax = \lambda Bx$ for some x having unit 2-norm. Show that there exists an orthogonal Q such that

$$
Q^TBQ = \begin{bmatrix} w & y & 0 \cdots 0 \\ \hline y & & \\ 0 & & \\ \vdots & & B_1 \\ 0 & & \end{bmatrix} , \quad Q^TAQ = \begin{bmatrix} \lambda w & \lambda y & 0 \cdots 0 \\ \hline \lambda y & & \\ 0 & & \\ \vdots & & A_1 \\ 0 & & \end{bmatrix} .
$$

Show how $Z = I - \alpha \epsilon_2 e_1^T$ can be determined such that if $Y = QZ$ then $Y^TAY = \text{diag}(\lambda w, A_1)$ and $Y^TBY = \text{diag}(w, B_1)$.

Notes and References for Sec. 8.6

The simultaneous reduction of two symmetric matrices to diagonal form is discussed in

A. Berman and A. Ben-Israel (1971). "A Note on Pencils of Hermitian or Symmetric Matrices," *SIAM J. Appl. Math. 21*, 51–54.

F. Uhlig (1973). "Simultaneous Block Diagonalization of Two Real Symmetric Matrices," *Lin. Alg. & Its Applic. 7*, 281–89.

F. Uhlig (1976). "A Canonical Form for a Pair of Real Symmetric Matrices That Generate a Nonsingular Pencil," *Lin. Alg. & Its Applic. 14*, 189–210.

K. N. Majindar (1979). "Linear Combinations of Hermitian and Real Symmetric Matrices," *Lin. Alg. & Its Applic. 25*, 95–105.

The perturbation theory that we presented for the symmetric-definite problem was taken from

G. W. Stewart (1979b). "Perturbation Bounds for the Definite Generalized Eigenvalue Problem," *Lin. Alg. & Its Applic. 23*, 69–86.

An excellent survey of computational methods is given in

G. W. Stewart (1976b). "A Bibliographical Tour of the Large Sparse Generalized Eigenvalue Problem," in *Sparse Matrix Compuations*, ed. J. R. Bunch and D. J. Rose, Academic Press, New York.

Some papers of particular interest include

C. R. Crawford (1973). "Reduction of a Band Symmetric Generalized Eigenvalue Problem," *Comm. Assoc. Comp. Mach. 16*, 41–44.

C. R. Crawford (1976). "A Stable Generalized Eigenvalue Problem," *SIAM J. Num. Anal. 13*, 854–60.

G. Fix and R. Heiberger (1972). "An Algorithm for the Ill-Conditioned Generalized Eigenvalue Problem," *SIAM J. Num. Anal. 9*, 78–88.

R. S. Martin and J. H. Wilkinson (1968c). "Reduction of the Symmetric Eigenproblem $Ax = \lambda Bx$ and Related Problems to Standard Form," *Numer. Math. 11*, 99–110.

G. Peters and J. H. Wilkinson (1969). "Eigenvalues of $Ax = \lambda Bx$ with Band Symmetric A and B," *Comp. J. 12*, 398–404.

A. Ruhe (1974). "SOR Methods for the Eigenvalue Problem with Large Sparse Matrices," *Math. Comp. 28*, 695–710.

The generalized SVD and some of its applications are discussed in

C. F. Van Loan (1976). "Generalizing the Singular Value Decomposition," *SIAM J. Num. Anal. 13*, 76–83.

C. C. Paige and M. Saunders (1981). "Towards a Generalized Singular Value Decomposition," *SIAM J. Num. Anal. 18*, 398–405.

A stable method for computing this decomposition was recently described in

G. W. Stewart (1983). "A Method for Computing the Generalized Singular Value Decomposition," in *Matrix Pencils*, ed. B. Kagstrom and A. Ruhe, Springer-Verlag, New York, pp. 207–20.

Lanczos Methods

§9.1 Derivation and Convergence Properties

§9.2 Practical Lanczos Procedures

§9.3 Applications to Linear Equations and Least Squares

In this chapter we develop the Lanczos method, a technique that is applicable to large, sparse, symmetric eigenproblems. The method involves tridiagonalizing the given matrix A. However, unlike the Householder approach, no intermediate (and full) submatrices are generated. Equally important, information about A's extremal eigenvalues tends to emerge long before the tridiagonalization is complete. This makes the Lanczos algorithm particularly useful in situations where a few of A's largest or smallest eigenvalues are desired.

The derivation and exact arithmetic properties of the method are presented in §9.1. Unfortunately, roundoff errors make the method somewhat difficult to use in practice. This shortcoming can be overcome in a number of ways, as we describe in §9.2. In the final section we show how the "Lanczos idea" can be applied to solve an assortment of singular value, least squares, and linear equations problems.

Reading Path

$$\S8.1 \rightarrow \S8.2 \rightarrow \S8.5 \nearrow \quad \S9.1 \longrightarrow \quad \S9.2 \longrightarrow \quad \S9.3$$
$$\S6.2 \nearrow \quad \S5.2, \rightarrow \S5.3 \nearrow \nearrow$$
$$\S6.5 \nearrow$$

Sec. 9.1. Derivation and Convergence Properties

Suppose $A \in \mathbb{R}^{n \times n}$ is large, sparse, and symmetric and assume that a few of its largest and/or smallest eigenvalues are desired. This problem can be solved by a method attributed to Lanczos (1950). The method generates a sequence of tridiagonal matrices $\{T_j\}$ with the property that the extremal eigenvalues of $T_j \in \mathbb{R}^{j \times j}$ are progressively better estimates of A's extremal eigenvalues. In this section we derive the technique and investigate its exact arithmetic properties.

The derivation of the Lanczos algorithm can proceed in several ways. So that its remarkable convergence properties do not come as a complete surprise, we prefer to lead into the technique by considering the optization of the Rayleigh quotient

$$r(x) = \frac{x^T A x}{x^T x} . \qquad\qquad x \neq 0$$

Recall from Theorem 8.1-2 that the maximum and minimum values of $r(x)$ are $\lambda_1(A)$ and $\lambda_n(A)$, respectively. Suppose $\{q_i\} \subset \mathbb{R}^n$ is a sequence of orthonormal vectors and define the scalars M_j and m_j by

$$M_j = \lambda_1(Q_j^T A Q_j) = \max_y \frac{y^T(Q_j^T A Q_j)y}{y^T y} = \max_y r(Q_j y) \leqslant \lambda_1(A)$$

$$m_j = \lambda_j(Q_j^T A Q_j) = \min_y \frac{y^T(Q_j^T A Q_j)y}{y^T y} = \min_y r(Q_j y) \geqslant \lambda_n(A)$$

where

$$Q_j = [q_1, \ldots, q_j].$$

The Lanczos algorithm can be derived by considering how to generate the q_j so that M_j and m_j are increasingly better estimates of $\lambda_1(A)$ and $\lambda_n(A)$.

Suppose $u_j \in \text{span}\{q_1, \ldots, q_j\}$ is such that $M_j = r(u_j)$. Since $r(x)$ increases most rapidly in the direction of the gradient

$$\nabla r(x) = \frac{2}{x^T x} [Ax - r(x)x] ,$$

we can ensure that $M_{j+1} > M_j$ if q_{j+1} is determined so

(9.1-1) $$\nabla r(u_j) \in \text{span}\{q_1, \ldots, q_{j+1}\}.$$

(This assumes $\nabla r(u_j) \neq 0$.) Likewise, if $v_j \in \text{span}\{q_1, \ldots, q_j\}$ satisfies $r(v_j) = m_j$, then it makes sense to require

(9.1-2) $$\nabla r(v_j) \in \text{span}\{q_1, \ldots, q_{j+1}\},$$

since $r(x)$ decreases most rapidly in the direction of $-\nabla r(x)$.

At first glance, the task of finding a single q_{j+1} that satisfies these two requirements appears impossible. However, since $\nabla r(x) \in \text{span}\{x, Ax\}$, it is clear that (9.1-1) and (9.1-2) can be simultaneously satisfied provided

$$\text{span}\{q_1, \ldots, q_j\} = \text{span}\{q_1, Aq_1, \ldots, A^{j-1}q_1\} \equiv \mathcal{K}(A, q_1, j)$$

and we choose q_{j+1} so

$$\mathcal{K}(A, q_1, j + 1) = \text{span}\{q_1, \ldots, q_{j+1}\}.$$

Thus, we are led to the problem of computing orthonormal bases for the *Krylov subspaces* $\mathcal{K}(A, q_1, j)$.

In order to do this efficiently, we exploit the connection between the tridiagonalization of A and the Q-R factorization of $K(A, q_1, n)$. (See §7.4.) Specifically, if $Q^TAQ = T$ is tridiagonal with $Qe_1 = q_1$, then

$$[q_1, Aq_1, \ldots, A^{n-1}q_1] = Q[e_1, Te_1, \ldots, T^{n-1}e_1]$$

is the Q-R factorization of $[q_1, \ldots, A^{n-1}q_1]$. Thus, the q_j can effectively be generated by tridiagonalizing A with an orthogonal matrix whose first column is q_1.

In §8.2 we discussed a tridiagonalization procedure based on Householder transformations. Unfortunately, this approach is usually impractical if A is large and sparse. Householder similarity transformations tend to destroy sparsity and as a result, unacceptably large, dense matrices may have to be dealt with during the reduction.

Loss of sparsity can sometimes be controlled by using Givens rather that Householder transformations. See Duff and Reid (1976). However, any method that computes T by successively "updating" A will not be useful in the majority of cases when A is sparse.

This suggests that we try to compute the elements of the tridiagonal matrix $T = Q^TAQ$ directly. Setting $Q = [q_1, \ldots, q_n]$ and

$$T = \begin{bmatrix} \alpha_1 & \beta_1 & & & \text{\Large 0} \\ \beta_1 & \alpha_2 & \ddots & & \\ & \ddots & \ddots & \ddots & \beta_{n-1} \\ & & \ddots & \ddots & \\ \text{\Large 0} & & \beta_{n-1} & & \alpha_n \end{bmatrix}$$

and equating columns in $AQ = QT$, we find

$$Aq_j = \beta_{j-1}q_{j-1} + \alpha_j q_j + \beta_j q_{j+1} \qquad \beta_0 q_0 \equiv 0$$

for $j = 1, \ldots, n - 1$. The orthonormality of the q_i implies

$$\alpha_j = q_j^T Aq_j.$$

Moreover, if

$$r_j = (A - \alpha_j I)q_j - \beta_{j-1}q_{j-1}$$

is nonzero, then $q_{j+1} = r_j/\beta_j$ where $\beta_j = \pm \|r_j\|_2$. Properly sequenced, these formulae define the *Lanczos iteration*:

$$r_0 = q_1, \beta_0 = 1, q_0 = 0, j = 0$$

(9.1-3)
$$\text{Do While } (\beta_j \neq 0)$$
$$q_{j+1} = r_j/\beta_j$$
$$j := j + 1$$
$$\alpha_j = q_j^T A q_j$$
$$r_j = (A - \alpha_j I)q_j - \beta_{j-1}q_{j-1}$$
$$\beta_j = \|r_j\|_2$$

There is no loss of generality in choosing the β_j to be positive. The q_i are called *Lanczos vectors*. With careful overwriting and use of the formula

$$\alpha_j = q_j^T(Aq_j - \beta_{j-1}q_{j-1}),$$

the whole process can be implemented with just a pair of n-vectors:

ALGORITHM 9.1-1:   *The Lanczos Algorithm.*   Given a symmetric $A \in \mathbb{R}^{n \times n}$ and $w \in \mathbb{R}^n$ having unit 2-norm, the following algorithm computes a j-by-j symmetric tridiagonal matrix T_j with the property that $\lambda(T_j) \subset \lambda(A)$. The diagonal and subdiagonal elements of T_j are stored in $\alpha_1, \ldots, \alpha_j$ and $\beta_1, \ldots, \beta_{j-1}$, respectively.

$$v_i := 0 \qquad (i = 1, \ldots, n)$$
$$\beta_0 := 1$$
$$j := 0$$
$$\text{Do While } (\beta_j \neq 0)$$
$$\quad \text{If } (j \neq 0) \text{ then}$$
$$\qquad \text{For } i = 1, \ldots, n$$
$$\qquad\qquad t := w_i, w_i := v_i/\beta_j, v_i := -\beta_j t$$
$$\quad v := Aw + v$$
$$\quad j := j + 1$$
$$\quad \alpha_j := w^T v$$
$$\quad v := v - \alpha_j w$$
$$\quad \beta_j := \|v\|_2$$

Note that A is not altered during the entire process. Thus, only a procedure for computing matrix-vector products involving A need be supplied. If sparsity is exploited in this procedure and only kn flops are involved in each call ($k \ll n$), then each Lanczos step requires approximately $(4 + k)n$ flops to execute.

The eigenvalues of T_j can be found using the symmetric QR algorithm or any of the special methods of §8.5, such as bisection.

The Lanczos vectors are generated in the n-vector w. If they are desired for later use, then special arrangements must be made for their storage. In the typical sparse matrix setting they would be stored on a disk or some other secondary storage device until required.

The iteration halts before complete tridiagonalization if q_1 is contained in a proper invariant subspace. This is one of several mathematical properties of the method that we summarize in the following result:

THEOREM 9.1-1. Let $A \in \mathbb{R}^{n \times n}$ be symmetric and $q_1 \in \mathbb{R}^n$ have unit 2-norm. Then the Lanczos iteration (9.1-3) runs until $j = m$, where $m =$ rank $[q_1, Aq_1, \ldots, A^{n-1}q_1]$. Moreover, for $j = 1, \ldots, m$ we have

(9.1-4)
$$AQ_j = Q_jT_j + r_je_j^T$$

where

$$T_j = \begin{bmatrix} \alpha_1 & \beta_1 & & & \mathbf{0} \\ \beta_1 & \alpha_2 & \ddots & & \\ & \ddots & \ddots & \ddots & \\ & & \ddots & \ddots & \beta_{j-1} \\ \mathbf{0} & & & \beta_{j-1} & \alpha_j \end{bmatrix}$$

and $Q_j = [q_1, \ldots, q_j]$ has orthonormal columns satisfying $R(Q_j) = \mathcal{K}(A, q_1, j)$.

Proof. The proof is by induction on j. Suppose the iteration has produced $Q_j = [q_1, \ldots, q_j]$ such that $R(Q_j) = \mathcal{K}(A, q_1, j)$ and $Q_j^TQ_j = I_j$. It is easy to see from (9.1-3) that (9.1-4) holds. Thus,

$$Q_j^TAQ_j = T_j + Q_j^Tr_je_j^T.$$

Since $\alpha_i = q_i^TAq_i$ for $i = 1, \ldots, j$ and

$$q_{i+1}^TAq_i = q_{i+1}^T(Aq_i - \alpha_iq_i - \beta_{i-1}q_{i-1}) = q_{i+1}^T(\beta_iq_{i+1}) = \beta_i$$

for $i = 1, \ldots, j - 1$, we have $Q_j^TAQ_j = T_j$. Consequently, $Q_j^Tr_j = 0$. If $r_j \neq 0$, then $q_{j+1} = r_j/\|r_j\|_2$ is orthogonal to $q_1, \ldots, q_j$ and

$$q_{j+1} \in \text{span}\{Aq_j, q_j, q_{j-1}\} \subset \mathcal{K}(A, q_1, j+1).$$

Thus, $Q_{j+1}^TQ_{j+1} = I_{j+1}$ and $R(Q_{j+1}) = \mathcal{K}(A, q_1, j+1)$.

On the other hand, if $r_j = 0$, then $AQ_j = Q_jT_j$. This says that $R(Q_j) = \mathcal{K}(A, q_1, j)$ is invariant. From this we conclude that $j = m = \dim[\mathcal{K}(A, q_1, n)]$. $\square$

Encountering a zero β_j in the Lanczos iteration is a welcome event in that it signals the computation of an exact invariant subspace. However, an exactly zero or even small β_j is a rarity in practice. Consequently, other explanations for the convergence of T_j's eigenvalues must be sought. The following result is a step in this direction.

THEOREM 9.1-2. Suppose that j steps of the Lanczos algorithm have been performed and that

$$S_j^T T_j S_j = \operatorname{diag}(\theta_1, \ldots, \theta_j)$$

is the Schur decomposition of the tridiagonal matrix T_j. If $Y_j \in \mathbb{R}^{n \times j}$ is defined by

$$Y_j = [y_1, \ldots, y_j] = Q_j S_j,$$

then for $i = 1, 2, \ldots, j$ we have

$$\| A y_i - \theta_i y_i \|_2 = |\beta_j| \ |s_{ji}|$$

where $S_j = (s_{pq})$.

Proof. Post-multiplying (9.1-4) by S_j gives

$$A Y_j = Y_j \operatorname{diag}(\theta_1, \ldots, \theta_j) + r_j e_j^T S_j,$$

i.e.,

$$A y_i = \theta_i y_i + r_j (e_j^T S e_i).$$

The proof is complete by taking norms and recalling that $\| r_j \|_2 = |\beta_j|$. $\square$

The theorem provides computable error bounds for T_j's eigenvalues:

$$\min_{\mu \in \lambda(A)} |\theta_i - \mu| \leqslant |\beta_j| \ |s_{ji}|. \qquad\qquad i = 1, \ldots, j$$

Note that in the terminology of §8.1, the (θ_i, y_i) are Ritz pairs for the subspace $R(Q_j)$.

Another way that T_j can be used to provide estimates of A's eigenvalues is described in Golub (1974) and involves the judicious construction of a rank-one matrix E such that $R(Q_j)$ is invariant for $A + E$. In particular, if we use the Lanczos method to compute $A Q_j = Q_j T_j + r_j e_j^T$ and set $E = \tau\, w w^T$, where $\tau = \pm 1$ and $w = a q_j + b r_j$, then it can be shown that

$$(A + E) Q_j = Q_j (T_j + \tau a^2 e_j e_j^T) + (1 + \tau a b) r_j e_j^T.$$

If $0 = 1 + \tau a b$, then the eigenvalues of the tridiagonal matrix

$$\tilde{T}_j = T_j + \tau a^2 e_j e_j^T$$

are also eigenvalues of $A + E$. We may then conclude from Corollary 8.1-5 that the intervals $[\lambda_i\, (\tilde{T}_j),\ \lambda_{i-1}\, (\tilde{T}_j)]$, where $i = 2, \ldots, j$, each contain an eigenvalue of A.

These bracketing intervals depend on the choice of τa^2. Suppose we have an approximate eigenvalue $\tilde{\lambda}$ of A. One possibility is to choose τa^2 so that

$$\det(\tilde{T}_j - \tilde{\lambda} I_j) = (\alpha_j + \tau a^2 - \tilde{\lambda}) p_{j-1}\, (\tilde{\lambda}) - \beta_{j-1}^2 p_{j-2}\, (\tilde{\lambda}) = 0$$

where the polynomials $p_i(x) = \det(T_i - xI_i)$ can be evaluated at $\bar{\lambda}$ using the three-term recurrence (8.5-2). (This assumes that $p_{j-1}(\bar{\lambda}) \neq 0$.) Eigenvalue estimation in this spirit is discussed in Lehmann (1963) and Householder (1968).

The preceding discussion indicates how eigenvalue estimates can be obtained via the Lanczos algorithm, but it reveals nothing about rate of convergence. Results of this variety constitute what is known as the *Kaniel-Paige Theory*, a sample of which follows.

Theorem 9.1-3. Let A be an n-by-n symmetric matrix with eigenvalues $\lambda_1 \geqslant \cdots \geqslant \lambda_n$ and corresponding orthonormal eigenvectors $z_1, \ldots, z_n$. If $\theta_1 \geqslant \cdots \geqslant \theta_j$ are the eigenvalues of the matrix T_j obtained after j steps of the Lanczos iteration, then

$$\lambda_1 \geqslant \theta_1 \geqslant \lambda_1 - \frac{(\lambda_1 - \lambda_n)\tan(\phi_1)^2}{[c_{j-1}(1 + 2\rho_1)]^2}$$

where $\cos(\phi_1) = |q_1^T z_1|$, $\rho_1 = (\lambda_1 - \lambda_2)/(\lambda_2 - \lambda_n)$, and $c_{j-1}(x)$ is Chebychev polynomial of degree $j - 1$.

Proof. From Theorem 8.1-2 we have

$$\theta_1 = \max_{y \neq 0} \frac{y^T T_j y}{y^T y} = \max_{y \neq 0} \frac{(Q_j y)^T A (Q_j y)}{(Q_j Y)^T (Q_j y)} = \max_{0 \neq w \in \mathcal{K}(A, q_1, j)} \frac{w^T A w}{w^T w}.$$

Since λ_1 is the maximum of $w^T A w / w^T w$ over *all* nonzero w, it follows that $\lambda_1 \geqslant \theta_1$. To obtain the lower bound for θ_1, note that

$$\theta_1 = \max_{p \in \mathcal{P}_{j-1}} \frac{q_1^T p(A) A p(A) q_1}{q_1^T p(A)^2 q_1},$$

where $\mathcal{P}_{j-1}$ is the set of all $j - 1$ degree polynomials. If

$$q_1 = \sum_{i=1}^{n} d_i z_i,$$

then

$$\frac{q_1^T p(A) A p(A) q_1}{q_1^T p(A)^2 q_1} = \frac{\displaystyle\sum_{i=1}^{n} d_i^2 p(\lambda_i)^2 \lambda_i}{\displaystyle\sum_{i=1}^{n} d_i^2 p(\lambda_i)^2}$$

$$\geqslant \lambda_1 - (\lambda_1 - \lambda_n) \frac{\displaystyle\sum_{i=2}^{n} d_i^2 p(\lambda_i)^2}{d_1^2 p(\lambda_1)^2 + \displaystyle\sum_{i=2}^{n} d_i^2 p(\lambda_i)^2}$$

We can make the lower bound tight by selecting a polynomial $p(x)$ that is large at $x = \lambda_1$ in comparison to its value at the remaining eigenvalues. One way of doing this is to set

$$p(x) = c_{j-1}\left[-1 + 2\,\frac{x - \lambda_n}{\lambda_2 - \lambda_n}\right],$$

where $c_{j-1}(z)$ is the $(j - 1)$-st Chebychev polynomial generated via the recursion

$$c_j(z) = 2zc_{j-1}(z) - c_{j-2}(z). \qquad\qquad c_0 = 1, c_1 = z$$

These polynomials are bounded by unity on $[-1, 1]$, but grow very rapidly outside this interval. By defining $p(x)$ this way, it follows that $|p(\lambda_i)|$ is bounded by unity for $i = 2, \ldots, n$, while $p(\lambda_1) = c_{j-1}(1 + 2\rho_1)$. Thus,

$$\theta_1 \geq \lambda_1 - (\lambda_1 - \lambda_n)\,\frac{1 - d_1^2}{d_1^2}\,\frac{1}{c_{j-1}^2(1 + 2\rho_1)}.$$

The desired lower bound is obtained by noting that $\tan(\phi_1)^2 = (1 - d_1^2)/d_1^2$. $\square$

An analogous result pertaining to θ_j follows immediately from this theorem:

COROLLARY 9.1-4. Using the same notation as the theorem,

$$\lambda_n \leq \theta_j \leq \lambda_n + \frac{(\lambda_1 - \lambda_n)\tan^2(\phi_n)}{c_{j-1}^2(1 + 2\rho_n)},$$

where $\rho_n = (\lambda_{n-1} - \lambda_n)/(\lambda_1 - \lambda_{n-1})$ and $\cos(\phi_n) = |q_1^T z_n|$.

Proof. Apply Theorem 9.1-3 with A replaced by $-A$. $\square$

It is worthwhile to compare θ_1 with the corresponding power method estimate of λ_1. (See §7.3.) For clarity, assume $\lambda_1 \geq \cdots \geq \lambda_n \geq 0$. After $j - 1$ power method steps applied to q_1, a vector is obtained in the direction of

$$v = A^{j-1}q_1 = \sum_{i=1}^{n} c_i \lambda_i^{j-1} z_i$$

along with an eigenvalue estimate

$$\gamma_1 = \frac{v^T A v}{v^T v}.$$

Using the proof and notation of Theorem 9.1-3, it is easy to show that

$$(9.1\text{-}5) \qquad \lambda_1 \geq \gamma_1 \geq \lambda_1 - (\lambda_1 - \lambda_n)\tan(\phi_1)^2 \left(\frac{\lambda_2}{\lambda_1}\right)^{2j-2}.$$

Table 9.1-1. L_{j-1}/R_{j-1}

λ_1/λ_2	$j = 5$	$j = 10$	$j = 15$	$j = 20$	$j = 25$
	1.1×10^{-4}	2.0×10^{-10}	3.9×10^{-16}	7.4×10^{-22}	1.4×10^{-27}
1.5	3.9×10^{-2}	6.8×10^{-4}	1.2×10^{-5}	2.0×10^{-7}	3.5×10^{-9}
	2.7×10^{-2}	5.5×10^{-5}	1.1×10^{-7}	2.1×10^{-10}	4.2×10^{-13}
1.1	4.7×10^{-1}	1.8×10^{-1}	6.9×10^{-2}	2.7×10^{-2}	1.0×10^{-2}
	5.6×10^{-1}	1.0×10^{-1}	1.5×10^{-2}	2.0×10^{-3}	2.8×10^{-4}
1.01	9.2×10^{-1}	8.4×10^{-1}	7.6×10^{-1}	6.9×10^{-1}	6.2×10^{-1}

(Hint: Set $p(x) = x^{j-1}$ in the proof.) Thus, we can compare the tightness of the lower bounds for θ_1 and γ_1 by comparing

$$L_{j-1} \equiv 1 \Big/ \left[c_{j-1}\left(2\,\frac{\lambda_1}{\lambda_2} - 1\right)\right]^2 \geq 1 \Big/ \left[c_{j-1}\left(1 + 2\,\rho_1\right)\right]^2$$

and

$$R_{j-1} = (\lambda_2/\lambda_1)^{2(j-1)}.$$

This is done in Table 9.1-1 for representative values of j and (λ_2/λ_1). The superiority of the Lanczos estimate is self-evident. This should be no surprise, since θ_1 is the maximum of $r(x) = x^T A x / x^T x$ over *all* of $\mathcal{K}(A, q_1, j)$, while $\gamma_1 = r(v)$ for a particular v in $\mathcal{K}(A, q_1, j)$, namely, $v = A^{j-1}q_1$.

We conclude with some remarks about error bounds for T_j's interior eigenvalues. The key idea in the proof of Theorem 9.1-3 is the use of the translated Chebychev polynomial. With this polynomial we amplified the component of q_1 in the direction z_1. A similar idea can be used to obtain bounds for an interior Ritz value θ_i. The bounds are not as satisfactory, however, because the "amplifying polynomial" has the form $q(x)\Pi_{k=1}^{i-1}(x - \lambda_k)$, where $q(x)$ is the $(j - i)$ degree Chebychev polynomial on $[\lambda_{i+1}, \lambda_n]$. For details, see Kaniel (1966), Paige (1971), or Scott (1978).

Problems

P9.1-1. Suppose $A \in \mathbb{R}^{n \times n}$ is skew-symmetric. Derive a Lanczos-like algorithm for computing a skew-symmetric tridiagonal matrix T_m such that $AQ_m = Q_m T_m$, where $Q_m^T Q_m = I_m$.

P9.1-2. Let $A \in \mathbb{R}^{n \times n}$ be symmetric and define $r(x) = x^T A x / x^T x$. Suppose $S \subset \mathbb{R}^n$ is a subspace with the property that $x \in S$ implies $\nabla r(x) \in S$. Show that S is invariant for A.

P9.1-3. Show that if a symmetric matrix $A \in \mathbb{R}^{n \times n}$ has a multiple eigenvalue, then the Lanczos iteration will terminate prematurely.

P9.1-4. Show that the index m in Theorem 9.1-1 is the dimension of the smallest invariant subspace for A that contains q_1.

P9.1-5. Let $A \in \mathbb{R}^{n \times n}$ be symmetric and consider the problem of determining an orthonormal sequence $q_1, q_2, \ldots$ with the property that once $Q_j = [q_1, \ldots, q_j]$ is known, q_{j+1} is chosen so as to minimize

$$\mu_j = \| (I - Q_{j+1}Q_{j+1}^T)AQ_j \|_F \; .$$

Show that if span $\{q_1, \ldots, q_j\} = \mathcal{K}(A, q_1, j)$, then it is possible to choose q_{j+1} so $\mu_j = 0$. Explain how this optimization problem once again leads to the Lanczos iteration.

Notes and References for Sec. 9.1

The classic reference for the Lanczos method is

C. Lanczos (1950). "An Iteration Method for the Solution of the Eigenvalue Problem of Linear Differential and Integral Operators," *J. Res. Nat. Bur. Stand. 45*, 255–82.

Although the convergence of the Ritz values is alluded to in this paper, for more details we refer the reader to

S. Kaniel (1966). "Estimates for Some Computational Techniques in Linear Algebra," *Math. Comp. 20*, 369–78.
C. C. Paige (1971). "The Computation of Eigenvalues and Eigenvectors of Very Large Sparse Matrices," Ph.D. thesis, London University.

The Kaniel-Paige theory set forth in these papers is nicely summarized in

D. Scott (1978). "Analysis of the Symmetric Lanczos Process," Electronic Research Laboratory Technical Report UCB/ERL M78/40, University of California.

See also Wilkinson (AEP, pp. 270 ff.), Parlett (SEP, chap. 13) and the more recent paper

Y. Saad (1980). "On the Rates of Convergence of the Lanczos and the Block Lanczos Methods," *SIAM J. Num. Anal. 17*, 687–706.

Of the several computational variants of the Lanczos method, Algorithm 9.1-1 is the most stable. For details, see

C. C. Paige (1972). "Computational Variants of the Lanczos Method for the Eigenproblem," *J. Inst. Math. Applic. 10*, 373–81.
J. Lewis (1977). "Algorithms for Sparse Matrix Eigenvalue Problems," Technical Report STAN-CS-77-595, Department of Computer Science, Stanford University.

The effect of an ill-chosen initial vector is thoroughly discussed in

D. S. Scott (1979). "How to Make the Lanczos Algorithm Converge Slowly," *Math. Comp. 33*, 239–47.

The connection between the Lanczos algorithm, orthogonal polynomials, and the theory of moments is surveyed in

G. H. Golub (1974). "Some Uses of the Lanczos Algorithm in Numerical Linear Algebra," in *Topics in Numerical Analysis*, ed. J.J.H. Miller, Academic Press, New York.

See also

A. S. Householder (1968). "Moments and Characteristic Roots II," *Numer. Math. 11*, 126-28.

N. J. Lehmann (1963). "Optimale Eigenwerteinschliessungen," *Numer. Math. 5*, 246-72.

We motivated our discussion of the Lanczos algorithm by discussing the inevitability of fill-in when Householder or Givens transformations are used to tridiagonalize. Actually, fill-in can sometimes be kept to an acceptable level if care is exercised. See

I. S. Duff (1974). "Pivot Selection and Row Ordering in Givens Reduction on Sparse Matrices," *Computing 13*, 239-48.

I. S. Duff and J. K. Reid (1976). "A Comparison of Some Methods for the Solution of Sparse Over-Determined Systems of Linear Equations," *J. Inst. Maths. Applic 17*, 267-80.

L. Kaufman (1979). "Application of Dense Householder Transformations to a Sparse Matrix," *ACM Trans. Math. Soft. 5*, 442-50.

Sec. 9.2. Practical Lanczos Procedures

Rounding errors greatly affect the behavior of Algorithm 9.1-1, the Lanczos iteration. The basic difficulty is caused by loss of orthogonality among the Lanczos vectors, a phenomenon that muddies the issue of termination and complicates the relationship between A's eigenvalues and those of the tridiagonal matrices T_j. This disturbing feature, coupled with the advent of Householder's perfectly stable method of tridiagonalization, explains why the Lanczos algorithm was disregarded by numerical analysts during the fifties and sixties. However, interest in the method was rejuvenated with the development of the Kaniel-Paige theory. Concurrently, the pressure to solve large, sparse eigenproblems increased with increased computer power. With many fewer than n iterations typically required to get good approximate extremal eigenvalues, the Lanczos method became attractive as a sparse matrix technique rather than as a competitor of the Householder approach.

The search for a practical, easy-to-use Lanczos procedure is rooted in the fundamental error analysis of the method by Paige (1971, 1976a, 1980). An examination of his results is the best way to motivate the several modified Lanczos procedures of this section.

After j steps of the algorithm we obtain the matrix of computed Lanczos vectors

$$\hat{Q}_j = [\hat{q}_1, \ldots, \hat{q}_j]$$

and the associated tridiagonal matrix

$$\hat{T}_j = \begin{bmatrix} \hat{\alpha}_1 & \hat{\beta}_1 & & & \mathbf{0} \\ \hat{\beta}_1 & \hat{\alpha}_2 & \ddots & & \\ & \ddots & \ddots & \hat{\beta}_{j-1} \\ \mathbf{0} & & \hat{\beta}_{j-1} & \hat{\alpha}_j \end{bmatrix}.$$

Paige (1971, 1976a) shows that if $\hat{r}_j$ is the computed analog of r_j, then

(9.2-1) $$A\hat{Q}_j = \hat{Q}_j\hat{T}_j + \hat{r}_j e_j^T + E_j,$$

where

(9.2-2) $$\|E_j\|_2 \cong \mathbf{u}\|A\|_2 .$$

This indicates that the important equation $AQ_j = Q_jT_j + r_je_j^T$ is satisfied to working precision, a perfectly pleasing result.

Unfortunately, the picture is much less rosy with respect to the orthogonality among the $\hat{q}_i$. (Normality is not an issue. The computed Lanczos vectors essentially have unit length.) If

$$\hat{\beta}_j = fl(\|\hat{r}_j\|_2)$$

and

$$\hat{q}_{j+1} = fl(\hat{r}_j/\hat{\beta}_j),$$

then a quick analysis shows that

$$\hat{\beta}_j\hat{q}_{j+1} = \hat{r}_j + w_j,$$

where $\|w_j\|_2 \cong \mathbf{u}\|r_j\|_2 \cong \mathbf{u}\|A\|_2$. Thus, we may conclude that

$$|\hat{q}_{j+1}^T\hat{q}_i| \cong \frac{|\hat{r}_j^T\hat{q}_i| + \mathbf{u}\|A\|_2}{|\hat{\beta}_j|}$$

for $i = 1, \ldots, j$. In other words, significant departures from orthogonality can be expected when $\hat{\beta}_j$ is small, *even* in the ideal situation when $\hat{r}_j^T\hat{Q}_j$ is zero. A small $\hat{\beta}_j$ implies cancellation in the calculation of $\hat{r}_j$. We stress that loss of orthogonality is due to this cancellation and is not the result of the gradual accumulation of roundoff error.

EXAMPLE 9.2-1. The matrix

$$A = \begin{bmatrix} 2.64 & -.48 \\ -.48 & 2.36 \end{bmatrix}$$

has eigenvalues $\lambda_1 = 3$ and $\lambda_2 = 2$. If the Lanczos algorithm is applied to this matrix with

$$\hat{q}_1 = \begin{bmatrix} .810 \\ -.586 \end{bmatrix}$$

and three-digit floating point arithmetic is performed, then

$$\hat{q}_2 \cong \begin{bmatrix} .707 \\ .707 \end{bmatrix} .$$

Loss of orthogonality occurs because $\text{span}\{\hat{q}_1\}$ is almost invariant for A. (The vector $x = (.8, -.6)^T$ is the eigenvector affiliated with λ_1.)

Further details of the Paige analysis will be given shortly. Suffice it to say now that loss of orthogonality always occurs in practice and with it, an apparent deterioration in the quality of T_j's eigenvalues:

THEOREM 9.2-1.   If (9.2-1) holds, then there exist eigenvalues $\mu_1, \ldots, \mu_j \in \lambda(A)$ such that

$$| \mu_i - \lambda_i(\hat{T}_j) | \leqslant \sqrt{2} \, \frac{\|\hat{r}_j\|_2 + \|E_j\|_2}{\sigma_j(\hat{Q}_j)}$$

for $i = 1, \ldots, j$.

Proof.   Apply Theorem 8.1-8 with $F_1 = \hat{r}_j e_j^T + E_j$. $\square$

An obvious way to prevent the denominator in the upper bound from going to zero is to orthogonalize each newly computed Lanczos vector against its predecessors. This leads directly to our first "practical" Lanczos procedure.

Lanczos with Complete Reorthogonalization

Let $r_0, \ldots, r_{j-1} \in \mathbb{R}^n$ be given and suppose that Householder matrices $P_0, \ldots, P_{j-1}$ have been computed such that $(P_0 \cdots P_{j-1})^T [r_0, \cdots, r_{j-1}]$ is upper triangular. Denote the first j columns of $(P_0 \cdots P_{j-1})$ by $[q_1, \cdots, q_j]$. Now suppose that we are given a vector $r_j \in \mathbb{R}^n$ and wish to compute a unit vector q_{j+1} in the direction of

$$w = r_j - \sum_{i=1}^{j} (q_i^T r_j) q_i \in \text{span}\{q_1, \ldots, q_j\}^{\perp}.$$

If a Householder matrix P_j is determined so $(P_0 \cdots P_j)^T [r_0, \ldots, r_j]$ is upper triangular, then it follows that the $j + 1$-st column of $P_0 \cdots P_j$ is the desired unit vector.

If we incorporate these Householder computations into the Lanczos process, we can produce Lanczos vectors that are orthogonal to working accuracy:

$$
\begin{aligned}
&r_0 := q_1 \quad \text{(given unit vector)}\\
&\text{Determine } P_0 = I - 2v_0 v_0^T / v_0^T v_0 \text{ so } P_0 r_0 = e_1.\\
&\alpha_1 := q_1^T A q_1\\
&\text{Do } j = 1, \ldots, n - 1\\
&\qquad r_j := (A - \alpha_j I) q_j - \beta_{j-1} q_{j-1} \qquad (\beta_0 q_0 \equiv 0)\\
&\qquad w := (P_{j-1} \cdots P_0) r_j\\
&\qquad \text{Determine } P_j = I - 2v_j v_j^T / v_j^T v_j \text{ such that}\\
&\qquad\qquad P_j w = (w_1, \ldots, w_j, \beta_j, 0, \ldots, 0)^T\\
&\qquad q_{j+1} := (P_0 \cdots P_j) e_{j+1}\\
&\qquad \alpha_{j+1} := q_{j+1}^T A q_{j+1}
\end{aligned}
$$

(9.2-3)

This is an example of a *complete reorthogonalization* Lanczos scheme. A thorough analysis may be found in Paige (1976b). The idea of using Householder matrices to enforce orthogonality appears in Golub, Underwood, and Wilkinson (1972).

That the computed q_i in (9.2-3) are orthogonal to working precision follows from the roundoff properties of Householder matrices. Note that by virtue of the definition of q_{j+1}, it makes no difference if $\beta_j = 0$. For this reason, the algorithm may safely run until $j = n - 1$. (In practice, however, one would terminate for a much smaller value of j.)

Of course, in any implementation of (9.2-3), one stores the Householder vectors v_j and never explicitly forms corresponding P_j. Since P_j has the form

$$
P_j = \begin{bmatrix} I_j & 0 \\ 0 & \tilde{P}_j \end{bmatrix} \begin{matrix} j \\ n-j \end{matrix}
$$

there is no need to compute the first j components of $w = (P_{j-1} \cdots P_0) r_j$. (In exact arithmetic, these components would be zero.)

Unfortunately, these economies make but a small dent in the computational overhead associated with complete reorthogonalization. The Householder calculations increase the work in the j-th Lanczos step by $O(jn)$ flops. Moreover, to compute q_{j+1}, Householder vectors $v_0, \cdots v_j$ must be accessed. For large n and j, this usually implies a great deal of swapping between primary and secondary store.

Thus, there is a high price associated with complete reorthogonalization. Fortunately, there are more effective courses of action to take, but these demand that we look more closely at precisely how orthogonality is lost.

Selective Orthogonalization

A remarkable, somewhat ironic consequence of the Paige (1971) error analysis is that loss of orthogonality goes hand in hand with convergence of a Ritz pair. To be precise, suppose the symmetric QR algorithm is applied to $\hat{T}_j$ and renders computed Ritz values $\hat{\theta}_1, \ldots, \hat{\theta}_j$ and a nearly orthogonal matrix of eigenvectors $\hat{S}_j = (\hat{s}_{pq})$. If

$$\hat{Y}_j = [\hat{y}_1, \ldots, \hat{y}_j] = fl(\hat{Q}_j\hat{S}_j),$$

then it can be shown that for $i = 1, \ldots, j$

$$(9.2\text{-}4) \qquad\qquad |\hat{q}_{j+1}^T \hat{y}_i| \cong \frac{\mathbf{u}\|A\|_2}{|\hat{\beta}_j| \ |\hat{s}_{ji}|}$$

and

$$(9.2\text{-}5) \qquad\qquad \|A\hat{y}_i - \theta_i\hat{y}_i\|_2 \cong |\beta_j|\|\hat{s}_{ji}|\,.$$

That is, the most recently computed Lanczos vector $\hat{q}_{j+1}$ tends to have a non-trivial (and unwanted) component in the direction of any converged Ritz vector. Consequently, instead of orthogonalizing $\hat{q}_{j+1}$ against all of the previously computed Lanczos vectors, we can achieve the same effect by orthogonalizing it against the (typically few) converged Ritz vectors.

The practical aspects of enforcing orthogonality in this way are discussed in Parlett and Scott (1979). In their scheme, known as *selective orthogonalization*, a computed Ritz pair $(\hat{\theta}, \hat{y})$ is called "good" if it satisfies

$$\|A\hat{y} - \hat{\theta}\hat{y}\|_2 \cong \sqrt{\mathbf{u}}\,\|A\|_2.$$

As soon as $\hat{q}_{j+1}$ is computed, it is orthogonalized against each good Ritz vector. This is much less costly than complete reorthogonalization, since there are usually many fewer good Ritz vectors than Lanczos vectors.

One way to implement selective orthogonalization is to diagonalize $\hat{T}_j$ at each step and then examine the $\hat{s}_{ji}$ in light of (9.2-4) and (9.2-5). A much more efficient approach is to estimate the loss-of-orthogonality measure $\|I_j - \hat{Q}_j^T\hat{Q}_j\|_2$ using the following result:

LEMMA 9.2-2. Suppose $S_+ = [S, d]$ where $S \in \mathbb{R}^{n \times j}$ and $d \in \mathbb{R}^n$. If

$$\|I_j - S^T S\|_2 \leqslant \mu \qquad \text{and} \qquad |1 - d^T d| \leqslant \delta,$$

then

$$\|I_{j+1} - S_+^T S_+\|_2 \leqslant \mu_+$$

where

$$\mu_+ = \tfrac{1}{2}[\mu + \delta + \sqrt{(\mu - \delta)^2 + 4\|S^T d\|_2^2}\,]$$

Proof.   See Kahan and Parlett (1974) or Parlett and Scott (1979). $\square$

Thus, if we have a bound for $\|I_j - \hat{Q}_j^T\hat{Q}_j\|_2$ we can generate a bound for $\|I_{j+1} - \hat{Q}_{j+1}^T\hat{Q}_{j+1}\|_2$ by applying the lemma with $S = \hat{Q}_j$ and $d = \hat{q}_{j+1}$. (In this case $\delta \cong \mathbf{u}$ and we assume that $\hat{q}_{j+1}$ has been orthogonalized against the set of currently good Ritz vectors.) It is possible to estimate the norm of $\hat{Q}_j^T\hat{q}_{j+1}$

from a simple recurrence that spares one the need for accessing $\hat{q}_1, \ldots, \hat{q}_j$. See Kahan and Parlett (1974) or Parlett and Scott (1979). The overhead is minimal, and when the bounds signal loss of orthogonality, it is time to contemplate the enlargement of the set of good Ritz vectors. Then and only then is $\hat{T}_j$ diagonalized.

Considerable effort has been spent in trying to develop a workable Lanczos procedure that does not involve any kind of orthogonality enforcement. Research in this direction focusses on the problem of "ghost" or "spurious" eigenvalues. These are multiple eigenvalues of $\hat{T}_j$ that correspond to simple eigenvalues of A. They arise because the iteration essentially restarts itself when orthogonality to a converged Ritz vector is lost. (By way of analogy, consider what would happen during orthogonal iteration (7.3-4) if we "forgot" to orthogonalize.)

The problem of identifying ghost eigenvalues and coping with their presence is discussed in Cullum and Willoughby (1979) and Parlett and Reid (1981). It is a particularly pressing problem in those applications where all of A's eigenvalues are desired, for then the above orthogonalization procedures are too expensive to implement.

Difficulties with the Lanczos iteration can be expected even if A has a genuinely multiple eigenvalue. This follows because the T_j are unreduced, and unreduced tridiagonal matrices cannot have multiple eigenvalues (cf §7.4). Our next practical Lanczos procedure attempts to circumvent this difficulty.

Block Lanczos

Like the simple power method, the Lanczos algorithm has a block analog. (See §7.3 for the discussion on simultaneous iteration.) Suppose $n = rp$ and consider the decomposition

$$(9.2-6) \qquad Q^T A Q = \overline{T} = \begin{bmatrix} M_1 & B_1^T & & & \mathbf{0} \\ B_1 & M_2 & B_2^T & & \\ & \ddots & \ddots & \ddots & \\ & & & & B_{r-1}^T \\ \mathbf{0} & & & B_{r-1} & M_r \end{bmatrix},$$

where

$$Q = [X_1, \ldots, X_r] \qquad\qquad X_i \in \mathbb{R}^{n \times p}$$

is orthogonal, each $M_i \in \mathbb{R}^{p \times p}$, and each $B_i \in \mathbb{R}^{p \times p}$ is upper triangular. Comparing blocks in $AQ = Q\overline{T}$ shows that

$$A X_j = X_{j-1} B_{j-1}^T + X_j M_j + X_{j+1} B_j \qquad\qquad X_0 B_0 \equiv 0$$

for $j = 1, 2, \ldots, r - 1$. From the orthogonality of Q it follows that

$$M_j = X_j^T A X_j. \qquad\qquad\qquad j = 1, 2, \ldots, r.$$

Moreover,

$$X_{j+1}B_j = R_j$$

represents the Q-R factorization of

$$R_j = AX_j - X_jM_j - X_{j-1}B_{j-1} \in \mathbb{R}^{n \times p}.$$

These observations suggest that the block tridiagonal matrix $\overline{T}$ in (9.2-6) can be generated as follows:

(9.2-7)
$$
\begin{aligned}
&X_1 \in \mathbb{R}^{n \times p} \text{ given, with } X_1^T X_1 = I_p. \\
&M_1 = X_1^T A X_1 \\
&\text{For } j = 1, \ldots, r-1 \\
&\quad R_j = AX_j - X_jM_j - X_{j-1}B_{j-1}^T \qquad (X_0B_0^T = 0) \\
&\quad X_{j+1}B_j = R_j \qquad \text{(Q-R factorization)} \\
&\quad M_{j+1} = X_{j+1}^T A X_{j+1}
\end{aligned}
$$

At the beginning of the j-th pass through the loop we have

$$A[X_1, \ldots, X_j] = [X_1, \ldots, X_j]\,\overline{T}_j + R_j\,[0, \ldots, 0, I_p],$$

where

$$
\overline{T}_j =
\begin{bmatrix}
M_1 & B_1^T & & \mathbf{0} \\
B_1 & M_2 & \ddots & \\
& \ddots & \ddots & B_{j-1}^T \\
\mathbf{0} & & B_{j-1} & M_j
\end{bmatrix}.
$$

Using an argument similar to the one used in the proof of Theorem 9.1-1, we can show that the X_j are mutually orthogonal provided none of the R_j are rank-deficient. However, if $\text{rank}(R_j) < p$ for some j, then it is possible to choose the columns of X_{j+1} such that $X_{j+1}^T X_i = 0$, for $i = 1, \ldots, j$. See Golub and Underwood (1977).

Because $\overline{T}_j$ has bandwidth p, it can be efficiently reduced to tridiagonal form using an algorithm of Schwartz (1968). Once tridiagonal form is achieved, the Ritz values can be obtained via the symmetric QR algorithm.

In order to intelligently decide when to use block Lanczos, it is necessary to understand how the block dimension affects convergence of the Ritz values. The following generalization of Theorem 9.1-3 sheds light on this issue:

THEOREM 9.2-3. Let A be an n-by-n symmetric matrix with eigenvalues $\lambda_1 \geqslant \cdots \geqslant \lambda_n$ and corresponding orthonormal eigenvectors $z_1, \ldots, z_n$. Let $\mu_1 \geqslant \cdots \geqslant \mu_p$ be the p largest eigenvalues of the matrix $\overline{T}_j$ obtained after j steps of the block Lanczos iteration (9.2-7). If $Z_1 = [z_1, \ldots, z_p]$ and $\cos(\theta_p) = \sigma_p(Z_1^T X_1) > 0$, then for $k = 1, \ldots, p$

$$\lambda_k \geqslant \mu_k \geqslant \lambda_k - \epsilon_k^2$$

where

$$\epsilon_k^2 = \frac{(\lambda_1 - \lambda_k)\tan^2(\theta_p)}{\left[c_{j-1}\left(\dfrac{1 + \gamma_k}{1 - \gamma_k}\right)\right]^2} \qquad\qquad \gamma_k = \frac{\lambda_k - \lambda_{p+1}}{\lambda_k - \lambda_n}$$

and $c_{j-1}(z)$ is the $(j - 1)$-st Chebychev polynomial.

Proof. See Underwood (1975). $\square$

Analogous inequalities can be obtained for $\overline{T}_j$'s smallest eigenvalues by applying the theorem with A replaced by $-A$.

Based on Theorem 9.2-3 and scrutiny of the block Lanczos iteration (9.2-7) we may conclude that

(a) the error bound for the Ritz values improve with increased p;

(b) the amount of work required to compute T_j's eigenvalues is proportional to p^2;

(c) the overhead associated with a Lanczos step does not increase much with increased p; and

(d) the block dimension should be at least as large as the largest multiplicity of any sought-after eigenvalue.

How to determine block dimension in the face of these trade-offs is discussed by Scott (1979) in detail.

Loss of orthogonality also plagues the block Lanczos algorithm. However, all of the orthogonality enforcement schemes described above can be applied in the block setting. See Scott (1979), Lewis (1977), and Ruhe (1979) for details.

s-Step Lanczos

The block Lanczos algorithm (9.2-7) can be used in an iterative fashion to calculate selected eigenvalues of A. To fix ideas, suppose we wish to calculate the p largest eigenvalues. If $X_1 \in \mathbb{R}^{n \times p}$ is a given matrix having orthonormal columns, we may proceed as follows:

1. Generate $X_2, \ldots, X_s \in \mathbb{R}^{n \times p}$ via the block Lanczos algorithm.

2. Form $\overline{T}_s = [X_1, \ldots, X_s]^T A [X_1, \ldots, X_s]$ an sp-by-sp, p-diagonal matrix.

3. Compute an orthogonal matrix $U = [u_1, \ldots, u_{sp}]$ such that $U^T \overline{T}_s U = \text{diag}(\theta_1, \ldots, \theta_{sp})$ where $\theta_1 \geqslant \cdots \geqslant \theta_{sp}$.

4. Set $X_1 := [X_1, \ldots, X_s][u_1, \ldots, u_p]$.
5. If $\| AX_1 - X_1 \overline{T}_s \|_F$ is still too large, go to 1.

This is the block analog of the *s-step Lanczos algorithm*, which has been extensively analyzed by Cullum and Donath (1974) and Underwood (1975).

The same idea can also be used to compute several of A's smallest eigenvalues or a mixture of both large and small eigenvalues. See Cullum (1978). The choice of the parameters s and p depends upon storage constraints as well as upon the factors we mentioned above in our discussion of block dimension. The block dimension p may be diminished as the good Ritz vectors emerge. However, this demands that orthogonality to the converged vectors be enforced. See Cullum and Donath (1974).

Problems

P9.2-1. Prove Lemma 9.2-2.

P9.2-2. Show that if $A \in \mathbb{R}^{n \times n}$ is symmetric, then

$$\max_{\substack{X^T X = I_k \\ Y^T Y = I_j}} \text{trace}(X^T A X - Y^T A Y) = \sum_{i=1}^{k} \lambda_i(A) - \sum_{i=1}^{j} \lambda_{n-i+1}(A).$$

Notes and References for Sec. 9.2

The behavior of the Lanczos method in the presence of roundoff error was originally reported in

C. C. Paige (1971). "The Computation of Eigenvalues and Eigenvectors of Very Large Sparse Matrices," Ph.D. thesis, University of London.

Important follow-up papers include

C. C. Paige (1976a). "Error Analysis of the Lancozs Algorithm for Tridiagonalizing a Symmetric Matrix," *J. Inst. Math. Applic. 18*, 341–49.
C. C. Paige (1980). "Accuracy and Effectiveness of the Lanczos Algorithm for the Symmetric Eigenproblem," *Lin. Alg. & Its Applic. 34*, 235–58.

Details pertaining to complete reorthogonalization may be found in

G. H. Golub, R. Underwood, and J. H. Wilkinson (1972). "The Lanczos Algorithm for the Symmetric $Ax = \lambda Bx$ Problem," Report STAN-CS-72-270, Department of Computer Science, Stanford University, Stanford, Calif.
C. C. Paige (1976b). "Practical Use of the Symmetric Lanczos Process with Reorthogonalization," *BIT 10*, 183–95.

The more efficient approach of maintaining orthogonality to the recently converged Ritz vectors is explored in

D. S. Scott (1978). "Analysis of the Symmetric Lanczos Process," UCB-ERL Technical Report M78/40, University of California, Berkeley.

B. N. Parlett and D. S. Scott (1979). "The Lanczos Algorithm with Selective Ortho-gonalization," *Math. Comp. 33*, 217–38.

Without any reorthogonalization it is necessary either to monitor the loss of ortho-gonality and quit at the appropriate instant or else to devise some scheme that will aid in the distinction between the ghost eigenvalues and the actual eigenvalues. See

W. Kahan and B. N. Parlett (1974). "An Analysis of Lanczos Algorithms for Symmet-ric Matrices," ERL-M467, University of California, Berkeley.

W. Kahan and B. N. Parlett (1976). "How Far Should You Go with the Lanczos Pro-cess?" in *Sparse Matrix Computations*, ed. J. Bunch and D. Rose, Academic Press, New York, pp. 131–44.

J. Lewis (1977). "Algorithms for Sparse Matrix Eigenvalue Problems," Report STAN-CS-77-595, Department of Computer Science, Stanford University, Stanford, Calif.

J. Cullum and R. A. Willoughby (1979). "Lanczos and the Computation in Specified Intervals of the Spectrum of Large, Sparse Real Symmetric Matrices," in *Sparse Matrix Proc., 1978*, ed. I. S. Duff and G. W. Stewart, SIAM Publications, Phila-delphia.

B. N. Parlett and J. K. Reid (1981). "Tracking the Progress of the Lanczos Algorithm for Large Symmetric Eigenproblems," *IMA J. Num. Anal. 1*, 135–55.

The original analysis of the block Lanczos algorithm appeared in

R. Underwood (1975). "An Iterative Block Lanczos Method for the Solution of Large Sparse Symmetric Eigenproblems," Report STAN-CS-75-496, Department of Computer Science, Stanford University, Stanford, Calif.

Subsequent papers include

G. H. Golub and R. Underwood (1977). "The Block Lanczos Method for Computing Eigenvalues," in *Mathematical Software III*, ed. J. Rice, Academic Press, New York, pp. 364–77.

J. Cullum and W. E. Donath (1974). "A Block Lanczos Algorithm for Computing the *Q* Algebraically Largest Eigenvalues and A Corresponding Eigenspace of Large Sparse Real Symmetric Matrices," *Proc. of the 1974 IEEE Conf. on Decision and Control*, Phoenix, Ariz. pp. 505–9.

J. Cullum (1978). "The Simultaneous Computation of a Few of the Algebraically Largest and Smallest Eigenvalues of a Large Sparse Symmetric Matrix," *BIT 18*, 265–75.

A. Ruhe (1979). "Implementation Aspects of Band Lanczos Algorithms for Computa-tion of Eigenvalues of Large Sparse Symmetric Matrices," *Math. Comp. 33*, 680–87.

A FORTRAN package for the block Lanczos algorithm is described in

D. S. Scott (1979). "Block Lanczos Software for Symmetric Eigenvalue Problems," Report ORNL/CSD-48, Oak Ridge National Laboratory, Union Carbide Corpora-tion, Oak Ridge, Tenn.

The block Lanczos algorithm generates a symmetric band matrix whose eigenvalues can be computed in any of several ways. One approach is described in

H. R. Schwartz (1968). "Tridiagonalization of a Symmetric Band Matrix," *Numer. Math. 12*, 231–41. See also HACLA, pp. 273–83.

In some applications it is necessary to obtain estimates of interior eigenvalues. The Lanczos algorithm, however, tends to find the extreme eigenvalues first. The following paper shows how a combination of the Lanczos method and inverse iteration can sometimes be used to obtain good estimates of interior eigenvalues:

A. K. Cline, G. H. Golub, and G. W. Platzman (1976). "Calculation of Normal Modes of Oceans Using a Lanczos Method," in *Sparse Matrix Computations*, ed. J. R. Bunch and D. J. Rose, Academic Press, New York, pp. 409–26.

Sec. 9.3. Applications to Linear Equations and Least Squares

In this section we show how the Lanczos iteration can be used to solve large sparse linear equation and least squares problems. Our aim is to give an overview without substantial editorial comment. The topics in this section are the focus of much current research. Readers interested in implementation details and numerical results should consult the appropriate references.

Symmetric Positive Definite Systems

Suppose $A \in \mathbb{R}^{n \times n}$ is symmetric and positive definite and consider the functional $\phi(x)$ defined by

$$\phi(x) = \tfrac{1}{2} x^T A x - x^T b,$$

where $b \in \mathbb{R}^n$. Since $\nabla \phi(x) = Ax - b$, it follows that $x = A^{-1}b$ is the unique minimizer of ϕ. Hence, an approximate minimizer of ϕ can be regarded as an approximate solution to $Ax = b$.

One way to produce a sequence $\{x_j\}$ that converges to x is to generate a sequence of orthonormal vectors $\{q_j\}$ and to let x_j minimize ϕ over $\text{span}\{q_1, \ldots, q_j\}$, where $j = 1, \ldots, n$. Let $Q_j = [q_1, \ldots, q_j]$. Since

$$x \in \text{span}\{q_1, \ldots, q_j\} \Rightarrow \phi(x) = \tfrac{1}{2} y^T (Q_j^T A Q_j) y - y^T (Q_j^T b)$$

for some $y \in \mathbb{R}^j$, it follows that

(9.3-1)
$$x_j = Q_j y_j$$

where

(9.3-2)
$$(Q_j^T A Q_j) y_j = Q_j^T b.$$

Note that $Ax_n = b$.

We now consider how this approach to solving $Ax = b$ can be made effective when A is large and sparse. There are two hurdles to overcome. First, the linear system (9.3-2) must be easily solved; second, we must be able to compute x_j *without* having to refer to $q_1, \ldots, q_j$ explicitly as (9.3-1) suggests. This would be intolerable for large j.

Suppose we use the Lanczos algorithm to generate the q_i. After j steps of that iteration we obtain the factorization

$$(9.3\text{-}3) \qquad AQ_j = Q_j T_j + r_j e_j^T$$

where

$$(9.3\text{-}4) \qquad T_j = Q_j^T A Q_j = \begin{bmatrix} \alpha_1 & \beta_1 & & & \mathbf{0} \\ \beta_1 & \alpha_2 & \ddots & & \\ & \ddots & \ddots & \ddots & \\ & & \ddots & \ddots & \beta_{j-1} \\ \mathbf{0} & & & \beta_{j-1} & \alpha_j \end{bmatrix}.$$

With this approach, (9.3-2) becomes a symmetric positive definite tridiagonal system which may be solved via the L-D-L^T factorization. (See Algorithm 5.3-7.) In particular, by setting

$$L_j = \begin{bmatrix} 1 & & & \mathbf{0} \\ \mu_2 & 1 & & \\ & \ddots & \ddots & \\ \mathbf{0} & & \mu_j & 1 \end{bmatrix} \qquad D_j = \begin{bmatrix} d_1 & & & \mathbf{0} \\ & d_2 & & \\ & & \ddots & \\ \mathbf{0} & & & d_j \end{bmatrix}$$

we find by comparing entries in

$$(9.3\text{-}5) \qquad T_j = L_j D_j L_j^T$$

that

$$d_1 = \alpha_1$$
$$\text{For } i = 2, \ldots, j$$
$$\mu_i = \beta_{i-1}/d_{i-1}$$
$$d_i = \alpha_i - \beta_{i-1}\mu_i$$

Note that we need only calculate

$$(9.3\text{-}6) \qquad \begin{aligned} \mu_j &= \beta_{j-1}/d_{j-1} \\ d_j &= \alpha_j - \beta_{j-1}\mu_j \end{aligned}$$

in order to obtain L_j and D_j from L_{j-1} and D_{j-1}.

As we mentioned above, it is critical to be able to compute x_j efficiently. To this end we define $C_j \in \mathbb{R}^{n \times j}$ and $p_j \in \mathbb{R}^j$ by the equations

$$(9.3\text{-}7) \qquad \begin{aligned} C_j L_j^T &= Q_j \\ L_j D_j p_j &= Q_j^T b \end{aligned}$$

and observe that

$$x_j = Q_j T_j^{-1} Q_j^T b = Q_j (L_j D_j L_j^T)^{-1} Q_j^T b = C_j p_j.$$

Let $C_j = [c_1, \ldots, c_j]$ be a column partitioning. It follows from (9.3-7) that

$$[c_1, \mu_2 c_1 + c_2, \ldots, \mu_j c_{j-1} + c_j] = [q_1, \ldots, q_j],$$

and therefore

$$C_j = [C_{j-1}, c_j] \qquad c_j = q_j - \mu_j c_{j-1}.$$

Also observe that if we set $p_j = [\rho_1, \ldots, \rho_j]^T$ in $L_j D_j p_j = Q_j^T b$, then that equation becomes

$$
\begin{bmatrix}
 & & & \vdots & \\
 & L_{j-1} D_{j-1} & & \vdots & \mathbf{0} \\
 & & & \vdots & \\
 \hdashline
 0 \cdots 0 & \mu_j d_{j-1} & \vdots & d_j
\end{bmatrix}
\begin{bmatrix}
\rho_1 \\
\rho_2 \\
\vdots \\
\rho_{j-1} \\
\rho_j
\end{bmatrix}
=
\begin{bmatrix}
q_1^T b \\
q_2^T b \\
\vdots \\
q_{j-1}^T b \\
q_j^T b
\end{bmatrix} .
$$

Since $L_{j-1} D_{j-1} p_{j-1} = Q_{j-1}^T b$, it follows that

$$p_j = \begin{bmatrix} p_{j-1} \\ \rho_j \end{bmatrix}, \qquad\qquad \rho_j = (q_j^T b - \mu_j d_{j-1} \rho_{j-1})/d_j$$

and thus,

$$x_j = C_j p_j = C_{j-1} p_{j-1} + \rho_j c_j = x_{j-1} + \rho_j c_j.$$

This is precisely the kind of recursive formula for x_j that we need. Together with (9.3-6) and (9.3-7) it enables us to make the transition from $(q_{j-1}, c_{j-1}, x_{j-1})$ to (q_j, c_j, x_j) with a minimal amount of work and storage.

A further simplification results if we set $q_1 = b/\beta_0$, where $\beta_0 = \| b \|_2$. For this choice of a Lanczos starting vector we see that $q_i^T b = 0$ for $i = 2, 3, \ldots$. It follows from (9.3-3) that

$$A x_j = A Q_j y_j = Q_j T_j y_j + r_j e_j^T y_j = Q_j Q_j^T b + r_j e_j^T y_j = b + r_j e_j^T y_j.$$

Thus, if $\beta_j = \| r_j \|_2 = 0$ in the Lanczos iteration, then $A x_j = b$. Moreover, since $\| A x_j - b \|_2 = \beta_j \, | e_j^T y_j |$, the iteration provides estimates of the current residual. Overall, we have the following:

ALGORITHM 9.3-1. Given $b \in \mathbb{R}^n$ and a symmetric positive definite $A \in \mathbb{R}^{n \times n}$, the following algorithm computes $x \in \mathbb{R}^n$ such that $A x = b$.

$$\beta_0 = \|b\|_2$$
$$q_1 = b/\beta_0$$
$$\alpha_1 = q_1^T A q_1$$
$$d_1 = \alpha_1$$
$$c_1 = q_1$$
$$x_1 = b/\alpha_1$$
For $j = 1, \ldots, n - 1$
$$\qquad r_j = (A - \alpha_j I)q_j - \beta_{j-1}q_{j-1} \qquad (\beta_0 q_0 \equiv 0)$$
$$\qquad \beta_j = \|r_j\|_2$$
$\qquad$ If $\beta_j = 0$
$\qquad\qquad$ *then*
$$\qquad\qquad\qquad \text{Set } x = x_j \text{ and quit}$$
$\qquad\qquad$ *else*
$$\qquad\qquad\qquad q_{j+1} = r_j/\beta_j$$
$$\qquad\qquad\qquad \alpha_{j+1} = q_{j+1}^T A q_{j+1}$$
$$\qquad\qquad\qquad \mu_{j+1} = \beta_j/d_j$$
$$\qquad\qquad\qquad d_{j+1} = \alpha_{j+1} - \mu_{j+1}\beta_j$$
$$\qquad\qquad\qquad \rho_{j+1} = -\mu_{j+1}d_j\rho_j/d_{j+1}$$
$$\qquad\qquad\qquad c_{j+1} = q_{j+1} - \mu_{j+1}c_j$$
$$\qquad\qquad\qquad x_{j+1} = x_j + \rho_{j+1}c_{j+1}$$
$$x = x_n$$

This algorithm requires one matrix-vector multiplication and $5n$ flops per iteration.

The numerical behavior of Algorithm 9.3-1 will be discussed in the next chapter, where it is rederived and identified as the widely known method of conjugate gradients.

Symmetric Indefinite Systems

A key feature in the above development is the idea of computing the L-D-L^T factorization of the tridiagonal matrices T_j. Unfortunately, this is potentially unstable if A, and consequently T_j, is not positive definite. A way around this difficulty proposed by Paige and Saunders (1975) is to develop the recursion for x_j via an *L-Q factorization* of T_j. In particular, at the j-th step of the iteration we will have Givens rotations $J_1, \ldots J_{j-1}$ such that

$$T_j J_1 \cdots J_{j-1} = L_j = \begin{bmatrix} d_1 & & & & & \\ e_2 & d_2 & & & \mathbf{0} & \\ f_3 & e_3 & d_3 & & & \\ & \ddots & \ddots & \ddots & & \\ \mathbf{0} & & & f_j & e_j & \bar{d}_j \end{bmatrix}.$$

Note that with this factorization, x_j is given by

$$x_j = Q_j y_j = Q_j T_j^{-1} Q_j^T b = W_j s_j,$$

where $W_j \in \mathbb{R}^{n \times j}$ and $s_j \in \mathbb{R}^j$ are defined by

$$W_j = Q_j J_1 \cdots J_{j-1} \text{ and } L_j s_j = Q_j^T b.$$

Scrutiny of these equations enables one to develop a formula for computing x_j from x_{j-1} and an easily computed multiple of w_j, the last column of W_j. For details, see Paige and Saunders (1975).

Bidiagonalization and the SVD

Suppose $U^T A V = B$ represents the bidiagonalization of $A \in \mathbb{R}^{m \times n}$ and that

(9.3-8)
$$U = [u_1, \ldots, u_m] \qquad\qquad U^T U = I_m$$
$$V = [v_1, \ldots, v_n] \qquad\qquad V^T V = I_n$$

and

(9.3-9)
$$B = \begin{bmatrix} \alpha_1 & \beta_1 & & 0 \\ & \ddots & \ddots & \\ 0 & & \ddots & \beta_{n-1} \\ & & & \alpha_n \\ \hline & & 0 & \end{bmatrix}.$$

Recall that this factorization serves as a front end for the SVD algorithm and may be computed using Householder transformations. See §6.5.

Unfortunately, if A is large and sparse, then we can expect large, dense submatrices to arise during the Householder bidiagonalization. Consequently, it would be nice to develop a means for computing B directly without any orthogonal updates of the matrix A.

Proceeding just as we did in §9.1, we compare columns in the equations $AV = UB$ and $A^T U = VB^T$. Doing this we find that

$$Av_j = \alpha_j u_j + \beta_{j-1} u_{j-1} \qquad\qquad \beta_0 u_0 \equiv 0$$
$$A^T u_j = \alpha_j v_j + \beta_j v_{j+1} \qquad\qquad \beta_n v_{n+1} \equiv 0$$

for $j = 1, \ldots, n$. Defining $r_j = Av_j - \beta_{j-1} u_{j-1}$ and $p_j = A^T u_j - \alpha_j v_j$, we may conclude from orthonormality that $\alpha_j = \pm \|r_j\|_2$, $u_j = r_j / \alpha_j$, $\beta_j = \pm \|p_j\|_2$, and $v_{j+1} = p_j / \beta_j$.

Properly sequenced, these equations define the Lanczos method for bidiagonalizing a rectangular matrix:

$v_1 \in \mathbb{R}^n$ given, with unit 2-norm

$$r_1 = A v_1$$

$$\alpha_1 = \| r_1 \|_2$$

For $j = 1, \ldots, n$

    If $\alpha_j = 0$

        then quit.

(9.3-10)   *else*

$$u_j = r_j / \alpha_j$$

$$p_j = A^T u_j - \alpha_j v_j$$

    $\beta_j = \| p_j \|_2$

    If $\beta_j = 0$

        then quit

        else

$$v_{j+1} = p_j / \beta_j$$

$$r_{j+1} = A v_{j+1} - \beta_j u_j$$

$$\alpha_{j+1} = \| r_{j+1} \|_2$$

This algorithm was first discussed in Paige (1974). It is essentially equivalent to applying the Lanczos tridiagonalization scheme to the symmetric matrix

$$C = \begin{bmatrix} 0 & A \\ A^T & 0 \end{bmatrix}.$$

In §8.3 we showed that

$$\lambda_i(C) = \sigma_i(A) = -\lambda_{n+m-i+1}(C)$$

for $i = 1, \ldots, n$. Because of this, it is not surprising that the large singular values of the bidiagonal matrix

$$B_j = \begin{bmatrix} \alpha_1 & \beta_1 & & & \\ & \alpha_2 & \ddots & & \mathbf{0} \\ & & \ddots & \ddots & \\ & \mathbf{0} & & \ddots & \beta_{j-1} \\ & & & & \alpha_j \end{bmatrix}$$

tend to be very good approximations to the large singular values of A. (The small singular values of A correspond to the interior eigenvalues of C and are not so well approximated.) The equivalent of the Kaniel-Paige theory for the Lanczos bidiagonalization may be found in Luk (1978) as well as in Golub, Luk, and Overton (1981). The analytic, algorithmic, and numerical developments of the previous two sections all carry over naturally to the bidiagonalization.

Least Squares

As detailed in §6.5, the full-rank LS problem $\min \| Ax - b \|_2$ can be solved via the bidiagonalization (9.3-8)–(9.3-9) In particular,

$$x_{LS} = V y_{LS} = \sum_{i=1}^{n} a_i v_i$$

where $y = (a_1, \ldots, a_n)^T$ solves the bidiagonal system $B y = (u_1^T b, \ldots, u_n^T b)^T$. Note that because B is *upper* bidiagonal, we cannot solve for y until the bidiagonalization is complete. Moreover, we are required to save the vectors v_1, $\ldots, v_n$, an unhappy circumstance if n is large.

The development of a sparse least squares algorithm based on the bidiagonalization can be accomplished more favorably if A is reduced to *lower* bidiagonal form:

$$U^T A V = B = \begin{bmatrix} \alpha_1 & & & & \\ \beta_1 & \alpha_2 & & \text{\huge 0} & \\ & \ddots & \ddots & & \\ & & \ddots & \alpha_n & \\ \text{\huge 0} & & & \beta_n \\ \hline & & \text{\huge 0} & \end{bmatrix},$$

where $V = [v_1, \ldots, v_n]$ and $U = [u_1, \ldots, u_m]$. It is straightforward to develop a Lanczos procedure for doing this, and the resulting algorithm is very similar to (9.3-10).

Let $V_j = [v_1, \ldots, v_j]$, $U_j = [u_1, \ldots, u_j]$, and

$$B_j = \begin{bmatrix} \alpha_1 & & & \\ \beta_1 & \alpha_2 & \text{\huge 0} & \\ & \ddots & \ddots & \\ & & \ddots & \alpha_j \\ \text{\huge 0} & & & \beta_j \end{bmatrix} \in \mathbb{R}^{(j+1) \times j}$$

and consider minimizing $\| Ax - b \|_2$ over all vectors of the form $x = V_j y, y \in \mathbb{R}^j$. Since

$$\| A V_j y - b \|_2 = \| U^T A V_j y - U^T b \|_2 = \| B_j y - U_{j+1}^T b \|_2 + \sum_{i=j+2}^{m} (u_i^T b)^2,$$

it follows that $x_j = V_j y_j$ is the minimizing x that we are after, where y_j minimizes the $(j+1)$-by-j LS problem $\min \| B_j y - U_{j+1}^T b \|_2$.

Since B_j is lower bidiagonal, it is easy to compute Jacobi rotations $J_1, \ldots, J_j$ such that

$$J_j \cdots J_1 B_j = \begin{bmatrix} R_j \\ 0 \end{bmatrix} \begin{matrix} j \\ 1 \end{matrix}$$

is upper bidiagonal. If

$$J_j \cdots J_1 U_{j+1}^T b = \begin{bmatrix} d_j \\ u \end{bmatrix} \begin{matrix} j \\ 1 \end{matrix},$$

then it follows that

$$x_j = V_j y_j = W_j d_j$$

where $V_j = W_j R_j$. Paige and Saunders (1978) show how x_j can be obtained from x_{j-1} via a simple recursion that involves the last column of W_j. The net result is a sparse LS algorithm that requires only a few n-vectors of storage to implement.

Unsymmetric Lanczos

We conclude this section with a brief discussion of the unsymmetric Lanczos algorithm. As it is pointed out in Wilkinson (AEP, pp. 388–405), the similarity reduction of a general matrix to tridiagonal form is inadvisable for reasons of stability. Despite this we feel that it is of interest to look at the unsymmetric Lanczos process, if only to highlight the added difficulties associated with the unsymmetric eigenvalue problem.

Suppose $A \in \mathbb{R}^{n \times n}$ and that a nonsingular matrix X exists so

$$X^{-1}AX = T = \begin{bmatrix} \alpha_1 & \gamma_1 & & & \mathbf{0} \\ \beta_1 & \alpha_2 & \ddots & & \\ & \ddots & \ddots & \ddots & \\ & & \ddots & \ddots & \gamma_{n-1} \\ \mathbf{0} & & & \beta_{n-1} & \alpha_n \end{bmatrix}.$$

With the column partitionings

$$x = [x_1, \ldots, x_n]$$

$$X^{-T} = Y = [y_1, \ldots, y_n]$$

we find upon comparing columns in $AX = XT$ and $A^T Y = YT^T$ that

$$Ax_j = \gamma_{j-1}x_{j-1} + \alpha_j x_j + \beta_j x_{j+1} \qquad\qquad \gamma_0 x_0 \equiv 0$$

and

$$A^T y_j = \beta_{j-1}y_{j-1} + \alpha_j y_j + \gamma_j y_{j+1} \qquad\qquad \beta_0 y_0 \equiv 0$$

for $j = 1, \ldots, n - 1$. These equations together with $Y^T X = I_n$ imply

$$\alpha_j = y_j^T A x_j$$

and

$$\beta_j x_{j+1} = r_j \equiv (A - \alpha_j I)x_j - \gamma_{j-1}x_{j-1}$$
$$\gamma_j y_{j+1} = p_j \equiv (A - \alpha_j I)^{\mathsf{T}}y_j - \beta_{j-1}y_{j-1}.$$

There is some flexibility in choosing the scale factors β_j and γ_j. A "canonical" choice is to set $\beta_j = \|r_j\|_2$ and $\gamma_j = x_{j+1}^{\mathsf{T}}p_j$ giving:

$x_1, y_1 \in \mathbb{R}^n$ given, and satisfy $x_1^{\mathsf{T}}x_1 = x_1^{\mathsf{T}}y_1 = 1$

For $j = 1, \ldots, n-1$
  $\alpha_j = y_j^{\mathsf{T}}Ax_j$
  $r_j = (A - \alpha_j I)x_j - \gamma_{j-1}x_{j-1}$　　$(\gamma_0 x_0 \equiv 0)$
  $\beta_j = \|r_j\|_2$

(9.3-11)　　If $\beta_j > 0$
    then $x_{j+1} = r_j/\beta_j$
    else quit
  $p_j = (A - \alpha_j I)^{\mathsf{T}}y_j - \beta_{j-1}y_{j-1}$　　$(\beta_0 y_0 \equiv 0)$
  $\gamma_j = x_{j+1}^{\mathsf{T}}p_j$
  If $\gamma_j \neq 0$
    then $y_{j+1} = p_j/\gamma_j$.
    else quit
$\alpha_n = x_n^{\mathsf{T}}Ay_n$

Defining $X_j = [x_1, \ldots, x_j]$, $Y_j = [y_1, \ldots, y_j]$, and T_j to be the leading j-by-j principal submatrix of T, it is easy to verify that

$$AX_j = X_jT_j + r_je_j^{\mathsf{T}}$$

and

$$A^{\mathsf{T}}Y_j = Y_jT_j^{\mathsf{T}} + p_je_j^{\mathsf{T}}.$$

Thus, whenever $\beta_j = \|r_j\|_2 = 0$, the columns of X_j define an invariant subspace for A. Termination in this regard is therefore a welcome event. Unfortunately, if $\gamma_j = x_{j+1}^{\mathsf{T}}p_j = 0$, then the iteration terminates without any invariant subspace information for either A or A^{T}.

This is but one of the severe difficulties associated with (9.3-11). Other problems include a lack of convergence of T_j's eigenvalues and near dependence among the x_i and among the y_i. We are unaware of any successful implementations of the unsymmetric Lanczos algorithm.

Problems

P9.3-1. Show that the vector x_j in Algorithm 9.3-1 satisfies

$$\|Ax_j - b\|_2 = \beta_j \,|e_n^{\mathsf{T}}y_j| . \qquad\qquad T_jy_j = Q_j^{\mathsf{T}}b$$

P9.3-2. Give an example of a starting vector for which the unsymmetric Lanczos iteration (9.3-11) breaks down without rendering any invariant subspace information. Use

$$A = \begin{bmatrix} 1 & 6 & 2 \\ 3 & 0 & 2 \\ 1 & 3 & 5 \end{bmatrix}.$$

Notes and References for Sec. 9.3

Most of the material in this section has been distilled from the following papers

C. C. Paige (1974a). "Bidiagonalization of Matrices and Solution of Linear Equations," *SIAM J. Num. Anal. 11*, 197–209.

C. C. Paige and M. A. Saunders (1975). "Solution of Sparse Indefinite Systems of Linear Equations," *SIAM J. Num. Anal. 12*, 617–29.

C. C. Paige and M. A. Saunders (1978). "A Bidiagonalization Algorithm for Sparse Linear Equations and Least Squares Problems," Report SOL 78-19, Department of Operations Research, Stanford University, Stanford, Calif.

C. C. Paige and M. A. Saunders (1982a). "LSQR: An Algorithm for Sparse Linear Equations and Sparse Least Squares," *ACM Trans. Math. Soft. 8*, 43–71.

C. C. Paige and M. A. Saunders (1982b). "Algorithm 583 LSQR: Sparse Linear Equations and Least Squares Problems," *ACM Trans. Math. Soft. 8*, 195–209.

Other applications of the Lanczos algorithm are described in

B. N. Parlett (1980a). "A New Look at the Lanczos Algorithm for Solving Symmetric Systems of Linear Equations," *Lin. Alg. & Its. Applic. 29*, 323–46.

O. Widlund (1978). "A Lanczos Method for a Class of Nonsymmetric Systems of Linear Equations," *SIAM J. Numer. Anal. 15*, 801–12.

G. H. Golub. F. T. Luk, and M. Overton (1981). "A Block Lanczos Method for Computing the Singular Values and Corresponding Singular Vectors of a Matrix," *ACM Trans. Math. Soft. 7*, 149–69.

J. M. Van Kats and H. A. Van der Vorst (1977). "Automatic Monitoring of Lanczos Schemes for Symmetric or Skew-Symmetric Generalized Eigenvalue Problems," Report TR 7, Academisch Computer Centru, Utrecht, The Netherlands.

G. H. Golub, R. Underwood, and J. H. Wilkinson (1972). "The Lanczos Algorithm for the Symmetric $Ax = \lambda Bx$ Problem," Report STAN-CS-72-270, Department of Computer Science, Stanford University, Stanford, Calif.

T. Ericsson and A. Ruhe (1980). "The Spectral Transformation Lanczos Method for the Numerical Solution of Large Sparse Generalized Symmetric Eigenvalue Problems," *Math. Comp. 35*, 1251–68.

F. T. Luk (1978). "Sparse and Parallel Matrix Computations," Report STAN-CS-78-685, Department of Computer Science, Stanford University, Stanford, Calif.

Iterative Methods for Linear Systems

§10.1 **The Standard Iterations**

§10.2 **Derivation and Properties of the Conjugate Gradient Method**

§10.3 **Practical Conjugate Gradient Procedures**

We concluded the previous chapter by showing how the Lanczos iteration could be used to solve various linear equation and least squares problems. The methods developed were suitable for large sparse problems because they did not require the factorization of the underlying matrix. In this section we continue the discussion of linear equation solvers that have this property.

The first section is a brisk review of the classical iterations: Jacobi, Gauss-Seidel, SOR, Chebychev semi-iterative, and so on. Our treatment of these methods is brief because our principal aim in this chapter is to highlight the method of conjugate gradients, a scheme for which there is not a great deal of expository literature. In §10.2 we carefully develop this important technique in a natural way from the method of steepest descent. Readers will recall that the conjugate gradient method has already been introduced via the Lanczos iteration in §9.3. The reason for deriving the method again is to motivate some of its practical variants, which are the subject of §10.3.

We warn the reader of an inconsistency in the notation of this chapter. In §10.1 methods are developed at the "(i, j) level" necessitating the use of superscripts: $x_i^{(k)}$ denotes the i-th component of a vector $x^{(k)}$. In the other sections, however, algorithmic developments can proceed without explicit mention of vector/matrix entries. Hence, in §10.2 and §10.3 we can dispense with superscripts and denote vector sequences by "$\{x_k\}$."

Reading Path

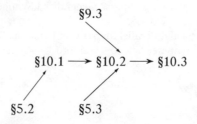

Sec. 10.1. The Standard Iterations

Most of the linear equation solvers of Chapters 4 and 5 require the factorization of the coefficient matrix A. Methods of this type are called *direct methods*. Direct methods can be impractical if A is large and sparse, because the sought-after factors can be dense. An exception to this occurs when A is banded (cf. §5.3). Yet in many band matrix problems even the band itself is sparse, making algorithms such as band Cholesky difficult to implement.

One reason for great interest in sparse linear equation solvers is the importance of being able to obtain numerical solutions to partial differential equations. Indeed, researchers in computational PDE's have been responsible for many of the sparse matrix techniques that are presently in general use.

Roughly speaking, there are two approaches to the sparse $Ax = b$ problem. One is to pick an appropriate direct method and adapt it to exploit A's sparsity. Typical adaptation strategies involve the intelligent use of data structures and special pivoting strategies that minimize fill-in. The literature in this area is vast, and the interested reader should consult the several articles in Bunch and Rose (1976) as well as the book by George and Liu (1981). We shall not delve further into this very active area of matrix computation research.

In contrast to the direct methods are the iterative methods. These methods generate a sequence of approximate solutions $\{x^{(k)}\}$ and usually involve the matrix A only multiplicatively. The evaluation of an iterative method invariably focuses on how quickly the iterates $x^{(k)}$ converge. In this section we present some basic iterative methods, discuss their practical implementation, and prove a few representative theorems concerned with their behavior.

Perhaps the simplest iterative scheme is the *Jacobi iteration*. It is defined for matrices that have nonzero diagonal elements. The method can be motivated by rewriting the 3-by-3 system $Ax = b$ as follows:

$$x_1 = (b_1 - a_{12}x_2 - a_{13}x_3)/a_{11}$$
$$x_2 = (b_2 - a_{21}x_1 - a_{23}x_3)/a_{22}$$
$$x_3 = (b_3 - a_{31}x_1 - a_{32}x_2)/a_{33}.$$

Here we are assuming that A has nonzero diagonal entries. Suppose $x^{(k)}$ is an approximation to $x = A^{-1}b$. A natural way to generate a new approximation $x^{(k+1)}$ is to compute

(10.1-1)
$$x_1^{(k+1)} = (b_1 - a_{12}x_2^{(k)} - a_{13}x_3^{(k)})/a_{11}$$
$$x_2^{(k+1)} = (b_2 - a_{21}x_1^{(k)} - a_{23}x_3^{(k)})/a_{22}$$
$$x_3^{(k+1)} = (b_3 - a_{31}x_1^{(k)} - a_{32}x_2^{(k)})/a_{33}.$$

This defines the Jacobi iteration for the case $n = 3$. For general n we have

For $i = 1, \ldots, n$

$$(10.1\text{-}2) \qquad x_i^{(k+1)} = \left(b_i - \sum_{j=1}^{i-1} a_{ij} x_j^{(k)} - \sum_{j=i+1}^{n} a_{ij} x_j^{(k)} \right) \Big/ a_{ii}$$

Note that in the Jacobi iteration one does not use the most recently available information when computing $x_i^{(k+1)}$. For example, $x_1^{(k)}$ is used in the calculation of $x_2^{(k+1)}$ even though $x_1^{(k+1)}$ is known. If we revise the Jacobi iteration so that we always use the most current estimate of the exact x_i, we obtain

For $i = 1, \ldots, n$

$$(10.1\text{-}3) \qquad x_i^{(k+1)} = \left(b_i - \sum_{j=1}^{i-1} a_{ij} x_j^{(k+1)} - \sum_{j=i+1}^{n} a_{ij} x_j^{(k)} \right) \Big/ a_{ii}$$

This defines what is called the *Gauss-Seidel iteration*.

For both the Jacobi and Gauss-Seiden iterations, the transition from $x^{(k)}$ to $x^{(k+1)}$ can be succinctly described in terms of the matrices L, D, and U defined by

$$L = \begin{bmatrix} 0 & & & & \\ a_{21} & 0 & & \mathbf{0} & \\ \vdots & \vdots & \ddots & & \\ \vdots & \vdots & & \ddots & \\ a_{n1} & a_{n2} & \cdots & a_{n,n-1} & 0 \end{bmatrix},$$

$$(10.1\text{-}4) \qquad D = \mathrm{diag}(a_{11}, \ldots, a_{nn}),$$

$$U = \begin{bmatrix} 0 & a_{12} & \cdots & & a_{1n} \\ & 0 & \cdots & & a_{2n} \\ & & \ddots & & \vdots \\ \mathbf{0} & & & \ddots & a_{n-1,n} \\ & & & & 0 \end{bmatrix}.$$

In particular, the Jacobi step has the form

$$M_J x^{(k+1)} = N_J x^{(k)} + b,$$

where $M_J = D$ and $N_J = -(L + U)$. On the other hand,

$$M_G x^{(k+1)} = N_G x^{(k)} + b$$

with $M_G = (D + L)$ and $N_G = -U$ defines Gauss-Seidel. The question of interest, therefore, is when iterations of the form $M x^{(k+1)} = N x^{(k)} + b$ converge to $x = A^{-1} b$.

THEOREM 10.1-1. Suppose $b \in \mathbb{R}^n$ and $A = M - N \in \mathbb{R}^{n \times n}$ is nonsingular. If M is nonsingular and the spectral radius of $M^{-1}N$ satisfies $\rho(M^{-1}N) < 1$, then the iterates $x^{(k)}$ defined by $Mx^{(k+1)} = Nx^{(k)} + b$ converge to $x = A^{-1}b$ for any starting vector $x^{(0)}$.

Proof. Let $e^{(k)} = x^{(k)} - x$ denote the error in the k-th iterate. Since $Mx = Nx + b$ it follows that $M(x^{(k+1)} - x) = N(x^{(k)} - x)$, and thus,

$$e^{(k+1)} = M^{-1}Ne^{(k)} = (M^{-1}N)^k e^{(0)}.$$

From Lemma 7.3-2 we know that $(M^{-1}N)^k \to 0$ if $\rho(M^{-1}N) < 1$. $\square$

This result is central in the area of iterative methods, where algorithmic development typically proceeds along the following lines:

(a) A *splitting* $A = M - N$ is proposed, where linear systems of the form $Mz = d$ are "easy" to solve.

(b) Classes of matrices are identified for which the *iteration matrix* $G = M^{-1}N$ satisfies $\rho(G) < 1$.

(c) Further results about $\rho(G)$ are established to gain intuition about how the error $e^{(k)}$ tends to zero.

Consider, for example, the Jacobi iteration

$$Dx^{(k+1)} = -(L + U)x^{(k)} + b.$$

One condition that guarantees $\rho(M_J^{-1}N_J) < 1$ is strict diagonal dominance. Indeed, if A has that property, then

$$\rho(M_J^{-1}N_J) \leqslant \|D^{-1}(L + U)\|_\infty = \max_i \sum_{j \neq i} |a_{ij}/a_{ii}| < 1.$$

Clearly, the "more dominant" the diagonal of A, the more rapid will be the convergence.

A more complicated spectral radius argument is needed to show that Gauss-Seidel converges for symmetric positive definite A:

THEOREM 10.1-2. If $A \in \mathbb{R}^{n \times n}$ is symmetric and positive definite, then the Gauss-Seidel iteration (10.1-3) converges for any $x^{(0)}$.

Proof. Write $A = L + D + L^T$ where $D = \text{diag}(a_{ii})$ and L is strictly lower triangular. In light of Theorem 10.1-1 our task is to show that the matrix $G = -(D + L)^{-1}L^T$ has eigenvalues that are inside the unit circle. Since D is positive definite we have

$$G_1 \equiv D^{1/2}GD^{-1/2} = -(I + L_1)^{-1}L_1^T,$$

where $L_1 = D^{-1/2}LD^{-1/2}$. Since G and G_1 have the same eigenvalues, we must verify that $\rho(G_1) < 1$. If $G_1x = \lambda x$ with $x^H x = 1$, then

$$-L_1^T x = \lambda(I + L_1)x$$

and thus,

$$-x^H L_1^T x = \lambda(1 + x^H L_1 x).$$

Letting $a + bi = x^H L_1 x$ we have

$$|\lambda|^2 = \left| \frac{-a + bi}{1 + a + bi} \right|^2 = \frac{a^2 + b^2}{1 + 2a + a^2 + b^2}.$$

However, since $D^{-1/2} A D^{-1/2} = I + L_1 + L_1^T$ is positive definite,

$$0 < 1 + x^H L_1 x + x^H L_1^T x = 1 + 2a$$

implying $|\lambda| < 1$. $\square$

This result is frequently applicable because many of the matrices that arise from discretized elliptic PDE's are symmetric positive definite. Numerous other results of this flavor appear in the literature. See Varga (1962), Young (1971) and Hageman and Young (1981).

We now focus on some practical details associated with the Gauss-Seidel iteration. With overwriting the method is particularly simple to implement:

ALGORITHM 10.1-1: *Gauss-Seidel Step.* Given $x \in \mathbb{R}^n$, $b \in \mathbb{R}^n$, and $A = L + D + U \in \mathbb{R}^{n \times n}$ with $D = \text{diag}(a_{ii})$ nonsingular, L strictly lower triangular and U strictly upper triangular, the following algorithm overwrites x with $(D + L)^{-1}(b - Ux)$.

For $i = 1, \ldots, n$

$$x_i := \left(b_i - \sum_{j=1}^{i-1} a_{ij} x_j - \sum_{j=i+1}^{n} a_{ij} x_j \right) \Big/ a_{ii}$$

This algorithm requires no more flops than there are entries in the matrix A. It makes no sense to be more precise about the work involved because the actual implementation depends greatly upon the structure of the problem at hand.

In order to stress this point we consider the application of Algorithm 10.1-1 to the problem

(10.1-6)

$$\begin{bmatrix} T & -I_N & & & \mathbf{0} \\ -I_N & T & -I_N & & \\ & \ddots & \ddots & \ddots & \\ & & & & -I_N \\ \mathbf{0} & & & -I_N & T \end{bmatrix} \begin{bmatrix} g_1 \\ g_2 \\ \vdots \\ \vdots \\ g_M \end{bmatrix} = \begin{bmatrix} f_1 \\ f_2 \\ \vdots \\ \vdots \\ f_M \end{bmatrix}, \qquad NM\text{-by-}NM$$

where

$$
T = \begin{bmatrix} 4 & -1 & & & \\ -1 & 4 & -1 & & \Large 0 \\ & \ddots & \ddots & \ddots & \\ & & & & -1 \\ \Large 0 & & & -1 & 4 \end{bmatrix}, \quad f_j = \begin{bmatrix} f(1,j) \\ f(2,j) \\ \vdots \\ f(N,j) \end{bmatrix}, \quad g_j = \begin{bmatrix} g(1,j) \\ g(2,j) \\ \vdots \\ g(N,j) \end{bmatrix}.
$$

This problem arises when the Poisson equation is discretized on a rectangle. It is easy to show that the matrix A is positive definite.

With the convention that $g(i, j) = 0$ whenever $i \in \{0, N + 1\}$ or $j \in \{0, M + 1\}$ we see that the Gauss-Seidel step takes on the form

For $j = 1, \ldots, M$
    For $i = 1, \ldots, N$
$$
g(i, j) := [f(i, j) + g(i - 1, j) + g(i + 1, j) + g(i, j - 1)
$$
$$
+ g(i, j + 1)]/4
$$

Note that in this problem no storage is required for the matrix A.

The Gauss-Seidel iteration is very attractive because of its simplicity. Unfortunately, if the spectral radius of $M_G^{-1} N_G$ is close to unity, then it may be prohibitively slow, for the error tends to zero like $\rho(M_G^{-1} N_G)^k$. In an effort to rectify this, let $\omega \in \mathbb{R}$ and consider the following modification of the Gauss-Seidel step:

For $i = 1, \ldots, n$

$$
x_i^{(k+1)} = \omega \left(b_i - \sum_{j=1}^{i-1} a_{ij} x_j^{(k+1)} - \sum_{j=i+1}^{n} a_{ij} x_j^{(k)} \right) \Big/ a_{ii} + (1 - \omega) x_i^{(k)}
$$

This defines the method of *successive over-relaxation* (SOR). In matrix terms, the SOR step is given by

(10.1-7)
$$
M_\omega x^{(k+1)} = N_\omega x^{(k)} + \omega b
$$

where

$$
M_\omega = D + \omega L
$$

and

$$
N_\omega = (1 - \omega)D - \omega U.
$$

For a few structured (but important) problems such as (10.1-6) the value of the *relaxation parameter* ω that minimizes $\rho(M_\omega^{-1} N_\omega)$ is known. Moreover, a significant reduction in $\rho(M_1^{-1} N_1) = \rho(M_G^{-1} N_G)$ can result. In more complicated problems, however, it may be necessary to perform a fairly sophisiticated eigenvalue analysis in order to determine an appropriate ω. A complete survey of "SOR theory" appears in Young (1971). Practical schemes for estimating the optimum ω are discussed in O'Leary (1976) and Wachpress (1966).

Another way to accelerate the convergence of an iterative method makes use of Chebychev polynomials. Suppose $x^{(1)}, \ldots, x^{(k)}$ have been generated via

$$Mx^{(j+1)} = Nx^{(j)} + b$$

and that we wish to determine coefficients $\nu_j(k), \ j = 0, \ldots, k$ such that

(10.1-9)
$$y^{(k)} = \sum_{j=0}^{k} \nu_j(k)x^{(j)}$$

represents an improvement over $x^{(k)}$. If $x^{(0)} = \cdots = x^{(k)} = x$, then it is reasonable to insist that $y^{(k)} = x$. Hence, we require

(10.1-10)
$$\sum_{j=0}^{k} \nu_j(k) = 1.$$

Subject to this constraint, we consider how to choose the $\nu_j(k)$ so that the error in $y^{(k)}$ is minimized.

Recalling from the proof of Theorem 10.1-1 that $x^{(k)} - x = (M^{-1}N)^k e^{(0)}$ where $e^{(0)} = x^{(0)} - x$, we see that

$$y^{(k)} - x = \sum_{j=0}^{k} \nu_j(k)[x^{(j)} - x] = \sum_{j=0}^{k} \nu_j(k)(M^{-1}N)^j e^{(0)}.$$

Working in the 2-norm we therefore obtain

(10.1-11)
$$\|y^{(k)} - x\|_2 \leqslant \|p_k(G)\|_2 \|e^{(0)}\|_2,$$

where $G = M^{-1}N$ and

$$p_k(z) = \sum_{j=0}^{k} \nu_j(k)z^j.$$

Note that the condition (10.1-10) implies $p_k(1) = 1$.

At this point we assume that G is symmetric with eigenvalues λ_i that satisfy

$$-1 < \alpha \leqslant \lambda_n \leqslant \cdots \leqslant \lambda_1 \leqslant \beta < 1.$$

It follows that

$$\|p_k(G)\|_2 = \max_{\lambda_i} |p_k(\lambda_i)| \leqslant \max_{\alpha \leqslant \lambda \leqslant \beta} |p_k(\lambda)|.$$

Thus, to make the norm of $p_k(G)$ small, we need a polynomial $p_k(z)$ that is small on $[\alpha, \beta]$ subject to the constraint that $p_k(1) = 1$.

Recall that the Chebychev polynomials $c_j(z)$ generated by the recursion

$$c_0(z) = 1$$

$$c_1(z) = z$$

$$c_j(z) = 2zc_{j-1}(z) - c_{j-2}(z)$$

satisfy $|c_j(z)| \leqslant 1$ on $[-1, 1]$ but grow rapidly off this interval. As a consequence, the polynomial

$$p_k(z) = \frac{c_k\left[-1 + 2\dfrac{z - \alpha}{\beta - \alpha}\right]}{c_k(\mu)},$$

where

$$\mu = -1 + 2\frac{1 - \alpha}{\beta - \alpha} = 1 + 2\frac{1 - \beta}{\beta - \alpha}$$

satisfies $p_k(1) = 1$ and tends to be small on $[\alpha, \beta]$. From the definition of $p_k(z)$ and equation (10.1-11) we see

$$\|y^{(k)} - x\|_2 \leqslant \frac{\|x - x^{(0)}\|_2}{|c_k(\mu)|}.$$

Thus, the larger μ is, the greater will be the acceleration of convergence.

In order for the above to be a practical acceleration procedure we need a more efficient method for calculating $y^{(k)}$ than (10.1-9). We have been tacitly assuming that n is large and thus, that the storage of $x^{(0)}, \ldots, x^{(k)}$ would be inconvenient or impossible.

Fortunately it is possible to derive a three-term recurrence among the $y^{(k)}$ by exploiting the three-term recurrence among the Chebychev polynomials. In particular, it can be shown that if

$$\omega_{k+1} = 2\frac{2 - \beta - \alpha}{\beta - \alpha}\frac{c_k(\mu)}{c_{k+1}(\mu)}$$

then

$$y^{(k+1)} = \omega_{k+1}(y^{(k)} - y^{(k-1)} + \gamma z^{(k)}) + y^{(k-1)}$$

(10.1-12) $$Mz^{(k)} = b - Ay^{(k)}$$

$$\gamma = 2/(2 - \alpha - \beta),$$

where $y^{(0)} = x^{(0)}$ and $y^{(1)} = x^{(1)}$. We refer to this scheme as the Chebychev semi-iterative method associated with $My^{(k+1)} = Ny^{(k)} + b$. For the acceleration to be effective we need good lower and upper bounds α and β. As in SOR, these parameters may be difficult to ascertain in all but a few structured problems.

Chebychev semi-iterative methods are extensively analyzed in Varga (1962, chap. 5) as well as in Golub and Varga (1961).

In deriving the Chebychev acceleration we assumed that the iteration matrix $G = M^{-1}N$ was symmetric. However, so long as G has real eigenvalues the analysis goes through. But even with this weaker assumption (10.1-12) cannot be applied to SOR because $M_\omega^{-1}N_\omega$ can have complex eigenvalues. Interestingly, it is possible to "symmetrize" SOR by coupling it with the *backward SOR* scheme

For $i = n, \ldots, 1$

$$x_i^{(k+1)} = \omega\left(b_i - \sum_{j=1}^{i-1} a_{ij}x_j^{(k)} - \sum_{j=i+1}^{n} a_{ij}x_j^{(k+1)}\right)\Big/a_{ii} + (1 - \omega)x_i^{(k)}$$

This can be described in matrix terms using (10.1-4). In particular, we have

$$\bar{M}_\omega x^{(k+1)} = \bar{N}_\omega x^{(k)} + \omega b,$$

where

(10.1-13) $\qquad \bar{M}_\omega = D + \omega U \quad$ and $\quad \bar{N}_\omega = (1 - \omega)D - \omega L.$

If A is symmetric ($U = L^T$), then $\bar{M}_\omega = M_\omega^T$ and $\bar{N}_\omega = N_\omega^T$, and we have the iteration

(10.1-14)
$$M_\omega x^{(k+1/2)} = N_\omega x^{(k)} + \omega b$$
$$M_\omega^T x^{(k+1)} = N_\omega^T x^{(k+1/2)} + \omega b.$$

It is clear that the iteration matrix for this method is given by

$$G = M_\omega^{-T}N_\omega^T M_\omega^{-1}N_\omega.$$

A calculation shows that

$$G = M^{-1}N \equiv (M_\omega D^{-1}M_\omega^T)^{-1}(N_\omega^T D^{-1}N_\omega).$$

It is easy to show that if D has positive diagonal entries, then G is similar to a symmetric matrix and hence, has real eigenvalues.

The iteration (10.1-14) is called the *symmetric successive over-relaxation* (SSOR) method. It is frequently used in conjunction with the Chebychev semi-iterative acceleration.

Problems

P10.1-1. Show that the Jacobi iteration can be written in the form

$$x^{(k+1)} = x^{(k)} + Hr^{(k)}$$

where $r^{(k)} = b - Ax^{(k)}$. Repeat for the Gauss-Seidel iteration.

P10.1-2. Show that if A is strictly diagonally dominant, then the Gauss-Seidel iteration converges.

P10.1-3. Show that the Jacobi iteration converges for 2-by-2 symmetric positive definite systems.

P10.1-4. Show that if $A = M - N$ is singular, then we can never have $\rho(M^{-1}N) < 1$ even if M is nonsigular.

P10.1-5. Compare $\rho(M_J^{-1}N_J)$ and $\rho(M_G^{-1}N_G)$ for the matrix

$$
A = \begin{bmatrix} 4 & -1 & -1 \\ -1 & 4 & -1 \\ -1 & -1 & 4 \end{bmatrix}.
$$

P10.1-6. What is the optimum SOR parameter for the example in P10.1-5?

P10.1-7. What is the SSOR iteration matrix for the example in P10.1-5?

Notes and References for Sec. 10.1

Three comprehensive volumes survey the area of iterative methods in far greater depth than our own treatment. They are

R. S. Varga (1962). *Matrix Iterative Analysis*, Prentice-Hall, Englewood Cliffs, N.J.

D. M. Young (1971). *Iterative Solution of Large Linear Systems*, Academic Press, New York.

L. A. Hageman and D. M. Young (1981). *Applied Iterative Methods*, Academic Press, New York.

We also highly recommend chapter 7 of

J. Ortega (1972). *Numerical Analysis: A Second Course*, Academic Press, New York.

The following books contain numerous articles on the different aspects of sparse matrix computation:

D. J. Rose and R. A. Willoughby, eds. (1972). *Sparse Matrices and Their Application*, Plenum Press, New York.

J. R. Bunch and D. J. Rose, eds. (1976). *Sparse Matrix Computations*, Academic Press, New York.

I. S. Duff and G. W. Stewart, eds. (1979). *Sparse Matrix Proceedings, 1978*, SIAM Publications, Philadelphia.

A. Bjorck, R. J. Plemmons, and H. Schneider (1981). *Large-Scale Matrix Problems*, North-Holland, New York.

These references describe some of the problems associated with the development of sparse matrix software. See also

J. A. George and J. Liu (1981). *Computer Solution of Large Sparse Positive Definite Systems*, Prentice-Hall, Englewood Cliffs, N.J.

As we mentioned, Young (1971) has the most comprehensive treatment of the SOR method. The object of "SOR theory" is to guide the user in choosing the relaxation parameter ω. In this setting, the ordering of equations and unknowns is critical. See

M.J.M. Bernal and J. H. Verner (1968). "On Generalizations of the Theory of Consistent Orderings for Successive Over-Relaxation Methods," *Numer. Math. 12*, 215-22.

R. A. Nicolaides (1974). "On a Geometrical Aspect of SOR and the Theory of Consistent Ordering for Positive Definite Matrices," *Numer. Math. 23*, 99-104.

D. M. Young (1970). "Convergence Properties of the Symmetric and Unsymmetric Over-Relaxation Methods," *Math. Comp. 24*, 793-807.

D. M. Young (1972). "Generalization of Property *A* and Consistent Ordering," *SIAM J. Num. Anal. 9*, 454-63.

Heuristic methods for estimating ω are also of interest. See

E. L. Wachpress (1966). *Iterative Solution of Elliptic Systems*, Prentice-Hall, Englewood Cliffs, N.J.

D. P. O'Leary (1976). "Hybrid Conjugate Gradient Algorithms," Report STAN-CS-76-548, Department of Computer Science, Stanford University (Ph.D. thesis).

An analysis of the Chebychev semi-iterative method appears in

G. H. Golub and R. S. Varga (1961). "Chebychev Semi-Iterative Methods, Successive Over-Relaxation Iterative Methods, and Second-Order Richardson Iterative Methods," Parts I and II, *Numer. Math. 3*, 147-56, 157-68.

This work is premised on the assumption that the underlying iteration matrix has real eigenvalues. How to proceed when this is not the case is discussed in

T. A. Manteuffel (1977). "The Tchebychev Iteration for Nonsymmetric Linear Systems," *Numer. Math. 28*, 307-27.

Sometimes it is possible to "symmetrize" an iterative method, thereby simplifying the acceleration process, since all the relevant eigenvalues are real. Such is the case of SSOR method discussed in

J. W. Sheldon (1955). "On the Numerical Solution of Elliptic Difference Equations," *Math. Tables Aids Comp. 9*, 101-12.

Sec. 10.2. Derivation and Properties of the Conjugate Gradient Method

A difficulty associated with the SOR, Chebychev semi-iterative, and related methods is that they depend upon parameters that are sometimes hard to choose properly. For example, for Chebychev acceleration to be successful we need good estimates of the largest and smallest eigenvalue of the underlying iteration matrix $M^{-1}N$. Unless this matrix is sufficiently structured, this may be analytically impossible and/or computationally expensive.

In this section we present a method without this difficulty—the Hestenes-Stiefel conjugate gradient method. We derived this method in §9.3, where we discussed how to use the Lanczos algorithm to solve symmetric linear systems. The reason for rederiving the conjugate gradient method is to set the stage for §10.3, where we show its connection with other iterative $Ax = b$ schemes and describe some of its useful generalizations.

The starting point in the derivation is to consider how we might go about minimizing the functional $\phi(x)$, defined by

$$\phi(x) = \tfrac{1}{2}x^TAx - x^Tb,$$

where $b \in \mathbb{R}^n$ and $A \in \mathbb{R}^{n \times n}$ is assumed to be positive definite and symmetric. The minimum value of ϕ is $-\tfrac{1}{2}b^TA^{-1}b$, achieved by setting $x = A^{-1}b$. Thus, minimizing ϕ and solving $Ax = b$ are equivalent problems.

One of the simplest strategies to minimize ϕ is the method of steepest descent. Recall that at a point x_k the function ϕ decreases most rapidly in the direction of the negative gradient $-\nabla\phi(x_k) = b - Ax_k$. We call

$$r_k = b - Ax_k$$

the *residual* of x_k. If the residual is nonzero, then there exists an $\alpha \in \mathbb{R}$ such that

$$\phi(x_k + \alpha r_k) < \phi(x_k).$$

In the method of steepest descent (with exact line search) the parameter α is chosen so as to minimize $\phi(x_k + \alpha r_k)$. This gives

(10.2-1)
$$
\begin{aligned}
&x_0 = 0 \\
&\text{For } k = 1, 2, \ldots \\
&\qquad r_{k-1} = b - Ax_{k-1} \\
&\qquad \text{If } r_{k-1} = 0 \\
&\qquad\qquad then \text{ quit} \\
&\qquad else \\
&\qquad\qquad \alpha_k = r_{k-1}^T r_{k-1} / r_{k-1}^T A r_{k-1} \\
&\qquad\qquad x_k = x_{k-1} + \alpha_k r_{k-1}
\end{aligned}
$$

The starting vector $x_0 = 0$ is not restrictive. If $\tilde{x}$ is a more appropriate initial guess, we merely apply (10.2-1) with b replaced by $b - A\tilde{x}$. The vectors $x_k + \tilde{x}$ are then the desired approximate solutions.

The global convergence of the steepest descent method follows from the inequality

$$\phi(x_k) + \frac{1}{2} b^TA^{-1}b \leqslant \left(1 - \frac{1}{\kappa_2(A)}\right)\left(\phi(x_{k-1}) + \frac{1}{2} b^TA^{-1}b\right)$$

Unfortunately, the speed of convergence may be prohibitively slow if $\kappa_2(A) = \lambda_1(A)/\lambda_n(A)$ is large. In this situation, the level curves of ϕ are very elongated hyperellipsoids and minimization corresponds to finding the lowest point on a relatively flat, steep-sided valley. In steepest descent, we are forced to traverse back and forth *across* the valley as depicted in Figure 10.2-1 rather than *down* the valley. Algebraically, the difficulty is that the gradient directions that arise are far too similar as the iteration progresses.

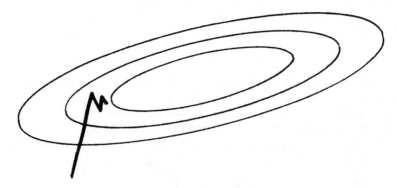

Figure 10.2-1.

This suggests that we successively minimize ϕ along a set of directions $\{p_1, p_2, \ldots\}$ that do not necessarily correspond to the residuals $\{r_0, r_1, \ldots\}$. It is easy to show that to minimize $\phi(x_{k-1} + \alpha p_k)$ with respect to α, we merely set

$$\alpha = \alpha_k \equiv p_k^T r_{k-1} / p_k^T A p_k.$$

With this choice, it can be shown that

(10.2-2) $$\phi(x_{k-1} + \alpha_k p_k) = \phi(x_{k-1}) - \tfrac{1}{2}(p_k^T r_{k-1})^2 / p_k^T A p_k.$$

Thus, to ensure a reduction in the size of ϕ we must insist that p_k not be orthogonal to r_{k-1}. This leads to the following minimization strategy:

$$x_0 = 0$$
For $k = 1, 2, \ldots$
$\quad r_{k-1} = b - A x_{k-1}$
$\quad$ If $r_{k-1} = 0$
$\quad\quad$ *then* quit
(10.2-3) $\quad\quad$ *else*
$\quad\quad\quad$ Choose a direction p_k such that $p_k^T r_{k-1} \neq 0$
$\quad\quad\quad \alpha_k = p_k^T r_{k-1} / p_k^T A p_k$
$\quad\quad\quad x_k = x_{k-1} + \alpha_k p_k$

Notice that x_k is a linear combination of $p_1, \ldots, p_k$. The problem, of course, is how to choose these vectors so as to guarantee global convergence and at the same time avoid the pitfalls of steepest descent.

A seemingly ideal approach would be to choose linearly independent p_i with the property that each x_k in (10.2-3) solves

(10.2-4)
$$\min_{x \in \text{span}\{p_1, \ldots, p_k\}} \phi(x).$$

This would guarantee not only global convergence but finite termination as well, i.e., $Ax_n = b$. Note that what we are seeking is a vector p_k such that when we solve the one-dimensional minimization problem

$$\min_\alpha \phi(x_{k-1} + \alpha p_k)$$

we also solve the k-dimensional minimization (10.2-4). This would appear to be an unrealizable goal. We show that it is not.

Define the matrix of search directions $P_k \in \mathbb{R}^{n \times k}$ by

$$P_k = [p_1, \ldots, p_k].$$

If $x \in R(P_k)$ then

$$x = P_{k-1}y + \alpha p_k$$

for some $y \in \mathbb{R}^{k-1}$ and $\alpha \in \mathbb{R}$. If x has this form then it is easy to show that

$$\phi(x) = \phi(P_{k-1}y) + \alpha y^T P_{k-1}^T A p_k + \frac{\alpha^2}{2} p_k^T A p_k - \alpha p_k^T b.$$

The presence of the "cross term" $\alpha y^T P_{k-1}^T A p_k$ complicates the minimization. If it were absent, then the minimization of ϕ over $R(P_k)$ would decouple into a minimization over $R(P_{k-1})$, whose solution x_{k-1} is assumed known, and a simple minimization involving the scalar α. One way to effect this decoupling is to insist that

$$P_{k-1}^T A p_k = 0.$$

If we do this and define $x_{k-1} \in R(P_{k-1})$ and $\alpha_k \in \mathbb{R}$ by

$$\phi(x_{k-1}) = \min_y \phi(P_{k-1}y)$$

and

$$\alpha_k = p_k^T b / p_k^T A p_k,$$

then

$$\min_{y,\alpha} \phi(P_{k-1}y + \alpha p_k) = \min_y \phi(P_{k-1}y) + \min_\alpha \left\{ \frac{\alpha^2}{2} p_k^T A p_k - \alpha p_k^T b \right\}$$

is solved by setting $P_{k-1}y = x_{k-1}$ and $\alpha = \alpha_k$. Since x_{k-1} is a linear combination of $p_1, \ldots, p_{k-1}$ it follows that $p_k^T A x_{k-1} = 0$ and so

$$\alpha_k = p_k^T r_{k-1} / p_k^T A p_k,$$

precisely the recipe for α_k appearing in (10.2-3).

Summarizing, if in each step of (10.2-3) we choose the search direction p_k to satisfy

(10.2-5) $p_k^T A p_i = 0,$ $i = 1, \ldots, k-1$

then the vector $x_k = x_{k-1} + \alpha_k p_k$ will minimize ϕ over span$\{p_1, \ldots, p_k\}$. If (10.2-5) holds then we say that p_k is A-*conjugate* to $p_1, \ldots, p_{k-1}$.

Suppose that we have used (10.2-3) and that $p_1, \ldots, p_{k-1}$ are nonzero and mutually A-conjugate, i.e., $p_i^T A p_j = 0$ for all $i \neq j$. Since

$$x_{k-1} \in \text{span}\{p_1, \ldots, p_{k-1}\},$$

it follows that for any $p \in \mathbb{R}^n$ we have

$$p^T r_{k-1} = p^T b - p^T A P_{k-1} z$$

for some $z \in \mathbb{R}^{k-1}$. If $p^T r_{k-1} = 0$ for every p that is A-conjugate to $p_1, \ldots, p_{k-1}$, then we must have $b \in \text{span}\{A p_1, \ldots, A p_{k-1}\}$. But this implies

$$x = A^{-1} b \in \text{span}\{p_1, \ldots, p_{k-1}\}$$

and thus, $x_{k-1} = x$, i.e., $r_{k-1} = 0$. Hence, if $r_{k-1} \neq 0$ then we can find a nonzero p_k that is A-conjugate to $p_1, \ldots, p_{k-1}$ and satisfies $p_k^T r_{k-1} \neq 0$. Since our aim is to bring about the swift reduction in the size of the residuals, it is natural to choose p_k to be the closest vector to r_{k-1} that is A-conjugate to $p_1, \ldots, p_{k-1}$. This defines "version zero" of the method of conjugate gradients:

(10.2-6)

$$
\begin{aligned}
&x_0 = 0 \\
&\text{For } k = 1, 2, \ldots, n \\
&\quad r_{k-1} = b - A x_{k-1} \\
&\quad \text{If } r_{k-1} = 0 \\
&\qquad then \\
&\qquad\quad \text{Set } x = x_{k-1} \text{ and quit.} \\
&\qquad else \\
&\qquad\quad \text{If } k = 1 \\
&\qquad\qquad then \\
&\qquad\qquad\quad p_1 = r_0 \\
&\qquad\qquad else \\
&\qquad\qquad\quad \text{Choose } p_k \text{ to be the closest vector in} \\
&\qquad\qquad\qquad \text{span}\{A p_1, \ldots, A p_{k-1}\}^\perp \text{ to } r_{k-1}. \\
&\qquad\quad \alpha_k = p_k^T r_{k-1} / p_k^T A p_k \\
&\qquad\quad x_k = x_{k-1} + \alpha_k p_k \\
&x = x_n
\end{aligned}
$$

The finite termination of this algorithm is guaranteed, since the p_k are nonzero and nonzero A-conjugate vectors are independent. (Either $r_{k-1} = 0$ for some $k \leqslant n$ or we compute x_n which minimizes ϕ over span$\{p_1, \ldots, p_n\} = R^n$.)

To make (10.2-6) an effective sparse $Ax = b$ solver, we need an efficient method for computing p_k. As a first step in this direction we characterize p_k as the solution to a certain LS problem.

LEMMA 10.2-1. For $k \geqslant 2$ the vectors p_k generated by (10.2-6) satisfy

$$p_k = r_{k-1} - AP_{k-1}z_{k-1},$$

where z_{k-1} solves

$$\min_z \|r_{k-1} - AP_{k-1}z\|_2.$$

Proof. Since p_k is the residual of the LS problem, it must satisfy $(AP_{k-1})^T p_k = 0$. Thus, $p_j^T A p_k = 0$ for $j = 1, \ldots, k - 1$. Moreover, since

$$p_k = [I - (AP_{k-1})(AP_{k-1})^+]r_{k-1}$$

is the orthogonal projection of r_{k-1} into $R(AP_{k-1})^\perp$, it is the closest vector in $R(AP_{k-1})^\perp$ to r_{k-1}. $\square$

With this result we can establish a number of important relationships between the residuals r_i and the search directions p_i.

LEMMA 10.2-2. If the conjugate gradient method (10.2-6) passes through k iterations, then for $j = 1, \ldots, k$ we have

(10.2-7) $$r_j = r_{j-1} - \alpha_j A p_j$$

(10.2-8) $$P_j^T r_j = 0$$

(10.2-9) $$\mathrm{span}\{p_1, \ldots, p_j\} = \mathrm{span}\{r_0, \ldots, r_{j-1}\}$$
$$= \mathrm{span}\{b, Ab, \ldots, A^{j-1}b\}.$$

Proof. Equation (10.2-7) follows by applying A to both sides of $x_j = x_{j-1} + \alpha_j p_j$ and using the definition of the residual.

To establish (10.2-8), observe that if $(P_j^T A P_j)y_j = P_j^T b$, then y_j minimizes

$$\phi(P_j y) = \tfrac{1}{2} y^T (P_j^T A P_j) y - y^T P_j^T b.$$

Consequently, $x_j = P_j y_j$ and $P_j^T r_j = P_j^T (b - A P_j y_j) = 0$.

Induction is necessary to prove (10.2-9). It is clearly true for $j = 1$. Suppose it holds for some j satisfying $1 \leqslant j < k$. From (10.2-7) it follows that

$r_j \in \text{span}\{b, Ab, \ldots, A^j b\}$ and thus $p_{j+1} = r_j - AP_j z_j \in \text{span}\{b, Ab, \ldots, A^j b\}$. Since $\text{span}\{r_0, \ldots, r_j\}$ and $\text{span}\{p_1, \ldots, p_{j+1}\}$ each have dimension $j + 1$, they must equal $\text{span}\{b, Ab, \ldots, A^j b\}$. $\square$

We next establish that the residuals r_i are mutually orthogonal.

THEOREM 10.2-3. After $k - 1$ steps of the conjugate gradient algorithm (10.2-6) the residuals $r_0, \ldots, r_{k-1}$ are mutually orthogonal.

Proof. From the equation $x_i = x_{i-1} + \alpha_i p_i$ it follows that $r_i = r_{i-1} - \alpha_i A p_i$. Thus, from Lemma 10.2-1 we have

$$p_j = r_{j-1} - [Ap_1, \ldots, Ap_{j-1}]z_{j-1} \in \text{span}\{r_0, \ldots, r_{j-1}\}$$

for $j = 1, \ldots, k$. Hence,

$$[p_1, \ldots, p_k] = [r_0, \ldots, r_{k-1}]T$$

for some upper triangular matrix $T \in \mathbb{R}^{k \times k}$. Since the p_i are independent, T is nonsingular. This implies

$$r_{j-1} \in \text{span}\{p_1, \ldots, p_j\}$$

for $j = 1, \ldots, k$. But from (10.2-8) we know that $P_{k-1}^T r_{k-1} = 0$. Since $r_0, \ldots, r_{k-2} \in \text{span}\{p_1, \ldots, p_{k-1}\}$, we must have $r_j^T r_{k-1} = 0$ for $j = 0, \ldots, k - 2$. $\square$

We now show that p_k can be expressed as a linear combination of p_{k-1} and r_{k-1}. Partitioning the vector z_{k-1} of Lemma 10.2-1 as

$$z_{k-1} = \begin{bmatrix} w \\ \mu \end{bmatrix} \begin{matrix} k-2 \\ 1 \end{matrix}$$

and using the identity $r_{k-1} = r_{k-2} - \alpha_{k-1} A p_{k-1}$, we see that

$$p_k = r_{k-1} - AP_{k-1}z_{k-1} = r_{k-1} - AP_{k-2}w - \mu A p_{k-1}$$

$$= \left(1 + \frac{\mu}{\alpha_{k-1}}\right) r_{k-1} + s_{k-1}$$

where

$$s_{k-1} = -\frac{\mu}{\alpha_{k-1}} r_{k-2} - AP_{k-2}w.$$

Because the r_i are mutually orthogonal, it follows that s_{k-1} and r_{k-1} are orthogonal. Thus, the LS problem of Lemma 10.2-1 boils down to choosing μ and w such that

$$\|p_k\|_2^2 = \left(1 + \frac{\mu}{\alpha_{k-1}}\right)^2 \|r_{k-1}\|_2^2 + \|s_{k-1}\|_2^2$$

is minimum. Since the 2-norm of $r_{k-2} - AP_{k-2}z$ is minimized by z_{k-2} giving residual p_{k-1}, it follows that s_{k-1} is a multiple of p_{k-1}. Consequently,

$$p_k \in \text{span}\{r_{k-1}, p_{k-1}\}.$$

Without loss of generality we may assume

$$p_k = r_{k-1} + \beta_k p_{k-1},$$

and since $p_{k-1}^T A p_k = 0$ and $p_{k-1}^T r_{k-1} = 0$, it follows that

$$\beta_k = -\frac{p_{k-1}^T A r_{k-1}}{p_{k-1}^T A p_{k-1}}.$$

and

$$\alpha_k = \frac{r_{k-1}^T r_{k-1}}{p_{k-1}^T A p_k}$$

This leads us to "version 1" of the conjugate gradient method:

$$
\begin{aligned}
&x_0 = 0 \\
&\text{For } k = 1, \ldots, n \\
&\quad r_{k-1} = b - A x_{k-1} \\
&\quad \text{If } r_{k-1} = 0 \\
&\quad\quad \text{then} \\
&\quad\quad\quad \text{quit} \\
&\quad\quad \text{else} \\
&\quad\quad\quad \text{If } k = 1 \\
&\quad\quad\quad\quad \text{then} \\
&\quad\quad\quad\quad\quad p_k = r_0 \\
&\quad\quad\quad\quad \text{else} \\
&\quad\quad\quad\quad\quad \beta_k = -p_{k-1}^T A r_{k-1} / p_{k-1}^T A p_{k-1} \\
&\quad\quad\quad\quad\quad p_k = r_{k-1} + \beta_k p_{k-1} \\
&\quad\quad\quad \alpha_k = r_{k-1}^T r_{k-1} / p_k^T A p_k \\
&\quad\quad\quad x_k = x_{k-1} + \alpha_k p_k \\
&x = x_n
\end{aligned}
$$

(10.2-10)

As it stands, this iteration seems to require two matrix-vector multiplications each time through the loop. However, by computing the residuals recursively via

$$r_k = r_{k-1} - \alpha_k A p_k$$

and substituting

(10.2-11) $$r_{k-1}^T r_{k-1} = -\alpha_{k-1} r_{k-1}^T A p_{k-1}$$

(10.2-12) $$r_{k-2}^T r_{k-2} = \alpha_{k-1} p_{k-1}^T A p_{k-1}$$

into the formula for β_k, we obtain the following more efficient algorithm:

$$x_0 = 0$$
$$r_0 = b$$
For $k = 1, \ldots, n$
    if $r_{k-1} = 0$
       then
            Set $x = x_{k-1}$ and quit.

(10.2-13)   *else*

$$\beta_k = r_{k-1}^T r_{k-1} / r_{k-2}^T r_{k-2} \qquad (\beta_1 \equiv 0)$$
$$p_k = r_{k-1} + \beta_k p_{k-1} \qquad\qquad (p_1 \equiv r_0)$$
$$\alpha_k = r_{k-1}^T r_{k-1} / p_k^T A p_k$$
$$x_k = x_{k-1} + \alpha_k p_k$$
$$r_k = r_{k-1} - \alpha_k A p_k$$

$$x = x_n$$

Iteration (10.2-13) is essentially the form of the conjugate gradient algorithm that appeared in the original paper by Hestenes and Stiefel (1952). The details associated with its practical implementation will be discussed in §10.3.

    In the remainder of this section we survey some of the method's exact arithmetic properties. We first establish the connection between the conjugate gradient and Lanczos iterations, i.e., between (10.2-13) and Algorithm 9.3-1. Define the matrix of residuals $R_k \in \mathbb{R}^{n \times k}$ by

$$R_k = [r_0, \ldots, r_{k-1}]$$

and the upper bidiagonal matrix $B_k \in \mathbb{R}^{k \times k}$ by

$$B_k = \begin{bmatrix} 1 & -\beta_2 & & & \mathbf{0} \\ & 1 & \ddots & & \\ & & \ddots & \ddots & \\ & & & \ddots & -\beta_k \\ \mathbf{0} & & & & 1 \end{bmatrix}.$$

From the equations $p_j = r_{j-1} + \beta_j p_{j-1}$, $j = 2, \ldots, k$, and $p_1 = r_0$ it follows that $R_k = P_k B_k$. Since the columns of $P_k = [p_1, \ldots, p_k]$ are A-conjugate, we see that

$$R_k^T A R_k = B_k^T \operatorname{diag}(p_1^T A p_1, \ldots, p_k^T A p_k) B_k$$

is tridiagonal. From (10.2-9) and Theorem 10.2-3 it follows that if

$$\Delta = \operatorname{diag}(\rho_0, \ldots, \rho_{k-1}), \qquad\qquad \rho_i = \|r_i\|_2$$

then the columns of $R_k \Delta^{-1}$ form an orthonormal basis for span$\{b, Ab, \ldots, A^{k-1}b\}$. Consequently, the columns of this matrix are essentially the Lanczos vectors of Algorithm 9.3-1, i.e., $q_i = \pm r_{i-1}/\rho_{i-1}$, where $i = 1, \ldots, k$. Moreover, the tridiagonal matrix associated with these Lanczos vectors is given by

$$(10.2\text{-}14) \qquad T_k = \Delta^{-1} B_k^T \operatorname{diag}(p_i^T A p_i) B_k \Delta^{-1}.$$

The diagonal and subdiagonal of this matrix involve quantities that are readily available during the conjugate gradient iteration. Thus, we can obtain good estimates of A's extremal eigenvalues (and condition number) as we generate the x_k in (10.2-13).

To obtain error bounds for the conjugate gradient iterates x_k observe that if $Ax = b$ then

$$(x - x_k)^T A (x - x_k) = b^T A^{-1} b + 2\phi(x_k).$$

Defining the "A-norm" on $\mathbb{R}^n$ by

$$\|w\|_A = w^T A w,$$

we see that the problem of minimizing ϕ is equivalent to the problem of minimizing the A-norm error. Since x_k minimizes ϕ over all vectors in span$\{b, \ldots, A^{k-1}b\}$, it follows that

$$(10.2\text{-}15) \qquad \|x - x_k\|_A = \min_{p \in \mathcal{P}_{k-1}} \|x - p(A)b\|_A,$$

where $\mathcal{P}_{k-1}$ denotes the set of all $k - 1$ degree polynomials. It can be shown that if $\kappa = \kappa(A)$ then

$$(10.2\text{-}16) \qquad \|x - x_k\|_A \leqslant \|x - x_0\|_A \left[\frac{1 - \sqrt{\kappa}}{1 + \sqrt{\kappa}}\right]^{2k}.$$

Although this result is pessimistic in practice, it has a useful role to play in the next section, where we introduce a way of effectively reducing the size of κ in the bound.

Problems

P10.2-1. Verify that the residuals in (10.2-1) satisfy $r_i^T r_j = 0$ whenever $j = i + 1$.

P10.2-2. Verify (10.2-2).

P10.2-3. Verify (10.2-11) and (10.2-12).

P10.2-4. Give formulae for the entries of the tridiagonal matrix T_k in (10.2-14).

P10.2-5. Show that if b is in a q-dimensional invariant subspace, then $Ax_q = b$ in (10.2-13).

Notes and References for Sec. 10.2

The conjugate gradient method is a member of a larger class of methods that are referred to as *conjugate direction* algorithms. In a conjugate direction algorithm the search directions are all B-conjugate for some suitably chosen matrix B. A discussion of these methods and their relation to matrix factorizations is given in

G. W. Stewart (1973). "Conjugate Direction Methods for Solving Systems of Linear Equations," *Numer. Math. 21*, 284–97.

The classic reference for the conjugate gradient method is

M. R. Hestenes and E. Stiefel (1952). "Methods of Conjugate Gradients for Solving Linear Systems," *J. Res. Nat. Bur. Stand. 49*, 409–36.

An exact arithmetic analysis of the method may be found in chapter 2 of

M. R. Hestenes (1980). *Conjugate Direction Methods in Optimization*, Springer-Verlag, Berlin.

See also

D. K. Faddeev and V. N. Faddeeva (1963). *Computational Methods of Linear Algebra*, W. H. Freeman & Co., San Francisco.
O. Axelsson (1977). "Solution of Linear Systems of Equations: Iterative Methods," in *Sparse Matrix Techniques: Copenhagen, 1976*, ed. V. A. Barker, Springer-Verlag, Berlin.

An ALGOL procedure is given in

T. Ginsburg (1971). "The Conjugate Gradient Method," in HACLA, pp. 57–69.

Sec. 10.3. Practical Conjugate Gradient Procedures

Rounding errors lead to a loss of orthogonality among the residuals r_i in the conjugate gradient algorithm (10.2-13). As a consequence, the property of finite termination which we labored so hard to establish does not apply in practice. This forces us to regard conjugate gradients in a new light, namely, as a genuinely iterative method. Thus, a practical conjugate gradient scheme would proceed along the following lines:

ALGORITHM 10.3-1. Given $b \in \mathbb{R}^n$, a symmetric positive definite $A \in \mathbb{R}^{n \times n}$, and a tolerance $\epsilon \geq \mathbf{u}$, the following algorithm computes a vector x such that $\|b - Ax\|_2 \cong \epsilon \|b\|_2$.

$x := 0,\ r := b,\ \rho_0 := \|r\|_2^2,\ k := 1$
Do While $(\sqrt{\rho_{k-1}} > \epsilon\|b\|_2)$
   If $k = 1$
      then
$$p := r$$
      else
$$\beta_k := \rho_{k-1}/\rho_{k-2},\ p := r + \beta_k p$$
$w := Ap,\ \alpha_k := \rho_{k-1}/p^\mathrm{T}w,\ x := x + \alpha_k p$
$r := r - \alpha_k w,\ \rho_k := \|r\|_2^2,\ k := k + 1$

This algorithm requires one matrix-vector multiplication and $5n$ flops per iteration. Four n-vectors of storage are essential: x, r, p, and w. The subscripting of the scalars is not necessary.

It is also possible to base the termination criteria on heuristic estimates of the error $A^{-1}r_k$ by approximating $\|A^{-1}\|_2$ with the reciprocal of the smallest eigenvalue of the tridiagonal matrix T_k given in (10.2-14).

It was Reid (1971b) who rejuvenated interest in the conjugate gradient algorithm by showing that it can be a very effective technique when used iteratively. While the iterative point of view permits us to tolerate loss of orthogonality, it does make the issue of rate of convergence extremely important. On that score, the bound (10.2-15) is somewhat distressing because it suggests a hopelessly slow rate of convergence for ill-conditioned problems.

An important way around this difficulty is to *pre-condition* A. This refers to finding a nonsingular symmetric matrix C such that $\tilde{A} = C^{-1}AC^{-1}$ has improved condition. We then can apply conjugate gradients (with improved convergence properties) to the transformed system

(10.3-1) $$\tilde{A}\tilde{x} = \tilde{b},$$

where $x = C^{-1}\tilde{x}$ and $\tilde{b} = C^{-1}b$. Using (10.2-13) this gives

$\tilde{x}_0 = 0,\ \tilde{r}_0 = \tilde{b}$
For $k = 1, \ldots, n$
   if $\tilde{r}_{k-1} = 0$
      then
        Set $\tilde{x} = \tilde{x}_{k-1}$ and quit.
(10.3-2)   *else*
$$\beta_k = \tilde{r}_{k-1}^\mathrm{T}\tilde{r}_{k-1}/\tilde{r}_{k-2}^\mathrm{T}\tilde{r}_{k-2} \qquad (\beta_1 \equiv 0)$$
$$\tilde{p}_k = \tilde{r}_{k-1} + \beta_k\tilde{p}_{k-1} \qquad\qquad (\tilde{p}_1 \equiv \tilde{r}_0)$$
$$\alpha_k = \tilde{r}_{k-1}^\mathrm{T}\tilde{r}_{k-1}/\tilde{p}_k^\mathrm{T}C^{-1}AC^{-1}\tilde{p}_k$$
$$\tilde{x}_k = \tilde{x}_{k-1} + \alpha_k\tilde{p}_k$$
$$\tilde{r}_k = \tilde{r}_{k-1} - \alpha_k C^{-1}AC^{-1}\tilde{p}_k$$
$\tilde{x} = \tilde{x}_n$

Here, the $\tilde{x}_k$ can be regarded as approximations to $\tilde{x}$. In order to simplify this iteration define

$$M = C^2$$
$$p_k = C^{-1}\tilde{p}_k$$
$$x_k = C^{-1}\tilde{x}_k$$
$$z_k = C^{-1}\tilde{r}_k$$
$$r_k = C\tilde{r}_k = b - Ax_k.$$

By substituting into (10.3-2) we obtain

(10.3-3)

$$x_0 = 0$$
$$r_0 = b$$
For $k = 1, \ldots, n$
    If $r_{k-1} = 0$
        then
            Set $x = x_{k-1}$ and quit.
        . *else*
            Solve $Mz_{k-1} = r_{k-1}$ for z_{k-1}.
            $\beta_k = z_{k-1}^T r_{k-1}/z_{k-2}^T r_{k-2}$ $(\beta_1 \equiv 0)$
            $p_k = z_{k-1} + \beta_k p_{k-1}$ $(p_1 \equiv z_0)$
            $\alpha_k = z_{k-1}^T r_{k-1}/p_k^T Ap_k$
            $x_k = x_{k-1} + \alpha_k p_k$
            $r_k = r_{k-1} - \alpha_k Ap_k$
    $x = x_n$

This is called the *preconditioned* conjugate gradient algorithm, and the symmetric positive definite matrix M is called the *preconditioner*. It can be shown that when the conjugate gradient algorithm is generalized in this way, the residuals and search directions satisfy

(10.3-4) $$r_j^T M^{-1} r_i = 0 \qquad\qquad i \neq j$$

(10.3-5) $$p_j^T (C^{-1}AC^{-1})p_i = 0 \qquad\qquad i \neq j$$

Clearly, for the preconditioned conjugate algorithm to be an effective sparse matrix technique we must be able to solve linear systems of the form $Mz = r$ easily. A good preconditioner can lead to very rapid convergence, often after $O(\sqrt{n})$ iterations. Thus, the question is how to choose M. Guidance in this matter may be obtained by rewriting the conjugate gradient method in the style of Concus, Golub, and O'Leary (1976):

$$x_{-1} = x_0 = 0$$

For $k = 1, \ldots, n$

$\quad r_{k-1} = b - Ax_{k-1}$

$\quad$ If $r_{k-1} = 0$

$\qquad$ *then*

$\qquad\qquad$ Set $x = x_{k-1}$ and quit.

$\quad$ *else*

$\qquad$ Solve $Mz_{k-1} = r_{k-1}$ for z_{k-1}.

$\qquad \gamma_{k-1} = z_{k-1}^T Mz_{k-1}/z_{k-1}^T Az_{k-1}$

(10.3-6) $\qquad$ If $k = 1$

$\qquad\qquad$ *then*

$\qquad\qquad\qquad \omega_1 = 1$

$\qquad\qquad$ *else*

$$\omega_k = \left(1 - \frac{\gamma_{k-1}}{\gamma_{k-2}} \frac{z_{k-1}^T Mz_{k-1}}{z_{k-2}^T Mz_{k-2}} \frac{1}{\omega_{k-1}}\right)^{-1}$$

$$x_k = x_{k-2} + \omega_k(\gamma_{k-1}z_{k-1} + x_{k-1} - x_{k-2})$$

$x = x_n$

Many other iterative methods have as their basic step

(10.3-7) $\qquad x_k = x_{k-2} + \omega_k(\gamma_k z_{k-1} + x_{k-1} - x_{k-2})$,

where

$$Mz_{k-1} = r_{k-1} = b - Ax_{k-1}.$$

For example, if we set $\omega_k = 1$ and $\gamma_k = 1$, then

$$x_k = M^{-1}(b - Ax_{k-1}),$$

i.e.,

$$Mx_k = Nx_{k-1} + b,$$

where

$$A = M - N.$$

Thus, the Jacobi, Gauss-Seidel, SOR, and SSOR methods of 10.1 have the form (10.3-7). So also does the Chebychev semi-iterative method (10.1-12). Thus, we can think of the scalars ω_k *and* γ_k in (10.3-7) as acceleration parameters that can be chosen to speed the convergence of the iteration

$$Mx_k = Nx_{k-1} + b.$$

Hence, any iterative method based on the splitting $A = M - N$ can be accelerated by the conjugate gradient algorithm so long as M (the preconditioner) is symmetric and positive definite.

Two common preconditioners when A is symmetric positive definite are

$$M = \text{diag}(a_{11}, \ldots, a_{nn})$$

and

$$M = (D + \omega L)D^{-1}(D + \omega L)^{\text{T}},$$

which, respectively, lead to acceleration of the Jacobi and SSOR iterations of §10.1.

Many elliptic PDE problems lead to symmetric positive definite matrices of the form

$$A = \begin{bmatrix} M_1 & F \\ F^{\text{T}} & M_2 \end{bmatrix} \begin{matrix} m \\ m \end{matrix},$$
$$\begin{matrix} m & m \end{matrix}$$

where systems of the form $M_1 z = d$ and $M_2 z = d$ are easy to solve. Setting

$$M = \begin{bmatrix} M_1 & 0 \\ 0 & M_2 \end{bmatrix}$$

leads to a very effective acceleration of the "block" Jacobi iteration. See Reid (1971b).

At other times it is possible to find a splitting $A = M - N$, where N is of low rank and M has some structured positive definite form. This occurs in certain elliptic problems with irregular boundaries. See the survey by Concus, Golub, and O'Leary (1976).

Another preconditioning strategy involves computing an *incomplete Cholesky factorization* of A. The idea behind this approach is to calculate a lower triangular matrix H with the property that H has some tractable sparsity structure and is somehow "close" to A's exact Cholesky factor G. The simplest way to do this is to set g_{ij} to zero in Cholesky factorization (Algorithm 5.2-1) every time the corresponding a_{ij} is zero, i.e.,

For $j = 1, \ldots, n$

$$h_{jj} = \left(a_{jj} - \sum_{i=1}^{j-1} h_{ij}^2 \right)^{1/2}$$

For $i = j + 1, \ldots, n$

(10.3-8)
$$\quad \text{if } a_{ij} = 0$$
$$\quad \quad \text{then}$$
$$\quad \quad \quad h_{ij} = 0$$
$$\quad \quad \text{else}$$

$$h_{ij} = \left(a_{ij} - \sum_{k=1}^{j-1} h_{ik} h_{jk} \right) \Big/ h_{jj}$$

In practice, A and H would be stored in an appropriate data structure and the looping in the above algorithm would take on a very special appearance.

Unfortunately, (10.3-8) is not always stable. To see why, recall that in the Cholesky decomposition we repeatedly perform the following calculation:

$$\begin{bmatrix} \alpha & v^{\mathrm{T}} \\ v & B \end{bmatrix} = \begin{bmatrix} \sqrt{\alpha} & 0 \\ w & I \end{bmatrix} \begin{bmatrix} 1 & 0 \\ 0 & B_1 \end{bmatrix} \begin{bmatrix} \sqrt{\alpha} & w^{\mathrm{T}} \\ 0 & I \end{bmatrix},$$

where $w^{\mathrm{T}} = v^{\mathrm{T}}/\sqrt{\alpha}$ and $B_1 = B - (vv^{\mathrm{T}}/\alpha)$. The next step is to likewise factor B_1, a matrix which can readily be shown to be positive definite. However, in the above incomplete Cholesky scheme, we go on to work with

$$B_1 = \mathrm{sparse}\left(B, B - \frac{vv^{\mathrm{T}}}{\alpha}\right),$$

where $\mathrm{sparse}(\cdot, \cdot)$ is defined by

$$C = \mathrm{sparse}(U, V) = (c_{ij}) \qquad c_{ij} = v_{ij} \text{ if } u_{ij} \neq 0, \text{ and zero}$$
otherwise.

Unfortunately, it is possible to construct an example in which B_1 is not positive definite.

Classes of positive definite matrices for which incomplete Cholesky is stable are identified in Manteuffel (1979).

Problems

P10.3-1. Give a detailed algorithm that performs incomplete Cholesky (10.3-8) to the positive definite matrix defined in (10.1-6).

P10.3-2. Introduce overwriting into the preconditioned conjugate gradient algorithm (10.3-3). How many n-vectors of storage does your algorithm require?

Notes and References for Sec. 10.3

The idea of using the conjugate gradient method as an iterative method was broached in

J. K. Reid (1971b). "On the Method of Conjugate Gradients for the Solution of Large Sparse Systems of Linear Equations," in *Large Sparse Sets of Linear Equations*, ed. J. K. Reid, Academic Press, New York, pp. 231-54.

Several authors have attempted to explain the algorithm's behavior in finite precision arithmetic. See

J. Cullum and R. A. Willoughby (1977). "The Equivalence of the Lanczos and the Conjugate Gradient Algorithms," IBM Research Report RE-6903.

H. Wozniakowski (1980). "Roundoff Error Analysis of a New Class of Conjugate Gradient Algorithms," *Lin. Alg. & Its Applic. 29*, 507-29.

A. Greenbaum (1981). "Behavior of the Conjugate Gradient Algorithm in Finite Precision Arithmetic," Report UCRL 85752, Lawrence Livermore Laboratory, Livermore, Calif.

See also the analysis in

A. Jennings (1977a). "Influence of the Eigenvalue Spectrum on the Convergence Rate of the Conjugate Gradient Method," *J. Inst. Math. Applic. 20*, 61-72.
G. W. Stewart (1975a). "The Convergence of the Method of Conjugate Gradients at Isolated Extreme Points in the Spectrum," *Numer. Math. 24*, 85-93.

The conjugate gradient algorithm can also be used to interpret the behavior of the Lanczos algorithm for eigenvalues. See

J. Cullum and R. Willoughby (1980). "The Lanczos Phenomena: An Interpretation Based on Conjugate Gradient Optimization," *Lin. Alg. & Its. Applic. 29*, 63-90.

Our discussion of the preconditioned conjugate gradient algorithm is patterned after

P. Concus, G. H. Golub, and D. P. O'Leary (1976). "A Generalized Conjugate Gradient Method for the Numerical Solution of Elliptic Partial Differential Equations," in *Sparse Matrix Computations*, ed. J. R. Bunch and D. J. Rose, Academic Press, New York.

Other references concerned with extensions of the basic method include

D. P. O'Leary (1980a). "The Block Conjugate Gradient Algorithm and Related Methods," *Lin. Alg. & Its Applic. 29*, 293-322.
J. K. Reid (1972). "The Use of Conjugate Gradients for Systems of Linear Equations Possessing Property *A*," *SIAM J. Num. Anal. 9*, 325-32.

The incomplete Cholesky factorization idea was initially discussed in

J. A. Meijerink and H. A. Van der Vorst (1977). "An Iterative Solution Method for Linear Equation Systems of Which the Coefficient Matrix is a Symmetric *M*-Matrix," *Math. Comp. 31*, 148-62.

For generalizations and analysis, see

T. A. Manteuffel (1979). "Shifted Incomplete Cholesky Factorization," in *Sparse Matrix Proceedings, 1978*, ed. I. S. Duff and G. W. Stewart, SIAM Publications, Philadelphia.

The conjugate gradient algorithm is based on the functional $\phi(x)$. However, the whole development goes through if the minimization of $\| Ax - b \|_2$ is considered instead. The resulting technique, called the *minimum residual variant*, is analyzed in

A. K. Cline (1976b). "Several Observations on the Use of Conjugate Gradient Methods," ICASE Report 76-22, NASA Langley Research Center, Hampton, Va.

It can be "derived" by replacing all the inner products in Algorithm 10.3-1 with A-inner products, i.e., $u^T v$ becomes $u^T A v$. The minimum residual variant sometimes brings about a slightly faster reduction in the norm of the residual.

Some representative papers concerned with the development of nonsymmetric conjugate gradient procedures include

D. M. Young and K. C. Jea (1980). "Generalized Conjugate Gradient Acceleration of Nonsymmetrizable Iterative Methods," *Lin. Alg. & Its. Applic. 34*, 159-94.

O. Axelsson (1980). "Conjugate Gradient Type Methods for Unsymmetric and Inconsistent Systems of Linear Equations," *Lin. Alg. & Its Applic. 29*, 1-66.

Finally, we mention that the method can be used when trying to compute an eigenvector of a large sparse symmetric matrix. See

A. Ruhe and T. Wiberg (1972). "The Method of Conjugate Gradients Used in Inverse Iteration," *BIT 12*, 543-54.

Functions of Matrices

> §11.1 **Eigenvalue Methods**
> §11.2 **Approximation Methods**
> §11.3 **The Matrix Exponential**

Computing a function $f(A)$ of an n-by-n matrix A is a frequently occurring problem in control theory and other application areas. Roughly speaking, if the scalar function $f(z)$ is defined on $\lambda(A)$, then $f(A)$ is "defined" by substituting "A" for "z" in the "formula" for $f(z)$. For example, if $f(z) = (1 + z)(1 - z)^{-1}$ and $1 \notin \lambda(A)$, then $f(A) = (I + A)(I - A)^{-1}$.

The computations get particularly interesting when the function f is transcendental. One approach in this more complicated situation is to compute an eigenvalue decomposition $A = YBY^{-1}$ and use the formula $f(A) = Yf(B)Y^{-1}$. If B is sufficiently simple then it is often possible to calculate $f(B)$ directly. This is illustrated in §11.1 for the cases when $A = YBY^{-1}$ represents the Jordan and Schur decompositions. Not surprisingly, reliance on the latter decomposition results in a more stable $f(A)$ procedure.

Another class of methods for the matrix function problem is to approximate the desired function $f(A)$ with an easy-to-calculate function $g(A)$. For example, g might be a truncated Taylor series approximate to f. Error bounds associated with the approximation of matrix functions are given in §11.2.

In the last section we discuss the special and very important problem of computing the matrix exponential e^A.

Reading Path

$$\text{Chapter 7} \longrightarrow \text{§11.1} \longrightarrow \text{§11.2} \longrightarrow \text{§11.3}$$

Sec. 11.1. Eigenvalue Methods

Given an n-by-n matrix A and a scalar function $f(z)$, there are several ways to define the *matrix function* $f(A)$. A very informal definition might be to substitute "A" for "z" in the formula for $f(z)$. For example, if $p(z) = 1 + z$ and $r(z) = [1 - (z/2)]^{-1}[1 + (z/2)]$ for $z \neq 2$, then it is certainly reasonable to define $p(A)$ and $r(A)$ by

$$p(A) = I + A$$

and

$$r(A) = [I - A/2]^{-1}[I + A/2]. \qquad\qquad 2 \notin \lambda(A)$$

"A-for-z" substitution also works for transcendental functions:

$$e^A = \sum_{k=0}^{\infty} A^k/k!$$

To make subsequent algorithmic developments precise, however, we need a more rigorous definition of $f(A)$.

Perhaps the most elegant definition is the following. Suppose $f(z)$ is analytic inside and on a closed contour Γ which encircles $\lambda(A)$. We define $f(A)$ to be the matrix

(11.1-1) $$f(A) = \frac{1}{2\pi i} \oint_\Gamma f(z)(zI - A)^{-1}dz.$$

This definition is immediately recognized as a matrix version of the Cauchy integral theorem. The integral is defined on an element-by-element basis:

$$f(A) = (f_{kj}) \Rightarrow f_{kj} = \frac{1}{2\pi i} \oint_\Gamma f(z)e_k^T(zI - A)^{-1}e_j dz.$$

Notice that the entries of $(zI - A)^{-1}$ are analytic on Γ and that $f(A)$ is defined whenever $f(z)$ analytic in a neighborhood of $\lambda(A)$.

Although fairly useless from the computational point of view, the definition (11.1-1) can be used to derive more practical characterizations of $f(A)$. For example, if $f(A)$ is defined and

$$A = XBX^{-1} = X \operatorname{diag}(B_1, \ldots, B_p)X^{-1}, \qquad B_i \in \mathbb{C}^{n_i \times n_i}$$

then it is easy to verify that

(11.1-2) $$f(A) = Xf(B)X^{-1} = X \operatorname{diag}[f(B_1), \ldots, f(B_p)]X^{-1}.$$

For the case when the B_i are Jordan blocks we obtain the following:

THEOREM 11.1-1. Let $X^{-1}AX = \operatorname{diag}(J_1, \ldots, J_p)$ be the Jordan canonical form (JCF) of $A \in \mathbb{C}^{n \times n}$ with

$$J_i = \begin{bmatrix} \lambda_i & 1 & & & \mathbf{0} \\ & \lambda_i & 1 & & \\ & & \ddots & \ddots & \\ & & & \ddots & 1 \\ \mathbf{0} & & & & \lambda_i \end{bmatrix} \in \mathbb{C}^{m_i \times m_i}$$

If $f(z)$ is analytic on an open set containing $\lambda(A)$, then

$$f(A) = X \operatorname{diag}[f(J_1), \ldots, f(J_p)]X^{-1}$$

where

$$f(J_i) = \begin{bmatrix} f(\lambda_i) & f^{(1)}(\lambda_i) & \cdots & \dfrac{f^{(m_i-1)}(\lambda_i)}{(m_i-1)!} \\ & f(\lambda_i) & \ddots & \vdots \\ & & \ddots & f^{(1)}(\lambda_i) \\ \text{\huge0} & & & f(\lambda_i) \end{bmatrix}.$$

Proof. In view of the remarks preceding the statement of the theorem, it suffices to examine $f(G)$ where

$$G = \lambda I + E \qquad E = (\delta_{i,j-1})$$

is a q-by-q Jordan block. Suppose $(zI - G)$ is nonsingular. Since

$$(zI - G)^{-1} = \sum_{k=0}^{q-1} \frac{E^k}{(z - \lambda)^{k+1}},$$

it follows from Cauchy's integral theorem that

$$f(G) = \sum_{k=0}^{q-1} \left[\frac{1}{2\pi i} \oint_\Gamma \frac{f(z)}{(z - \lambda)^{k+1}} \, dz \right] E^k = \sum_{k=0}^{q-1} \frac{f^{(k)}(\lambda)}{k!} E^k.$$

The theorem follows from the observation that $E^k = (\delta_{i,j-k})$. $\square$

COROLLARY 11.1-2. If $A \in \mathbb{C}^{n \times n}$, $A = X \operatorname{diag}(\lambda_1, \ldots, \lambda_n)X^{-1}$, and $f(A)$ is defined, then $f(A) = X \operatorname{diag}[f(\lambda_1), \ldots, f(\lambda_n)]X^{-1}$.

These results illustrate the close connection between $f(A)$ and the eigensystem of A. Unfortunately, the JCF approach to the matrix function problem has dubious computational merit unless A is diagonalizable with a well-conditioned matrix of eigenvectors. Indeed, rounding errors of order $\mathbf{u} \kappa_2(X)$ can be expected to contaminate the computed result, since a linear system involving the matrix X must be solved.

EXAMPLE 11.1-1. If

$$A = \begin{bmatrix} 1 + 10^{-5} & 1 \\ 0 & 1 - 10^{-5} \end{bmatrix},$$

then any matrix of eigenvectors is a column scaled version of

$$X = \begin{bmatrix} 1 & -1 \\ 0 & 2(1 - 10^{-5}) \end{bmatrix}$$

and has a 2-norm condition number of order 10^5. Using a computer with machine precision $\mathbf{u} \cong 10^{-7}$ we find

$$\text{fl}[X^{-1} \text{diag}(\exp(1 + 10^{-5}), \exp(1 - 10^{-5}))X] = \begin{bmatrix} 2.718307 & 2.750000 \\ 0.000000 & 2.718254 \end{bmatrix}$$

while

$$e^A = \begin{bmatrix} 2.718309 & 2.718282 \\ 0.000000 & 2.718255 \end{bmatrix}.$$

The example suggests that whenever possible, ill-conditioned similarity transformations should be avoided when computing $f(A)$. This can be accomplished by computing the Schur decomposition $A = QTQ^H$ and then using the formula

$$f(A) = Qf(T)Q^H.$$

For this to be effective, we need an algorithm for computing functions of upper triangular matrices. Unfortunately, an explicit expression for $f(T)$ is very complicated, as the following theorem shows.

THEOREM 11.1-3. Let $T = (t_{ij})$ be an n-by-n upper triangular matrix with $\lambda_i = t_{ii}$ and assume $f(T)$ is defined. If $f(T) = (f_{ij})$, then for $i > j, f_{ij} = 0$; for $i = j, f_{ii} = f(\lambda_i)$; and for $i < j$,

$$f_{ij} = \sum_{(s_0, \ldots, s_k) \in S_{ij}} t_{s_0, s_1} \cdot t_{s_1, s_2} \cdots t_{s_{k-1}, s_k} f[\lambda_{s_0}, \ldots, \lambda_{s_k}],$$

where S_{ij} is the set of all strictly increasing sequences of integers that start at i and end at j and where $f[\lambda_{s_0}, \ldots, \lambda_{s_k}]$ is the k-th order divided difference of f at $\lambda_{s_0}, \ldots, \lambda_{s_k}$.

Proof. See Descloux (1963), Davis (1973), or Van Loan (1975b). $\square$

Computing $f(T)$ via Theorem 11.1-3 would require about 2^n flops. Fortunately, Parlett (1974a) has derived an elegant recursive method for determining the strictly upper triangular portion of the matrix $F = f(T)$. It requires only $n^3/3$ flops and can be derived from the following commutivity result:

$$(11.1-3) \qquad\qquad FT = TF.$$

Indeed, by comparing (i, j) entries in this equation we find

$$\sum_{k=i}^{j} f_{ik} t_{kj} = \sum_{k=i}^{j} t_{ik} f_{kj}, \qquad\qquad j > i$$

and thus, if t_{ii} and t_{jj} are distinct,

(11.1-4) $$f_{ij} = t_{ij} \frac{f_{jj} - f_{ii}}{t_{jj} - t_{ii}} + \frac{\displaystyle\sum_{k=i+1}^{j-1} [t_{ik} f_{kj} - f_{ik} t_{kj}]}{t_{jj} - t_{ii}}.$$

From this we conclude that f_{ij} is a linear combination of its neighbors to its left and below in the matrix F. For example, the entry f_{25} depends upon f_{22}, $f_{23}, f_{24}, f_{55}, f_{45}$, and f_{35}:

$$F = \begin{bmatrix} x & x & x & x & x & x \\ & \boxed{x \quad x \quad x \quad x} \to x & & & & x \\ & & x & x & \boxed{x} & x \\ & \mathbf{0} & & x & \boxed{x} & x \\ & & & & \boxed{x} & x \\ & & & & & x \end{bmatrix}.$$

Because of this, the entire upper triangular portion of F can be computed one superdiagonal at a time beginning with the diagonal, $f(t_{11}), \ldots, f(t_{nn})$. The complete procedure is as follows:

ALGORITHM 11.1-1. Given an upper triangular $T \in \mathbb{C}^{n \times n}$ with distinct eigenvalues $t_{11}, \ldots, t_{nn}$ and a function $f(z)$, the following algorithm computes $F = f(T)$ assuming that it is defined:

For $i = 1, \ldots, n$
    $f_{ii} := f(t_{ii})$
For $p = 1, 2, \ldots, n - 1$
    For $i = 1, \ldots, n - p$
        $j := i + p$
        $f_{ij} := [t_{ij}(f_{jj} - f_{ii}) + \sum_{k=i+1}^{j-1} (t_{ik} f_{kj} - f_{ik} t_{kj})]/(t_{jj} - t_{ii})$

This algorithm requires $n^3/3$ flops. Once $F = f(T)$ is calculated, then $f(A)$ is given by the formula $f(A) = QFQ^H$. Clearly, most of the work in computing $f(A)$ by this approach is in the computation of the Schur decomposition $A = QTQ^H$.

EXAMPLE 11.1-2. If

$$T = \begin{bmatrix} 1 & 2 & 3 \\ 0 & 3 & 4 \\ 0 & 0 & 5 \end{bmatrix}$$

and $f(z) = (1 + z)/z$ then $F = (f_{ij}) = f(T)$ is defined by

$f_{11} = (1 + 1)/1 = 2$
$f_{22} = (1 + 3)/3 = \frac{4}{3}$
$f_{33} = (1 + 5)/5 = \frac{6}{5}$
$f_{12} = t_{12}(f_{22} - f_{11})/(t_{22} - t_{11}) = -\frac{2}{3}$
$f_{23} = t_{23}(f_{33} - f_{22})/(t_{33} - t_{22}) = -4/15$
$f_{13} = [t_{13}(f_{33} - f_{11}) + (t_{12}f_{23} - f_{12}t_{23})]/(t_{33} - t_{11}) = -1/15.$

If A has close or multiple eigenvalues, then Algorithm 11.1-1 leads to poor results. In this case it is advisable to use a block version of the algorithm. We outline this approach below; for more details the reader should consult Parlett (1974a).

The first step is to choose Q in the Schur decomposition such that close or multiple eigenvalues are clustered in blocks $T_{11}, \ldots, T_{pp}$ along the diagonal of T. In particular, we must compute a partitioning

$$T = \begin{bmatrix} T_{11} & T_{12} & \cdots & T_{1p} \\ & T_{22} & \cdots & T_{2p} \\ & & \ddots & \vdots \\ \mathbf{0} & & & T_{pp} \end{bmatrix} \qquad F = \begin{bmatrix} F_{11} & F_{12} & \cdots & F_{1p} \\ & F_{22} & \cdots & F_{2p} \\ & & \ddots & \vdots \\ \mathbf{0} & & & F_{pp} \end{bmatrix}$$

where $\lambda(T_{ii}) \cap \lambda(T_{jj}) = \emptyset$, $i \neq j$. The actual determination of the block sizes can be done using the methods of §7.6.

Next we compute the submatrices $F_{ii} = f(T_{ii})$ for $i = 1, \ldots, p$. Since the eigenvalues of T_{ii} are presumably close, these calculations require special methods. (Some possibilities are discussed in the next two sections.) Once the diagonal blocks of F are known, the blocks in the strict upper triangle of F can be found recursively, as in the scalar case. To derive the governing equations, we equate (i, j) blocks in $FT = TF$ for $i < j$ and obtain the following generalization of (11.1-4):

$$(11.1-5) \quad F_{ij}T_{jj} - T_{ii}F_{ij} = T_{ij}F_{jj} - F_{ii}T_{ij} + \sum_{k=i+1}^{j-1} (T_{ik}F_{kj} - F_{ik}T_{kj}).$$

This is a linear system whose unknowns are the elements of the block F_{ij} and whose right-hand side is "known" if we compute the F_{ij} one block super-diagonal at a time. We can solve (11.1-5) using the Bartels-Stewart algorithm (Algorithm 7.6-2).

The block variant of Algorithm 11.1-1 described here is useful when computing real functions of real matrices. After computing the real Schur form $A = QTQ^T$, the block algorithm can be invoked in order to handle the 2-by-2 bumps along the diagonal of T.

Problems

P11.1-1. Using the definition (11.1-1) show that (a) $Af(A) = f(A)A$; (b) $f(A)$ is upper triangular if A is upper triangular; (c) $f(A)$ is Hermitian if A is Hermitian.

P11.1-2. Rewrite Algorithm 11.1-1 so that $f(T)$ is computed "column-at-a-time".

P11.1-3. Suppose $A = X \operatorname{diag}(\lambda_i)X^{-1}$ where $X = [x_1, \ldots, x_n]$ and $X^{-1} = [y_1, \ldots, y_n]^H$. Show that if $f(A)$ is defined, then

$$f(A) = \sum_{k=1}^{n} f(\lambda_i)x_i y_i^H.$$

P11.1-4. Show that if

$$T = \begin{bmatrix} T_{11} & T_{12} \\ 0 & T_{22} \end{bmatrix} \begin{matrix} p \\ q \end{matrix}$$
$$\quad\;\; p \quad\;\; q$$

and $f(T)$ is defined, then it has the form

$$f(T) = \begin{bmatrix} F_{11} & F_{12} \\ 0 & F_{22} \end{bmatrix} \begin{matrix} p \\ q \end{matrix}$$
$$\quad\;\; p \quad\;\; q$$

where $F_{11} = f(T_{11})$ and $F_{22} = f(T_{22})$.

Notes and References for Sec. 11.1

The following texts have a better-than-average treatment of the matrix function problem:

R. Bellman (1969). *Introduction to Matrix Analysis*, McGraw-Hill, New York.

F. Gantmacher (1959). *The Theory of Matrices*, vols. 1–2, Chelsea Publishing Co., New York.

L. Mirsky (1955). *An Introduction to Linear Algebra*, Oxford University Press, London.

The contour integral representation of $f(A)$ given in the text is useful in functional analysis because of its generality. See

N. Dunford and J. Schwartz (1958). *Linear Operators, Part I*, Interscience, New York.

As we discussed, other definitions of $f(A)$ are possible. However, for the matrix functions typically encountered in practice, all these definitions are equivalent. See

R. F. Rinehart (1955). "The Equivalence of Definitions of a Matrix Function," *Amer. Math. Monthly 62*, 395-414.

Various aspects of the Jordan representation are detailed in

J. S. Frame (1964a). "Matrix Functions and Applications, Part II," *IEEE Spectrum 1 (April)*, 102-8.
J. S. Frame (1964b). "Matrix Functions and Applications, Part IV," *IEEE Spectrum 1 (June)*, 123-31.

The following are concerned with the Schur decomposition and its relationship to the $f(A)$ problem:

C. Davis (1973). "Explicit Functional Calculus," *Lin. Alg. & Its Applic. 6*, 193-99.
J. Descloux (1963). "Bounds for the Spectral Norm of Functions of Matrices," *Numer. Math. 5*, 185-90.
C. F. Van Loan (1975b). "A Study of the Matrix Exponential," Numerical Analysis Report no. 7, University of Manchester, England.

Algorithm 11.1-1 and the various computational difficulties that arise when it is applied to a matrix having close or repeated eigenvalues are discussed in

B. N. Parlett (1974a). "Computation of Functions of Triangular Matrices," Memorandum no. ERL-M481, Electronics Research Laboratory, College of Engineering, University of California, Berkeley.
B. N. Parlett (1976). "A Recurrence among the Elements of Functions of Triangular Matrices," *Lin. Alg. & Its Applic. 14*, 117-21.

A compromise between the Jordan and Schur approaches to the $f(A)$ problem results if A is reduced to block diagonal form as described in §7.6. See

B. Kagstrom (1977). "Numerical Computation of Matrix Functions," Department of Information Processing Report UMINF-58.77, University of Umea, Umea, Sweden.

Sec. 11.2. Approximation Methods

We now consider a class of methods for computing matrix functions which prima facie do not involve eigenvalues. These techniques are based on the idea that if $g(z)$ approximates $f(z)$ on $\lambda(A)$, then $f(A)$ approximates $g(A)$, e.g.,

$$e^A \cong I + A + \frac{A^2}{2!} + \cdots + \frac{A^q}{q!}.$$

The Jordan representation of matrix functions (Theorem 11.1-1) can be used to bound the error in the approximant $g(A)$:

THEOREM 11.2-1. Let $X^{-1}AX = \text{diag}(J_1, \ldots, J_p)$ be the JCF of $A \in \mathbb{C}^{n \times n}$ with

$$
J_i = \begin{bmatrix}
\lambda_i & 1 & \cdots & 0 \\
& \lambda_i & \cdots & 0 \\
& & \ddots & \vdots \\
& & & 1 \\
\mathbf{0} & & & \lambda_i
\end{bmatrix} \in \mathbb{C}^{m_i \times m_i}.
$$

If $f(z)$ and $g(z)$ are analytic on an open set containing $\lambda(A)$, then

$$
\|f(A) - g(A)\|_2 \leq \kappa_2(X) \max_{\substack{1 \leq i \leq p \\ 0 \leq r \leq m_i - 1}} m_i \frac{|f^{(r)}(\lambda_i) - g^{(r)}(\lambda_i)|}{r!}.
$$

Proof.   Defining $h(z) = f(z) - g(z)$ we have

$$
\|f(A) - g(A)\|_2 = \|X \, \text{diag}[h(J_1), \ldots, h(J_p)] X^{-1}\|_2
$$
$$
\leq \kappa_2(X) \max_{1 \leq i \leq p} \|h(J_i)\|_2.
$$

Using Theorem 11.1-1 and equation (2.2-10) we conclude that

$$
\|h(J_i)\|_2 \leq m_i \max_{0 \leq r \leq m_i - 1} \frac{|h^{(r)}(\lambda_i)|}{r!},
$$

thereby proving the theorem. $\square$

An analogous result can be obtained via the Schur decomposition:

THEOREM 11.2-2. Let $Q^H A Q = T = \text{diag}(\lambda_i) + N$ be the Schur decomposition of $A \in \mathbb{C}^{n \times n}$, with N being the strictly upper triangular portion of T. If $f(z)$ and $g(z)$ are analytic on a closed convex set Ω whose interior contains $\lambda(A)$, then

$$
\|f(A) - g(A)\|_F \leq \sum_{r=0}^{n-1} \delta_r \frac{\| \, |N|^r \|_F}{r!}
$$

where

$$
\delta_r = \sup_{z \in \Omega} |f^{(r)}(z) - g^{(r)}(z)|.
$$

Proof. Let $h(z) = f(z) - g(z)$ and set $H = (h_{ij}) = h(A)$. Let $S_{ij}^{(r)}$ denote the set of strictly increasing integer sequences $(s_0, \ldots, s_r)$ with the property that $s_0 = i$ and $s_r = j$. Notice that

$$S_{ij} = \bigcup_{r=1}^{j-i} S_{ij}^{(r)}$$

and so from Theorem 11.1-3 we obtain the following for all $i < j$:

$$h_{ij} = \sum_{r=1}^{j-1} \sum_{s \in S_{ij}^{(r)}} n_{s_0, s_1} n_{s_1, s_2} \cdots n_{s_{r-1}, s_r} h[\lambda_{s_0}, \ldots, \lambda_{s_r}].$$

Now since Ω is convex and h analytic, we have

(11.2-1) $$|h[\lambda_{s_0}, \ldots, \lambda_{s_r}]| \leqslant \sup_{z \in \Omega} \frac{|h^{(r)}(z)|}{r!} = \frac{\delta_r}{r!}.$$

Furthermore, if $|N|^r = (n_{ij}^{(r)})$ for $r \geqslant 1$, then it can be shown that

(11.2-2) $$n_{ij}^{(r)} = \begin{cases} 0 & j < i + r \\ \sum_{s \in S_{ij}^{(r)}} |n_{s_0, s_1} n_{s_1, s_2} \cdots n_{s_{r-1}, s_r}| & j \geqslant i + r \end{cases}$$

The theorem now follows by taking absolute values in the expression for h_{ij} and then using (11.2-1) and (11.2-2). $\square$

EXAMPLE 11.2-1. Suppose

$$A = \begin{bmatrix} -.01 & 1 & 1 \\ 0 & 0 & 1 \\ 0 & 0 & .01 \end{bmatrix}$$

and that $f(z) = e^z$ and $g(z) = 1 + z + z^2/2$. A calculation shows that

$$\|f(A) - g(A)\| \cong 10^{-5}$$

in either the Frobenius norm or the 2-norm. Since $\kappa_2(X) \cong 10^7$, the error predicted by Theorem 11.2-1 is $O(1)$, rather pessimistic. On the other hand, the error predicted by the Schur decomposition approach is $O(10^{-2})$.

The bounds in the above theorems suggest that there is more to approximating $f(A)$ than just approximating $f(z)$ on the spectrum of A. In particular, we see that if the eigensystem of A is ill-conditioned and/or A's departure from normality is large, then the discrepancy between $f(A)$ and $g(A)$ may be considerably larger than $\max_{z \in \lambda(A)} |f(z) - g(z)|$. Thus, even though approximation methods avoid eigenvalue computations, they appear to be influenced by the structure of A's eigensystem, a point that we pursue further in the next section.

A popular way of approximating a matrix function such as e^A is through the truncation of its Taylor series. The conditions under which a matrix function $f(A)$ has a Taylor series representation are easily established.

THEOREM 11.2-3.   If $f(z)$ has a power series representation

$$f(z) = \sum_{k=0}^{\infty} c_k z^k$$

on an open disk containing $\lambda(A)$, then

$$f(A) = \sum_{k=0}^{\infty} c_k A^k.$$

Proof.   We prove the theorem for the case when A is diagonalizable. (In P11.2-1 we give a hint as to how to proceed without this assumption.) Suppose

$$X^{-1}AX = D = \text{diag}(\lambda_1, \ldots, \lambda_n).$$

Using Corrollary 11.1-2 we have

$$f(A) = X \, \text{diag}[f(\lambda_1), \ldots, f(\lambda_n)]X^{-1}$$

$$= X \, \text{diag}\left[\sum_{k=0}^{\infty} c_k \lambda_1^k, \ldots, \sum_{k=0}^{\infty} c_k \lambda_n^k \right] X^{-1}$$

$$= X \left[\sum_{k=0}^{\infty} c_k D^k \right] X^{-1} = \sum_{k=0}^{\infty} c_k (XDX^{-1})^k = \sum_{k=0}^{\infty} c_k A^k. \quad \square$$

Several important transcendental matrix functions have particularly simple series representations:

$$\log(I - A) = \sum_{k=1}^{\infty} A^k/k \qquad\qquad |\lambda| < 1, \lambda \in \lambda(A)$$

$$\sin(A) = \sum_{k=0}^{\infty} (-1)^k A^{2k+1}/(2k + 1)!$$

$$\cos(A) = \sum_{k=0}^{\infty} (-1)^k A^{2k}/(2k)!$$

The following theorem bounds the errors which arise when matrix functions such as these are approximated via truncated Taylor series:

THEOREM 11.2-4.   If $f(z)$ has a Taylor series

$$f(z) = \sum_{k=0}^{\infty} \alpha_k z^k$$

on an open disk containing the eigenvalues of $A \in \mathbb{C}^{n \times n}$, then

$$\left\| f(A) - \sum_{k=0}^{q} \alpha_k A^k \right\|_2 \leqslant \frac{n}{(q+1)!} \max_{0 \leqslant s \leqslant 1} \left\| f^{(q+1)}(As) \right\|_2.$$

Proof. Define the matrix $E(s)$ by

(11.2-3)
$$f(As) = \sum_{k=0}^{q} \alpha_k (As)^k + E(s). \qquad 0 \leqslant s \leqslant 1$$

If $f_{ij}(s)$ is the (i, j) entry of $f(As)$, then it is necessarily analytic and so

(11.2-4)
$$f_{ij}(s) = \sum_{k=0}^{q} \frac{f_{ij}^{(k)}(0)}{k!} s^k + \frac{f_{ij}^{(q+1)}(\xi_{ij})}{(q+1)!} s^{q+1}$$

where ξ_{ij} satisfies $0 \leqslant \xi_{ij} \leqslant s \leqslant 1$. By comparing powers of s in (11.2-3) and (11.2-4) we conclude that $e_{ij}(s)$, the (i, j) entry of $E(s)$, has the form

$$e_{ij}(s) = \frac{f_{ij}^{(q+1)}(\xi_{ij})}{(q+1)!} s^{q+1}.$$

Now $f_{ij}^{(q+1)}(s)$ is the (i, j) entry of $f^{(q+1)}(As)$ and therefore

$$|e_{ij}(s)| \leqslant \max_{0 \leqslant s \leqslant 1} \frac{|f_{ij}^{(q+1)}(s)|}{(q+1)!} \leqslant \max_{0 \leqslant s \leqslant 1} \frac{\|f^{(q+1)}(As)\|_2}{(q+1)!}.$$

The theorem now follows from the observation that $\|E(t)\|_2 \leqslant n \max_{i,j} |e_{ij}(t)|$. $\square$

EXAMPLE 11.2-2. If

$$A = \begin{bmatrix} -49 & 24 \\ -64 & 31 \end{bmatrix},$$

then

$$e^A = \begin{bmatrix} -0.735759 & 0.551819 \\ -1.471518 & 1.103638 \end{bmatrix}.$$

For $q = 59$ Theorem 11.2-4 predicts that

$$\left\| e^A - \sum_{k=0}^{q} A^k / k! \right\|_2 \leqslant \frac{n}{(q+1)!} \cdot \max_{0 \leqslant s \leqslant 1} \|e^{As}\|_2 \leqslant 10^{-60}.$$

If, however, $\beta = 16$, $t = 6$, chopped arithmetic is used we find

$$fl\left[\sum_{k=0}^{59} A^k / k! \right] = \begin{bmatrix} -22.25880 & -1.432766 \\ -61.49931 & -3.474280 \end{bmatrix}.$$

The problem is that some of the partial sums have large elements. For example, $I + \cdots + A^{17}/17!$ has entries of order 10^7. Since the machine precision is approximately 10^{-7}, rounding errors larger than the norm of the solution are sustained.

The example highlights a shortcoming of truncated Taylor series approximation: it tends to be worthwhile only near the origin. The problem can sometimes be circumvented through a change of scale. For example, by repeated application of the *double angle* formulae

$$\cos(2A) = 2\cos(A)^2 - I$$

$$\sin(2A) = 2\sin(A)\cos(A)$$

it is possible to "build up" the sine and cosine of a matrix from suitably truncated Taylor series approximates:

$S_0 = $ Taylor approximate to $\sin(A/2^k)$
$C_0 = $ Taylor approximate to $\cos(A/2^k)$
For $j = 1, \ldots, k$
    $S_j = 2S_{j-1}C_{j-1}$
    $C_j = 2C_{j-1}^2 - I$

Here k is a positive integer chosen so that, say, $\|A\|_\infty \leqslant 2^k$. See Serbin and Blalock (1979).

Since the approximation of transcendental matrix functions so often involves the evaluation of polynomials, it is worthwhile to look at the details of computing

$$p(A) = b_0 I + b_1 A + \cdots + b_q A^q$$

where the scalars $b_0, \ldots, b_q \in \mathbb{R}$ are given. The most obvious approach is to invoke Horner's scheme:

$S := b_q A + b_{q-1} I$
For $k = q - 2, \ldots, 0$
    $S := AS + b_k I$

whereupon $S = p(A)$. This requires about $(q - 1)n^3$ flops. However, unlike the scalar case, this summation process is not optimal. To see why, suppose $q = 9$ and observe that

$$p(A) = A^3[A^3[b_9 A^3 + (b_8 A^2 + b_7 A + b_6 I)]$$
$$+ (b_5 A^2 + b_4 A + b_3 I)] + b_2 A^2 + b_1 A + b_0 I.$$

Thus, $S = p(A)$ can be evaluated in only $4n^3$ flops:

$$A_2 := A^2$$
$$A_3 := AA_2$$
$$S := b_9 A_3 + b_8 A_2 + b_7 A + b_6 I$$
$$S := A_3 S + b_5 A_2 + b_4 A + b_3 I$$
$$S := A_3 S + b_2 A_2 + b_1 A + b_0 I.$$

In general, if s is any integer satisfying $1 \leqslant s \leqslant \sqrt{q}$ then

(11.2-5) $$p(A) = \sum_{k=0}^{r} B_k (A^s)^k \qquad r = \text{floor}(q/s)$$

where

$$B_k = \begin{cases} b_{sk+s-1} A^{s-1} + \cdots + b_{sk+1} A + b_{sk} I & k = 0, \ldots, r-1 \\ b_q A^{q-sr} + \cdots + b_{sr+1} A + b_{sr} I & k = r \end{cases}$$

Once $A^2, \ldots, A^s$ are computed, Horner's rule can be applied to (11.2-5) and the net result is that $p(A)$ can be computed in approximately $(s + r - 1)n^3$ flops. By choosing $s = \text{floor}(\sqrt{q})$, we see that this flop count has an approximately minimal value of $n^3 \sqrt{q}$. This technique is discussed in Paterson and Stockmeyer (1973). Van Loan (1978b) shows how the procedure can be implemented without storage arrays for $A^2, \ldots, A^s$.

The problem of raising a matrix to a given power deserves special mention. Suppose it was required to compute A^{13}. Noting that $A^4 = (A^2)^2$, $A^8 = (A^4)^2$, and $A^{13} = A^8 A^4 A$, we see that this can be accomplished in $5n^3$ flops. In general we have

ALGORITHM 11.2-1: *Binary Powering.* Given $A \in \mathbb{R}^{n \times n}$ and a positive integer s, the following algorithm computes $Y = A^s$.

Let $s = \sum_{k=0}^{t} \beta_k 2^k$ be the binary expansion of s, with $\beta_t \neq 0$.

$W := A$
$Y := A^{\beta_0}$
For $k = 1, 2, \ldots, t$
    $W := W^2$
    If $\beta_k = 1$, then $Y := YW$

This algorithm requires at most $2 \cdot \text{floor}[\log_2(s)]n^3$ flops. If s is an integral power of 2, then only $\log_2(s)n^3$ flops are needed.

We conclude this section with some remarks on the integration of matrix functions. Suppose $f(At)$ is defined for all $t \in [a, b]$ and that we wish to compute

$$F = \int_a^b f(At)\,dt.$$

As in (11.1-1) the integration is on an element-by-element basis.

Ordinary quadrature rules can be applied to F. For example, with Simpson's rule we have

(11.2-6)
$$F \cong \tilde{F} = \frac{h}{3} \sum_{k=0}^{m} w_k f[A(a + kh)]$$

where m is even, $h = (b - a)/m$, and

$$w_k = \begin{cases} 1 & k = 0, m \\ 4 & k \text{ odd} \\ 2 & k \text{ even}, k \neq 0, m. \end{cases}$$

If $(d^4/dz^4)f(zt) = f^{(4)}(zt)$ is continuous for $t \in [a, b]$ and if $f^{(4)}(At)$ is defined on this same interval, then it can be shown that $F = \tilde{F} + E$ where

(11.2-7)
$$\|E\|_2 \leqslant \frac{nh^4(b - a)}{180} \max_{a \leqslant t \leqslant b} \|f^{(4)}(At)\|_2.$$

Let f_{ij}, $\tilde{f}_{ij}$, and e_{ij} denote the (i, j) entries of F, $\tilde{F}$, and E, respectively. Under the above assumptions we can apply the standard error bounds for Simpson's rule and obtain

$$|e_{ij}| \leqslant \frac{h^4(b - a)}{180} \max_{a \leqslant t \leqslant b} |e_i^T f^{(4)}(At)e_j|.$$

The inequality (11.2-7) now follows because $\|E\|_2 \leqslant n \max |e_{ij}|$ and

$$\max_{a \leqslant t \leqslant b} |e_i^T f^{(4)}(At)e_j| \leqslant \max_{a \leqslant t \leqslant b} \|f^{(4)}(At)\|_2.$$

Of course, in the practical application of (11.2-6), the evaluations $f[A(a + kh)]$ normally have to be approximated. Thus, the overall error involves the error in approximating $f(A(a + kh))$ as well as the Simpson rule error.

Problems

P11.2-1. (a) Suppose $G = \lambda I + E$ is a p-by-p Jordan block, where $E = \{\delta_{i,\,j-1}\}$. Show that

$$(\lambda I + E)^k = \sum_{j=0}^{\min\{p-1,k\}} \binom{k}{j} \lambda^{k-j} E^j.$$

(b) Use (a) and Theorem 11.1-1 to prove Theorem 11.2-3.

P11.2-2. Verify (11.2-2).

P11.2-3. Show that if $\|A\|_2 < 1$, then $\log(I + A)$ exists and

$$\|\log(I + A)\|_2 \leqslant \frac{\|A\|_2}{1 - \|A\|_2}.$$

P11.2-4. Let A be an n-by-n symmetric positive definite matrix. (a) Show that there exists a unique symmetric positive definite X such that $A = X^2$. (b) Denote the matrix X in part (a) by $\sqrt{A}$. Show that if $X_0 = I$ and

$$X_{k+1} = \tfrac{1}{2}(X_k + AX_k^{-1}) \qquad\qquad k = 0, 1, \ldots$$

then $X_k \to \sqrt{A}$ quadratically.

P11.2-5. Specialize Algorithm 11.2-1 to the case when A is symmetric. Repeat for the case when A is upper triangular. In both instances, give the associated flop counts.

P11.2-6. Show that $X(t) = C_1 \cos(t\sqrt{A}) + C_2\sqrt{A^{-1}} \sin(t\sqrt{A})$ solves

$$\ddot{X}(t) = -AX(t)$$

$$X(0) = C_1$$

$$\dot{X}(0) = C_2,$$

where A is symmetric positive definite.

P11.2-7. Using Theorem 11.2-4 bound the error in the approximations

$$\sin(A) \cong \sum_{k=0}^{q} (-1)^k A^{2k+1}/(2k + 1)!$$

$$\cos(A) \cong \sum_{k=0}^{q} (-1)^k A^{2k}/(2k)!$$

Notes and References for Sec. 11.2

The optimality of Horner's rule for polynomial evaluation is discussed in

D. Knuth (1969). *The Art of Computer Programming*, vol. 2, *Seminumerical Algorithms*, Addison-Wesley, Reading, Mass.

M. S. Paterson and L. J. Stockmeyer (1973). "On The Number of Nonscalar Multiplications Necessary to Evaluate Polynomials," *SIAM J. Comp. 2*, 60–66.

The Horner evaluation of matrix polynomials is analyzed in

C. F. Van Loan (1978b). "A Note on the Evaluation of Matrix Polynomials," *IEEE Trans. Auto. Cont. AC-24*, 320–21.

The Newton and Lagrange representations for $f(A)$ and their relationship in other matrix function definitions is discussed in

R. F. Rinehart (1955). "The Equivalence of Definitions of a Matrix Function," *Amer. Math. Monthly 62*, 395-414.

The "double angle" method for computing the cosine of matrix is analyzed in

S. Serbin and S. Blalock (1979). "An Algorithm for Computing the Matrix Cosine," *SIAM J. Sci. Stat. Comp. 1*, 198-204.

Sec. 11-3. The Matrix Exponential

One of the most frequently computed matrix functions is the exponential

$$(11.3-1) \qquad e^{At} = \sum_{k=0}^{\infty} (At)^k/k!.$$

Numerous algorithms for computing e^{At} have been proposed, but most of them are of dubious numerical quality, as is pointed out in the survey article by Moler and Van Loan (1978). In order to illustrate what the computational difficulties are, we present a brief perturbation analysis of the matrix exponential problem and then use it to assess one of the better e^{At} algorithms.

The starting point in the discussion is the initial value problem

$$(11.3-2) \qquad \begin{aligned} \dot{X}(t) &= AX(t) \qquad\qquad A, X(t) \in \mathbb{R}^{n\times n} \\ X(0) &= I, \end{aligned}$$

which has the unique solution $X(t) = e^{At}$. This characterization of the matrix exponential can be used to establish the identity

$$(11.3-3) \qquad e^{(A+E)t} - e^{At} = \int_0^t e^{A(t-s)} E e^{(A+E)s} ds,$$

whereupon

$$(11.3-4) \qquad \frac{\|e^{(A+E)t} - e^{At}\|_2}{\|e^{At}\|_2} \leqslant \frac{\|E\|_2}{\|e^{At}\|_2} \int_0^t \|e^{A(t-s)}\|_2 \|e^{(A+E)s}\|_2 ds.$$

Further simplifications result if we bound the norms of the exponentials that appear in the integrand. One way of doing this is through the Schur decomposition. If

$$Q^H A Q = \text{diag}(\lambda_i) + N \qquad\qquad A \in \mathbb{C}^{n\times n}$$

is the Schur decomposition of A, then it can be shown that

$$(11.3-5) \qquad \|e^{At}\|_2 \leqslant e^{\alpha(A)t} M_S(t),$$

where

$$(11.3-6) \qquad \alpha(A) = \max\{Re(\lambda) \mid \lambda \in \lambda(A)\}$$

$$M_S(t) = \sum_{k=0}^{n-1} \|Nt\|_2^k/k!.$$

With a little manipulation, these results can be used to establish the inequality

$$\frac{\|e^{(A+E)t} - e^{At}\|_2}{\|e^{At}\|_2} \leqslant t\|E\|_2 M_S(t)^2 e^{tM_S(t)\|E\|_2}.$$

Notice that $M_S(t) \equiv 1$ if and only if A is normal, suggesting that the matrix exponential problem is "well-behaved" if A is normal. This observation is confirmed by the behavior of the *matrix exponential condition number* $\nu(A, t)$, defined by

$$\nu(A, t) = \max_{\|E\|_2 \leqslant 1} \left\| \int_0^t e^{A(t-s)} E e^{As} ds \right\|_2 \frac{\|A\|_2}{\|e^{At}\|_2}.$$

This quantity, discussed in Van Loan (1977a), measures the sensitivity of the map $A \to e^{At}$ in that for a given t, there is a matrix E for which

$$\frac{\|e^{(A+E)t} - e^{At}\|_2}{\|e^{At}\|_2} \cong \nu(A, t) \frac{\|E\|_2}{\|A\|_2}.$$

Thus, if $\nu(A, t)$ is large, small changes in A can induce relatively large changes in e^{At}. Unfortunately, it is difficult to precisely characterize those A for which $\nu(A, t)$ is large. (This is in contrast to the linear equation problem $Ax = b$, where the ill-conditioned A are neatly described in terms of SVD.) One thing we can say, however, is that $\nu(A, t) \geqslant t\|A\|_2$, with equality holding for all non-negative t if and only if A is normal.

Bearing in mind these sensitivity properties of the map $A \to e^{At}$, we now present and analyze the method of *scaling and squaring with diagonal Padé approximation*. Following the discussion in §11.2, if $g(z) \cong e^z$ then $g(A) \cong e^A$. A very useful class of approximants for this purpose are the Padé functions defined by

$$R_{pq}(z) = D_{pq}(z)^{-1} N_{pq}(z),$$

where

$$N_{pq}(z) = \sum_{k=0}^{p} \frac{(p + q - k)! \, p!}{(p + q)! \, k! \, (p - k)!} z^k$$

and

$$D_{pq}(z) = \sum_{k=0}^{q} \frac{(p + q - k)! \, q!}{(p + q)! \, k! \, (q - k)!} (-z)^k$$

Notice that $R_{po}(z) = 1 + z + \cdots + z^p/p!$ is the p-th order Taylor polynomial.

Unfortunately, the Padé approximants are good only near the origin, as the following identity reveals:

(11.3-7)

$$e^A = R_{pq}(A) + \frac{(-1)^q}{(p+q)!} A^{p+q+1} D_{pq}(A)^{-1} \int_0^1 u^p (1-u)^q e^{A(1-u)} \, du.$$

However, this problem can be overcome by exploiting the fact that $e^A = (e^{A/m})^m$. In particular, we scale A by m such that $R_{pq}(A/m)$ is a suitably accurate approximation to $e^{A/m}$ and then raise the resulting matrix to the m-th power. Typically, m is chosen to be a power of 2 so that when Algorithm 11.2-1 is applied, it amounts to repeated squaring. The success of the overall procedure depends on the accuracy of the approximant

$$F_{pq} = \left[R_{pq}\!\left(\frac{A}{2^j}\right) \right]^{2^j}.$$

In Moler and Van Loan (1978) it is shown that if

(11.3-8)
$$\frac{\|A\|_\infty}{2^j} \leqslant \frac{1}{2}$$

then there exists an $E \in \mathbb{R}^{n \times n}$ such that

$$F_{pq} = e^{A+E}$$

$$AE = EA$$

(11.3-9)

$$\|E\|_\infty \leqslant \epsilon \|A\|_\infty$$

$$\epsilon = 2^{3-(p+q)} \frac{p! \, q!}{(p+q)! \, (p+q+1)!}.$$

We make two remarks concerning the selection of p and q. First, from the easily established inequality

(11.3-10)
$$\frac{\|e^A - F_{pq}\|_\infty}{\|e^A\|_\infty} \leqslant \epsilon \|A\|_\infty e^{\epsilon \|A\|_\infty}$$

we see that p and q can be determined according to some relative error tolerance. However, if A itself has errors of order $\delta \|A\|_\infty$ then it makes little sense to choose $\epsilon < \delta$.

The other comment pertaining to the selection of F_{pq} is based on the observation that F_{pq} requires about $[j + \max\{p, q\}]n^3$ flops. Therefore, for a given amount of work the error bound in (11.3-10) is smallest when $p = q$.

The overall method is summarized as follows:

ALGORITHM 11.3-1. Given $\delta > 0$ and $A \in \mathbb{R}^{n \times n}$, the following algorithm computes $F = e^{A+E}$ where $\|E\|_\infty \leqslant \delta \|A\|_\infty$.

$j := \max\{0, 1 + \text{floor}(\log_2(\|A\|_\infty))\}$
$A := A/2^j$
Let q be the smallest non-negative integer such that

$$2^{3-2q}\, \frac{(q!)^2}{(2q)!\,(2q+1)!} \leqslant \delta$$

$D := I, N := I, X := I$
For $k = 1, \ldots, q$

$$c := \frac{(2q-k)!\,q!}{(2q)!\,k!\,(q-k)!}$$

$X := AX$
$N := N + cX$
$D := D + (-1)^k cX$
Solve $DF = N$ for F using Gaussian elimination with partial pivoting.
For $k = 1, \ldots, j$
$\quad F := F^2$

This algorithm requires about $(q + j + \frac{1}{3})n^3$ flops.

The special Horner techniques of §11.2 can be applied to quicken the computation of $D = D_{qq}(AA)$ and $N = N_{qq}(A)$. For example, if $q = 8$ we have

$$N_{qq}(A) = U + AV$$

and

$$D_{qq}(A) = U - AV,$$

where

$$U = c_0 I + c_2 A^2 + (c_4 I + c_6 A^2 + c_8 A^4)A^4$$

and

$$V = c_1 I + c_3 A^2 + (c_5 I + c_7 A^2)A^4.$$

Clearly, N and D can be found in $5n^3$ flops, whereas the method described in the algorithm requires $7n^3$ flops.

The roundoff error properties of Algorithm 11.3-1 have essentially been analyzed by Ward (1977). The bounds he derives are computable and incorporated in a FORTRAN subroutine.

There are some interesting questions pertaining to the stability of the scaling and squaring approach. A potential difficulty arises during the squaring process if A is a matrix whose exponential grows before it decays, as depicted in Figure 11.3-1. If

(11.3-12) $$G = R_{qq}\left(\frac{A}{2^j}\right) \cong e^{A/2^j}$$

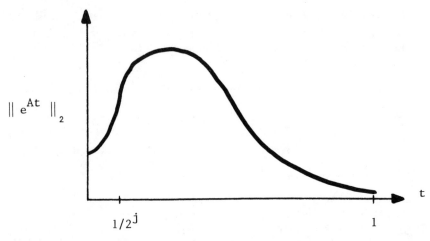

Figure 11.3-1.

then it can be shown that rounding errors of order

$$\hat{\epsilon} = \mathbf{u} \| G \|_2^2 \| G^2 \|_2 \| G^4 \|_2 \| G^8 \|_2 \cdots \| G^{2^{j-1}} \|_2$$

can be expected to contaminate the computed G^{2^j}. Now if $\| e^{At} \|_2$ has a hump as depicted above, then it may be the case that

$$\hat{\epsilon} \gg \mathbf{u} \| G^{2^j} \|_2 \cong \| e^A \|_2,$$

thus ruling out the possibility of small relative errors.

If A is normal, then so is the matrix G and therefore $\| G^m \|_2 = \| G \|_2^m$ for all positive integers. Thus,

$$\hat{\epsilon} = \mathbf{u} \| G^{2^j} \|_2 \cong \mathbf{u} \| e^A \|_2.$$

Conclusion: the problems of the hump disappear and the algorithm can essentially be guaranteed to produce small relative error when A is normal. On the other hand, it is more difficult to draw conclusions about the method when A is non-normal because the connection between $\nu(A, t)$ and the hump phenomena is unclear. Judging from limited experiments, however, it seems that Algorithm 11.3-1 fails to produce a relatively accurate e^A only when $\nu(A, 1)$ is correspondingly large.

Problems

P11.3-1. Show that $e^{(A+B)t} = e^{At} e^{Bt}$ for all t if and only if $AB = BA$. (Hint: Express both sides as a power series in t and compare the coefficient of "t".)

P11.3-2. Show that $\lim_{t \to \infty} e^{At} = 0$ if and only if $\alpha(A) < 0$, where $\alpha(A)$ is defined by (11.3-6).

P11.3-3. Show that if A is skew-symmetric then both e^A and $R_{11}(A)$ are orthogonal. Are there any other values of p and q for which $R_{pq}(A)$ is orthogonal?

P11.3-4. Use (11.3-9) to prove (11.3-10).

P11.3-5. Let $\mu_2(A)$ denote the most positive eigenvalue of the Hermitian matrix $(A + A^H)/2$.

(a) For arbitrary $v \in \mathbb{C}^n$, define $\phi_v(t) = \|e^{At}v\|_2^2$. Show that

$$\frac{d}{dt}\phi_v(t) \leqslant 2\mu_2(A)\phi_v(t).$$

(b) Show that $\|e^{At}\|_2 \leqslant e^{\mu_2(A)t}$ for $t \geqslant 0$.

(c) Show that $\sup_{t \geqslant 0}\|e^{At}\|_2 = 1$ if and only if $\mu_2(A) \leqslant 0$.

(d) Show $\|e^{At}\|_2 \geqslant e^{-\mu_2(-A)t}$ for $t \geqslant 0$.

(e) Show $\mu_2(A + E) \leqslant \mu_2(A) + \|E\|_2$.

(f) Using the fact that $\|e^{At}\|_2 \geqslant e^{\alpha(A)t}$ show

$$\frac{\|e^{(A+E)t} - e^{At}\|_2}{\|e^{At}\|_2} \leqslant t\|E\|_2 \exp[\|E\|_2 + \mu_2(A) - \alpha_2(A)]t.$$

P11.3-6. Show that if A is nonsingular, then there exists a matrix X such that $A = e^X$. Is X unique?

P11.3-7. Prove the identity (11.3-3).

P11.3-8. Suppose $T = D + N$ is an n-by-n upper triangular matrix with N its strictly upper triangular portion. Show that

$$e^{Tt} = \sum_{k=0}^{n-1} T_k(t)$$

where $T_0(t) = e^{Dt}$ and

$$T_k(t) = \int_0^t e^{D(t-s)}NT_{k-1}(s)ds.$$

Use this result to establish (11.3-5).

P11.3-9. Show that if

$$C = \begin{bmatrix} -A^T & P \\ 0 & A \end{bmatrix} \begin{matrix} n \\ n \end{matrix} \quad \text{and} \quad e^{C\Delta} = \begin{bmatrix} F_{11} & F_{12} \\ 0 & F_{21} \end{bmatrix},$$

$$\begin{matrix} n & n \end{matrix}$$

then $F_{11}^T F_{12} = \int_0^\Delta e^{A^T t} P e^{At} dt.$

Notes and References for Sec. 11.3

Much of what appears in this section and an extensive bibliography may be found in the survey article

C. B. Moler and C. F. Van Loan (1978). "Nineteen Dubious Ways to Compute the Exponential of a Matrix," *SIAM Review 20*, 801–36.

Scaling and squaring with Padé approximants (Algorithm 11.3-1) and a careful implementation of Parlett's Schur decomposition method (Algorithm 11.1-1) were found to be among the less dubious of the nineteen methods scrutinized. Various aspects of Padé approximation of the matrix exponential are discussed in

W. Fair and Y. Luke (1970). "Padé Approximations to the Operator Exponential," *Numer. Math. 14*, 379–82.

C. F. Van Loan (1977a). "On the Limitation and Application of Padé Approximation to the Matrix Exponential," in *Padé and Rational Approximation*, ed. E. B. Saff and R. S. Varga, Academic Press, New York.

R. C. Ward (1977). "Numerical Computation of the Matrix Exponential with Accuracy Estimate," *SIAM J. Num. Anal. 14*, 600–14.

A. Wragg (1973). "Computation of the Exponential of a Matrix I: Theoretical Considerations," *J. Inst. Math. Applic. 11*, 369–75.

A. Wragg (1975). "Computation of the Exponential of a Matrix II: Practical Considerations," *J. Inst. Math. Applic. 15*, 273–78.

A proof of equation (11.3-7) for the scalar case appears in

R. S. Varga (1961). "On Higher-Order Stable Implicit Methods for Solving Parabolic Partial Differential Equations," *J. Math. Phys. 40*, 220–31.

There are many applications in control theory calling for the computation of the matrix exponential. In the linear optimal regulator problem, for example, various integrals involving the matrix exponential are required. See

E. S. Armstrong and A. K. Caglayan (1976). "An Algorithm for the Weighting Matrices in the Sample-Data Optimal Linear Regulator Problem," NASA Technical Note, TN D-8372.

J. Johnson and C. L. Phillips (1971). "An Algorithm for the Computation of the Integral of the State Transition Matrix," *IEEE Trans. Auto. Cont. AC-16*, 204–5.

C. Van Loan (1978a). "Computing Integrals Involving the Matrix Exponential," *IEEE Trans. Auto. Cont. AC-23*, 395–404.

An understanding of the map $A \rightarrow \exp(At)$ and its sensitivity is helpful when assessing the performance of algorithms for computing the matrix exponential. Work in this direction includes

B. Kagstrom (1977). "Bounds and Perturbation Bounds for the Matrix Exponential," *BIT 17*, 39–57.

C. F. Van Loan (1977b). "The Sensitivity of the Matrix Exponential," *SIAM J. Num. Anal. 14*, 971–81.

The computation of a logarithm of a matrix is an important area demanding much more work. These calculations arise in various "system identification" problems. See

B. Singer and S. Spilerman (1976). "The Representation of Social Processes by Markov Models," *Amer. J. Sociology 82*, 1–54.

B. W. Helton (1968). "Logarithms of Matrices," *Proc. Amer. Math. Soc. 19*, 733–36.

Special Topics

In this final chapter we discuss an assortment of problems that have attracted our attention over the years. If there is an underlying theme it is the versatility of the matrix methods that we have presented throughout the book.

We first consider least squares minimization with constraints. Two types of constraints are considered, quadratic inequality and linear equality. The singular value decomposition and Q-R factorization have a central role to play in the solution of these problems.

The next two sections are also concerned with variations on the standard LS problem $\min \| Ax - b \|_2$. In §12.2 we consider how b might be approximated by some subset of A's columns, a course of action that is sometimes appropriate when A is rank-deficient. In the next section we consider a variation of ordinary regression known as total least squares that has appeal when A is contaminated with error. More applications of the SVD are considered in §12.4, where various subspace calculations are considered.

Some variations of the symmetric eigenvalue problem are discussed in §12.5, while in the last section we investigate a topic of great interest in nonlinear equations and optimization: how to update the Q-R factorization of a matrix that has been altered by a rank-one matrix.

Reading Paths

$$\S8.6 \;\rightarrow\; \S12.1$$

$$\S6.1 \;\rightarrow\; \S6.2 \;\rightarrow\; \S6.5 \begin{array}{l} \nearrow \;\S12.2 \\ \rightarrow\; \S12.3 \\ \searrow \;\S12.4 \end{array}$$

$$\S8.1 \;\rightarrow\; \S8.2 \;\rightarrow\; \S8.3 \searrow$$

$$\S6.2 \;\rightarrow\; \S12.5$$

$$\S9.1 \;\nearrow$$

$$\S5.2 \searrow$$

$$\S12.6$$

$$\S6.2 \;\nearrow$$

Sec. 12.1. Some Constrained Least Squares Problems

Least squares minimization with a quadratic inequality constraint—the LSQI problem—is a technique that can be used whenever the solution to the ordinary LS problem needs to be regularized. A simple LSQI problem that arises when attempting to fit a function to noisy data is

$$\text{minimize } \| Ax - b \|_2$$

(12.1-1)

$$\text{subject to } \| Bx \|_2 \leqslant \alpha,$$

where $A \in \mathbb{R}^{m \times n}$, $b \in \mathbb{R}^m$, $B \in \mathbb{R}^{n \times n}$ (nonsingular), and $\alpha \geqslant 0$. The constraint, which defines a hyperellipsoid in $\mathbb{R}^n$, is usually chosen in this setting to damp out excessive oscillation in the fitting function. This can be done, for example, if B is a discretized second derivative operator.

More generally, we have the problem

$$\text{minimize } \| Ax - b \|_2 \qquad\qquad A \in \mathbb{R}^{m \times n}, b \in \mathbb{R}^m$$

(12.1-2)

$$\text{subject to } \| Bx - d \|_2 \leqslant \alpha, \qquad B \in \mathbb{R}^{p \times n}, d \in \mathbb{R}^p, \alpha \geqslant 0$$

which clearly has a solution if and only if $\min \| Bx - d \|_2 \leqslant \alpha$. Assuming that there is a solution, then either

$$\| BA^+ b - d \|_2 \leqslant \alpha \text{ and } x = A^+ b \text{ solves (12.1-2)},$$

or

$\|BA^+b - d\|_2 > \alpha$ and the solution satisfies the generalized normal equation $(A^TA + \lambda B^TB)x = (A^Tb + \lambda B^Td)$ where λ is chosen such that $\|Bx - d\|_2 = \alpha$.

Paraphrasing this alternative, either the unconstrained solution is feasible, or the solution to the constrained problem satisfies $\|Bx - d\|_2 = \alpha$. In the latter case, the method of Lagrange multipliers leads to the generalized normal equation. (Hint: Take the gradient of $\|Ax - b\|_2^2 + \lambda(\|Bx - d\|_2^2 - \alpha^2)$ with respect to x.)

The generalized singular value decomposition of §8.6 sheds light on the solvability of the LSQI problem. Indeed, if

$$
\text{(12.1-3)} \quad
\begin{aligned}
U^TAX &= \text{diag}(\alpha_1, \ldots, \alpha_n) & U^TU &= I_m \\
V^TBX &= \text{diag}(\beta_1, \ldots, \beta_q) & V^TV &= I_p, q = \min\{p, n\}
\end{aligned}
$$

is the generalized singular value decomposition of A and B, then (12.1-2) transforms to

$$\text{minimize } \|D_A y - \tilde{b}\|_2$$

$$\text{subject to } \|D_B y - \tilde{d}\|_2 \leqslant \alpha,$$

where $\tilde{b} = U^Tb$, $\tilde{d} = V^Td$, and $y = X^{-1}x$. The simple form of the objective function

$$\text{(12.1-4)} \qquad \|D_A y - \tilde{b}\|_2^2 = \sum_{i=1}^{n} (\alpha_i y_i - \tilde{b}_i)^2 + \sum_{i=n+1}^{m} \tilde{b}_i^2$$

and the constraint equation

$$\text{(12.1-5)} \qquad \|D_B y - \tilde{d}\|_2^2 = \sum_{i=1}^{r} (\beta_i y_i - \tilde{d}_i)^2 + \sum_{i=r+1}^{q} \tilde{d}_i^2 \leqslant \alpha^2$$

$$r = \text{rank}(B)$$

facilitate the analysis of the LSQI problem.

To begin with, the problem has a solution if and only if

$$\sum_{i=r+1}^{q} \tilde{d}_i^2 \leqslant \alpha^2.$$

Moreover, if we have equality, then consideration of (12.1-4) and (12.1-5) shows that the vector y defined by

$$
\text{(12.1-6)} \qquad y_i = \begin{cases}
\tilde{d}_i/\beta_i & i = 1, \ldots, r \\
\tilde{b}_i/\alpha_i & i = r + 1, \ldots, n; \alpha_i \neq 0 \\
0 & i = r + 1, \ldots, n; \alpha_i = 0
\end{cases}
$$

solves the LSQI problem. Hence, we hereafter assume

$$(12.1\text{-}7) \qquad \sum_{i=r+1}^{q} \tilde{d}_i^2 < \alpha^2.$$

Now the vector $y \in \mathbb{R}^n$, defined by

$$y_i = \begin{cases} \tilde{b}_i/\alpha_i & \alpha_i \neq 0 \\ 0 & \alpha_i = 0 \end{cases} \qquad i = 1, \ldots, n$$

minimizes $\| D_A y - \tilde{b} \|_2$. If this vector is also feasible, then we have a solution to (12.1-2). (This is not necessarily the solution of minimum 2-norm, however.) We therefore assume that

$$(12.1\text{-}8) \qquad \sum_{\substack{i=1 \\ \alpha_i \neq 0}}^{q} \left[\beta_i \frac{\tilde{b}_i}{\alpha_i} - \tilde{d}_i \right]^2 > \alpha^2.$$

This implies that the solution to the LSQI problem occurs on the boundary of the feasible set. Thus, our remaining goal is to

$$\text{minimize } \| D_A y - \tilde{b} \|_2$$

$$\text{subject to } \| D_B y - \tilde{d} \|_2 = \alpha.$$

To solve this problem we use the method of Lagrange multipliers. Defining

$$h(\lambda, y) = \| D_A y - \tilde{b} \|_2^2 + \lambda \{ \| D_B y - \tilde{d} \|_2^2 - \alpha^2 \},$$

we see that the equations $\frac{\partial h}{\partial y_i} = 0$ and $1, \ldots, n$ lead to the linear system

$$(D_A^\mathsf{T} D_A + \lambda D_B^\mathsf{T} D_B) y = D_A^\mathsf{T} \tilde{b} + \lambda D_B^\mathsf{T} \tilde{d}.$$

Assuming that the matrix of coefficients is nonsingular, this has a solution $y(\lambda)$ where

$$y_i(\lambda) = \frac{\alpha_i \tilde{b}_i + \lambda \beta_i \tilde{d}_i}{\alpha_i^2 + \lambda \beta_i^2}. \qquad i = 1, \ldots, n$$

The Lagrange parameter λ is resolved by the constraint

$$\phi(\lambda) \equiv \| D_B y(\lambda) - \tilde{d} \|_2^2 = \alpha^2,$$

which can be expressed as follows

$$\phi(\lambda) = \sum_{i=1}^{r} \left[\alpha_i \frac{\beta_i \tilde{b}_i - \alpha_i \tilde{d}_i}{\alpha_i^2 + \lambda \beta_i^2} \right]^2 + \sum_{i=r+1}^{q} \tilde{d}_i^2 = \alpha^2.$$

Equations of this type are referred to as *secular equations*.

From (12.1-8) we see that $\phi(0) > \alpha^2$. Now $\phi(\lambda)$ is monotone decreasing for $\lambda > 0$, and therefore, (12.1-8) implies the existence of a unique positive λ^* for which $\phi(\lambda^*) = \alpha^2$. It is easy to show that this is the desired root. It can be found through the application of any standard root-finding technique, such as Newton's method. The solution of the original LSQI problem is then $x = Xy(\lambda^*)$.

For the simple but important case when $B = I_n$ and $d = 0$ we have the following overall procedure:

ALGORITHM 12.1-1. Given $A \in \mathbb{R}^{m \times n}$ $(m \geqslant n)$, $b \in \mathbb{R}^m$, and $\alpha > 0$, the following algorithm computes a vector $x \in \mathbb{R}^n$ such that $\|Ax - b\|_2$ is minimum, subject to the constraint that $\|x\|_2 \leqslant \alpha$.

Compute the SVD of A, $A = U \operatorname{diag}(\sigma_i) V^\mathsf{T}$.
   Save $V = [v_1, \ldots, v_n]$
   $b := U^\mathsf{T} b$
   $r := \operatorname{rank}(A)$

If $\sum\limits_{i=1}^{r} (b_i/\sigma_i)^2 \leqslant \alpha^2$

   then

$$x := \sum_{i=1}^{r} (b_i/\sigma_i) v_i$$

   else

   Find $\lambda^* > 0$ such that $\sum\limits_{i=1}^{r} (\sigma_i b_i/(\sigma_i^2 + \lambda^*))^2 = \alpha^2$.

$$x := \sum_{i=1}^{r} (\sigma_i b_i/(\sigma_i^2 + \lambda^*)) v_i$$

This algorithm requires $mn^2 + \frac{17}{3} n^3$ flops.

EXAMPLE 12.1-1. The secular equation for the problem

$$\min_{\|x\|_2=1} \left\| \begin{bmatrix} 2 & 0 \\ 0 & 1 \\ 0 & 0 \end{bmatrix} \begin{bmatrix} x_1 \\ x_2 \end{bmatrix} - \begin{bmatrix} 4 \\ 2 \\ 3 \end{bmatrix} \right\|_2$$

is

$$\left[\frac{8}{\lambda + 4} \right]^2 + \left[\frac{2}{\lambda + 1} \right]^2 = 1.$$

This has solution $\lambda^* = 4.57132$, giving solution $x = (.93334, .35898)^\mathsf{T}$.

We know that underlying the problem solved by Algorithm 12.1-1 is the equivalent to the Lagrange multiplier problem of determining $\lambda > 0$ such that

(12.1-9)
$$(A^TA + \lambda I)x = A^Tb$$

and

$$\|x\|_2 = \alpha.$$

Now (12.1-9) is precisely the normal equation formulation for the *ridge regression* problem

$$\min_x \left\| \begin{bmatrix} A \\ \sqrt{\lambda}I \end{bmatrix} x - \begin{bmatrix} b \\ 0 \end{bmatrix} \right\|_2^2 = \min_x \|Ax - b\|_2^2 + \lambda\|x\|_2^2.$$

In the general ridge regression problem one has some criteria for selecting the ridge parameter λ, e.g., $\|x(\lambda)\| = \alpha$ for some given α. We describe a λ-selection procedure described in Golub, Heath, and Wahba (1977).

Set $D_k = I - e_k e_k^T = \text{diag}(1, \ldots, 1, 0, 1, \ldots, 1) \in \mathbb{R}^{m \times m}$ and let $x_k(\lambda)$ solve

(12.1-10)
$$\min_x \|D_k(Ax - b)\|_2^2 + \lambda\|x\|_2^2.$$

Thus, $x_k(\lambda)$ is the solution to the ridge regression problem with the k-th row of A and k-th component of b deleted—i.e., the k-th experiment is ignored. Now consider choosing λ so as to minimize the *cross-validation weighted square error* $C(\lambda)$ defined by

$$C(\lambda) = \frac{1}{m} \sum_{k=1}^{m} w_k [a_k^T x_k(\lambda) - b_k]^2.$$

Here, $w_1, \ldots, w_m$ are given non-negative weights and a_k^T is the k-th row of A. Noting that

$$\|Ax_k(\lambda) - b\|_2^2 = \|D_k[Ax_k(\lambda) - b]\|_2^2 + (a_k^T x_k(\lambda) - b_k)^2,$$

we see that $[a_k^T x_k(\lambda) - b_k]^2$ is the increase in the sum of squares resulting when the k-th row is "reinstated." Minimizing $C(\lambda)$ is tantamount to choosing λ such that the final model is not overly dependent on any one experiment.

A more rigorous analysis will make this statement precise and also suggest a method for minimizing $C(\lambda)$. Assuming that $\lambda > 0$, an algebraic manipulation shows that

(12.1-11)
$$x_k(\lambda) = x(\lambda) + \frac{a_k^T x(\lambda) - b_k}{1 - z_k^T a_k} z_k,$$

where

$$z_k = (A^TA + \lambda I)^{-1}a_k$$
$$x(\lambda) = (A^TA + \lambda I)^{-1}A^Tb.$$

Applying $-a_k^T$ to (12.1-11) and then adding b_k to each side of the resulting equation gives:

$$(12.1\text{-}12) \qquad b_k - a_k^T x_k(\lambda) = \frac{e_k^T[I - A(A^TA + \lambda I)^{-1}A^T]b}{e_k^T[I - A(A^TA + \lambda I)^{-1}A^T]e_k}.$$

Noting that the residual $r = (r_1, \ldots, r_m)^T = b - Ax(\lambda)$ is given by

$$r = [I - A(A^TA + \lambda I)^{-1}A^T]b,$$

we see

$$C(\lambda) = \frac{1}{m} \sum_{k=1}^{m} w_k \left[\frac{r_k}{\partial r_k / \partial b_k}\right]^2.$$

The quotient $\partial r_k / (\partial r_k / \partial b_k)$ may be regarded as an inverse measure of the "impact" of the k-th observation b_k on the model. When $\partial r_k / \partial b_k$ is small, this says that the error in the model's prediction of b_k is somewhat independent of b_k. The tendency for this to be true is lessened by basing the model on the λ^* that minimizes $C(\lambda)$.

The actual determination of λ^* is simplified by computing the SVD of A. Indeed, if

$$U^TAV = \text{diag}(\sigma_1, \ldots, \sigma_n)$$

with $\sigma_1 \geqslant \cdots \geqslant \sigma_n$ and $\tilde{b} = U^Tb$, then it can be shown from (12.1-12) that

$$C(\lambda) = \frac{1}{m} \sum_{k=1}^{m} w_k \left[\frac{\tilde{b}_k - \sum_{j=1}^{r} u_{kj}\tilde{b}_j \left(\dfrac{\sigma_j^2}{\sigma_j^2 + \lambda}\right)}{1 - \sum_{j=1}^{r} u_{kj}^2 \left(\dfrac{\sigma_j^2}{\sigma_j^2 + \lambda}\right)}\right]^2.$$

The minimization of this expression is discussed in Golub, Heath, and Wahba (1979).

We conclude the section by considering the least square problem with linear equality constraints:

$$(12.1\text{-}13) \qquad \min_{Bx=d} \|Ax - b\|_2$$

where $A \in \mathbb{R}^{m \times n}$, $B \in \mathbb{R}^{p \times n}$, $b \in \mathbb{R}^m$, $d \in \mathbb{R}^p$, and $\text{rank}(B) = p$. Note that this is a special case of (12.1-2). However, it is simpler to approach (12.1-13) directly rather than through Lagrange multipliers.

Assume for clarity that both A and B have full rank. Let

$$Q^TB^T = \begin{bmatrix} R \\ 0 \end{bmatrix} \begin{matrix} p \\ n-p \end{matrix}$$

be the Q-R factorization of B^T and set

$$AQ = [A_1, \ A_2 \ ] \qquad Q^T x = \begin{bmatrix} y \\ z \end{bmatrix} \begin{matrix} p \\ n-p \end{matrix} .$$
$$\hspace{2.5cm} p \quad n-p$$

It is clear that with these transformations (12.1-13) becomes

$$\min_{\substack{[R^T \ 0][\begin{smallmatrix} y \\ z \end{smallmatrix}]=d}} \|A_1 y + A_2 z - b\|_2.$$

Thus, y is determined from the equation $R^T y = d$. The vector z is obtained by solving the *un*constrained LS problem

$$\min_z \|A_2 z - (b - A_1 y)\|_2.$$

Combining the above, we see that $x = Q \begin{bmatrix} y \\ z \end{bmatrix}$ solves (12.1-13).

   An interesting way to obtain an approximate solution to (12.1-13) is to solve the unconstrained LS problem

(12.1-14)
$$\min_x \left\| \begin{bmatrix} A \\ \lambda B \end{bmatrix} x - \begin{bmatrix} b \\ \lambda d \end{bmatrix} \right\|_2$$

for large λ. The generalized singular value decomposition sheds light on the quality of the approximation. Let

$$U^T A X = \text{diag}(\alpha_1, \ \ldots, \ \alpha_n) = D_A \in \mathbb{R}^{m \times n}$$

$$V^T B X = \text{diag}(\beta_1, \ \ldots, \ \beta_p) = D_B \in \mathbb{R}^{p \times n}$$

be the GSVD of (A, B) and assume that both matrices have full rank. If $U = [u_1, \ \ldots, \ u_m]$, $V = [v_1, \ \ldots, \ v_p]$, and $X = [x_1, \ \ldots, \ x_n]$, then it is easy to show that

(12.1-15)
$$x = \sum_{i=1}^{p} \frac{v_i^T d}{\beta_i} x_i + \sum_{i=p+1}^{n} \frac{u_i^T b}{\alpha_i} x_i$$

is the exact solution to (12.1-13), while

(12.1-16)
$$x(\lambda) = \sum_{i=1}^{p} \frac{\alpha_i u_i^T b + \lambda^2 \beta_i^2 v_i^T d}{\alpha_i^2 + \lambda^2 \beta_i^2} x_i + \sum_{i=p+1}^{n} \frac{u_i^T b}{\alpha_i} x_i$$

solves (12.1-14). Since

(12.1-17)
$$x(\lambda) - x = \sum_{i=1}^{p} \frac{\alpha_i(\beta_i u_i^T b - \alpha_i v_i^T d)}{\beta_i(\alpha_i^2 + \lambda^2 \beta_i^2)} x_i,$$

it follows that $x(\lambda) \rightarrow x$.

   The appeal of this approach to (12.1-13) is that no special subroutines are required: an ordinary LS solver will do. However, for large values of λ numerical problems can arise, and it is necessary to take precautions. See Powell and Reid (1968) and Van Loan (1982a).

EXAMPLE 12.1-2. The problem

$$\min_{x_1=x_2} \left\| \begin{bmatrix} 1 & 2 \\ 3 & 4 \\ 5 & 6 \end{bmatrix} \begin{bmatrix} x_1 \\ x_2 \end{bmatrix} - \begin{bmatrix} 7 \\ 1 \\ 3 \end{bmatrix} \right\|_2$$

has solution $x = (.3407821, .3407821)^{\mathsf{T}}$. This can be approximated by solving

$$\min_{x} \left\| \begin{bmatrix} 1 & 2 \\ 3 & 4 \\ 5 & 6 \\ 1000 & -1000 \end{bmatrix} \begin{bmatrix} x_1 \\ x_2 \end{bmatrix} - \begin{bmatrix} 7 \\ 1 \\ 3 \\ 0 \end{bmatrix} \right\|_2,$$

which has solution $x = (.3407810, .3407829)^{\mathsf{T}}$.

Problems

P12.1-1. Show that if $N(A) \cap N(B) \neq \{0\}$, then (12.1-2) cannot have a unique solution. Give an example which shows that the converse is not true. (Hint: A^+b feasible.)

P12.1-2. Let $p_0(x), \ldots, p_n(x)$ be given polynomials and $(x_0, y_0), \ldots, (x_m, y_m)$ a given set of coordinate pairs with $x_i \in [a, b]$. It is desired to find a polynomial

$$p(x) = \sum_{k=0}^{n} a_k p_k(x)$$

such that

$$\sum_{i=0}^{m} [p(x_i) - y_i]^2$$

is minimized subject to the constraint that

$$\int_a^b [p''(x)]^2 dx \cong h \sum_{i=0}^{N} \left[\frac{p(z_{i-1}) - 2p(z_i) + p(z_{i+1})}{h^2} \right]^2 \leq \alpha^2,$$

where $z_i = a + ih$ and $b = a + Nh$. Show that this leads to an LSQI problem of the form (12.1-1).

P12.1-3. Suppose $Y = [y_1, \ldots, y_k]$ where $y_i \in \mathbb{R}^m$ has the property that

$$Y^{\mathsf{T}} Y = \operatorname{diag}(d_1^2, \ldots, d_k^2). \qquad d_1 \geqslant d_2 \geqslant \cdots \geqslant d_k \geqslant 0.$$

Show that if $Y = QR$ is the Q-R decomposition of Y, then R is diagonal with $|r_{ii}| = d_i$.

P12.1-4. (a) Show that if $(A^TA + \lambda I)x = A^Tb$, $\lambda > 0$, and $\|x\|_2 = \alpha$, then $z = (Ax - b)/\lambda$ solves the *dual equations*

$$(AA^T + \lambda I)z = -b$$

$$\|A^Tz\|_2 = \alpha$$

(b) Show that if $(AA^T + \lambda I)z = -b$, $\|A^Tz\|_2 = \alpha$, then $x = -A^Tz$ satisfies $(A^TA + \lambda I)x = A^Tb$, $\|x\|_2 = \alpha$.

P12.1-5. Suppose $A \in \mathbb{R}^{m \times 1}$ is given by $A^T = (1, \ldots, 1)^T$ and let $b \in \mathbb{R}^m$. Show that the cross-validation technique described in the text with unit weights leads to an optimal λ given by

$$\lambda = \left[\left(\frac{\bar{b}}{s} \right)^2 - \frac{1}{m} \right]^{-1}$$

where $\bar{b} = (b_1 + \cdots + b_m)/m$ and $s^2 = \Sigma_{i=1}^m (b_i - \bar{b})^2/(m - 1)$.

P12.1-6. Establish equations (12.1-15) through (12.1-17).

Notes and References for Sec. 12.1

Roughly speaking, regularization is a technique for transforming an ill-conditioned problem into a stable one. Quadratically constrained least squares is an important example. See

L. Elden (1977*a*). "Algorithms for the Regularization of Ill-Conditioned Least Squares Problems," *BIT 17*, 134–45.
L. Elden (1977*b*). "Numerical Analysis of Regularization and Constrained Least Square Methods," Ph.D. thesis, Linkoping Studies in Science and Technology, Dissertation no. 20, Linkoping, Sweden.

Our treatment of cross-validation is patterned after

G. H. Golub, M. Heath, and G. Wahba (1979). "Generalized Cross-Validation as a Method for Choosing a Good Ridge Parameter," *Technometrics 21*, 215–23.

A complete analysis of the LSQI problem is given in

W. Gander (1981). "Least Squares with a Quadratic Constraint," *Numer. Math. 36*, 291–307.

See also

G. E. Forsythe and G. H. Golub (1965). "On the Stationary Values of a Second-Degree Polynomial on the Unit Sphere," *SIAM J. App. Math. 13*, 1050–68.
L. Elden (1980). "Perturbation Theory for the Least Squares Problem with Linear Equality Constraints," *SIAM J. Num. Anal. 17*, 338–50.

K. Schittkowski and J. Stoer (1979). "A Factorization Method for the Solution of Constrained Linear Least Squares Problems Allowing for Subsequent Data Changes," *Numer. Math. 31*, 431–63.

Solving the LS problem with linear equality constraints via (12.1-14) is discussed in Lawson and Hansen (SLE, chap. 22). The problems associated with the technique are analyzed in

M.J.D. Powell and J. K. Reid (1968). "On Applying Householder's Method to Linear Least Squares Problems," *Proc. IFIP Congress*, pp. 122–26.

C. Van Loan (1982a). "A Generalized SVD Analysis of Some Weighting Methods for Equality-Constrained Least Squares," in *Proceedings of the Conference on Matrix Pencils*, ed. B. Kagstrom and A. Ruhe, Springer-Verlag, New York. (Forthcoming.)

Sec. 12.2. Subset Selection Using the Singular Value Decomposition

As described in §6.5, the rank-deficient LS problem $\min \| Ax - b \|_2$ can be approached by approximating the minimum norm solution

$$x_{LS} = \sum_{i=1}^{r} \frac{u_i^T b}{\sigma_i} v_i$$

with

$$x_{\hat{r}} = \sum_{i=1}^{\hat{r}} \frac{u_i^T b}{\sigma_i} v_i, \qquad\qquad \hat{r} \leqslant r$$

where

$$(12.2\text{-}1) \qquad\qquad A = U\Sigma V^T = \sum_{i=1}^{r} \sigma_i u_i v_i^T$$

is the SVD of A. Here, $\hat{r}$ is some numerically determined estimate of r. Note that $x_{\hat{r}}$ minimizes $\| A_{\hat{r}} x - b \|_2$ where

$$A_{\hat{r}} = \sum_{i=1}^{\hat{r}} \sigma_i u_i v_i^T$$

is the closest matrix to A that has rank $\hat{r}$. See Corollary 2.3-3.

Replacing A by $A_{\hat{r}}$ in the LS problem amounts to filtering out the small singular values and can make a great deal of sense in those situations where A is derived from noisy data. In other applications, however, rank deficiency implies redundancy among the factors that compose the underlying model. In this case, the model-builder may not be interested in a predictor such as $A_{\hat{r}} x_{\hat{r}}$ that involves all n redundant factors. Instead, a predictor Ay may be sought where y has at most $\hat{r}$ nonzero components. The position of the nonzero entries determines which columns of A, i.e., which factors in the model, are to be used in approximating the observation vector b. How to pick these columns is the problem of *subset selection*.

QR with column pivoting is one method of subset selection that we are already acquainted with. Indeed, if we use Algorithm 6.4-1 to compute

$$y = \Pi \begin{bmatrix} R_{11}^{-1} c \\ 0 \end{bmatrix},$$

where

$$Q^T A \Pi = \begin{bmatrix} R_{11} & R_{12} \\ 0 & R_{22} \end{bmatrix} \begin{matrix} \hat{r} \\ m - \hat{r} \end{matrix}, \qquad Q^T b = \begin{bmatrix} c \\ d \end{bmatrix} \begin{matrix} r \\ m - \hat{r} \end{matrix}$$

and $\Pi = [e_{p_1}, \ldots, e_{p_n}]$, then Ay involves only columns $p_1, \ldots, p_{\hat{r}}$ of A.

However, in settings that involve near rank deficiency, it is advisable to use the SVD. A subset selection procedure that is based on this decomposition has been proposed by Golub, Klema, and Stewart (1976). Their method proceeds as follows:

(a) Compute the SVD $A = U \Sigma V^T$ and use it to determine $\hat{r}$.

(b) Calculate a permutation matrix P such that the columns of the matrix $B_1 \in \mathbb{R}^{m \times \hat{r}}$ in $AP = [B_1, B_2]$ are "sufficiently independent."

(c) Predict b with Ay where $y = P \begin{bmatrix} z \\ 0 \end{bmatrix}$ and $z \in \mathbb{R}^{\hat{r}}$ minimizes $\| B_1 z - b \|_2$.

The key step is (b). Since

$$\min_{z \in \mathbb{R}^{\hat{r}}} \| B_1 z - b \|_2 = \| Ay - b \|_2 \geqslant \min \| Ax - b \|_2,$$

it can be argued that the permutation P should be chosen to make the residual $(I - B_1 B_1^+)b$ as small as possible. Unfortunately, such a solution can lead to an unstable solution. For example, if

$$A = \begin{bmatrix} 1 & 1 & 0 \\ 1 & 1 + \epsilon & 0 \\ 0 & 0 & 1 \end{bmatrix}, \qquad b = \begin{bmatrix} 1 \\ -1 \\ 0 \end{bmatrix},$$

$\hat{r} = 2$, and $P = I$, then $\min \| B_1 z - b \|_2 = 0$, but $\| B_1^+ b \|_2 = O(1/\epsilon)$. On the other hand, any proper subset involving the third column of A is strongly independent, but renders a much worse residual.

The example shows that there can be a trade-off between the independence of the chosen columns and the norm of the residual that they render. How to proceed in the face of this trade-off requires additional mathematical machinery in the form of useful bounds on $\sigma_{\hat{r}}(B_1)$, the smallest singular value of B_1.

THEOREM 12.2-1. Let the SVD of $A \in \mathbb{R}^{m \times n}$ be given by (12.2-1), and define the matrix $B_1 \in \mathbb{R}^{m \times \hat{r}}$, $\hat{r} \leqslant \operatorname{rank}(A)$, by

$$AP = [\underset{\hat{r}}{B_1}, \underset{n-\hat{r}}{B_2}]$$

where $P \in \mathbb{R}^{n \times n}$ is a permutation. If

$$(12.2-2) \qquad P^T V = \begin{bmatrix} \tilde{V}_{11} & \tilde{V}_{12} \\ \tilde{V}_{21} & \tilde{V}_{22} \end{bmatrix} \begin{matrix} \hat{r} \\ n-\hat{r} \end{matrix}$$
$$\qquad\qquad \underset{\hat{r}}{} \quad \underset{n-\hat{r}}{}$$

and $\tilde{V}_{11}$ is nonsingular, then

$$\frac{\sigma_{\hat{r}}(A)}{\| \tilde{V}_{11}^{-1} \|_2} \leqslant \sigma_{\hat{r}}(B_1) \leqslant \sigma_{\hat{r}}(A).$$

Proof. The upper bound follows from the minimax characterization of singular values given in §8.3.

To establish the lower bound, partition the diagonal matrix of singular values as follows:

$$\Sigma = \begin{bmatrix} \Sigma_1 & 0 \\ 0 & \Sigma_2 \end{bmatrix} \begin{matrix} \hat{r} \\ m-\hat{r} \end{matrix}.$$
$$\qquad \underset{\hat{r}}{} \quad \underset{n-\hat{r}}{}$$

If $w \in \mathbb{R}^{\hat{r}}$ is a unit vector with the property that $\| B_1 w \|_2 = \sigma_{\hat{r}}(B_1)$, then

$$\sigma_{\hat{r}}(B_1)^2 = \| B_1 w \|_2^2 = \| U \Sigma V^T P \binom{w}{0} \|_2^2 = \| \Sigma_1 \tilde{V}_{11}^T w \|_2^2 + \| \Sigma_2 \tilde{V}_{12}^T w \|_2^2.$$

The theorem now follows because

$$\| \Sigma_1 \tilde{V}_{11}^T w \|_2 \geqslant \sigma_{\hat{r}}(A)/\| \tilde{V}_{11}^{-1} \|_2. \qquad \square$$

This result suggests that in the interest of obtaining a sufficiently independent subset of columns, we choose the permutation P such that the resulting $\tilde{V}_{11}$ submatrix is as well-conditioned as possible. A heuristic solution to this problem can be obtained by computing the QR with column-pivoting factorization of the matrix $[V_{11}^T, V_{21}^T]$, where

$$V = \begin{bmatrix} V_{11} & V_{12} \\ V_{21} & V_{22} \end{bmatrix} \begin{matrix} \hat{r} \\ n-\hat{r} \end{matrix}.$$
$$\qquad \underset{\hat{r}}{} \quad \underset{n-\hat{r}}{}$$

is a partitioning of the matrix V in (12.2-1). In particular, if we use Algorithm 6.4-1 to compute

$$Q^T[V_{11}^T, V_{21}^T]P = [\underset{r}{R_{11}}, \underset{n-r}{R_{12}}]$$

where Q is orthogonal, P is a permutation matrix, and R_{11} is upper triangular, then (12.2-2) implies

$$\begin{bmatrix} \tilde{V}_{11} \\ \tilde{V}_{21} \end{bmatrix} = P^T \begin{bmatrix} V_{11} \\ V_{21} \end{bmatrix} = \begin{bmatrix} R_{11}^T Q^T \\ R_{12}^T Q^T \end{bmatrix} .$$

Note that R_{11} is nonsingular and that $\| \tilde{V}_{11}^{-1} \|_2 = \| R_{11}^{-1} \|_2$. Heuristically, column pivoting tends to produce a well-conditioned R_{11}, and so the overall process tends to produce a well-conditioned $\tilde{V}_{11}$.

We return to the discussion of the trade-off between column independence and norm of residual. In particular, to assess the above method of subset selection, we need to examine the residual of the vector y that it computes:

$$r_y = b - Ay = b - B_1 z = (I - B_1 B_1^+)b.$$

To this end, it is appropriate to compare r_y with

$$r_{x_{\hat{r}}} = b - A x_{\hat{r}}$$

since we are regarding A as a rank-$\hat{r}$ matrix and since $x_{\hat{r}}$ solves the nearest rank-$\hat{r}$ LS problem, namely, $\min \| A_{\hat{r}} x - b \|_2$.

THEOREM 12.2-2. If r_y and $r_{x_{\hat{r}}}$ are defined as above and if $\tilde{V}_{11}$ is the leading r-by-r submatrix of $P^T V$, then

$$\| r_{x_{\hat{r}}} - r_y \|_2 \leqslant \frac{\sigma_{\hat{r}+1}(A)}{\sigma_{\hat{r}}(A)} \| \tilde{V}_{11}^{-1} \|_2 \| b \|_2 .$$

Proof. It is easy to verify that

$$r_{x_{\hat{r}}} = (I - U_1 U_1^T)b$$

and

$$r_y = (I - Q_1 Q_1^T)b$$

where

$$U = [\underset{\hat{r}}{U_1}, \underset{m-\hat{r}}{U_2}]$$

is a partitioning of the matrix U in (12.2-1) and where

$$Q_1 = B_1 (B_1^T B_1)^{-1/2}.$$

From §2.4 we obtain

$$\| r_{x_{\hat{r}}} - r_y \|_2 \leqslant \| U_1 U_1^T - Q_1 Q_1^T \|_2 \| b \|_2 = \| U_2^T Q_1 \|_2 \| b \|_2 ,$$

while Theorem 12.2-1 permits us to conclude that

$$\|U_2^T Q_1\|_2 \leqslant \|U_2^T B_1\|_2 \|(B_1^T B_1)^{-1/2}\|_2 \leqslant \sigma_{\hat{r}+1}(A) \cdot \frac{1}{\sigma_{\hat{r}}(B_1)}$$

$$\leqslant \frac{\sigma_{\hat{r}+1}(A)}{\sigma_{\hat{r}}(A)} \|\tilde{V}_{11}^{-1}\|_2. \quad \square$$

Noting that

$$\|r_{x_{\hat{r}}} - r_y\|_2 = \|B_1 y - \sum_{i=1}^{\hat{r}} (u_i^T b) u_i\|_2,$$

we see that Theorem 12.2-2 sheds light on how well $B_1 y$ can predict the "stable" component of b, i.e., $U_1^T b$. Any attempt to approximate $U_2^T b$ can lead to a large norm solution. Moreover, the theorem says that if $\sigma_{\hat{r}+1}(A) \ll \sigma_{\hat{r}}(A)$, then any reasonably independent subset of columns produces essentially the same-sized residual. On the other hand, if there is no well-defined gap in the singular values, then the determination of $\hat{r}$ becomes difficult and the entire subset selection problem more complicated.

These observations together with Theorem 12.2-1 form the basis of the following algorithm proposed by Golub, Klema, and Stewart (1976):

ALGORITHM 12.2-1. Given $A \in \mathbb{R}^{m \times n}$, $b \in \mathbb{R}^m$, and a method for computing an integer $\hat{r}$ that approximates rank(A), the following algorithm computes a permutation P and a vector $z \in \mathbb{R}^{\hat{r}}$ such that the first $\hat{r}$ columns of AP are independent and such that $\|AP \binom{z}{0} - b\|_2$ is minimized.

Compute the SVD $U^T A V = \text{diag}(\sigma_1, \ldots, \sigma_n)$ and save V.

Determine $\hat{r} \leqslant$ rank(A) and partition $V = \begin{bmatrix} V_{11} & V_{12} \\ V_{21} & V_{22} \end{bmatrix} \begin{matrix} \hat{r} \\ n-\hat{r} \end{matrix}$.
$\qquad\qquad\qquad\qquad\qquad\qquad\quad \hat{r} \quad\; n-\hat{r}$

Use QR with column pivoting (Algorithm 6.4-1) to compute

$$Q^T [V_{11}^T, V_{21}^T] P = [R_{11}, R_{12}]$$

and set $AP = [\underset{\hat{r}}{B_1}, \underset{n-\hat{r}}{B_2}]$

Determine $z \in \mathbb{R}^{\hat{r}}$ such that $\|b - B_1 z\|_2 = \min$.

This algorithm requires $mn^2 + 6n^3$ flops.

EXAMPLE 12.2-1. Let

$$A = \begin{bmatrix} 3 & 4 & 1 \\ 7 & 4 & -3 \\ 2 & 5 & 3 \\ -1 & 4 & 5 \end{bmatrix}, \quad b = \begin{bmatrix} 1 \\ 1 \\ 1 \\ 1 \end{bmatrix},$$

Rank$(A) = 2$, and

$$x_{LS} = \begin{bmatrix} .0815 \\ .1545 \\ .0730 \end{bmatrix}.$$

If Algorithm 12.2-1 is applied, we find $P = [e_2, e_1, e_3]$ and solution

$$x = \begin{bmatrix} .0845 \\ .2275 \\ .0000 \end{bmatrix}.$$

Note: $\| b - Ax_{LS} \|_2 \cong \| b - Ax \|_2 = .1966$.

Problems

P12.2-1. Suppose $A \in \mathbb{R}^{m \times n}$ and that $\| u^T A \|_2 = \sigma$ with $u^T u = 1$. Show that if $u^T(Ax - b) = 0$ for $x \in \mathbb{R}^n$ and $b \in \mathbb{R}^m$, then $\| x \| \geqslant |u^T b| / \sigma$.

P12.2-2. Show that if $B_1 \in \mathbb{R}^{m \times k}$ is comprised of k columns from $A \in \mathbb{R}^{m \times n}$, then $\sigma_k(B_1) \leqslant \sigma_k(A)$.

P12.2-3. In equation (12.2-2) we know that the matrix

$$P^T V = \begin{bmatrix} \tilde{V}_{11} & \tilde{V}_{12} \\ \tilde{V}_{21} & \tilde{V}_{22} \end{bmatrix} \begin{matrix} \hat{r} \\ n - \hat{r} \end{matrix}$$
$$\quad\quad\; \hat{r} \quad\; n - \hat{r}$$

is orthogonal. Thus, $\| \tilde{V}_{11}^{-1} \|_2 = \| \tilde{V}_{22}^{-1} \|_2$ from Theorem 2.4-1. Show how to compute P by applying the QR with column pivoting algorithm to $[V_{22}^T, V_{12}^T]$. (For $\hat{r} > n/2$, this procedure would be more economical than the technique discussed in the text.)

Notes and References for Sec. 12.2

The material in this section is derived from

G. H. Golub, V. Klema and G. W. Stewart (1976). "Rank Degeneracy and Least Squares Problems," Technical Report TR-456, Department of Computer Science, University of Maryland, College Park.

A detailed treatment of numerical rank is given in this reference. The literature on subset selection is vast; we refer the reader to

H. Hotelling (1957). "The Relations of the Newer Multivariate Statistical Methods to Factor Analysis," *Brit. J. Stat. Psych.* **10**, 69–79.

Sec. 12.3. Total Least Squares

The problem of minimizing $\|D(Ax - b)\|_2$ where $A \in \mathbb{R}^{m \times n}$, and $D = \text{diag}(d_1, \ldots, d_m)$ is nonsingular can be recast as follows:

(12.3-1)
$$\text{minimize } \|Dr\|_2 \qquad\qquad r \in \mathbb{R}^m$$
$$\text{subject to } b + r \in R(A)$$

In this problem there is a tacit assumption that the errors are confined to the "observation" b. When error is also present in the "data" A, then it may be more natural to consider the problem

(12.3-2)
$$\text{minimize } \|D[E, r]T\|_F \qquad\qquad E \in \mathbb{R}^{m \times n}, \quad r \in \mathbb{R}^m$$
$$\text{subject to } b + r \in R(A + E)$$

where $D = \text{diag}(d_1, \ldots, d_m)$ and $T = \text{diag}(t_1, \ldots, t_{n+1})$ are nonsingular. This problem, discussed in Golub and Van Loan (1980), is referred to as the *total least squares* (TLS) problem.

If a minimizing $[E_0, r_0]$ can be found for (12.3-2), then any x satisfying $(A + E_0)x = b + r_0$ is called a TLS solution. It should be realized however, that (12.3-2) may fail to have a solution altogether. For example, if

$$A = \begin{bmatrix} 1 & 0 \\ 0 & 0 \\ 0 & 0 \end{bmatrix}, \quad b = \begin{bmatrix} 1 \\ 1 \\ 1 \end{bmatrix},$$

$D = I_3, T = I_3$, and

$$E_\epsilon = \begin{bmatrix} 0 & 0 \\ 0 & \epsilon \\ 0 & \epsilon \end{bmatrix},$$

then for all $\epsilon > 0$, $b \in R(A + E_\epsilon)$. Thus, there is no smallest value of $\|[E, r]\|_F$ for which $b + r \in R(A + E)$.

A generalization of (12.3-2) results if we allow multiple right-hand sides. In particular, if $B \in \mathbb{R}^{m \times k}$ then we have the problem

(12.3-3)
$$\text{minimize } \|D[E, R]T\|_F \qquad\qquad E \in \mathbb{R}^{m \times n}, R \in \mathbb{R}^{m \times k}$$
$$\text{subject to } R(B + R) \subset R(A + E),$$

where $D = \text{diag}(d_1, \ldots, d_m)$ and $T = \text{diag}(t_1, \ldots, t_{n+k})$ are nonsingular. If $[E_0, R_0]$ solves (12.3-3), then any $X \in \mathbb{R}^{n \times k}$ satisfying $(A + E_0)X = (B + R_0)$ is said to be a TLS solution to (12.3-3). The following theorem gives conditions for the uniqueness and existence of such an X:

THEOREM 12.3-1.   Let A, B, D, and T be as above and assume $m \geqslant n + k$. Let $C = D[A, B]T$ have SVD $U^T C V = \text{diag}(\sigma_1, \ldots, \sigma_{n+k}) = \Sigma$ where

$$
U = \underset{n \quad k}{[U_1, U_2]} \; m \; , \qquad
V = \begin{bmatrix} V_{11} & V_{12} \\ V_{12} & V_{22} \end{bmatrix} \begin{matrix} n \\ k \end{matrix} \; , \qquad
\Sigma = \begin{bmatrix} \Sigma_1 & 0 \\ 0 & \Sigma_2 \end{bmatrix} \begin{matrix} n \\ k \end{matrix} \; .
$$

$$
 \quad \qquad \phantom{V = \begin{bmatrix} V_{11} \end{bmatrix}} \underset{n \quad k}{} \qquad \phantom{\Sigma = \begin{bmatrix} \Sigma_1 \end{bmatrix}} \underset{n \quad k}{}
$$

If $\sigma_n > \sigma_{n+1}$ then the matrix $[E_0, R_0]$ defined by

$$
(12.3\text{-}4) \quad D[E_0, R_0]T = -U_2 \Sigma_2 [V_{12}^T, V_{22}^T] = -C \begin{bmatrix} V_{12} \\ V_{22} \end{bmatrix} [V_{12}^T, V_{22}^T]
$$

solves (12.3-3) and

$$
(12.3\text{-}5) \qquad X = -T_1 V_{12} V_{22}^{-1} T_2^{-1} \qquad
\begin{aligned}
T_1 &= \text{diag}(t_1, \ldots, t_n) \\
T_2 &= \text{diag}(t_{n+1}, \ldots, t_{n+k})
\end{aligned}
$$

exists and is the unique X satisfying $(A + E_0)X = B + R_0$.

Proof. $R(B + R) \subset R(A + E)$ implies that there exists an $X \in \mathbb{R}^{n \times k}$ such that $(A + E)X = B + R$, i.e.,

$$
\{D[A, B]T + D[E, R]T\}T^{-1} \begin{bmatrix} X \\ -I_k \end{bmatrix} = 0.
$$

Thus, the matrix in curly brackets has, at most, rank n. By following the argument in Corollary 2.3-3 it can be shown that

$$
\|D[E, R]T\|_F^2 \geqslant \sum_{i=n+1}^{n+k} \sigma_i^2,
$$

and that equality results by setting $[E, R] = [E_0, R_0]$. The condition $\sigma_n > \sigma_{n+1}$ ensures that $[E_0, R_0]$ is the unique minimizer.

If V_{22} is nonsingular, the matrix X defined by (12.3-5) solves the TLS problem (12.3-3):

$$
\{D[A, B]T + D[E_0, R_0]T\}T^{-1} \begin{bmatrix} X \\ -I_k \end{bmatrix}
$$

$$
= -U_1 \Sigma_1 [V_{11}^T, V_{21}^T] \begin{bmatrix} V_{12} \\ V_{22} \end{bmatrix} V_{22}^{-1} T_2^{-1} = 0
$$

The uniqueness of X follows from the condition that $\sigma_n > \sigma_{n+1}$. This same condition guarantees that V_{22} is nonsingular. To see this, define A_1 and B_1 by $[A_1, B_1] = D[A, B]T$. If $V_{22}x = 0$, then it follows from the equation

$$[A_1, B_1] \begin{bmatrix} V_{12} \\ V_{22} \end{bmatrix} = U_2 \Sigma_2$$

that $A_1(V_{12}x) = U_2\Sigma_2 x$. Hence,

$$\|A_1(V_{12}x)\|_2 = \|\Sigma_2 x\|_2 \leqslant \sigma_{n+1}\|x\|_2.$$

But from the easily established inequality $\sigma_n < \sigma_n(A_1)$ we conclude that $x = 0$. $\square$

If $\sigma_n = \sigma_{n+1}$ then the TLS problem may still have a solution, although it may not be unique. In this case it may be desirable to single out a "minimal norm" solution. To this end, consider the τ-norm defined on $\mathbb{R}^{n \times k}$ by $\|Z\|_\tau = \|T_1^{-1}ZT_2\|_2$. If X is given by (12.3-5), then from Theorem 2.4-1 we have

$$\|X\|_\tau^2 = \|V_{12}V_{22}^{-1}\|_2^2 = \frac{1 - \sigma_k(V_{22})^2}{\sigma_k(V_{22})^2}.$$

This suggests choosing V in Theorem 12.3-1 so that $\sigma_k(V_{22})$ is maximized.

We show how this can be accomplished for the important case $k = 1$. Suppose $\sigma_{n-p} > \sigma_{n-p+1} = \cdots = \sigma_{n+1}$, and let $v_{n+1-p}, \ldots, v_{n+1}$ be the last $p + 1$ columns of V. If $\tilde{Q}$ is a Householder matrix such that

$$[v_{n+1-p}, \ldots, v_{n+1}]\tilde{Q} = \begin{array}{c} \left[\begin{array}{c|c} \times & z \\ \hline 0 & \alpha \end{array}\right] \begin{array}{c} n \\ 1 \end{array} \\ \begin{array}{cc} p & 1 \end{array} \end{array}$$

then $\begin{bmatrix} z \\ \alpha \end{bmatrix}$ has the largest $(n+1)$-st component of all the vectors in span$\{v_{n+1-p}, \ldots, v_{n+1}\}$. If $\alpha = 0$, the TLS problem has no solution. Otherwise $x = -T_1 z/(t_{n+1}\cdot\alpha)$. Moreover,

$$\begin{bmatrix} I_{n-1} & 0 \\ 0 & \tilde{Q} \end{bmatrix} U^\mathsf{T}(D[A, b]T)V \begin{bmatrix} I_{n-p} & 0 \\ 0 & \tilde{Q} \end{bmatrix} = \Sigma,$$

and so

$$D[E_0, r_0]T = -D[A, b]T \begin{bmatrix} z \\ \alpha \end{bmatrix} [z^\mathsf{T} \quad \alpha].$$

Overall, we have the following algorithm:

ALGORITHM 12.3-1.   Given $A \in \mathbb{R}^{m \times n}$ ($m > n$), $b \in \mathbb{R}^m$, and nonsingular $D = \mathrm{diag}(d_1, \ldots, d_m)$ and $T = \mathrm{diag}(t_1, \ldots, t_{n+1})$, the following algorithm computes (if possible) a vector $x \in \mathbb{R}^n$ such that $(A + E_0)x = (b + r_0)$ and $\|D[E_0, r_0]T\|_F$ is minimal.

Compute the SVD $U^T D[A, b] TV = \text{diag}(\sigma_1, \ldots, \sigma_{n+1})$. Save V.
Define an index p by

$$\sigma_1 \geqslant \cdots \geqslant \sigma_{n-p} > \sigma_{n-p+1} = \cdots = \sigma_{n+1}$$

Compute a Householder matrix P such that entries $(n + 1, j)$, $j = n - p + 1, \ldots, n$, of VP are zero. $V := VP$

If $v_{n+1,n+1} = 0$
    then
        quit
    else
        $x_i := -t_i v_{i,n+1}/(t_{n+1} v_{n+1,n+1})$ $i = 1, \ldots, n$

This algorithm requires $mn^2 + \frac{17}{3} n^3$ flops.

EXAMPLE 12.3-1. The TLS problem

$$\min_{(a+e)x=b+r} \| [e, r] \|_F,$$

where

$$a = \begin{bmatrix} 1 \\ 2 \\ 3 \\ 4 \end{bmatrix} \qquad b = \begin{bmatrix} 2.01 \\ 3.99 \\ 5.80 \\ 8.30 \end{bmatrix}$$

has solution $x = 2.0212$, $e = (-.0045, -.0209, -.1048, .0855)^T$, and $r = (.0022, .0103, .0519, -.4023)^T$. Note: for this data $x_{LS} = 2.0197$.

It can be shown that the TLS solution x minimizes

$$\psi(x) = \sum_{i=1}^{m} d_i^2 \frac{|a_i^T x - b_i|^2}{x^T T_1^{-2} x + t_{n+1}^{-2}},$$

where a_i^T is i-th row of A and b_i is the i-th component of b. A geometrical interpretation of the TLS problem is made possible by this observation. Indeed,

$$\frac{|a_i^T x - b_i|^2}{x^T T_1^{-2} x + t_{n+1}^{-2}}$$

is the square of the distance from $\begin{bmatrix} a_i \\ b_i \end{bmatrix} \in \mathbb{R}^{n+1}$ to the nearest point in the subspace

$$P_x = \left\{ \begin{bmatrix} a \\ b \end{bmatrix} \mid a \in \mathbb{R}^n, b \in \mathbb{R}, b = x^T a \right\},$$

where distance in $\mathbb{R}^{n+1}$ is measured by the norm $\| z \| = \| Tz \|_2$. A great deal has been written about this kind of fitting. See Pearson (1901) and Madansky (1959).

Problems

P12.3-1. Let $A \in \mathbb{R}^{m \times n}$ $(m > n)$ and $b \in \mathbb{R}^m$ be given. (a) For fixed $x \in \mathbb{R}^n$ and $r \in \mathbb{R}^m$, show that

$$E = E(r, x) = \frac{(b - Ax + r)x^T}{x^T x}$$

solves

$$\min_{(A+E)x=b+r} \| E \|_F .$$

(b) For fixed $x \in \mathbb{R}^n$, show that $r = r_x = (Ax - b)/(1 + x^T x)$ solves

$$\min_r \| [E(r, x), r] \|_F^2 .$$

(c) Show

$$\| [E(r_x, x), r_x] \|_F^2 = \frac{\left\| [A, b] \begin{bmatrix} x \\ -1 \end{bmatrix} \right\|_2^2}{\left\| \begin{bmatrix} x \\ -1 \end{bmatrix} \right\|_2^2} .$$

(d) Show how one is led to Algorithm 12.3-1 when the expression in (c) is minimized over x.

P12.3-2. Repeat P12.3-1, now using the weighted Frobenius norm $\| SCT \|_F$. ($C \in \mathbb{R}^{m \times (n+1)}$, $S = \text{diag}(s_i) \in \mathbb{R}^{m \times m}$, $T = \text{diag}(t_i) \in R^{(n+1) \times (n+1)}$.)

P12.3-3. Consider the TLS problem (12.3-2) with nonsingular D and T. (a) Show that if $\text{rank}(A) < n$, then (12.3-2) has a solution if and only if $b \in R(A)$. (b) Show that if $\text{rank}(A) = n$, then (12.3-2) has no solution if $A^T D^2 b = 0$ and $|t_{n+1}| \| Db \|_2 \geqslant \sigma_n(DAT_1)$, where $T_1 = \text{diag}(t_1, \ldots, t_n)$.

P12.3-4. Show that if $C = D[A, b]T = [A_1, d]$ and $\sigma_n(C) > \sigma_{n+1}(C)$, then the TLS solution x satisfies $[A_1^T A_1 - \sigma_{n+1}(C)^2 I]x = A_1^T d$.

Notes and References for Sec. 12.3

All of the material in this section is taken from

G. H. Golub and C. F. Van Loan (1980). "An Analysis of the Total Least Squares Problem," *SIAM J. Num. Anal. 17*, 883–93.

Other references concerned with least squares fitting when there are errors in the data matrix include:

W. G. Cochrane (1968). "Errors of Measurement in Statistics," *Technometrics 10*, 637-66.

R. F. Gunst, J. T. Webster, and R. L. Mason (1976). "A Comparison of Least Squares and Latent Root Regression Estimators," *Technometrics 18*, 75-83.

I. Linnik (1961). *Method of Least Squares and Principles of the Theory of Observations*, Pergamon Press, New York.

A. Madansky (1959). "The Fitting of Straight Lines When Both Variables Are Subject to Error," *J. Amer. Stat. Assoc. 54*, 173-205.

K. Pearson (1901). "On Lines and Planes of Closest Fit to Points in Space," *Phil. Mag. 2*, 559-72.

A. Van der Sluis and G. W. Veltkamp (1979). "Restoring Rank and Consistency by Orthogonal Projection," *Lin. Alg. & Its Applic. 28*, 257-78.

G. W. Stewart (1977c). "Sensitivity Coefficients for the Effects of Errors in the Independent Variables in a Linear Regression," Technical Report TR-571, Department of Computer Science, University of Maryland.

Sec. 12.4. Comparing Subspaces Using the Singular Value Decomposition

It is sometimes necessary in practice to investigate the relationship between two given subspaces. How close are they? Do they intersect? Can one be "rotated" into the other? And so on.

In this section we show how questions like these can be answered using the singular value decomposition.

Rotation of Subspaces

Suppose $A \in \mathbb{R}^{m \times p}$ is a data matrix obtained by performing a certain set of experiments. If the same set of experiments is performed again, then a different data matrix, $B \in \mathbb{R}^{m \times p}$, is obtained. In the *orthogonal Procrustes problem* the possibility that B can be rotated into A is explored by solving the following problem:

$$\text{(12.4-1)} \qquad \begin{array}{c} \text{minimize } \|A - BQ\|_F \\ \text{subject to } Q^T Q = I_p \end{array}$$

Note that if $Q \in \mathbb{R}^{p \times p}$ is orthogonal, then

$$\|A - BQ\|_F^2 = \text{trace}(A^T A) + \text{trace}(B^T B) - 2\,\text{trace}(Q^T B^T A),$$

and thus, (12.4-1) is equivalent to the problem of maximizing $\text{trace}(Q^T B^T A)$.

The maximizing Q can be found by calculating the SVD of $B^T A$. Indeed, if

$$U^T(B^T A)V = \Sigma = \text{diag}(\sigma_1, \ldots, \sigma_p)$$

is the SVD of this matrix and we define the orthogonal matrix Z by

$$Z = V^T Q^T U,$$

then

$$\text{trace}(Q^TB^TA) = \text{trace}(Q^TU\Sigma V^T) = \text{trace}(Z\Sigma) = \sum_{i=1}^{p} z_{ii}\sigma_i \leqslant \sum_{i=1}^{p} \sigma_i.$$

Clearly, the upper bound is attained by setting $Q = UV^T$, for then $Z = I_p$. This gives the following algorithm:

ALGORITHM 12.4-1. Given A and B in $\mathbb{R}^{m \times p}$, the following algorithm finds an orthogonal $Q \in \mathbb{R}^{p \times p}$ such that $\|A - BQ\|_F$ is minimum

$C := B^TA$
Compute the SVD $U^TCV = \Sigma$. Save U and V.
$Q := UV^T$

This algorithm requires $mp^2 + 12p^3$ flops.

EXAMPLE 12.4-1. If

$$A = \begin{bmatrix} 1 & 2 \\ 3 & 4 \\ 5 & 6 \\ 7 & 8 \end{bmatrix} \quad \text{and} \quad B = \begin{bmatrix} 1.2 & 2.1 \\ 2.9 & 4.3 \\ 5.2 & 6.1 \\ 6.8 & 8.1 \end{bmatrix},$$

then

$$Q = \begin{bmatrix} .9999 & -.0126 \\ .0126 & .9999 \end{bmatrix}$$

minimizes $\|A - BQ\|_F$ over all orthogonal matrices Q. The minimum value is .4661.

Intersection of Null Spaces

Let $A \in \mathbb{R}^{m \times n}$ and $B \in \mathbb{R}^{p \times n}$ be given, and consider the problem of finding an orthonormal basis for $N(A) \cap N(B)$. One approach is to apply the SVD algorithm to $\begin{bmatrix} A \\ B \end{bmatrix}$, since

$$\begin{bmatrix} A \\ B \end{bmatrix} x = 0 \iff x \in N(A) \cap N(B).$$

However, a more economical procedure results if we exploit the following theorem.

THEOREM 12.4-1. Let $\{z_1, \ldots, z_r\}$ be an orthonormal basis for $N(A)$ where $A \in \mathbb{R}^{m \times n}$. Define $Z = [z_1, \ldots, z_r]$ and let $\{w_1, \ldots, w_q\}$ be an orthonormal basis for $N(BZ)$ where $B \in \mathbb{R}^{p \times n}$. If $W = [w_1, \ldots, w_q]$, then the columns of ZW form an orthonormal basis for $N(A) \cap N(B)$.

Proof. Since $AZ = 0$ and $(BZ)W = 0$ we clearly have $R(ZW) \subset N(A) \cap N(B)$. Now suppose x is in both $N(A)$ and $N(B)$. It follows that $x = Za$ for some $0 \neq a \in \mathbb{R}^r$. But since $0 = Bx = BZa$, we must have $a = Wb$ for some $b \in \mathbb{R}^q$. Thus, $x = ZWb \in R(ZW)$. $\square$

When the SVD algorithm is used to compute the orthonormal basis in this theorem, the following algorithm results:

ALGORITHM 12.4-2. Given $A \in \mathbb{R}^{m \times n}$ and $B \in \mathbb{R}^{p \times n}$, the following algorithm computes a matrix $Y = [y_1, \ldots, y_s]$ having orthonormal columns that span $N(A) \cap N(B)$.

> Compute the SVD $U_A^T A V_A = \Sigma_A$. Save V_A and set $r = \text{rank}(A)$.
> If $r = n$
> > *then*
> > > $s := 0$
> > > quit
> > *else*
> > > Partition $V_A = [V_{1A}, V_{2A}]$, $V_{2A} \in \mathbb{R}^{m \times (n-r)}$.
> > > $C := BV_{2A}$
> > > Compute the SVD $U_C^T C V_C = \Sigma_C$. Save V_C and set $q = \text{rank}(C)$.
> > > If $q = n - r$
> > > > *then*
> > > > > $s := 0$
> > > > > quit
> > > > *else*
> > > > > Partition $V_C = [V_{1C}, V_{2C}]$, $V_{2C} \in \mathbb{R}^{m \times (n-r-q)}$
> > > > > $s := n - r - q$
> > > > > $Y := V_{2A} V_{2C}$

The amount of work required by this algorithm depends upon the relative sizes of m, n, p, and r.

We mention that a computer implementation of this algorithm would require a means for deciding when a computed singular value $\hat{\sigma}_i$ is negligible. The use of a tolerance δ for this purpose (e.g. $\hat{\sigma}_i < \delta \Rightarrow \hat{\sigma}_i = 0$) implies that the columns of the computed $\hat{Y}$ "almost" define the common null space of A and B in the sense that

$$\| A\hat{Y} \|_2 \cong \| B\hat{Y} \|_2 \cong \delta.$$

EXAMPLE 12.4-2. If

$$
A = \begin{bmatrix} 1 & -1 & 1 \\ 1 & -1 & 1 \\ 1 & -1 & 1 \end{bmatrix} \quad \text{and} \quad B = \begin{bmatrix} 4 & 2 & 0 \\ 2 & 1 & 0 \\ 6 & 3 & 0 \end{bmatrix},
$$

then $N(A) \cap N(B) = \text{span}\{x\}$, where $x = (1, -2, -3)^T$. Applying Algorithm 12.4-2 we find

$$
V_{2A} = \begin{bmatrix} -.8165 & .0000 \\ -.4082 & .7071 \\ .4082 & .7071 \end{bmatrix}
$$

and

$$
V_{2C} = \begin{bmatrix} -.3273 \\ -.9449 \end{bmatrix},
$$

and thus,

$$
V_{2A} V_{2C} \cong \begin{bmatrix} .2673 \\ -.5345 \\ -.8018 \end{bmatrix} \cong .2673 \begin{bmatrix} 1 \\ -2 \\ -3 \end{bmatrix}.
$$

Angles between Subspaces

Let F and G be subspace in $\mathbb{R}^m$ whose dimensions satisfy

$$
p = \dim(F) \geqslant \dim(G) = q \geqslant 1.
$$

The *principal angles* $\theta_1, \ldots, \theta_q \in [0, \pi/2]$ between F and G are defined recursively by

$$
\cos(\theta_k) = \max_{u \in F} \max_{v \in G} u^T v = u_k^T v_k,
$$

subject to

$$
\|u\|_2 = \|v\|_2 = 1
$$
$$
u^T u_i = 0 \qquad\qquad i = 1, \ldots, k - 1
$$
$$
v^T v_i = 0 \qquad\qquad i = 1, \ldots, k - 1
$$

Note that $0 \leqslant \theta_1 \leqslant \cdots \leqslant \theta_q \leqslant \pi/2$. The vectors $\{u_1, \ldots, u_q\}$ and $\{v_1, \ldots, v_q\}$ are called the *principal vectors* of the subspace pair (F, G).

Principal angles and vectors arise in many important statistical applications. The smallest principal angle is related to our earlier notion of distance between equidimensional subspaces:

$$\text{dist}(F, G) = \sqrt{1 - \cos(\theta_1)^2} = \sin(\theta_1).$$

If the columns of $Q_F \in \mathbb{R}^{m \times p}$ and $Q_G \in \mathbb{R}^{m \times q}$ define orthonormal bases for F and G respectively, then

$$\max_{\substack{u \in F \\ \|u\|_2 = 1}} \max_{\substack{v \in G \\ \|v\|_2 = 1}} u^T v = \max_{\substack{y \in \mathbb{R}^p \\ \|y\|_2 = 1}} \max_{\substack{z \in \mathbb{R}^q \\ \|z\|_2 = 1}} y^T(Q_F^T Q_G)z.$$

From the minimax characterization of singular values given in §8.3, it follows that if

$$Y^T(Q_F^T Q_G)Z = \text{diag}(\sigma_1, \ldots, \sigma_q)$$

is the SVD of $Q_F^T Q_G$, then we may define the u_k, v_k, and θ_k by

$$[u_1, \ldots, u_p] = Q_F Y$$

$$[v_1, \ldots, v_q] = Q_G Z$$

$$\cos(\theta_k) = \sigma_k. \qquad\qquad k = 1, \ldots, q$$

Typically, the spaces F and G are defined as the ranges of given matrices $A \in \mathbb{R}^{m \times p}$ and $B \in \mathbb{R}^{m \times q}$. When this is the case, the desired orthonormal bases can be obtained by computing the Q-R factorizations of these two matrices.

ALGORITHM 12.4-3. Given $A \in \mathbb{R}^{m \times p}$ and $B \in \mathbb{R}^{m \times q}$ $(p \geqslant q)$ each with linearly independent columns, the following algorithm computes $\{\cos(\theta_k), u_k, v_k\}_{k=1}^q$ where the θ_k are the principal angles of the subspace pair $\{R(A), R(B)\}$ and u_k and v_k are the associated principal vectors.

Use Algorithm 6.2-1 to compute the decompositions:

$$A = Q_A R_A \qquad Q_A^T Q_A = I_p \qquad R_A \in \mathbb{R}^{p \times p}$$

$$B = Q_B R_B \qquad Q_B^T Q_B = I_q \qquad R_B \in \mathbb{R}^{q \times q}$$

$C := Q_A^T Q_B$
Compute the SVD $Y^T C Z = \text{diag}[\cos(\theta_k)]$
$[u_1, \ldots, u_p] := Q_A Y$
$[v_1, \ldots, v_q] := Q_B Z$

This algorithm requires about $2m(q^2 + p^2) + pq(m + q) + 6q^3$ flops.

The idea of using the SVD to compute the principal angles and vectors is due to Bjorck and Golub (1973). The problem of rank deficiency in A and B is also treated in this paper.

Intersection of Subspaces

Algorithm 12.4-3 can also be used to compute an orthonormal basis for $R(A) \cap R(B)$ where $A \in \mathbb{R}^{m \times p}$ and $B \in \mathbb{R}^{m \times q}$.

THEOREM 12.4-2. Let $\{\cos(\theta_k), u_k, v_k\}_{k=1}^{q}$ be prescribed by Algorithm 12.4-3. If the index s is defined by

$$1 = \cos(\theta_1) = \cdots = \cos(\theta_s) > \cos(\theta_{s+1}),$$

then

$$R(A) \cap R(B) = \text{span}\{u_1, \ldots, u_s\} = \text{span}\{v_1, \ldots, v_s\}.$$

Proof. The proof follows from the observation that if $\cos(\theta_k) = 1$, then $u_k = v_k$. $\square$

EXAMPLE 12.4-3. If

$$A = \begin{bmatrix} 1 & 2 \\ 3 & 4 \\ 5 & 6 \end{bmatrix} \quad \text{and} \quad B = \begin{bmatrix} 1 & 5 \\ 3 & 7 \\ 5 & -1 \end{bmatrix},$$

then the cosines of the principal angles between $R(A)$ and $R(B)$ are 1.000 and .856.

Problems

P12.4-1. Show that if A and B are m-by-p matrices, with $p \leqslant m$, then

$$\min_{Q^T Q = I_p} \|A - BQ\|_F = \sum_{i=1}^{p} \sigma_i(A)^2 - 2\sigma_i(B^T A) + \sigma_i(B)^2.$$

P12.4-2. Extend Algorithm 12.4-2 to the problem of computing an orthonormal basis for $N(A_1) \cap \cdots \cap N(A_s)$.

P12.4-3. Extend Algorithm 12.4-3 to handle the case when A and B are rank-deficient.

P12.4-4. Relate the principal angles and vectors between $R(A)$ and $R(B)$ to the eigenvalues and eigenvectors of the generalized eigenvalue problem

$$\begin{bmatrix} 0 & A^T B \\ B^T A & 0 \end{bmatrix} \begin{bmatrix} y \\ z \end{bmatrix} = \sigma \begin{bmatrix} A^T A & 0 \\ 0 & B^T B \end{bmatrix} \begin{bmatrix} y \\ z \end{bmatrix}.$$

Notes and References for Sec. 12.4

The problem of minimizing $\|A - BQ\|_F$ over all orthogonal matrices arises in psychometrics; see

B. Green (1952). "The Orthogonal Approximation of an Oblique Structure in Factor Analysis," *Psychometrika 17*, 429-40.

P. Schonemann (1966). "A Generalized Solution of the Orthogonal Procrustes Problem," *Psychometrika 31*, 1-10.

I. Y. Bar-Itzhack (1975). "Iterative Optimal Orthogonalization of the Strapdown Matrix," *IEEE Trans. Aerospace and Electronic Systems 11*, 30-37.

When $B = I$, this problem amounts to finding the closest orthogonal matrix of A. An iterative algorithm for computing this matrix is given in

A. Bjorck and C. Bowie (1971). "An Interative Algorithm for Computing the Best Estimate of an Orthogonal Matrix," *SIAM J. Num. Anal. 8*, 358-64.

If A is reasonably close to being orthogonal itself, then Bjorck and Bowie's technique is more efficient than the SVD algorithm. It would be interesting to extend their iteration to the case when B is not the identity.

Using the SVD to solve the canonical correlation problem was originally proposed in

A. Bjorck and G. H. Golub (1973). "Numerical Methods for Computing Angles between Linear Subspaces," *Math. Comp. 27*, 579-94.

Sec. 12.5. Some Modified Eigenvalue Problems

In this section some variations of the standard eigenvalue problem are considered. The discussion will underscore the wide applicability of the matrix methods that have been described throughout the book.

Stationary Values of a Constrained Quadratic Form

Let $A \in \mathbb{R}^{n \times n}$ be symmetric and recall that the gradient of $r(x) = x^T A x / x^T x$ is zero if and only if x is an eigenvector of A. (See §8.1.) The stationary or critical values of $r(x)$ are therefore the eigenvalues of A.

In certain applications it is necessary to find the stationary values of $r(x)$ subject to the constraint

$$C^T x = 0$$

where $C \in \mathbb{R}^{n \times p}$ with $n \geq p$. Suppose

$$Q^T C Z = \begin{bmatrix} S & 0 \\ 0 & 0 \end{bmatrix} \begin{matrix} r \\ n-r \end{matrix} \qquad r = \text{rank}(C)$$
$$ \begin{matrix} r & p-r \end{matrix}$$

is a complete orthogonal decomposition of C. Define $B \in \mathbb{R}^{n \times n}$ by

$$Q^T A Q = B = \begin{bmatrix} B_{11} & B_{12} \\ B_{21} & B_{22} \end{bmatrix} \begin{matrix} r \\ n-r \end{matrix}$$
$$ \begin{matrix} r & n-r \end{matrix}$$

and set

$$y = Q^Tx = \begin{bmatrix} u \\ v \end{bmatrix} \begin{matrix} r \\ n-r \end{matrix}.$$

Since $C^Tx = 0$ transforms to $S^Tu = 0$, the original problem becomes one of finding the stationary values of

$$r(y) = \frac{y^TBy}{y^Ty},$$

subject to the constraint that $u = 0$. But this amounts merely to finding the stationary values (eigenvalues) of the $(n - r)$-by-$(n - r)$ symmetric matrix B_{22}.

ALGORITHM 12.5-1.   Given the n-by-n symmetric matrix A and $C \in \mathbb{R}^{n \times p}$ $(n \geqslant p)$ having rank r, the following algorithm computes the stationary values $\tilde{\lambda}_1, \ldots, \tilde{\lambda}_{n-r}$ of the functional x^TAx/x^Tx subject to the constraint $C^Tx = 0$.

Compute the complete orthogonal decomposition

$$Q^TCZ = \begin{bmatrix} S & 0 \\ 0 & 0 \end{bmatrix} \begin{matrix} r \\ n-r \end{matrix} \qquad Q = [Q_1, \ Q_2]$$
$$\begin{matrix} r & p-r \end{matrix} \qquad\qquad\qquad \begin{matrix} r & n-r \end{matrix}$$

Compute the eigenvalues of $Q_2^TAQ_2$:

$$U^T(Q_2^TAQ_2)U = \mathrm{diag}(\tilde{\lambda}_1, \ldots, \tilde{\lambda}_{n-r})$$

The work required by this algorithm depends upon how the complete orthogonal decomposition is computed. If the SVD is used and $p = r$, then about $np^2 + 13p^3$ flops are needed.

An Inverse Eigenvalue Problem

Consider the problem of finding the stationary values of x^TAx/x^Tx subject to the constraint $c^Tx = 0$ where $A \in \mathbb{R}^{n \times n}$ is symmetric and $0 \neq c \in \mathbb{R}^n$. From the above it is easy to show that the desired stationary values $\tilde{\lambda}_1, \ldots, \tilde{\lambda}_{n-1}$ interlace the eigenvalues λ_i of A:

$$\lambda_n \leqslant \tilde{\lambda}_{n-1} \leqslant \lambda_{n-1} \leqslant \cdots \leqslant \lambda_2 \leqslant \tilde{\lambda}_1 \leqslant \lambda_1.$$

Now suppose that A has distinct eigenvalues and that we are *given* $\tilde{\lambda}_1, \ldots, \tilde{\lambda}_{n-1}$ satisfying

$$\lambda_n < \tilde{\lambda}_{n-1} < \lambda_{n-1} < \cdots < \lambda_2 < \tilde{\lambda}_1 < \lambda_1.$$

We seek to determine a unit vector $c \in \mathbb{R}^n$ such that the $\tilde{\lambda}_i$ are the stationary values of x^TAx subject to $x^Tx = 1$ and $c^Tx = 0$.

In order to determine the properties that c must have, we use the method of Lagrange multipliers. Equating the gradient of

$$\phi(x, \lambda, \mu) = x^TAx - \lambda(x^Tx - 1) + 2\mu x^Tc$$

to zero we obtain the important equation

$$(A - \lambda I)x = -\mu c.$$

This implies $\lambda = x^TAx$, i.e., λ is a stationary value. Thus, $A - \lambda I$ is non-singular and so

$$x = -\mu(A - \lambda I)^{-1}c.$$

Applying c^T to both sides and substituting the eigenvalue decomposition $Q^TAQ = \text{diag}(\lambda_i)$ we obtain

$$0 = \sum_{i=1}^{n} \frac{d_i^2}{(\lambda_i - \lambda)} \qquad\qquad d = Q^Tc$$

or

$$p(\lambda) \equiv \sum_{i=1}^{n} d_i^2 \prod_{\substack{j=1 \\ j \neq i}}^{n} (\lambda_j - \lambda) = 0.$$

Notice that $1 = \|c\|_2^2 = \|d\|_2^2 = d_1^2 + \cdots + d_n^2$ is the coefficient of $(-\lambda)^{n-1}$. Since $p(\lambda)$ is a polynomial having zeros $\tilde\lambda_1, \ldots, \tilde\lambda_{n-1}$, we must have

$$p(\lambda) = \prod_{j=1}^{n-1} (\tilde\lambda_j - \lambda).$$

It follows from these two formulas for $p(\lambda)$ that

$$d_k^2 = \frac{\displaystyle\prod_{j=1}^{n-1} (\tilde\lambda_j - \lambda_k)}{\displaystyle\prod_{\substack{j=1 \\ j \neq k}}^{n-1} (\lambda_j - \lambda_k)}.$$

Since there are two possible signs for d_k, there are 2^n different solutions to this problem.

Eigenvalues of a Matrix Modified by a Rank-One Matrix.

Given $\sigma \in \mathbb{R}$, $u \in \mathbb{R}^n$, and $D = \text{diag}(d_1, \ldots, d_n)$ satisfying $d_1 \geqslant \cdots \geqslant d_n$, it is sometimes necessary to compute the eigenvalues $\lambda_1 \geqslant \cdots \geqslant \lambda_n$ of

$$C = D + \sigma uu^T.$$

Using the minimax characterization of eigenvalues, it can be shown that

(i) if $\sigma \geqslant 0$

$$d_i \leqslant \lambda_i \leqslant d_{i-1}, \qquad\qquad i = 2, \ldots, n$$
$$d_2 \leqslant \lambda_1 \leqslant d_1 + \sigma u^T u;$$

(ii) if $\sigma \leqslant 0$

$$d_{i+1} \leqslant \lambda_i \leqslant d_i, \qquad\qquad i = 1, \ldots, n-1$$
$$d_n + \sigma u^T u \leqslant \lambda_n \leqslant d_n.$$

One approach to computing $\lambda(C)$ is to apply Newton's method to the function

$$p(\lambda) = \det(D + \sigma u u^T - \lambda I).$$

It can be shown that if $p_n(\lambda)$ is defined by

(12.5-2)
$$\begin{aligned}
&r_1(\lambda) = 1 \\
&p_1(\lambda) = (d_1 - \lambda) + \sigma u_1^2 \\
&\text{For } k = 2, \ldots, n \\
&\quad r_k(\lambda) = (d_{k-1} - \lambda) r_{k-1}(\lambda) \\
&\quad p_k(\lambda) = (d_k - \lambda) p_{k-1}(\lambda) + \sigma u_k^2 r_k(\lambda)
\end{aligned}$$

then $p_n(\lambda) = p(\lambda)$. Therefore, the evaluation of both $p(\lambda)$ and $p'(\lambda)$ is straightforward. The interlacing properties can be used to obtain good starting values.

An alternative method for finding $\lambda(C)$ leads to a generalized tridiagonal eigenvalue problem. Let P be a permutation matrix such that if $v = Pu$ then

$$v_1 = \cdots = v_s = 0 \quad \text{and} \quad 0 < |v_{s+1}| \leqslant \cdots \leqslant |v_n|.$$

Furthermore, let K be the bidiagonal matrix

$$K = \begin{bmatrix} 1 & r_1 & & & \\ & 1 & r_2 & & \text{\Large 0} \\ & & \ddots & \ddots & \\ \text{\Large 0} & & & 1 & r_{n-1} \\ & & & & 1 \end{bmatrix} \qquad r_i = \begin{cases} 0 & i = 1, \ldots, s-1 \\ -v_i/v_{i+1} & i = s, \ldots, n-1 \end{cases}$$

Notice that $Kv = v_n e_n$ and thus the equation

$$(D + \sigma u u^T)x = \lambda x$$

transforms to

$$[KPDP^T K^T + v_n^2 e_n e_n^T] y = \lambda K K^T y. \qquad\qquad x = P K^T y$$

The matrices $KPDP^T K^T + \sigma v_n^2 e_n e_n^T$ and KK^T are both symmetric and tri-diagonal. A straightforward generalization of the bisection scheme of §8.5 can be used to find the desired eigenvalues.

Another Inverse Eigenvalue Problem

Consider the problem of finding a tridiagonal matrix

$$
T = \begin{bmatrix}
\alpha_1 & \beta_1 & & & \text{\LARGE 0} \\
\beta_1 & \alpha_2 & \cdot & & \\
& \cdot & \cdot & \cdot & \\
& & \cdot & \cdot & \beta_{n-1} \\
\text{\LARGE 0} & & & \beta_{n-1} & \alpha_n
\end{bmatrix}
$$

such that T and its leading $n-1$ order principal submatrix $\tilde{T}$ have pre-scribed eigenvalues. In other words, given $\lambda(T) = \{\lambda_1, \ldots, \lambda_n\}$ and $\lambda(\tilde{T}) = \{\tilde{\lambda}_1, \ldots, \tilde{\lambda}_{n-1}\}$ satisfying

$$\lambda_1 > \tilde{\lambda}_1 > \lambda_2 > \cdots > \lambda_{n-1} > \tilde{\lambda}_{n-1} > \lambda_n,$$

find T.

For these relationships to hold, there must be an orthogonal Q such that $Q^T T Q = \Lambda = \text{diag}(\lambda_1, \ldots, \lambda_n)$ and such that the stationary values of

$$\frac{x^T T x}{x^T x} \quad \text{subject to } e_1^T x = 0$$

are given by $\tilde{\lambda}_1, \ldots, \tilde{\lambda}_{n-1}$. In other words, the $\tilde{\lambda}_i$ are stationary values of

$$\frac{y^T \Lambda y}{y^T y} \quad \text{subject to } d^T y = 0$$

where $d = Q^T e_1$, the first column of Q.

From the above we know that

$$
d_k^2 = \frac{\displaystyle\prod_{j=1}^{n-1} (\tilde{\lambda}_j - \lambda_k)}{\displaystyle\prod_{\substack{j=1 \\ j \neq k}}^{n-1} (\lambda_j - \lambda_k)}. \qquad k = 1, \ldots, n
$$

Consider applying the Lanczos iteration (9.1-3) with $A = \Lambda$ and $q_1 = d$. If the algorithm runs for n iterations, we generate an orthogonal matrix Q with the property that $Q^T \Lambda Q = T$ is tridiagonal. Since the first column of Q is d, it follows that T has the desired properties.

Problems

P12.5-1. Let $A \in \mathbb{R}^{m \times n}$ and consider the problem of finding the stationary values of

$$R(x, y) = \frac{y^T A x}{\|y\|_2 \|x\|_2} \qquad y \in \mathbb{R}^m, x \in \mathbb{R}^n$$

subject to the constraints

$$C^T x = 0 \qquad C \in \mathbb{R}^{n \times p} \qquad n \geqslant p$$

$$D^T y = 0 \qquad D \in \mathbb{R}^{m \times q} \qquad m \geqslant q$$

Show how to solve this problem by first computing complete orthogonal decompositions of C and D and then computing the SVD of a certain submatrix of a transformed A.

Notes and References for Sec. 12.5

Many of the problems discussed in this section appear in the following survey article:

G. H. Golub (1973). "Some Modified Matrix Eigenvalue Problems," *SIAM Review* 15, 318-44.

References for the stationary value problem include:

G. E. Forsythe and G. H. Golub (1965). "On the Stationary Values of a Second-Degree Polynomial on the Unit Sphere," *SIAM J. App. Math. 13*, 1050-68.

G. H. Golub and R. Underwood (1970). "Stationary Values of the Ratio of Quadratic Forms Subject to Linear Constraints," *Z. Angew. Math. Phys. 21*, 318-26.

Using the QR algorithm to determine the Gauss-type quadrature rules is discussed in

G. H. Golub and J. H. Welsch (1969). "Calculation of Gauss Quadrature Rules," *Math. Comp. 23*, 221-30.

We discussed certain inverse eigenvalue problems earlier in §9.3 in connection with the Lanczos algorithm. Other inverse eigenvalue problems include finding a diagonal matrix D so that AD (or sometimes $A + D$) has prescribed eigenvalues. See

S. Friedland (1975). "On Inverse Multiplicative Eigenvalue Problems for Matrices," *Lin. Alg. & Its Applic. 12*, 127-38.

The use of the Lanczos algorithm to solve various inverse eigenvalue problems is discussed in

D. L. Boley and G. H. Golub (1978). "The Matrix Inverse Eigenvalue Problem for Periodic Jacobi Matrices," in *Proc. Fourth Symposium on Basic Problems of Numerical Mathematics*, Prague, pp. 63-76.

Sec. 12.6. Updating the Q-R Factorization

In many applications it is necessary to re-factor a given matrix $A \in \mathbb{R}^{m \times n}$ after it has been altered in some minimal sense. For example, given that we have the Q-R factorization of A, we may need to calculate the Q-R factorization of a matrix $\bar{A}$ that is obtained by

(a) adding a rank-one matrix to A,

(b) appending a row (or column) to A, or

(c) deleting a row (or column) from A.

In a situations like these it is much more efficient to "update" A's Q-R factorization than to start from scratch by treating $\bar{A}$ as a matrix about which we know nothing. In this section we illustrate this point for the modifications (a), (b), and (c).

Before beginning, we mention that there are also techniques for updating the factorizations $PA = LU$, $A = GG^{T}$, and $A = LDL^{T}$. Updating these factorizations, however, can be quite delicate because of pivoting requirements and because when we tamper with a positive definite matrix the result may not be positive definite. See Gill, Golub, Murray, and Saunders (1974) and Stewart (1979c).

Rank-One Changes

Much of the work on updating matrix factorizations was prompted by the development of quasi-Newton methods for nonlinear equations and optimization. The simplest of these is Broyden's method, which can be used to find a zero of the function $f: \mathbb{R}^{n} \rightarrow \mathbb{R}^{n}$ as follows:

> Given: $x_0 \in \mathbb{R}^n$ and $B_0 \in \mathbb{R}^{n \times n}$.
> For $k = 0, 1, \ldots$
> > Solve $B_k s_k = -f(x_k)$ for s_k.
> > $x_{k+1} = x_k + s_k$.
> > $y_k = f(x_{k+1}) - f(x_k)$
> > Set $B_{k+1} = B_k + (y_k - B_k s_k)s_k^{T}/s_k^{T}s_k$.

Here, each B_k can be regarded as an approximation to the Jacobian of f at x_k and is assumed to be nonsingular.

To implement Broyden's method it is necessary to solve a linear system at each step. If the Q-R factorization is used and we fail to exploit the rank-one connection between B_{k-1} and B_k, then $O(n^3)$ flops per iteration are required. This prompts us to ask: Is there an $O(n^2)$ way to calculate the Q-R factorization of B_k, given that we have the Q-R factorization of B_{k-1}?

The answer to this question turns out to be affirmative. Suppose we have the Q-R factorization $QR = B \in \mathbb{R}^{m \times n}$ and that we need to compute the Q-R factorization

$$B + uv^T = Q_1 R_1$$

where $u, v \in \mathbb{R}^n$ are given. Observe that

(12.6-1) $$B + uv^T = Q(R + wv^T)$$

where

$$w = Q^T u.$$

Suppose that we have computed Givens rotations $J_k = J(k, k + 1, \theta_k)$ for $k = n - 1, \ldots, 1$ such that

$$J_1^T \cdots J_{n-1}^T w = \begin{bmatrix} \alpha \\ 0 \\ \vdots \\ 0 \end{bmatrix}$$

where $\alpha = \pm \| w \|_2$. (For details, see Algorithm 6.3-1.) If these same Givens rotations are applied to R, it can be shown that

(12.6-2) $$H = J_1^T \cdots J_{n-1}^T R$$

is upper Hessenberg. For example, in the $n = 4$ case,

$$R = \begin{bmatrix} x & x & x & x \\ & x & x & x \\ & & x & x \\ & & & x \end{bmatrix} \quad \text{and} \quad w = \begin{bmatrix} x \\ x \\ x \\ x \end{bmatrix}$$

are updated as follows:

$$R := J_3^T R = \begin{bmatrix} x & x & x & x \\ & x & x & x \\ & & x & x \\ & & x & x \end{bmatrix} \qquad w := J_3^T w = \begin{bmatrix} x \\ x \\ x \\ 0 \end{bmatrix}$$

$$R := J_2^T R = \begin{bmatrix} x & x & x & x \\ & x & x & x \\ & & x & x & x \\ & & & x & x \end{bmatrix} \qquad w := J_2^T w = \begin{bmatrix} x \\ x \\ 0 \\ 0 \end{bmatrix}$$

$$H := J_1^T R = \begin{bmatrix} x & x & x & x \\ x & x & x & x \\ & x & x & x \\ & & x & x \end{bmatrix} \qquad w := J_1^T w = \begin{bmatrix} x \\ 0 \\ 0 \\ 0 \end{bmatrix}.$$

Consequently,

(12.6-3) $$(J_1^T \cdots J_{n-1}^T)(B + wv^T) = H + \alpha e_1 v^T = H_1$$

is another upper Hessenberg matrix.

In §7.4 we showed how to compute the Q-R factorization of an upper Hessenberg matrix in $O(n^2)$ flops. In particular, we can find Givens rotations $G_k = J(k, k + 1, \phi_k)$ for $k = 1, \ldots, n - 1$ such that

(12.6-4) $$G_{n-1}^T \cdots G_1^T H_1 = R_1$$

is upper triangular. Combining (12.6-1) through (12.6-4) we obtain the Q-R factorization

$$B + uv^T = Q_1 R_1,$$

where

$$Q_1 = Q J_{n-1} \cdots J_1 G_1 \cdots G_{n-1}.$$

A careful assessment of the work reveals that $13n^2$ flops are required:

Computing $w = Q^T u$:	n^2
Computing H and accumulating the J_k into Q:	$6n^2$
Computing R_1 and accumulating the G_k into Q:	$6n^2$
	$13n^2$

The amount of work can be somewhat reduced if modified Householder transformations are used instead of Givens rotations. The technique readily extends to the case when B is rectangular. It can also be generalized to compute the Q-R factorization of $B + UV^T$ where $\text{rank}(UV^T) = p > 1$.

Appending or Deleting a Column

Assume that we have the Q-R factorization

(12.6-5) $$QR = A = [a_1, \ldots, a_n] \qquad a_i \in \mathbb{R}^m$$

and partition the upper triangular matrix $R \in \mathbb{R}^{m \times n}$ as follows:

$$R = \begin{bmatrix} R_{11} & v & R_{13} \\ 0 & r_{kk} & w^T \\ 0 & 0 & R_{33} \end{bmatrix} \begin{matrix} k-1 \\ 1 \\ m-k \end{matrix} \ .$$
$$\quad\ \ k-1 \quad 1 \quad n-k$$

Now suppose that we want to compute the Q-R factorization of

$$\bar{A} = [a_1, \ldots, a_{k-1}, a_{k+1}, \ldots, a_n] \in \mathbb{R}^{m \times (n-1)}.$$

Note that $\bar{A}$ is just A with its k-th column deleted and that

$$Q^T \bar{A} = \begin{bmatrix} R_{11} & R_{12} \\ 0 & w^T \\ 0 & R_{33} \end{bmatrix} = H$$

is upper Hessenberg, e.g.,

$$H = \begin{bmatrix} x & x & x & x & x \\ 0 & x & x & x & x \\ 0 & 0 & x & x & x \\ 0 & 0 & x & x & x \\ 0 & 0 & 0 & x & x \\ 0 & 0 & 0 & 0 & x \\ 0 & 0 & 0 & 0 & 0 \end{bmatrix}. \qquad m = 7, n = 6, k = 3$$

Clearly, the unwanted subdiagonal elements $h_{k+1,k}, \ldots, h_{n,n-1}$ can be zeroed by a sequence of Givens rotations:

$$J_{n-1}^T \cdots J_k^T H = R_1 = \Delta.$$

Here, $J_i = J(i, i+1, \theta_i)$ for $i = k, \ldots, n-1$. Thus, if

$$Q_1 = Q J_k \cdots J_{n-1},$$

then $\bar{A} = Q_1 R_1$ is the Q-R factorization of $\bar{A}$.

The above update procedure can be executed in $O(n^2)$ flops and is very useful in certain least squares problems. For example, one may wish to examine the significance of the k-th factor in the underlying model by deleting the k-th column of the corresponding data matrix and solving the resulting LS problem.

In a similar vein, it is useful to be able to compute efficiently the solution to the LS problem after a column has been appended to A. Suppose we have the Q-R factorization (12.6-5) and now wish to compute the Q-R factorization of

$$\bar{A} = [a_1, \ldots, a_k, z, a_{k+1}, \ldots, a_n]$$

where $z \in \mathbb{R}^m$ is given. Note that if

$$w = Q^T z$$

then

$$Q^T \bar{A} = [Q^T a_1, \ldots, Q^T a_k, w, Q^T a_{k+1}, \ldots, Q^T a_n] = \bar{R}$$

is upper triangular except for the presence of a "spike" in its k-th column, e.g.,

$$\bar{R} = \begin{bmatrix} x & x & x & x & x & x \\ 0 & x & x & x & x & x \\ 0 & 0 & x & x & x & x \\ 0 & 0 & 0 & x & x & x \\ 0 & 0 & 0 & x & 0 & x \\ 0 & 0 & 0 & x & 0 & 0 \\ 0 & 0 & 0 & x & 0 & 0 \end{bmatrix}. \qquad m = 7, n = 5, k = 3$$

A calculation shows that if Givens rotations $J_{m-1}, \ldots, J_{k+1}$ are determined so that

$$J_{k+2}^T \cdots J_{m-1}^T w = \begin{bmatrix} w_1 \\ \vdots \\ w_{k+1} \\ \bar{w}_{k+2} \\ 0 \\ 0 \end{bmatrix}$$

then $J_{k+2}^T \cdots J_{m-1}^T)\bar{R} = H$ is upper Hessenberg. In the example above, H is calculated as follows:

$$H := J_6 \bar{R} = \begin{bmatrix} x & x & x & x & x & x \\ 0 & x & x & x & x & x \\ 0 & 0 & x & x & x & x \\ 0 & 0 & 0 & x & x & x \\ 0 & 0 & 0 & x & x & x \\ 0 & 0 & 0 & x & 0 & 0 \\ 0 & 0 & 0 & 0 & 0 & 0 \end{bmatrix}$$

$$H := J_5 H = \begin{bmatrix} x & x & x & x & x & x \\ 0 & x & x & x & x & x \\ 0 & 0 & x & x & x & x \\ 0 & 0 & 0 & x & x & x \\ 0 & 0 & 0 & x & x & x \\ 0 & 0 & 0 & 0 & x & x \\ 0 & 0 & 0 & 0 & 0 & 0 \end{bmatrix}.$$

Clearly, an appropriate sequence of Givens rotations $G_{k+1}, \ldots, G_n$ can be chosen so $G_n^T \cdots G_{k+1}^T H = R_1$ is upper triangular. Thus,

$$\bar{A} = (Q J_{m-1} \cdots J_{k+2} G_{k+1} \cdots G_n) R_1$$

is the Q-R factorization of $\bar{A}$. This update requires $O(mn)$ flops.

Appending or Deleting a Row

Suppose we have the Q-R factorization $QR = A \in \mathbb{R}^{m \times n}$ and now wish to obtain the Q-R factorization of

$$\bar{A} = \begin{bmatrix} w^T \\ A \end{bmatrix},$$

where $w \in \mathbb{R}^n$. Note that

$$\mathrm{diag}(1, Q^T) \bar{A} = \begin{bmatrix} w^T \\ R \end{bmatrix} = H$$

is upper Hessenberg. Thus, Givens rotations $J_1, \ldots, J_n$ could be determined so $J_n^T \cdots J_1^T H = R_1$ is upper triangular. It follows that

$$\bar{A} = Q_1 R_1$$

is the desired Q-R factorization, where $Q_1 = \mathrm{diag}(1, Q) J_1 \cdots J_n$.

No essential complications result if the new row is added between rows k and $k + 1$ of A. We merely apply the above with A replaced by PA and Q replaced by PQ where

$$P = \begin{bmatrix} 0 & I_{m-k} \\ I_k & 0 \end{bmatrix}.$$

Upon completion we set $Q_1 := \mathrm{diag}(1, P^T) Q_1$.

Lastly, we consider how to update the Q-R factorization $QR = A \in \mathbb{R}^{m \times n}$ when the first row of A is deleted. In particular, we wish to compute the Q-R factorization of the submatrix A_1 in

$$A = \begin{bmatrix} z^{\mathrm{T}} \\ A_1 \end{bmatrix} \begin{matrix} 1 \\ m-1 \end{matrix}.$$

(The procedure is similar when an arbitrary row is deleted.)

Let q^{T} be the first row of Q and compute Givens rotations $J_1, \ldots, J_{m-1}$ such that

$$J_1^{\mathrm{T}} \cdots J_{m-1}^{\mathrm{T}} q = \alpha e_1,$$

where $\alpha = \pm 1$. Note that

$$H = J_1^{\mathrm{T}} \cdots J_{m-1}^{\mathrm{T}} R = \begin{bmatrix} v^{\mathrm{T}} \\ R_1 \end{bmatrix} \begin{matrix} 1 \\ m-1 \end{matrix}$$

is upper Hessenberg and that

$$QJ_{m-1} \cdots J_1 = \begin{bmatrix} \alpha & 0 \\ 0 & Q_1 \end{bmatrix},$$

where $Q_1 \in \mathbb{R}^{(m-1) \times (m-1)}$, is orthogonal. Thus,

$$A = \begin{bmatrix} z^{\mathrm{T}} \\ A_1 \end{bmatrix} = (QJ_{m-1} \cdots J_1)(J_1^{\mathrm{T}} \cdots J_{m-1}^{\mathrm{T}} R) = \begin{bmatrix} \alpha & 0 \\ 0 & Q_1 \end{bmatrix} \begin{bmatrix} v^{\mathrm{T}} \\ R_1 \end{bmatrix},$$

from which we conclude that $A_1 = Q_1 R_1$ is the desired Q-R factorization.

Problems

P12.6-1. Suppose we have the Q-R factorization for $A \in \mathbb{R}^{m \times n}$ and now wish to solve $\min_x \| (A + uv^{\mathrm{T}})x - b \|_2$ where $u, b \in \mathbb{R}^m$ and $v \in \mathbb{R}^n$ are given. Give an algorithm for solving this problem that requires $O(mn)$ flops. Assume that Q must be updated.

P12.6-2. Suppose we have the Q-R factorization $QR = A \in \mathbb{R}^{m \times n}$. Give an algorithm for computing the Q-R factorization of the matrix $\bar{A}$ obtained by deleting the k-th row of A. Your algorithm should require $O(mn)$ flops.

Notes and References for Sec. 12.6

Numerous aspects of the updating problem are presented in

P. E. Gill, G. H. Golub, W. Murray, and M. A. Saunders (1974). "Methods for Modifying Matrix Factorizations," *Math. Comp. 28*, 505–35.

Further references include

D. Goldfarb (1976). "Factorized Variable Metric Methods for Unconstrained Optimization," *Math. Comp. 30*, 796–811.
P. E. Gill, W. Murray, and M. A. Saunders (1975). "Methods for Computing and Modifying the LDV Factors of a Matrix," *Math. Comp. 29*, 1051–77.

J. Daniel, W. B. Gragg, L. Kaufman, and G. W. Stewart (1976). "Reorthogonalization and Stable Algorithms for Updating the Gram-Schmidt QR Factorization," *Math. Comp. 30*, 772-95.

FORTRAN programs for updating the Q-R and Cholesky factorizations are included in LINPACK (chap. 10). The stability of downdating the Cholesky factorization is analyzed in

G. W. Stewart (1979c). "The Effects of Rounding Error on an Algorithm for Downdating a Cholesky Factorization," *J. Inst. Math. Applic. 23*, 203-13.

The role of updating techniques in the area of nonlinear equations and optimization is surveyed in

J. E. Dennis and R. B. Schnabel (1983). *Quasi-Newton Methods for Nonlinear Problems*, Prentice-Hall, Englewood Cliffs, N.J.

Bibliography

J. O. Aasen (1971). "On the Reduction of a Symmetric Matrix to Tridiagonal Form," *BIT 11*, 233–42.

N. N. Abdelmalck (1971). "Roundoff Error Analysis for Gram-Schmidt Method and Solution of Linear Least Squares Problems," *BIT 11*, 345–68.

E. L. Allgower (1973). "Exact Inverses of Certain Band Matrices," *Numer. Math. 21*, 279–84.

A. R. Amir-Moez (1965). *Extremal Properties of Linear Transformations and Geometry of Unitary Spaces*, Texas Tech University Mathematics Series, no. 243, Lubbock, Tex.

N. Anderson and I. Karasalo (1975). "On Computing Bounds for the Least Singular Value of a Triangular Matrix," *BIT 15*, 1–4.

P. Anderson and G. Loizou (1973). "On the Quadratic Convergence of an Algorithm which Diagonalizes a Complex Symmetric Matrix," *J. Inst. Math. Applic. 12*, 261–71.

P. Anderson and G. Loizou (1976). "A Jacobi-Type Method for Complex Symmetric Matrices (Handbook)," *Numer. Math. 25*, 347–63.

E. S. Armstrong and A. K. Caglayan (1976). "An Algorithm for the Weighting Matrices in the Sample-Data Optimal Linear Regulator Problem," NASA Technical Note, TN D-8372.

O. Axelsson (1977). "Solution of Linear Systems of Equations: Iterative Methods," in *Sparse Matrix Techniques: Copenhagen, 1976*, ed. V. A. Barker, Springer-Verlag, Berlin.

O. Axelsson (1980). "Conjugate Gradient Type Methods for Unsymmetric and Inconsistent Systems of Linear Equations," *Lin. Alg. & Its Applic. 29*, 1–66.

I. Y. Bar-Itzhack (1975). "Iterative Optimal Orthogonalization of the Strapdown Matrix," *IEEE Trans. Aerospace and Electronic Systems 11*, 30–37.

S. Barnett and C. Storey (1968). "Some Applications of the Lyapunov Matrix Equation," *J. Inst. Math. Applic. 4*, 33–42.

I. Barrodale and C. Phillips (1975). "Algorithm 495: Solution of an Overdetermined System of Linear Equations in the Chebychev Norm," *ACM Trans. Math. Soft. 1*, 264–70.

I. Barrodale and F. D. K. Roberts (1973). "An Improved Algorithm for Discrete L_1 Linear Approximation," *SIAM J. Num. Anal. 10*, 839–48.

R. H. Bartels, A. R. Conn, and C. Charalambous (1978). "On Cline's Direct Method for Solving Overdetermined Linear Systems in the L_∞ Sense," *SIAM J. Num. Anal. 15*, 255–70.

445

R. H. Bartels, A. R. Conn, and J. W. Sinclair (1978). "Minimization Techniques for Piecewise Differentiable Functions: The L_1 Solution to an Overdetermined Linear System," *SIAM J. Num. Anal. 15*, 224–41.

R. H. Bartels and G. W. Stewart (1972). "Solution of the Equation $AX + XB = C$," *Comm. Assoc. Comp. Mach. 15*, 820–26.

W. Barth, R. S. Martin, and J. H. Wilkinson (1967). "Calculation of the Eigenvalues of a Symmetric Tridiagonal Matrix by the Method of Bisection," *Numer. Math. 9*, 386–93. See also HACLA, pp. 249–56.

V. Barwell and J. A. George (1976). "A Comparison of Algorithms for Solving Symmetric Indefinite Systems of Linear Equations," *ACM Trans. Math. Soft. 2*, 242–51.

K. J. Bathe and E. L. Wilson (1973). "Solution Methods for Eigenvalue Problems in Structural Mechanics," *Int. J. Numer. Meth. Eng. 6*, 213–26.

F. L. Bauer (1963). "Optimally Scaled Matrices," *Numer. Math. 5*, 73–87.

F. L. Bauer (1965). "Elimination with Weighted Row Combinations for Solving Linear Equations and Least Squares Problems," *Numer. Math 7*, 338–52. See also HACLA, pp. 119–33.

F. L. Bauer and C. T. Fike (1960). "Norms and Exclusion Theorems," *Numer. Math. 2*, 137–44.

F. L. Bauer and C. Reinsch (1968). "Rational QR Transformation with Newton Shift for Symmetric Tridiagonal Matrices," *Numer. Math. 11*, 264–72. See also HACLA, pp. 257–65.

F. L. Bauer and C. Reinsch (1970). "Inversion of Positive Definite Matrices by the Gauss-Jordan Method," in HACLA, pp. 45–49.

C. Bavely and G. W. Stewart (1979). "An Algorithm for Computing Reducing Subspaces by Block Diagonalization," *SIAM J. Num. Anal. 16*, 359–67.

R. Bellman (1970). *Introduction to Matrix Analysis*, 2nd ed., McGraw-Hill, New York.

E. Beltrami (1873). "Sulle Funzioni Bilineari," *Giornale di Mathematiche 11*, 98–106.

C. F. Bender and I. Shavitt (1970). "An Iterative Procedure for the Calculation of the Lowest Real Eigenvalue and Eigenvector of a Non-Symmetric Matrix," *J. Comp. Physics 6*, 146–49.

A. Berman and A. Ben-Israel (1971). "A Note on Pencils of Hermitian or Symmetric Matrices," *SIAM J. Appl. Math. 21*, 51–54.

M.J.M. Bernal and J. H. Verner (1968). "On Generalizations of the Theory of Consistent Orderings for Successive Over-Relaxation Methods," *Numer. Math. 12*, 215–22.

A. Bjorck (1967a). "Iterative Refinement of Linear Least Squares Solution I," *BIT 7*, 257–78.

A. Bjorck (1967b). "Solving Linear Least Squares Problems by Gram-Schmidt Orthogonalization," *BIT 7*, 1–21.

A. Bjorck (1968). "Iterative Refinement of Linear Least Squares Solution II," *BIT 8*, 8–30.

A. Bjorck and C. Bowie (1971). "An Iterative Algorithm for Computing the Best Estimate of an Orthogonal Matrix," *SIAM J. Num. Anal. 8*, 358–64.

A. Bjorck and T. Elfving (1973). "Algorithms for Confluent Vandermonde Systems," *Numer. Math. 21*, 130–37.

A. Bjorck and G. H. Golub (1967). "Iterative Refinement of Linear Least Squares Solutions by Householder Transformation," *BIT 7*, 322-37.

A. Bjorck and G. H. Golub (1973). "Numerical Methods for Computing Angles between Linear Subspaces," *Math. Comp. 27*, 579-94.

A. Bjorck and V. Pereyra (1970). "Solution of Vandermonde Systems of Equations," *Math. Comp. 24*, 893-903.

A. Bjorck, R. J. Plemmons, and H. Schneider (1981). *Large-Scale Matrix Problems*, North-Holland, New York.

J. M. Blue (1978). "A Portable FORTRAN Program to Find the Eclidean Norm of a Vector," *ACM Trans. Math. Soft. 4*, 15-23.

Z. Bohte (1975). "Bounds for Rounding Errors in the Gaussian Elimination for Band Systems," *J. Inst. Math. Applic. 16*, 133-42.

D. L. Boley and G. H. Golub (1978). "The Matrix Inverse Eigenvalue Problem for Periodic Jacobi Matrices," in *Proc. Fourth Symposium on Basic Problems of Numerical Mathematics*, Prague, pp. 63-76.

J. Boothroyd and P. J. Eberlein (1968). "Solution to the Eigenproblem by a Norm-Reducing Jacobi-Type Method (Handbook),' *Numer. Math. 11*, 1-12. See also HACLA, pp. 327-38.

H. J. Bowdler, R. S. Martin, G. Peters, and J. H. Wilkinson (1966). "Solution of Real and Complex Systems of Linear Equations," *Numer. Math. 8*, 217-34. See also HACLA, pp. 93-110.

H. Bowdler, R. S. Martin, C. Reinsch, and J. H. Wilkinson (1968). "The QR and QL Algorithms for Symmetric Matrices," *Numer. Math. 11*, 293-306. See also HACLA, pp. 227-40.

K. W. Brodlie and M.J.D. Powell (1975). "On the Convergence of Cyclic Jacobi Methods," *J. Inst. Math. Applic. 15*, 279-87.

C. G. Broyden (1973). "Some Condition Number Bounds for the Gaussian Elimination Process," *J. Inst. Math. Applic. 12*, 273-86.

A. Buckley (1974). "A Note on Matrices $A = 1 + H$, H Skew-Symmetric," *Z. Angew. Math. Mech 54*, 125-26.

A. Buckley (1977). "On the Solution of Certain Skew-Symmetric Linear Systems," *SIAM J. Num. Anal. 14*, 566-70.

J. R. Bunch (1971a). "Analysis of the Diagonal Pivoting Method," *SIAM J. Num. Anal. 8*, 656-80.

J. R. Bunch (1971b). "Equilibration of Symmetric Matrices in the Max-Norm," *J. Assoc. Comp. Mach. 18*, 566-72.

J. R. Bunch (1974). "Partial Pivoting Strategies for Symmetric Matrices," *SIAM J. Num. Anal. 11*, 521-28.

J. R. Bunch (1976). "Block Methods for Solving Sparse Linear Systems," in *Sparse Matrix Computations*, ed. J. R. Bunch and D. J. Rose, Academic Press, New York.

J. R. Bunch and L. Kaufman (1977). "Some Stable Methods for Calculating Inertia and Solving Symmetric Linear Systems," *Math. Comp. 31*, 162-79.

J. R. Bunch, L. Kaufman, and B. N. Parlett (1976). "Decomposition of a Symmetric Matrix," *Numer. Math. 27*, 95-109.

J. R. Bunch and B. N. Parlett (1971). "Direct Methods for Solving Symmetric Indefinite Systems of Linear Equations," *SIAM J. Num. Anal. 8*, 639-55.

J. R. Bunch and D. J. Rose, eds. (1976). *Sparse Matrix Computations*, Academic Press, New York.

O. Buneman (1969). "A Compact Non-Interative Poisson Solver," Report 294, Stanford University Institute for Plasma Research, Stanford, Calif.

P. A. Businger (1968). "Matrices Which Can be Optimally Scaled," *Numer. Math.* *12*, 346–48.

P. A. Businger (1969). "Reducing a Matrix to Hessenberg Form," *Math. Comp. 23*, 819–21.

P. A. Businger (1971a). "Monitoring the Numerical Stability of Gaussian Elimination," *Numer. Math. 16*, 360–61.

P. A. Businger (1971b). "Numerically Stable Deflation of Hessenberg and Symmetric Tridiagonal Matrices," *BIT 11*, 262–70.

P. A. Businger and G. H. Golub (1965). "Linear Least Squares Solutions by Householder Transformations," *Numer. Math. 7*, 269–76. See also HACLA, pp. 111–18.

P. A. Businger and G. H. Golub (1969). "Algorithm 358: Singular Value Decomposition of a Complex Matrix," *Comm. Assoc. Comp. Mach. 12*, 564–65.

B. L. Buzbee and F. W. Dorr (1974). "The Direct Solution of the Biharmonic Equation on Rectangular Regions and the Poisson Equation on Irregular Regions," *SIAM J. Num. Anal. 11*, 753–63.

B. L. Buzbee, F. W. Dorr, J. A. George, and G. H. Golub (1971). "The Direct Solution of the Discrete Poisson Equation on Irregular Regions," *SIAM J. Num. Anal. 8*, 722–36.

B. L. Buzbee, G. H. Golub, and C. W. Nielson (1970). "On Direct Methods for Solving Poisson's Equations," *SIAM J. Num. Anal. 7*, 627–56.

S. P. Chan and B. N. Parlett (1977). "Algorithm 517: A Program for Computing the Condition Numbers of Matrix Eigenvalues without Computing Eigenvectors," *ACM Trans. Math. Soft. 3*, 186–203.

T. F. Chan (1982a). "An Improved Algorithm for Computing the Singular Value Decomposition," *ACM Trans. Math. Soft. 8*, 72–83.

T. F. Chan (1982b). "Algorithm 581: An Improved Algorithm for Computing the Singular Value Decomposition," *ACM Trans. Math. Soft. 8*, 84–88.

A. K. Cline (1973). "An Elimination Method for the Solution of Linear Least Squares Problems," *SIAM J. Num. Anal. 10*, 283–89.

A. K. Cline (1976a). "A Descent Method for the Uniform Solution to Overdetermined Systems of Equations," *SIAM J. Num. Anal. 13*, 293–309.

A. K. Cline (1976b). "Several Observations on the Use of Conjugate Gradient Methods," ICASE Report 76-22. NASA Langley Research Center, Hampton, Va.

A. K. Cline, A. R. Conn, and C. Van Loan (1982). "Generalizing the LINPACK Condition Estimator," in *Numerical Analysis*, ed. J. P. Hennart, Lecture Notes in Mathematics no. 909, Springer-Verlag, New York.

A. K. Cline, G. H. Golub, and G. W. Platzman (1976). "Calculation of Normal Modes of Oceans Using a Lanczos Method," in *Sparse Matrix Computations*, ed. J. R. Bunch and D. J. Rose, Academic Press, New York, pp. 409–26.

A. K. Cline, C. B. Moler, G. W. Stewart, and J. H. Wilkinson (1979). "An Estimate for the Condition Number of a Matrix," *SIAM J. Num. Anal. 16*, 368–75.

R. E. Cline and R. J. Plemmons (1976). "l_2-Solutions to Undetermined Linear Systems," *SIAM Review 18*, 92–106.

M. Clint and A. Jennings (1970). "The Evaluation of Eigenvalues and Eigenvectors of Real Symmetric Matrices by Simultaneous Iteration," *Com. J. 13*, 76–80.

M. Clint and A. Jennings (1971). "A Simultaneous Iteration Method for the Unsymmetric Eigenvalue Problem," *J. Inst. Math. Applic. 8*, 111-21.

W. G. Cochrane (1968). "Errors of Measurement in Statistics," *Technometrics 10*, 637-66.

A. M. Cohen (1974). "A Note on Pivot Size in Gaussian Elimination," *Lin. Alg. & Its Applic. 8*, 361-68.

P. Concus and G. H. Golub (1973). "Use of Fast Direct Methods for the Efficient Numerical Solution of Nonseparable Elliptic Equations," *SIAM J. Num. Anal. 10*, 1103-20.

P. Concus, G. H. Golub, and D. P. O'Leary (1976). "A Generalized Conjugate Gradient Method for the Numerical Solution of Elliptic Partial Differential Equations," in *Sparse Matrix Computations*, ed. J. R. Bunch and D. J. Rose, Academic Press, New York.

S. D. Conte and C. de Boor (1980). *Elementary Numerical Analysis: An Algorithmic Approach*, 3rd ed., McGraw-Hill, New York.

J. E. Cope and B. W. Rust (1979). "Bounds on Solutions of Systems with Inaccurate Data," *SIAM J. Num. Anal. 16*, 950-63.

R. W. Cottle (1974). "Manifestations of the Schur Complement," *Lin. Alg. & Applic. 8*, 189-211.

C. R. Crawford (1973). "Reduction of a Band Symmetric Generalized Eigenvalue Problem," *Comm. Assoc. Comp. Mach. 16*, 41-44.

C. R. Crawford (1976). "A Stable Generalized Eigenvalue Problem," *SIAM J. Num. Anal. 13*, 854-60.

C. W. Cryer (1968). "Pivot Size in Gaussian Elimination," *Numer. Math. 12*, 335-45.

J. Cullum (1978). "The Simultaneous Computation of a Few of the Algebraically Largest and Smallest Eigenvalues of a Large Sparse Symmetric Matrix," *BIT 18*, 265-75.

J. Cullum and W. E. Donath (1974). "A Block Lanczos Algorithm for Computing the Q Algebraically Largest Eigenvalues and A Corresponding Eigenspace of Large, Sparse Real Symmetric Matrices," *Proc. of the 1974 IEEE Conf. on Decision and Control*, Phoenix, Ariz., pp. 505-9.

J. Cullum and R. A. Willoughby (1977). "The Equivalence of the Lanczos and the Conjugate Gradient Algorithms," IBM Research Report RC-6903.

J. Cullum and R. A. Willoughby (1979). "Lanczos and the Computation in Specified Intervals of the Spectrum of Large, Sparse Real Symmetric Matrices," in *Sparse Matrix Proc., 1978*, ed. I. S. Duff and G. W. Stewart, SIAM Publications, Philadelphia.

J. Cullum and R. Willoughby (1980). "The Lanczos Phenomena: An Interpretation Based on Conjugate Gradient Optimization," *Lin. Alg. & Its Applic. 29*, 63-90.

J.J.M. Cuppen (1981). "A Divide and Conquer Method for the Symmetric Eigenproblem," *Numer. Math. 36*, 177-95.

E. Cuthill (1972). "Several Strategies for Reducing the Bandwidth of Matrices," in *Sparse Matrices and Their Applications*, ed. D. J. Rose and R. A. Willoughby, Plenum Press, New York.

G. Cybenko (1978). "Error Analysis of Some Signal Processing Algorithms," Ph.D. thesis, Princeton University.

G. Cybenko (1980). "The Numerical Stability of the Levinson-Durbin Algorithm for Toeplitz Systems of Equations," *SIAM J. Sci & Stat. Comp. 1*, 303-10.

J. Daniel, W. B. Gragg, L. Kaufman, and G. W. Stewart (1976). "Reorthogonalization and Stable Algorithms for Updating the Gram-Schmidt QR Factorization," *Math. Comp. 30*, 772-95.

C. Davis (1973). "Explicit Functional Calculus," *Lin. Alg. & Its Applic. 6*, 193-99.

C. Davis and W. M. Kahan (1970). "The Rotation of Eigenvectors by a Perturbation, III," *SIAM J. Num. Anal. 7*, 1-46.

A. Dax and S. Kaniel (1977). "Pivoting Techniques for Symmetric Gaussian Elimination," *Numer. Math. 28*, 221-42.

C. deBoor and A. Pinkus (1977). "A Backward Error Analysis for Totally Positive Linear Systems," *Numer. Math. 27*, 485-90.

T. J. Dekker and J. F. Traub (1971). "The Shifted QR Algorithm for Hermitian Matrices," *Lin. Alg. & Its Applic. 4*, 137-54.

J. E. Dennis and R. Schnable (1983). *Quasi-Newton Methods for Nonlinear Problems*, Prentice-Hall, Englewood Cliffs..

J. Descloux (1963). "Bounds for the Spectral Norm of Functions of Matrices," *Numer. Math. 5*, 185-90.

M. A. Diamond and D.L.V. Ferreira (1976). "On a Cyclic Reduction Method for the Solution of Poisson's Equation." *SIAM J. Num. Anal. 13*, 54-70.

J. Dongarra, J. R. Bunch, C. B. Moler, and G. W. Stewart (1978). *LINPACK Users Guide*, SIAM Publications, Philadelphia.

F. W. Dorr (1970). "The Direct Solution of the Discrete Poisson Equation on a Rectangle," *SIAM Review 12*, 248-63.

F. W. Dorr (1973). "The Direct Solution of the Discrete Poisson Equation in $O(n^2)$ Operations," *SIAM Review, 15*, 412-15.

A. Dubrulle (1970). "A Short Note on the Implicit QL Algorithm for Symmetric Tridiagonal Matrices," *Numer. Math. 15*, 450.

A. Dubrulle, R. S. Martin, and J. H. Wilkinson (1968). "The Implicit QL Algorith," *Numer. Math. 12*, 377-83. See also HACLA, pp. 241-48.

I. S. Duff (1974). "Pivot Selection and Row Ordering in Givens Reduction on Sparse Matrices," *Computing 13*, 239-48.

I. S. Duff (1977). "A Survey of Sparse Matrix Research," *Proc. IEEE 65*, 500-35.

I. S. Duff and J. K. Reid (1975). "On the Reduction of Sparse Matrices to Condensed Forms by Similarity Transformations," *J. Inst. Math. Applic. 15*, 217-24.

I. S. Duff and J. K. Reid (1976). "A Comparison of Some Methods for the Solution of Sparse Over-Determined Systems of Linear Equations," *J. Inst. Math. Applic. 17*, 267-80.

I. S. Duff and G. W. Stewart, eds. (1979). *Sparse Matrix Proceedings, 1978*, SIAM Publications, Philadelphia.

N. Dunford and J. Schwartz (1958). *Linear Operators, Part I*, Interscience, New York.

J. Durbin (1960). "The Fitting of Time Series Models," *Rev. Inst. Int. Stat. 28*, 233-43.

P. J. Eberlein (1965). "On Measures of Non-normality for Matrices," *Amer. Math. Soc. Monthly 72*, 995-96.

P. J. Eberlein (1970). "Solution to the Complex Eigenproblem by a Norm-Reducing Jacobi-Type Method," *Numer. Math. 14*, 232-45. See also HACLA, pp. 404-17.

P. J. Eberlein (1971). "On the Diagonalization of Complex Symmetric Matrices," *J. Inst. Math. Applic. 7*, 377-83.

C. Eckart and G. Young (1939). "A Principal Axis Transformation for Non-Hermitian Matrices," *Bull. Amer. Math. Soc. 45*, 118-21.

L. Elden (1977a). "Algorithms for the Regularization of Ill-Conditioned Least Squares Problems," *BIT 17*, 134-45.

L. Elden (1977b). "Numerical Analysis of Regularization and Constrained Least Square Methods," Ph.D. thesis, Linkoping Studies in Science and Technology, Dissertation no. 20, Linkoping, Sweden.

L. Elden (1980). "Perturbation Theory for the Least Squares Problem with Linear Equality Constraints," *SIAM J. Num. Anal. 17*, 338-50.

W. Enwright (1979). "On the Efficient and Reliable Numerical Solution of Large Linear Systems of O.D.E.'s," *IEEE Trans. Auto. Cont. AC-24*, 905-8.

I. Erdelyi (1967). "On the Matrix Equation $Ax = \lambda Bx$," *J. Math. Anal. and Applic. 17*, 119-32.

T. Ericsson and A. Ruhe (1980). "The Spectral Transformation Lanczos Method for the Numerical Solution of Large Sparse Generalized Symmetric Eigenvalue Problems," *Math. Comp. 35*, 1251-68.

A. M. Erisman and J. K. Reid (1974). "Monitoring the Stability of the Triangular Factorization of a Sparse Matrix," *Numer. Math. 22*, 183-86.

D. K. Faddeev and V. N. Faddeeva (1963). *Computational Methods of Linear Algebra*, W. H. Freeman & Co., San Francisco.

W. Fair and Y. Luke (1970). "Padé Approximations to the Operator Exponential," *Numer. Math. 14*, 379-82.

D. G. Feinold and R. S. Varga (1962). "Block Diagonally Dominant Matrices and Generalizations of the Gershgorin Circle Theorem," *Pacific J. Math. 12*, 1241-50.

T. Fenner and G. Loizou (1974). "Some New Bounds on the Condition Numbers of Optimally Scaled Matrices," *J. Assoc. Comp. Mach. 1*, 514-24.

C. Fischer and R. A. Usmani (1969). "Properties of Some Tridiagonal Matrices and Their Application to Boundary Value Problems," *SIAM J. Num. Anal. 6*, 127-42.

G. Fix and R. Heiberger (1972). "An Algorithm for the Ill-Conditioned Generalized Eigenvalue Problem," *SIAM J. Num. Anal. 9*, 78-88.

R. Fletcher (1976). "Factorizing Symmetric Indefinite Matrices," *Lin. Alg. & Its Applic. 14*, 257-72.

G. E. Forsythe (1960). "Crout with Pivoting," *Comm. Assoc. Comp. Mach. 3*, 507-8.

G. E. Forsythe and G. H. Golub (1965). "On the Stationary Values of a Second-Degree Polynomial on the Unit Sphere," *SIAM J. App. Math. 13*, 1050-68.

G. E. Forsythe and P. Henrici (1960). "The Cyclic Jacobi Method for Computing the Principal Values of a Complex Matrix," *Trans. Amer. Math. Soc. 94*, 1-23.

G. E. Forsythe, M. A. Malcolm, and C. B. Moler (1977). *Computer Methods for Mathematical Computations*, Prentice-Hall, Englewood Cliffs, N.J.

G. E. Forsythe and C. B. Moler (1967). *Computer Solution of Linear Algebraic Systems*, Prentice-Hall, Englewood Cliffs, N.J.

L. Fox (1964). *An Introduction to Numerical Linear Algebra*, Oxford University Press, Oxford.

J. S. Frame (1964a). "Matrix Functions and Applications, Part II," *IEEE Spectrum 1 (April)*, 102-8.

J. S. Frame (1964b). "Matrix Functions and Applications, Part IV," *IEEE Spectrum 1 (June)*, 123-31.

J.G.F. Francis (1961). "The QR Transformation: A Unitary Analogue to the LR Transformation, Parts I and II," *Comp. J. 4*, 265–72, 332–45.

S. Friedland (1975). "On Inverse Multiplicative Eigenvalue Problems for Matrices," *Lin. Alg. & Its Applic. 12*, 127–38.

S. Friedland (1977). "Inverse Eigenvalue Problems," *Lin. Alg. & Its Applic. 17*, 15–52.

C. E. Froberg (1965). "On Triangularization of Complex Matrices by Two-Dimensional Unitary Transformations," *BIT 5*, 230–34.

G. Galimberti and V. Pereyra (1970). "Numerical Differentiation and the Solution of Multidimensional Vandermonde Systems," *Math. Comp. 24*, 357–64.

G. Galimberti and V. Pereyra (1971). "Solving Confluent Vandermonde Systems of Hermite Type," *Numer. Math. 18*, 44–60.

W. Gander (1981). "Least Squares with a Quadratic Constraint," *Numer. Math. 36*, 291–307.

F. R. Gantmacher (1959). *The Theory of Matrices*, vols. 1–2, Chelsea, New York.

B. S. Garbow, J. M. Boyle, J. J. Dongarra, and C. B. Moler (1972). *Matrix Eigensystem Routines: EISPACK Guide Extension*, Springer-Verlag, New York.

W. Gautschi (1975a). "Norm Estimates for Inverses of Vandermonde Matrices," *Numer. Math. 23*, 337–47.

W. Gautschi (1975b). "Optimally Conditioned Vandermonde Matrices," *Numer. Math. 24*, 1–12.

M. Gentleman (1973a). "Error Analysis of QR Decompositions by Givens Transformations," *Lin. Alg. & Its Appl. 10*, 189–97.

M. Gentleman (1973b). "Least Squares Computations by Givens Transformations without Square Roots," *J. Inst. Math. Appl. 12*, 329–36.

J. A. George (1973). "Nested Dissection of a Regular Finite Element Mesh," *SIAM J. Num. Anal. 10*, 345–63.

J. A. George (1974). "On Block Elimination for Sparse Linear Systems," *SIAM J. Num. Anal. 11*, 585–603.

J. A. George and M. T. Heath (1980). "Solution of Sparse Linear Least Squares Problems Using Givens Rotations," *Lin. Alg. & Its Applic. 34*, 69–83.

J. A. George and J. W. Liu (1981). *Computer Solution of Large Sparse Positive Definite Systems*, Prentice-Hall, Englewood Cliffs, N.J.

N. E. Gibbs and W. G. Poole, Jr. (1974). "Tridiagonalization by Permutations," *Comm. Assoc. Comp. Mach. 17*, 20–24.

N. E. Gibbs, W. G. Poole, and P. K. Stockmeyer (1976a). "A Comparison of Several Bandwidth and Profile Reduction Algorithms," *ACM Trans. Math. Soft. 2*, 322–30.

N. E. Gibbs, W. G. Poole, Jr., and P. K. Stockmeyer (1976b). "An Algorithm for Reducing the Bandwidth and Profile of a Sparse Matrix," *SIAM J. Num. Anal. 13*, 236–50.

P. E. Gill, G. H. Golub, W. Murray, and M. A. Saunders (1974). "Methods for Modifying Matrix Factorizations," *Math. Comp. 28*, 505–35.

P. E. Gill and W. Murray (1976). "The Orthogonal Factorization of a Large Sparse Matrix," in *Sparse Matrix Computations*, ed. J. R. Bunch and D. J. Rose, Academic Press, New York, pp. 177–200.

P. E. Gill, W. Murray, and M. A. Saunders (1975). "Methods for Computing and Modifying the LDV Factors of a Matrix," *Math. Comp. 29*, 1051–77.

T. Ginsburg (1971). "The Conjugate Gradient Method," in HACLA, pp. 57-69.

W. Givens (1958). "Computation of Plane Unitary Rotations Transforming a General Matrix to Triangular Form," *SIAM J. App. Math. 6*, 26-50.

I. C. Gohberg and M. G. Krein (1969). *Introduction to the Theory of Linear Non-Self-Adjoint Operators*, Amer. Math. Soc., Providence, R.I.

H. H. Goldstine and L. P. Horowitz (1959). "A Procedure for the Diagonalization of Normal Matrices," *J. Assoc. Comp. Mach. 6*, 176-95.

D. Goldfarb (1976). "Factorized Variable Metric Methods for Unconstrained Optimization," *Math. Comp. 30*, 796-811.

G. H. Golub (1965). "Numerical Methods for Solving Linear Least Squares Problems," *Numer. Math. 7*, 206-16.

G. H. Golub (1969). "Matrix Decompositions and Statistical Computation," in *Statistical Computation*, ed. R. C. Milton and J. A. Nelder, Academic Press, New York, pp. 365-97.

G. H. Golub (1973). "Some Modified Matrix Eigenvalue Problems," *SIAM Review 15*, 318-44.

G. H. Golub (1974). "Some Uses of the Lanczos Algorithm in Numerical Linear Algebra," in *Topics in Numerical Analysis*, ed. J.J.H. Miller, Academic Press, New York.

G. H. Golub, M. Heath, and G. Wahba (1979). "Generalized Cross-Validation as a Method for Choosing a Good Ridge Parameter," *Technometrics 21*, 215-23.

G. H. Golub and W. Kahan (1965). "Calculating the Singular Values and Pseudo-Inverse of a Matrix," *SIAM J. Num. Anal. Ser. B 2*, 205-24.

G. H. Golub, V. Klema, and G. W. Stewart (1976). "Rank Degeneracy and Least Squares Problems," Technical Report TR-456, Department of Computer Science, University of Maryland, College Park.

G. H. Golub, F. T. Luk, and M. Overton (1981). "A Block Lanczos Method for Computing the Singular Values and Corresponding Singular Vectors of a Matrix," *ACM Trans. Math. Soft. 7*, 149-69.

G. H. Golub, S. Nash, and C. Van Loan (1979). "A Hessenberg-Schur Method for the Matrix Problem $AX + XB = C$," *IEEE Trans. Auto. Cont. AC-24*, 909-913.

G. H. Golub and V. Pereyra (1973). "The Differentiation of Pseudo-Inverses and Nonlinear Least Squares Problems Whose Variables Separate," *SIAM J. Num. Anal. 10*, 413-32.

G. H. Golub and V. Pereyra (1976). "Differentiation of Pseudo-Inverses, Separable Nonlinear Least Squares Problems and Other Tales," in *Generalized Inverses and Applications*, ed. M. Z. Nashed, Academic Press, New York, pp. 303-24.

G. H. Golub and C. Reinsch (1970). "Singular Value Decomposition and Least Squares Solutions," *Numer. Math. 14*, 403-20. See also HACLA, pp. 134-51.

G. H. Golub and W. P. Tang (1981). "The Block Decomposition of a Vandermonde Matrix and Its Applications," *BIT 21*, 505-17.

G. H. Golub and R. Underwood (1970). "Stationary Values of the Ratio of Quadratic Forms Subject to Linear Constraints," *Z. Angew. Math. Phys. 21*, 318-26.

G. H. Golub and R. Underwood (1977). "The Block Lanczos Method for Computing Eigenvalues," in *Mathematical Software III*, ed. J. Rice, Academic Press, New York, pp. 364-77.

G. H. Golub, R. Underwood, and J. H. Wilkinson (1972). "The Lanczos Algorithm for the Symmetric $Ax = \lambda Bx$ Problem," Report STAN-CS-72-270, Department of Computer Science, Stanford University, Stanford, Calif.

G. H. Golub and C. F. Van Loan (1979). "Unsymmetric Positive Definite Linear Systems," *Lin. Alg. & Its Applic. 28*, 85–98.

G. H. Golub and C. F. Van Loan (1980). "An Analysis of the Total Least Squares Problem," *SIAM J. Num. Anal. 17*, 883–93.

G. H. Golub and J. M. Varah (1974). "On a Characterization of the Best l_2-scaling of a Matrix," *SIAM J. Num. Anal. 11*, 472–79.

G. H. Golub and R. S. Varga (1961). "Chebychev Semi-Iterative Methods, Successive Over-Relaxation Iterative Methods, and Second-Order Richardson Iterative Methods," Parts I and II, *Numer. Math. 3*, 147–56, 157–68.

G. H. Golub and J. H. Welsch (1969). "Calculation of Gauss Quadrature Rules," *Math. Comp. 23*, 221–30.

G. H. Golub and J. H. Wilkinson (1966). "Note on the Iterative Refinement of Least Squares Solution," *Numer. Math. 9*, 139–48.

G. H. Golub and J. H. Wilkinson (1976). "Ill-Conditioned Eigensystems and the Computation of the Jordan Canonical Form," *SIAM Review 18*, 578–619.

A. R. Gourlay (1970). "Generalization of Elementary Hermitian Matrices," *Comp. J. 13*, 411–12.

B. Green (1952). "The Orthogonal Approximation of an Oblique Structure in Factor Analysis," *Psychometrika 17*, 429–40.

A. Greenbaum (1981). "Behavior of the Conjugate Gradient Algorithm in Finite Precision Arithmetic," Report UCRL 85752, Lawrence Livermore Laboratory, Livermore, Calif.

R. G. Grimes and J. G. Lewis (1981). "Condition Number Estimation for Sparse Matrices," *SIAM J. Sci. & Stat. Comp. 2*, 384–88.

R. F. Gunst, J. T. Webster, and R. L. Mason (1976). "A Comparison of Least Squares and Latent Root Regression Estimators," *Technometrics 18*, 75–83.

K. K. Gupta (1972). "Solution of Eigenvalue Problems by Sturm Sequence Method," *Int. J. Num. Meth. Eng. 4*, 379–404.

L. A. Hageman and D. M. Young (1981). *Applied Iterative Methods*, Academic Press, New York.

HACLA. See Wilkinson and Reinsch (1971).

P. Halmos (1958). *Finite Dimensional Vector Spaces*, Van Nostrand, New York.

S. Hammarling (1974). "A Note on Modifications to the Givens Plane Rotation," *J. Inst. Math. Appl. 13*, 215–18.

E. R. Hansen (1962). "On Quasicyclic Jacobi Methods," *ACM J. 9*, 118–35.

E. R. Hansen (1963). "On Cyclic Jacobi Methods," *SIAM J. Applied Math. 11*, 448–59.

R. J. Hanson and C. L. Lawson (1969). "Extensions and Applications of the Householder Algorithm for Solving Linear Least Squares Problems," *Math. Comp. 23*, 787–812.

M. T. Heath (1978). "Numerical Algorithms for Nonlinearly Constrained Optimization," Report STAN-CS-78-656, Department of Computer Science, Stanford University (Ph.D. thesis).

D. Heller (1976). "Some Aspects of the Cyclic Reduction Algorithm for Block Tridiagonal Linear Systems," *SIAM J. Num. Anal. 13*, 484–96.

B. W. Helton (1968). "Logarithms of Matrices," *Proc. Amer. Math. Soc. 19*, 733–36.

P. Henrici (1958). "On the Speed of Convergence of Cyclic and Quasicyclic Jacobi

Methods for Computing the Eigenvalues of Hermitian Matrices," *SIAM J. Applied Math. 6*, 144-62.

P. Henrici (1962). "Bounds for Iterates, Inverses, Spectral Variation, and Fields of Values of Non-Normal Matrices," *Numer. Math. 4*, 24-40.

P. Henrici and K. Zimmermann (1968). "An Estimate for the Norms of Certain Cyclic Jacobi Operators," *Lin. Alg. & Its Applic. 1*, 489-501.

M. R. Hestenes (1980). *Conjugate Direction Methods in Optimization*, Springer-Verlag, Berlin.

M. R. Hestenes and E. Stiefel (1952). "Methods of Conjugate Gradients for Solving Linear Systems," *J. Res. Nat. Bur. Stand. 49*, 409-36.

D. Hoaglin (1977). "Mathematical Software and Exploratory Data Analysis," in *Mathematical Software III*, ed. John Rice, Academic Press, New York, pp. 139-59.

R. W. Hockney (1965). "A Fast Direct Solution of Poisson's Equation Using Fourier Analysis," *J. Assoc. Comp. Mach. 12*, 95-113.

W. Hoffmann and B. N. Parlett (1978). "A New Proof of Global Convergence for the Tridiagonal QL Algorithm," *SIAM J. Num. Anal. 15*, 929-37.

H. Hotelling (1957). "The Relations of the Newer Multivariate Statistical Methods to Factor Analysis," *Brit. J. Stat. Psych. 10*, 69-79.

A. S. Householder (1958). "Unitary Triangularization of a Nonsymmetric Matrix," *J. Assoc. Comp. Mach. 5*, 339-42.

A. S. Householder (1968). "Moments and Characteristic Roots II," *Numer. Math. 11*, 126-28.

A. S. Householder (1974). *The Theory of Matrices in Numerical Analysis*, Dover Publications, New York.

C. P. Huang (1975). "A Jacobi-Type Method for Triangularizing an Arbitrary Matrix," *SIAM J. Num. Anal. 12*, 566-70.

C. P. Huang (1981). "On the Convergence of the QR Algorithm with Origin Shifts for Normal Matrices," *IMA J. Num. Anal. 1*, 127-33.

T. E. Hull and J. R. Swenson (1966). "Tests of Probabilistic Models for Propagation of Roundoff Errors," *Comm. Assoc. Comp. Mach. 9*, 108-13.

Y. Ikebe (1979). "On Inverses of Hessenberg Matrices," *Lin. Alg. & Its Applic. 24*, 93-97.

C.G.J. Jacobi (1846). "Uber ein Leichtes Verfahren Die in der Theorie der Sacular-storungen Vorkommendern Gleichungen Numerisch Aufzulosen," *Crelle's J. 30*, 51-94.

A. Jennings (1977a). "Influence of the Eigenvalue Spectrum on the Convergence Rate of the Conjugate Gradient Method," *J. Inst. Math. Applic. 20*, 61-72.

A. Jennings (1977b). *Matrix Computation for Engineers and Scientists*, John Wiley & Sons, New York.

A. Jennings and D.R.L. Orr (1971). "Application of the Simultaneous Iteration Method to Undamped Vibration Problems," *Inst. J. Numer. Meth. Eng. 3*, 13-24.

A. Jennings and M. R. Osborne (1977). "Generalized Eigenvalue Problems for Certain Unsymmetric Band Matrices," *Lin. Alg. & Its Applic. 29*, 139-50.

A. Jennings and W. J. Stewart (1975). "Simultaneous Iteration for the Partial Eigensolution of Real Matrices," *J. Inst. Math. Applic. 15*, 351-62.

L. S. Jennings and M. R. Osborne (1974). "A Direct Error Analysis for Least Squares," *Numer. Math. 22*, 322-32.

P. S. Jenson (1972). "The Solution of Large Symmetric Eigenproblems by Sectioning," *SIAM J. Num. Anal. 9*, 534-45.

J. Johnson and C. L. Phillips (1971). "An Algorithm for the Computation of the Integral of the State Transition Matrix," *IEEE Trans. Auto. Cont. AC-16*, 204-5.

R. L. Johnston (1971). "Gershgorin Theorems for Partitioned Matrices," *Lin. Alg. & Its Applic. 4*, 205-20.

B. Kagstrom (1977a). "Bounds and Perturbation Bounds for the Matrix Exponential," *BIT 17*, 39-57.

B. Kagstrom (1977b). "Numerical Computation of Matrix Functions," Department of Information Processing Report UMINF-58.77, University of Umea, Umea, Sweden.

B. Kagstrom and A. Ruhe (1980a). "An Algorithm for Numerical Computation of the Jordan Normal Form of a Complex Matrix," *ACM Trans. Math. Soft. 6*, 398-419.

B. Kagstrom and A. Ruhe (1980b). "Algorithm 560 JNF: An Algorithm for Numerical Computation of the Jordan Normal Form of a Complex Matrix," *ACM Trans. Math. Soft. 6*, 437-43.

W. Kahan (1966). "Numerical Linear Algebra," *Canadian Math. Bull. 9*, 757-801.

W. Kahan (1967). "Inclusion Theorems for Clusters of Eigenvalues of Hermitian Matrices," Computer Science Report, University of Toronto.

W. Kahan (1975). "Spectra of Nearly Hermitian Matrices," *Proc. Amer. Math. Soc. 48*, 11-17.

W. Kahan and B. N. Parlett (1974). "An Analysis of Lanczos Algorithms for Symmetric Matrices," ERL-M467, University of California, Berkeley.

W. Kahan and B. N. Parlett (1976). "How Far Should You Go with the Lanczos Process?" in *Sparse Matrix Computations*, ed. J. Bunch and D. Rose, Academic Press, New York, pp. 131-44.

S. Kaniel (1966). "Estimates for Some Computational Techniques in Linear Algebra," *Math. Comp. 20*, 369-78.

I. Karasalo (1974). "A Criterion for Truncation of the QR Decomposition Algorithm for the Singular Linear Least Squares Problem," *BIT 14*, 156-66.

T. Kato (1966). *Perturbation Theory for Linear Operators*, Springer-Verlag, New York.

L. Kaufman (1974). "The LZ Algorithm to Solve the Generalized Eigenvalue Problem," *SIAM J. Num. Anal. 11*, 997-1024.

L. Kaufman (1977). "Some Thoughts on the QZ Algorithm for Solving the Generalized Eigenvalue Problem," *ACM Trans. Math. Soft. 3*, 65-75.

L. Kaufman (1979). "Application of Dense Householder Transformations to a Sparse Matrix," *ACM Trans. Math. Soft. 5*, 442-50.

D. Knuth (1969). *The Art of Computer Programming*, vol. 2, *Seminumerical-Algorithms*, Addison-Wesley, Reading, Mass.

S. Kourouklis and C. C. Paige (1981). "A Constrained Least Squares Approach to the General Gauss-Markov Linear Model," *J. Amer. Stat. Assoc. 76*, 620-25.

V. N. Kublanovskaya (1961). "On Some Algorithms for the Solution of the Complete Eigenvalue Problem," *USSR Comp. Math. Phys. 3*, 637-57.

V. N. Kublanovskaja and V. N. Fadeeva (1964). "Computational Methods for the Solution of a Generalized Eigenvalue Problem," *Amer. Math. Soc. Transl. 2*, 271-90.

C. D. La Budde (1964). "Two Classes of Algorithms for Finding the Eigenvalues and Eigenvectors of Real Symmetric Matrices," *J. Assoc. Comp. Mach. 11*, 53-58.

J. Lambiotte and R. G. Voigt (1975). "The Solution of Tridiagonal Linear Systems on the CDC-STAR 100 Computer," *ACM Trans. Math. Soft. 1*, 308–29.

P. Lancaster (1970). "Explicit Solution of Linear Matrix Equations," *SIAM Review 12*, 544–66.

C. Lanczos (1950). "An Interation Method for the Solution of the Eigenvalue Problem of Linear Differential and Integral Operators," *J. Res. Nat. Bur. Stand. 45*, 255–82.

J. Larson and A. Sameh (1978). "Efficient Calculation of the Effects of Roundoff Errors," *ACM Trans. Math. Soft. 4*, 228–36.

A. Laub (1981). "Efficient Multivariable Frequency Response Computations," *IEEE Trans. Auto. Cont. AC-26*, 407–8.

C. L. Lawson and R. J. Hanson (1969). "Extensions and Applications of the Householder Algorithm for Solving Linear Least Squares Problems," *Math Comp. 23*, 787–812.

C. L. Lawson and R. J. Hanson (1974). *Solving Least Squares Problems*, Prentice-Hall, Englewood Cliffs, N.J.

C. L. Lawson, R. J. Hanson, F. T. Krogh, and O. R. Kincaid (1979). "Basic Linear Algebra Subprograms for FORTRAN Usage," *ACM Trans. Math. Soft. 5*, 308–23.

N. J. Lehmann (1963). "Optimale Eigenwerteinschliessungen," *Numer. Math. 5*, 246–72.

F. Lemeire (1973). "Bounds for Condition Numbers of Triangular and Trapezoid Matrices," *BIT 15*, 58–64.

S. J. Leon (1980). *Linear Algebra with Applications*. Macmillan, New York.

N. Levinson (1947). "The Weiner RMS Error Criterion in Filter Design and Prediction," *J. Math. Phys. 25*, 261–78.

J. Lewis (1977). "Algorithms for Sparse Matrix Eigenvalue Problems," Technical Report STAN-CS-77-595, Department of Computer Science, Stanford University, Stanford, Calif.

I. Linnik (1961). *Method of Least Squares and Principles of the Theory of Observations*, Pergamon Press, New York.

G. Loizou (1969). "Nonnormality and Jordan Condition Numbers of Matrices," *J. Assoc. Comp. Mach. 16*, 580–84.

G. Loizou (1972). "On the Quadratic Convergence of the Jacobi Method for Normal Matrices," *Comp. J. 15*, 274–76.

M. Lotkin (1956). "Characteristic Values of Arbitrary Matrices," *Quart. Appl. Math. 14*, 267–75.

F. T. Luk (1978). "Sparse and Parallel Matrix Computations," Report STAN-CS-78-685, Department of Computer Science, Stanford University, Stanford, Calif.

F. T. Luk (1980). "Computing the Singular Value Decomposition on the ILLIAC IV," *ACM Trans. Math. Soft. 6*, 524–39.

C. McCarthy and G. Strang (1973). "Optimal Conditioning of Matrices," *SIAM J. Num. Anal. 10*, 370–88.

S. F. McCormick (1972). "A General Approach to One-Step Iterative Methods with Application to Eigenvalue Problems," *J. Comput. Sys. Sci. 6*, 354–72.

W. M. McKeeman (1962). "Crout with Equilibration and Iteration," *Comm. Assoc. Comp. Mach. 5*, 553–55.

A. Madansky (1959). "The Fitting of Straight Lines When Both Variables Are Subject to Error," *J. Amer. Stat. Assoc. 54*, 173–205.

K. N. Majindar (1979). "Linear Combinations of Hermitian and Real Symmetric Matrices," *Lin. Alg. & Its Applic. 25*, 95–105.

J. Makhoul (1975). "Linear Predication: A Tutorial Review," *Proc. IEEE 63* (4), 561–80.

M. A. Malcolm and J. Palmer (1974). "A Fast Method for Solving a Class of Tridiagonal Systems of Linear Equations," *Comm. Assoc. Comp. Mach. 17*, 14–17.

T. A. Manteuffel (1977). "The Tchebychev Iteration for Nonsymmetric Linear Systems," *Numer. Math. 28*, 307–27.

T. A. Manteuffel (1979). "Shifted Incomplete Cholesky Factorization," in *Sparse Matrix Proceedings, 1978*, ed. I. S. Duff and G. W. Stewart, SIAM Publications, Philadelphia.

M. Marcus and H. Minc (1964). *A Survey of Matrix Theory and Matrix Inequalities*, Allyn and Bacon, Boston.

J. Markel and A. Gray (1974). "Fixed-Point Truncation Arithmetic Implementation of a Linear Predication Autocorrelation Vocoder," *IEEE Trans. ASSP 22*, 273–81.

J. Markel and A. Gray (1976). *Linear Prediction of Speech*, Springer-Verlag, Berlin and New York.

R. S. Martin, G. Peters, and J. H. Wilkinson (1965). "Symmetric Decomposition of a Positive Definite Matrix," *Numer. Math. 7*, 362–83. See also HACLA, pp. 9–30.

R. S. Martin, G. Peters, and J. H. Wilkinson (1966). "Iterative Refinement of the Solution of a Positive Definite System of Equations," *Numer. Math. 8*, 203–16. See also HACLA, pp. 31–44.

R. S. Martin, G. Peters, and J. H. Wilkinson (1970). "The QR Algorithm for Real Hessenberg Matrices," *Numer. Math. 14*, 219–31. See also HACLA, pp. 359–71.

R. S. Martin, C. Reinsch, and J. H. Wilkinson (1970). "The QR Algorithm for Band Symmetric Matrices," *Numer. Math. 16*, 85–92. See also HACLA, pp. 266–272.

R. S. Martin and J. H. Wilkinson (1965). "Symmetric Decomposition of Positive Definite Band Matrices," *Numer. Math. 7*, 355–61. See also HACLA, pp. 50–56.

R. S. Martin and J. H. Wilkinson (1967). "Solution of Symmetric and Unsymmetric Band Equations and the Calculation of Eigenvalues of Band Matrices," *Numer. Math 9*, 279–301. See also HACLA, pp. 70–92.

R. S. Martin and J. H. Wilkinson (1968a). "Householder's Tridiagonalization of a Symmetric Matrix," *Numer. Math. 11*, 181–95. See also HACLA, pp. 212–26.

R. S. Martin and J. H. Wilkinson (1968b). "The Modified LR Algorithm for Complex Hessenberg Matrices," *Numer. Math. 12*, 369–76. See also HACLA, pp. 396–403.

R. S. Martin and J. H. Wilkinson (1968c). "Reduction of the Symmetric Eigenproblem $Ax = \lambda Bx$ and Related Problems to Standard Form," *Numer. Math. 11*, 99–110.

R. S. Martin and J. H. Wilkinson (1968d). "Similarity Reduction of a General Matrix to Hessenberg Form," *Numer. Math. 12*, 349–68. See also HACLA, pp. 339–58.

J. A. Meijerink and H. A. Van der Vorst (1977). "An Iterative Solution Method for Linear Equation Systems of Which the Coefficient Matrix is a Symmetric M-Matrix," *Math. Comp. 31*, 148–62.

W. Miller and D. Spooner (1978). "Software for Roundoff Analysis, II," *ACM Trans. Math. Soft. 4*, 369–90.

L. Mirsky (1955). *An Introduction to Linear Algebra*, Oxford University Press, London.

C. B. Moler (1967). "Iterative Refinement in Floating Point," *J. Assoc. Comp. Mach.* *14*, 316-71.

C. B. Moler (1980). "MATLAB User's Guide," Technical Report CS81-1, Department of Computer Science, University of New Mexico, Albuquerque.

C. B. Moler and G. W. Stewart (1973). "An Algorithm for Generalized Matrix Eigenvalue Problems," *SIAM J. Num. Anal. 10*, 241-56.

C. B. Moler and C. F. Van Loan (1978). "Nineteen Dubious Ways to Compute the Exponential of a Matrix," *SIAM Review 20*, 801-36.

D. Mueller (1966). "Householder's Method for Complex Matrices and Hermitian Matrices," *Numer. Math. 8*, 72-92.

F. D. Murnaghan and A. Wintner (1931). "A Canonical form for Real Matrices under Orthogonal Transformations," *Proc. Nat. Acad. Sci. 17*, 417-20.

J. C. Nash (1975). "A One-Sided Transformation Method for the Singular Value Decomposition and Algebraic Eigenproblem," *Comp. J. 18*, 74-76.

M. Z. Nashed (1976). *Generalized Inverses and Applications*, Academic Press, New York.

R. A. Nicolaides (1974). "On a Geometrical Aspect of SOR and the Theory of Consistent Ordering for Positive Definite Matrices," *Numer. Math. 23*, 99-104.

B. Noble and J. W. Daniel (1977). *Applied Linear Algebra*, Prentice-Hall, Englewood Cliffs, N.J.

D. P. O'Leary (1976). "Hybrid Conjugate Gradient Algorithms," Report STAN-CS-76-548, Department of Computer Science, Stanford University (Ph.D. thesis).

D. P. O'Leary (1980a). "The Block Conjugate Gradient Algorithm and Related Methods," *Lin. Alg. & Its Applic. 29*, 293-322.

D. P. O'Leary (1980b). "Estimating Matrix Condition Numbers," *SIAM J. Sci. & Stat. Comp. 1*, 205-9.

A. V. Oppenheim (1978). *Applications of Digital Signal Processing*, Prentice-Hall, Englewood Cliffs, N.J.

J. M. Ortega (1972). *Numerical Analysis: A Second Course*, Academic Press, New York.

E. E. Osborne (1960). "On Preconditioning of Matrices," *J. Assoc. Comp. Mach. 7*, 338-45.

M.H.C. Paardekooper (1971). "An Eigenvalue Algorithm for Skew Symmetric Matrices," *Numer. Math. 17*, 189-202.

C. C. Paige (1971). "The Computation of Eigenvalues and Eigenvectors of Very Large Sparse Matrices," Ph.D. thesis, London University.

C. C. Paige (1972). "Computational Variants of the Lanczos Method for the Eigenproblem," *J. Inst. Math. Applic. 10*, 373-81.

C. C. Paige (1973). "An Error Analysis of a Method for Solving Matrix Equations," *Math. Comp. 27*, 355-59.

C. C. Paige (1974a). "Bidiagonalization of Matrices and Solution of Linear Equations," *SIAM J. Num. Anal. 11*, 197-209.

C. C. Paige (1974b). "Eigenvalues of Perturbed Hermitian Matrices," *Lin. Alg. & Its Applic. 8*, 1-10.

C. C. Paige (1976a). "Error Analysis of the Lanczos Algorithm for Tridiagonalizing a Symmetric Matrix," *J. Inst. Math. Applic. 18*, 341-49.

C. C. Paige (1976b). "Practical Use of the Symmetric Lanczos Process with Reorthogonalization," *BIT 10*, 183-95.

C. C. Paige (1979a). "Computer Solution and Perturbation Analysis of Generalized Least Squares Problems," *Math. Comp. 33*, 171-84.

C. C. Paige (1979b). "Fast Numerically Stable Computations for Generalized Linear Least Squares Problems," *SIAM J. Num. Anal. 16*, 165-71.

C. C. Paige (1980). "Accuracy and Effectiveness of the Lanczos Algorithm for the Symmetric Eigenproblem," *Lin. Alg. & Its Applic. 34*, 235-58.

C. C. Paige (1981). "Properties of Numerical Algorithms Related to Computing Controllability," *IEEE Trans. Auto. Cont. AC-26*, 130-38.

C. C. Paige and M. A. Saunders (1975). "Solution of Sparse Indefinite Systems of Linear Equations," *SIAM J. Num. Anal. 12*, 617-29.

C. C. Paige and M. A. Saunders (1978). "A Bidiagonalization Algorithm for Sparse Linear Equations and Least Squares Problems," Report SOL 78-19, Department of Operations Research, Stanford University, Stanford, Calif.

C. C. Paige and M. Saunders (1981). "Towards a Generalized Singular Value Decomposition," *SIAM J. Num. Anal. 18*, 398-405.

C. C. Paige and M. A. Saunders (1982a). "LSQR: An Algorithm for Sparse Linear Equations and Sparse Least Squares," *ACM Trans. Math. Soft. 8*, 43-71.

C. C. Paige and M. A. Saunders (1982b). "Algorithm 583 LSQR: Sparse Linear Equations and Least Squares Problems," *ACM Trans. Math. Soft. 8*, 195-209.

B. N. Parlett (1965). "Convergence of the Q-R Algorithm," *Numer. Math. 7*, 187-93. (Correction in *Numer. Math. 10*, 163-64.)

B. N. Parlett (1966). "Singular and Invariant Matrices under the QR Algorithm," *Math. Comp. 20*, 611-15.

B. N. Parlett (1967). "Canonical Decomposition of Hessenberg Matrices," *Math. Comp. 21*, 223-27.

B. N. Parlett (1968). "Global convergence of the Basic QR Algorithm on Hessenberg Matrices," *Math. Comp. 22*, 803-17.

B. N. Parlett (1971). "Analysis of Algorithms for Reflections in Bisectors," *SIAM Review 13*, 197-208.

B. N. Parlett (1974a). "Computation of Functions of Triangular Matrices," Memorandum no. ERL-M481, Electronics Research Laboratory, College of Engineering, University of California, Berkeley.

B. N. Parlett (1974b). "The Rayleigh Quotient Iteration and Some Generalizations for Nonnormal Matrices," *Math. Comp. 28*, 679-93.

B. N. Parlett (1976). "A Recurrence among the Elements of Functions of Triangular Matrices," *Lin. Alg. & Its Applic. 14*, 117-21.

B. N. Parlett (1980a). "A New Look at the Lanczos Algorithm for Solving Symmetric Systems of Linear Equations," *Lin. Alg. & Its Applic. 29*, 323-46.

B. N. Parlett (1980b). *The Symmetric Eigenvalue Problem*, Prentice-Hall, Englewood Cliffs, N.J.

B. N. Parlett and W. G. Poole (1973). "A Geometric Theory for the QR, LU, and Power Iterations," *SIAM J. Num. Anal. 10*, 389-412.

B. N. Parlett and J. K. Reid (1970). "On the Solution of a System of Linear Equations Whose Matrix is Symmetric But Not Definite," *BIT 10*, 386-97.

B. N. Parlett and J. K. Reid (1981). "Tracking the Progress of the Lanczos Algorithm for Large Symmetric Eigenproblems," *IMA J. Num. Anal. 1*, 135-55.

B. N. Parlett and C. Reinsch (1969). "Balancing a Matrix for Calculation of Eigenvalues and Eigenvectors," *Numer. Math. 13*, 292-304. See also HACLA, pp. 315-26.

B. N. Parlett and D. S. Scott (1979). "The Lanczos Algorithm with Selective Orthogonalization," *Math. Comp. 33*, 217-38.

M. S. Paterson and L. J. Stockmeyer (1973). "On the Number of Nonscalar Multiplications Necessary to Evaluate Polynomials," *SIAM J. Comp. 2*, 60-66.

K. Pearson (1901). "On Lines and Planes of Closest Fit to Points in Space," *Phil. Mag. 2*, 559-72.

G. Peters and J. H. Wilkinson (1969). "Eigenvalues of $Ax = \lambda Bx$ with Band Symmetric A and B," *Comp. J. 12*, 398-404.

G. Peters and J. H. Wilkinson (1970a). "$Ax = \lambda Bx$ and the Generalized Eigenproblem," *SIAM J. Num. Anal. 7*, 479-92.

G. Peters and J. H. Wilkinson (1970b). "The Least Squares Problem and Pseudo-Inverses," *Comp. J. 13*, 309-16.

G. Peters and J. H. Wilkinson (1971). "The Calculation of Specified Eigenvectors by Inverse Iteration," in HACLA, pp. 418-39.

G. Peters and J. H. Wilkinson (1979). "Inverse Iteration, Ill-Conditioned Equations, and Newton's Method," *SIAM Review 21*, 339-60.

J. L. Phillips (1971). "The Triangular Decomposition of Hankel Matrices," *Math. Comp. 25*, 599-602.

R. J. Plemmons (1974). "Linear Least Squares by Elimination and MGS," *J. Assoc. Comp. Mach. 21*, 581-85.

D. A. Pope and C. Tompkins (1957). "Maximizing Functions of Rotations: Experiments Concerning Speed of Diagonalization of Symmetric Matrices Using Jacobi's Method," *J. Assoc. Comp. Mach. 4*, 459-66.

M.J.D. Powell and J. K. Reid (1968). "On Applying Householder's Method to Linear Least Squares Problems," *Proc. IFIP Congress*, pp. 122-26.

J. K. Reid (1967). "A Note on the Least Squares Solution of a Band System of Linear Equations by Householder Reductions," *Comp. J. 10*, 188-89.

J. K. Reid (1971a). "A Note on the Stability of Gaussian Elimination," *J. Inst. Math. Applic. 8*, 374-75.

J. K. Reid (1971b). "On the Method of Conjugate Gradients for the Solution of Large Sparse of Linear Equations," in *Large Sparse Sets of Linear Equations*, ed. J. K. Reid, Academic Press, New York, pp. 231-54.

J. K. Reid (1972). "The Use of Conjugate Gradients for Systems of Linear Equations Possessing Property A," *SIAM J. Num. Anal. 9*, 325-32.

C. Reinsch and F. L. Bauer (1968). "Rational QR Transformation with Newton's Shift for Symmetric Tridiagonal Matrices," *Numer. Math. 11*, 264-72. See also HACLA, pp. 257-65.

J. R. Rice (1966a). "Experiments on Gram-Schmidt Orthogonalization," *Math. Comp. 20*, 325-28.

J. Rice (1966b). "A Theory of Condition," *SIAM J. Num. Anal. 3*, 287-310.

R. F. Rinehart (1955). "The Equivalence of Definitions of a Matrix Function," *Amer. Math. Monthly 62*, 395-414.

J. Rissanen (1973). "Algorithms for Triangular Decomposition of Block Hankel and Toeplitz Matrices with Application to Factoring Positive Matrix Polynomials," *Math. Comp. 27*, 147-54.

H. H. Robertson (1977). "The Accuracy of Error Estimates for Systems of Linear Algebraic Equations," *J. Inst. Math. Applic. 20*, 409-14.

G. Rodrigue (1973). "A Gradient Method for the Matrix Eigenvalue Problem $Ax = \lambda Bx$," *Numer. Math. 22*, 1-16.

D. J. Rose (1969). "An Algorithm for Solving a Special Class of Tridiagonal Systems of Linear Equations," *Comm. Assoc. Comp. Mach. 12*, 234-36.

D. J. Rose and R. A. Willoughby, eds. (1972). *Sparse Matrices and Their Applications*, Plenum Press, New York.

A. Ruhe (1967). "On the Quadratic Convergence of the Jacobi Method for Normal Matrices," *BIT 7*, 305-13.

A. Ruhe (1968). "On the Quadratic Convergence of a Generalization of the Jacobi Method to Arbitrary Matrices," *BIT 8*, 210-31.

A. Ruhe (1969). "The Norm of a Matrix after a Similarity Transformation," *BIT 9*, 53-58.

A. Ruhe (1970a). "An Algorithm for Numerical Determination of the Structure of a General Matrix," *BIT 10*, 196-216.

A. Ruhe (1970b). "Perturbation Bounds for Means of Eigenvalues and Invariant Subspaces," *BIT 10*, 343-54.

A. Ruhe (1970c). "Properties of a Matrix with a Very Ill-Conditioned Eigenproblem," *Numer. Math. 15*, 57-60.

A. Ruhe (1974). "SOR Methods for the Eigenvalue Problem with Large Sparse Matrices," *Math. Comp. 28*, 695-710.

A. Ruhe (1975). "On the Closeness of Eigenvalues and Singular Values for Almost Normal Matrices," *Lin. Alg. & Its Applic. 11*, 87-94.

A. Ruhe (1978). "A Note on the Efficient Solution of Matrix Pencil Systems," *BIT 18*, 276-81.

A. Ruhe (1979). "Implementation Aspects of Band Lanczos Algorithms for Computation of Eigenvalues of Large Sparse Symmetric Matrices," *Math. Comp. 33*, 680-87.

A. Ruhe and T. Wiberg (1972). "The Method of Conjugate Gradients Used in Inverse Iteration," *BIT 12*, 543-54.

H. Rutishauser (1958). "Solution of Eigenvalue Problems with the LR Transformation," *Nat. Bur. Stand. App. Math. Ser. 49*, 47-81.

H. Rutishauser (1966). "The Jacobi Method for Real Symmetric Matrices," *Numer. Math. 9*, 1-10. See also HACLA, pp. 202-11.

H. Rutishauser (1969). "Computation Aspects of F. L. Bauer's Simultaneous Iteration Method," *Numer. Math. 13*, 4-13.

H. Rutishauser (1970). "Simultaneous Iteration Method for Symmetric Matrices," *Numer. Math. 16*, 205-23. See also HACLA, pp. 284-302.

Y. Saad (1980). "On the Rates of Convergence of the Lanczos and the Block Lanczos Methods," *SIAM J. Num. Anal. 17*, 687-706.

A. Sameh (1971). "On Jacobi and Jacobi-like Algorithms for a Parallel Computer," *Math. Comp. 25*, 579-90.

A. Sameh, J. Lermit, and K. Noh (1975). "On the Intermediate Eigenvalues of Symmetric Sparse Matrices," *BIT 12*, 543-54.

K. Schittkowski and J. Stoer (1979). "A Factorization Method for the Solution of Constrained Linear Least Squares Problems Allowing for Subsequent Data Changes," *Numer. Math. 31*, 431-63.

P. Schonemann (1966). "A Generalized Solution of the Orthogonal Procrustes Problem," *Psychometrika 31*, 1–10.

A. Schonhage (1964). "On the Quadratic Convergence of the Jacobi Process," *Numer. Math. 6*, 410–12.

A. Schonhage (1979). "Arbitrary Perturbations of Hermitian Matrices," *Lin. Alg. & Its Applic. 24*, 143–49.

I. Schur (1909). "On the Characteristic Roots of a Linear Substitution with an Application to the Theory of Integral Equations," *Math. Ann. 66*, 488–510 (German).

H. R. Schwartz (1968). "Tridiagonalization of a Symmetric Band Matrix," *Numer. Math 12*, 231–41. See also HACLA, pp. 273–83.

H. R. Schwartz (1974). "The Method of Coordinate Relaxation for $(A - \lambda B)x = 0$," *Numer. Math. 23*, 135–52.

D. S. Scott (1978). "Analysis of the Symmetric Lanczos Process," UCB-ERL Technical Report M78/40, University of California, Berkeley.

D. S. Scott (1979a). "Block Lanczos Software for Symmetric Eigenvalue Problems," Report ORNL/CSD-48, Oak Ridge National Laboratory, Union Carbide Corporation, Oak Ridge, Tenn.

D. S. Scott (1979b). "How to Make the Lanczos Algorithm Converge Slowly," *Math. Comp. 33*, 239–47.

J. J. Seaton (1969). "Diagonalization of Complex Symmetric Matrices Using a Modified Jacobi Method," *Comp. J. 12*, 156–57.

S. Serbin and S. Blalock (1979). "An Algorithm for Computing the Matrix Cosine," *SIAM J. Sci. Stat. Comp. 1*, 198–204.

J. W. Sheldon (1955). "On the Numerical Solution of Elliptic Difference Equations," *Math. Tables Aids Comp. 9*, 101–12.

B. Singer and S. Spilerman (1976). "The Representation of Social Processes by Markov Models," *Amer. J. Sociology 82*, 1–54.

R. Skeel (1981). "Effect of Equilibration on Residual Size for Partial Pivoting," *SIAM J. Num. Anal. 18*, 449–55.

B. T. Smith, J. M. Boyle, Y. Ikebe, V. C. Klema, and C. B. Moler (1970). *Matrix Eigensystem Routines: EISPACK Guide*, 2nd ed. Springer-Verlag, New York.

R. A. Smith (1967). "The Condition Numbers of the Matrix Eigenvalue Problem," *Numer. Math. 10*, 232–40.

F. Smithies (1970). *Integral Equations*, Cambridge University Press, Cambridge.

D. Stevenson (1981). "A Proposed Standard for Binary Floating Point Arithmetic," *Computer 14 (March)*, 51–62.

G. W. Stewart (1969). "Accelerating the Orthogonal Iteration for the Eigenvalues of a Hermitian Matrix," *Numer. Math. 13*, 362–76.

G. W. Stewart (1970). "Incorporating Origin Shifts into the QR Algorithm for Symmetric Tridiagonal Matrices," *Comm. Assoc. Comp. Mach. 13*, 365–67.

G. W. Stewart (1971). "Error Bounds for Approximate Invariant Subspaces of Closed Linear Operators," *SIAM J. Num. Anal. 8*, 796–808.

G. W. Stewart (1972). "On the Sensitivity of the Eigenvalue Problem $Ax = \lambda Bx$," *SIAM J. Num. Anal. 9*, 669–86.

G. W. Stewart (1973a). "Conjugate Direction Methods for Solving Systems of Linear Equations," *Numer. Math. 21*, 284–97.

G. W. Stewart (1973b). "Error and Perturbation Bounds for Subspaces Associated with Certain Eigenvalue Problems," *SIAM Review 15*, 727–64.

G. W. Stewart (1973c). *Introduction to Matrix Computations*. Academic Press, New York.

G. W. Stewart (1974). "The Numerical Treatment of Large Eigenvalue Problems," *Proc. IFIP Congress 74*, North-Holland, pp. 666-72.

G. W. Stewart (1975a). "The Convergence of the Method of Conjugate Gradients at Isolated Extreme Points in the Spectrum," *Numer. Math. 24*, 85-93.

G. W. Stewart (1975b). "Gershgorin Theory for the Generalized Eigenvalue Problem $Ax = \lambda Bx$," *Math. Comp. 29*, 600-606.

G. W. Stewart (1975c). "Methods of Simultaneous Iteration for Calculating Eigenvectors of Matrices," in *Topics in Numerical Analysis II*, ed. J. H. Miller, Academic Press, New York, pp. 185-96.

G. W. Stewart (1976a). "Algorithm 406: HQR3 and EXCHNG: FORTRAN Subroutines for Calculating and Ordering and Eigenvalues of a Real Upper Hessenberg Matrix," *ACM Trans. Math. Soft. 2*, 275-80.

G. W. Stewart (1976b). "A Bibliographical Tour of the Large Sparse Generalized Eigenvalue Problem," in *Sparse Matrix Computations*, ed. J. R. Bunch and D. J. Rose, Academic Press, New York.

G. W. Stewart (1976c). "The Economical Storage of Plane Rotations," *Numer. Math. 25*, 137-38.

G. W. Stewart (1976d). "Simultaneous Iteration for Computing Invariant Subspaces of Non-Hermitian Matrices," *Numer. Math. 25*, 123-36.

G. W. Stewart (1977a). "On the Perturbation of Pseudo-Inverses, Projections, and Linear Least Squares Problems," *SIAM Review 19*, 634-62.

G. W. Stewart (1977b). "Perturbation Bounds for the QR Factorization of a Matrix," *SIAM J. Num. Anal. 14*, 509-18.

G. W. Stewart (1977c). "Sensitivity Coefficients for the Effects of Errors in the Independent Variables in a Linear Regression," Technical Report TR-571, Department of Computer Science, University of Maryland.

G. W. Stewart (1978). "Perturbation Theory for the Generalized Eigenvalue Problem," in *Recent Advances in Numerical Analysis*, ed. C. deBoor and G. H. Golub, Academic Press, New York.

G. W. Stewart (1979a). "A Note on the Perturbation of Singular Values," *Lin. Alg. & Its Applic. 28*, 213-16.

G. W. Stewart (1979b). "Perturbation Bounds for the Definite Generalized Eigenvalue Problem," *Lin. Alg. & Its Applic. 23*, 69-86.

G. W. Stewart (1979c). "The Effects of Rounding Error on an Algorithm for Downdating a Cholesky Factorization," *J. Inst. Math. Applic. 23*, 203-13.

G. W. Stewart (1980). "The Efficient Generation of Random Orthogonal Matrices with an Application to Condition Estimators," *SIAM J. Num. Anal. 17*, 403-9.

G. W. Stewart (1983). "A Method for Computing the Generalized Singular Value Decomposition," in *Matrix Pencils*, ed. B. Kagstrom and A. Ruhe, Springer-Verlag, New York, pp. 207-20.

H. S. Stone (1973). "An Efficient Parallel Algorithm for the Solution of a Tridiagonal Linear System of Equations," *J. Assoc. Comp. Mach. 20*, 27-38.

H. S. Stone (1975). "Parallel Tridiagonal Equation Solvers," *ACM Trans. Math. Soft. 1*, 289-307.

G. Strang (1976). *Linear Algebra and Its Applications*, Academic Press, New York.

P. N. Swarztrauber and R. A. Sweet (1973). "The Direct Solution of the Discrete Poisson Equation on a Disk," *SIAM J. Num. Anal. 10*, 900–907.

R. A. Sweet (1974). "A Generalized Cyclic Reduction Algorithm," *SIAM J. Num. Anal. 11*, 506–20.

R. A. Sweet (1977). "A Cyclic Reduction Algorithm for Solving Block Tridiagonal Systems of Arbitrary Dimension," *SIAM J. Num. Anal. 14*, 706–20.

H. J. Symm and J. H. Wilkinson (1980). "Realistic Error Bounds for a Simple Eigenvalue and Its Associated Eigenvector," *Numer. Math. 35*, 113–26.

G. L. Thompson and R. L. Weil (1970). "Reducing the Rank of $A - \lambda B$," *Proc. Amer. Math. Soc. 26*, 548–54.

G. L. Thompson and R. L. Weil (1972). "Roots of Matrix Pencils $Ay = \lambda By$: Existence, Calculations, and Relations to Game Theory," *Lin. Alg. & Its Applic. 5*, 207–26.

W. F. Trench (1964). "An Algorithm for the Inversion of Finite Toeplitz Matrices," *J. SIAM 12*, 515–22.

W. F. Trench (1974). "Inversion of Toeplitz Band Matrices," *Math. Comp. 28*, 1089–95.

N. K. Tsao (1975). "A Note on Implementing the Householder Transformation," *SIAM J. Num. Anal. 12*, 53–58.

H. W. Turnbull and A. C. Aitken (1961). *An Introduction to the Theory of Canonical Matrices*, Dover, New York.

F. Uhlig (1973). "Simultaneous Block Diagonalization of Two Real Symmetric Matrices," *Lin. Alg. & Its Applic. 7*, 281–89.

F. Uhlig (1976). "A Canonical Form for a Pair of Real Symmetric Matrices That Generate a Nonsingular Pencil," *Lin. Alg. & Its Applic. 14*, 189–210.

R. Underwood (1975). "An Iterative Block Lanczos Method for the Solution of Large Sparse Symmetric Eigenproblems," Report STAN-CS-75-496, Department of Computer Science, Stanford University, Stanford, Calif.

J. Vandergraft (1971). "Generalized Rayleigh Methods with Applications to Finding Eigenvalues of Large Matrices," *Lin. Alg. & Its Applic. 4*, 353–68.

A. van der Sluis (1969). "Condition Numbers and Equilibration Matrics," *Numer. Math. 14*, 14–23.

A. van der Sluis (1970). "Condition, Equilibration, and Pivoting in Linear Algebraic Systems," *Numer. Math. 15*, 74–86.

A. van der Sluis (1975a). "Perturbations of Eigenvalues of Nonnormal Matrices," *Comm. Assoc. Comp. Mach. 18*, 30–36.

A. van der Sluis (1975b). "Stability of the Solutions of Linear Least Squares Problem," *Numer. Math. 23*, 241–54.

A. van der Sluis and G. W. Veltkamp (1979). "Restoring Rank and Consistency by Orthogonal Projection," *Lin. Alg. & Its Applic. 28*, 257–78.

H. Van de Vel (1977). "Numerical Treatment of a Generalized Vandermonde System of Equations," *Lin. Alg. & Its Applic. 17*, 149–74.

P. Van Dooren (1979). "The Computation of Kronecker's Canonical Form of a Singular Pencil," *Lin. Alg. & Its Applic. 27*, 103–40.

J. M. Van Kats and H. A. Van der Vorst (1977). "Automatic Monitoring of Lanczos Schemes for Symmetric or Skew-Symmetric Generalized Eigenvalue Problems," Report TR 7, Academisch Computer Centru, Utrecht, The Netherlands.

H.P.M. van Kempen (1966). "On Quadratic Convergence of the Special Cyclic Jacobi Method," *Numer. Math. 9*, 19–22.

C. F. Van Loan (1973). "Generalized Singular Values with Algorithms and Applications," Ph.D. thesis, University of Michigan, Ann Arbor.

C. F. Van Loan (1975a). "A General Matrix Eigenvalue Algorithm," *SIAM J. Num. Anal. 12*, 819–34.

C. F. Van Loan (1975b). "A Study of the Matrix Exponential," Numerical Analysis Report no. 7, University of Manchester, England.

C. F. Van Loan (1976). "Generalizing the Singular Value Decomposition," *SIAM J. Num. Anal. 13*, 76–83.

C. F. Van Loan (1977a). "On the Limitation and Application of Padé Approximation to the Matrix Exponential," in *Padé and Rational Approximation*, ed. E. B. Saff and R. S. Varga, Academic Press, New York.

C. F. Van Loan (1977b). "The Sensitivity of the Matrix Exponential," *SIAM J. Num. Anal. 14*, 971–81.

C. F. Van Loan (1978a). "Computing Integrals Involving the Matrix Exponential," *IEEE Trans. Auto. Cont. AC-23*, 395–404.

C. F. Van Loan (1978b). "A Note on the Evaluation of Matrix Polynomials." *IEEE Trans. Auto. Cont. AC-24*, 320–21.

C. F. Van Loan (1982a). "A Generalized SVD Analysis of Some Weighting Methods for Equality-Constrained Least Squares," in *Proceedings of the Conference on Matrix Pencils*, ed. B. Kagstrom and A. Ruhe, Springer-Verlag, New York.

C. F. Van Loan (1982b). "Using the Hessenberg Decomposition in Control Theory," in *Algorithms and Theory in Filtering and Control*, ed. D. C. Sorenson and R. J. Wets, Mathematical Programming Study no. 18, North Holland, Amsterdam, pp. 102–11.

J. Varah (1968a). "The Calculation of the Eigenvectors of a General Complex Matrix by Inverse Iteration," *Math. Comp. 22*, 785–91.

J. Varah (1968b). "Rigorous Machine Bounds for the Eigensystem of a General Complex Matrix," *Math. Comp. 22*, 793–801.

J. Varah (1970). "Computing Invariant Subspaces of a General Matrix When the Eigensystem is Poorly Determined," *Math. Comp. 24*, 137–49.

J. Varah (1979). "On the Separation of Two Matrices," *SIAM J. Num. Anal. 16*, 216–22.

J. M. Varah (1972). "On the Solution of Block-Tridiagonal Systems Arising from Certain Finite-Difference Equations," *Math. Comp. 26*, 859–68.

J. M. Varah (1973). "On the Numerical Solution of Ill-Conditioned Linear Systems with Applications to Ill-Posed Problems," *SIAM J. Num. Anal. 10*, 257–67.

J. M. Varah (1975). "A Lower Bound for the Smallest Singular Value of a Matrix," *Lin. Alg. & Its Applic. 11*, 1–2.

R. S. Varga (1961). "On Higher-Order Stable Implicit Methods for Solving Parabolic Partial Differential Equations," *J. Math. Phys. 40*, 220–31.

R. S. Varga (1962). *Matrix Iterative Analysis*, Prentice-Hall, Englewood Cliffs, N.J.

R. S. Varga (1970). "Minimal Gershgorin Sets for Partitioned Matrices," *SIAM J. Num. Anal. 7*, 493–507.

R. S. Varga (1976). "On Diagonal Dominance Arguments for Bounding $\| \|A^{-1}\| \|$," *Lin. Alg. & Its Applic. 14*, 211–17.

W. J. Vetter (1975). "Vector Structures and Solutions of Linear Matrix Equations," *Lin. Alg. & Its Applic. 10*, 181–88.

E. L. Wachpress (1966). *Iterative Solution of Elliptic Systems*, Prentice-Hall, Englewood Cliffs, N.J.

R. C. Ward (1975). "The Combination Shift QZ Algorithm," *SIAM J. Num. Anal. 12*, 835–53.

R. C. Ward (1977). "Numerical Computation of the Matrix Exponential with Accuracy Estimate," *SIAM J. Num. Anal. 14*, 600–614.

R. C. Ward and L. J. Gray (1978). "Eigensystem Computation for Skew-Symmetric and a Class of Symmetric Matrices," *ACM Trans. Math. Soft. 4*, 278–85.

G. A. Watson (1973). "An Algorithm for the Inversion of Block Matrices of Toeplitz Form," *J. Assoc. Comp. Mach. 20*, 409–15.

P. A. Wedin (1972). "Perturbation Bounds in Connection with the Singular Value Decomposition," *BIT 12*, 99–111.

P. A. Wedin (1973a). "On the Almost Rank-Deficient Case of the Least Squares Problem," *BIT 13*, 344–54.

P. A. Wedin (1973b). "Perturbation Theory for Pseudo-Inverses," *BIT 13*, 217–32.

O. Widlund (1978). "A Lanczos Method for a Class of Nonsymmetric Systems of Linear Equations," *SIAM J. Numer. Anal. 15*, 801–12.

J. H. Wilkinson (1961). "Error Analysis of Direct Methods of Matrix Inversion," *J. Assoc. Comp. Mach. 8*, 281–330.

J. H. Wilkinson (1962). "Note on the Quadratic Convergence of the Cyclic Jacobi Process," *Numer. Math 4*, 296–300.

J. H. Wilkinson (1963). *Rounding Errors in Algebraic Processes*, Prentice-Hall, Englewood Cliffs, N.J.

J. H. Wilkinson (1965a). *The Algebraic Eigenvalue Problem*, Clarendon Press, Oxford.

J. H. Wilkinson (1965b). "Convergence of the LR, QR, and Related Algorithms," *Comp. J. 8*, 77–84.

J. H. Wilkinson (1968a). "Almost Diagonal Matrices with Multiple or Clase Eigenvalues," *Lin. Alg. & Its Applic. 1*, 1–12.

J. H. Wilkinson (1968b). "Global Convergence of Tridiagonal QR Algorithm with Origin Shifts," *Lin. Alg. & Its Applic. 1*, 409–20.

J. H. Wilkinson (1968c). "A Priori Error Analysis of Algebraic Processes," *Proc. International Congress Math.* (Moscow: Izdat. Mir, 1968), pp. 629–39.

J. H. Wilkinson (1971). "Modern Error Analysis," *SIAM Review 13*, 548–68.

J. H. Wilkinson (1972). "Note on Matrices with a Very Ill-Conditioned Eigenproblem," *Numer. Math. 19*, 176–78.

J. H. Wilkinson (1977). "Some Recent Advances in Numerical Linear Algebra," in *The State of the Art in Numerical Analysis*, ed. D.A.H. Jacobs, Academic Press, New York, pp. 1–53.

J. H. Wilkinson (1978). "Linear Differential Equations and Kronecker's Canonical Form," in *Recent Advances in Numerical Analysis*, ed. C. deBoor and G. H. Golub, Academic Press, New York, pp. 231–65.

J. H. Wilkinson (1979). "Kronecker's Canonical Form and the QZ Algorithm," *Lin. Alg. & Its Applic. 28*, 285–303.

J. H. Wilkinson and C. Reinsch, eds. (1971). *Handbook for Automatic Computation, vol. 2, Linear Algebra*, Springer-Verlag, New York. (Here abbreviated as HACLA.)

H. Wimmer and A. D. Ziebur (1972). "Solving the Matrix Equation $\Sigma f_p(A)Xg_p(A)$," *SIAM Review 14*, 318-23.

H. Wozniakowski (1980). "Roundoff Error Analysis of a New Class of Conjugate Gradient Algorithms," *Lin. Alg. & Its Applic. 29*, 507-29.

A. Wragg (1973). "Computation of the Exponential of a Matrix I: Theoretical Considerations," *J. Inst. Math. Applic. 11*, 369-75.

A. Wragg (1975). "Computation of the Exponential of a Matrix II: Practical Considerations," *J. Inst. Math. Applic. 15*, 273-78.

J. M. Yohe (1979). "Software for Interval Arithmetic: A Reasonable Portable Package," *ACM Trans. Math. Soft. 5*, 50-63.

D. M. Young (1970). "Convergence Properties of the Symmetric and Unsymmetric Over-Relaxation Methods," *Math. Comp. 24*, 793-807.

D. M. Young (1971). *Iterative Solution of Large Linear Systems*, Academic Press, New York.

D. M. Young (1972). "Generalization of Property A and Consistent Ordering," *SIAM J. Num. Anal. 9*, 454-63.

D. M. Young and K. C. Jea (1980). "Generalized Conjugate Gradient Acceleration of Nonsymmetrizable Iterative Methods," *Lin Alg. & Its Applic. 34*, 159-94.

S. Zohar (1969). "Toeplitz Matrix Inversion: The Algorithm of W. F. Trench," *J. Assoc. Comp. Mach. 16*, 592-601.

Index

469